The Development of Biological Systematics

The Development of Biological Systematics:

Antoine-Laurent de Jussieu, Nature, and the Natural System

Peter F. Stevens

COLUMBIA UNIVERSITY PRESS

NEW YORK

Columbia University Press
New York Chichester, West Sussex

Library of Congress Cataloging-in-Publication Data
Stevens, Peter F.
The development of biological systemaitics: Antoine-Laurent de Jussieu, nature, and the natural system / Peter F. Stevens.
p. cm.
Includes bibliographical references and index.
ISBN 0-231-06440-3
1. Biology—Classification—History. 2. Botany—Classification—History. 3. Jussieu, Antione Laurent de, 1748–1836. I. Title.
QH83.S76 1994
574'.012—dc20 94-27231
CIP

Casebound editions of Columbia University Press books are printed on permanent and durable acid-free paper.
Printed in the United States of America
c 10 9 8 7 6 5 4 3 2 1

To E. A. K.

[Ces exemples] montrent que tous nos efforts sont impuissants, en présence des relations multiples qu'affectent de toutes parts les êtres qui nous entourent. C'est la lutte, dont parle le grand botaniste Geothe, de l'homme contre la nature infie. On est assuré toujours de trouver l'homme surpassé.

–Henri Baillon, *Étude general du famille des Euphorbiacées* (1850)

CONTENTS

FIGURES

Acknowledgments

I have incurred more intellectual debts in the course of the rather lengthy gestation of this work than I can ever hope to repay. Some of my most valuable leads have come from jotted notes which surfaced sometimes years after they had been made—long enough for me to have forgotten the circumstances that had prompted me to make them. My first acknowledgments, no less heartfelt for being so vague, are thus to the numerous people who, in this fashion, have affected my path of inquiry.

I owe thanks to the librarians and staffs of libraries in three continents who have patiently helped me deal with my chronic incapacity to find correctly filed books. I am particularly grateful to Judith Warnement, librarian of the combined libraries of the Harvard University Herbaria, to her two predecessors, Lenore Dickinson and Barbara Callahan, and to Jean Cargill, archivist in the libraries; to Eva Jonas, librarian of the Museum of Comparative Zoology, Harvard University, and Dana Fisher, administrative assistant. I am also grateful to Sylvia Fitzgerald, librarian at the Royal Botanic Gardens Kew, and Leonora Thompson and Cheryl Piggott, archivists there; Hervé Burdet, curator of the library at the Conservatoire et jardin Botaniques de la Ville de Genève, Switzerland; Bernadette Callery, research librarian at the New

York Botanical Garden; and the staff at Houghton and Widener libraries at Harvard, the Boston Public Library, the library of the Muséum National d'Histoire Naturelle, Paris, and the Black Mountain Library, C.S.I.R.O., Canberra. I am particularly grateful to Mme J. P. Weber-de Candolle and the late M. Roger de Candolle for their kindness and hospitality during my visits to the Fondation Auguste de Candolle, in Geneva.

Most of the illustrations are taken from books held by the library of the Gray Herbarium of Harvard University. The reproduction in figure 15 was supplied by the library of the Museum of Comparative Zoology, Harvard University, while figure 6 is based on a copy of a drawing held by the Conservatoire et jardin Botaniques, Geneva. I am grateful to the directors of all these institutions for permission to use their material.

My colleagues in the Harvard University Herbaria have accommodated with grace my idiosyncrasies; to them I owe special thanks. I have been greatly helped by discussions that developed from talks I presented at meetings of the International Botanical Congress (Berlin) and the Australian Systematic Botanical Society (Canberra), and at Acadia, Harvard, McGill, and Toronto Universities, the New York Botanical Garden, and the Royal Botanic Gardens, Kew. I am particularly grateful to the students of Biology 164, History of Botanical Systematics, who helped me clarify my thoughts while I was working out some of my ideas in the fall of 1991.

I thank Mark Barrow, Janet Browne, Mark Coode, Quentin Cronk, Richard Drayton, Brent Elliott, Hunter Dupree, Marion Filipiuk, Paul Groff, Joy Harvey, Jason Koh, Chris Kraus, David Mabberley, Gordon McOuat, Ernst Mayr, Bob O'Hara, Dorinda Outram, Duncan Porter, Susan Rosa, Betsy Shaw, Emma Spary, Sonia Sultan, and Liz Taylor for information, discussion, and much other valuable input. The late Dick Eyde and the late Andrew Kanis are also gratefully acknowledged. I spent much and pleasurable time talking with David Hendler about system and method of the eighteenth century. Judy West and the other inhabitants of the Australian National Herbarium provided me a very pleasant safe haven for work during the final stages of manuscript writing, and I learned a lot from them that found its way into the book. Jean Cargill checked nearly all the references in the endnotes; Linda Fahey reformatted most of the bibliography; and Chris Dick, Diane Ferguson, and Santiago Madriñan helped in various ways at the end—to all my thanks. I am grateful, as well, to Réjane Bernier and Denis Lamy, who kindly read drafts of the manuscript, as did three readers for Columbia University Press, and Polly Winsor also made helpful comments on some chapters. Many thanks are due to the editors and staff of Columbia University Press for their help and patience: Ed Lugenbeel encouraged me by quietly assuming that I would finish, even when this outcome was less than clear to me, and I owe a special debt to Connie Barlow for working so hard to make the manuscript understandable. I am particularly grateful to Arthur Cain, Alain Cuerrier, Toby Kellogg, and Jim Larson; only they know how much this book owes to their help.

Finally, because a familiarity with Antoine-Laurent de Jussieu's early writings is essential for understanding his whole work, I am particularly grateful to Susan Rosa at the University of Wisconsin (Milwaukee) for successfully tackling the daunting task of translating the lengthy and at times obscure Introduction to Jussieu's *Genera plantarum* and to Susan Kelley at the Arnold Arboretum of Harvard University for her translation of Jussieu's memoir on the relationship between the characters of plants and their virtues.

Preface

My active interest in the history of systematic or classificatory biology developed from attempts to teach it. Like all systematists I had used the older primary literature extensively in the course of my day-to-day work, and I had read secondary literature on the development of the discipline. I soon realized, however, that there was much I did not understand.

For instance, I found that the frequent comments that Darwin's *On the Origin of Species* had little effect on the practice of systematics were accompanied by only the barest hints as to what systematics in the early nineteenth century entailed—beyond the description of groups based on characters that were presumably the essences, or a reflection of the essences, of the groups. Yet when I reread the primary literature with the aim of trying to understand why systematists throughout the nineteenth century were doing what they were doing, I found that little mention was made of essences. In an essay that I wrote in 1984 for the journal *Taxon*, I concluded that the metaphors and analogies used by botanical systematists had indeed changed little throughout the nineteenth century. But I was not satisfied that this was the whole story. These metaphors and analogies were connected with classifications, and it seemed to me that the real problem was that we knew little about how and why systematists in the late eighteenth and early nineteenth centuries produced their classifications. In particular, little was known about the work of Antoine-Laurent de Jussieu, who published his seminal work, his great *Genera plantarum*, in 1789. I realized that to consider his work a classification represented a fundamental misunderstanding of his aims. Yet Jussieu is the person who effectively codified the so-called natural system in botany; his work was lauded by botanists and zoologists alike and served as a basis for subsequent botanical studies in systematics for the next two centuries. I decided to learn more about Jussieu, and this book is the fruit of that effort.

In the course of my research I came to realize that Jussieu's work was underlain by a commitment to the principle of continuity—that there are no gaps in nature. A similar commitment is found in many of his successors, too, and this necessitated a reevaluation of the history of biological systematics. This belief in continuity represents a radical departure from the essentialist thinking attributed to Jussieu's predecessors, particularly Linnaeus. Darwin, of course, dealt the death blow to essentialist thinking in his *On the Origin of Species* in 1859, although by all accounts essentialism died very slowly, especially in the systematic community. Buffon's ideas of a properly historical natural history represent an earlier challenge to Linnaeus's ideas, but even in German-speaking countries they seem to have been largely without lasting effect. Jussieu's commitment to the principle of continuity represents an implicit rejection that higher-level groups, at least, had essences or even definitions. If Jussieu did indeed bring such thinking to the discipline of systematics well before Darwin, the history of science as currently written does not give Jussieu his due.

This book is an attempt to document the extent to which a belief in continuity is evident in Jussieu's work, to explore how botanists in the nineteenth century related Jussieu's work to their own, and to examine how a better understanding of Jussieu ought to affect our portrayal of the history of systematics—and perhaps even our modern practice of systematics. I have restricted this history very largely to botanists and botanical systematists for several reasons. First, our understanding of the history of biology, and of biology itself, is skewed because of an overemphasis on zoologists and zoology—more specifically, ornithology. The history of botanists and botany deserves far more attention. The need for a greater botanical focus is even more pressing because ideas of classification and relationships were better developed for plants than for animals in the late eighteenth century, and they served as a springboard for the remarkable advances zoological systematists made in the first half of the nineteenth century.

Within botany, I focus on flowering plants—as did systematists themselves, particularly in the first part of the period under study. I discuss genera and families and how their circumscriptions and relationships were determined. Again, this stance is taken because historical studies of systematics have tended to focus on the development of species concepts, ideas of variation, and associated topics. Finally, I emphasize practice as much as theory; the names of many of the botanists I cite are not, therefore, widely known. Indeed, this study is interesting in part precisely because major ideas of life and nature have had so little obvious effect on the work of those who so laboriously classified living things, and because in few of even the more important systematic studies of the whole period from 1789 to 1859, and on up to 1959, can the reader understand what was being attempted, other than an inventory of nature, and far less how and why. Although the botanists on whom I was working by and large dismissed theory as of no importance for their work, I found that they had nevertheless unknowingly embraced Jussieu's ideas of the living world. The system they helped build formed the comparative framework for botany in the nineteenth and twentieth centuries, but the practitioners might well have repudiated Jussieu's ideas if they had been aware of them.

I began my research well aware that I would find little help in the literature in the form of previous attempts to catalog and interpret the development of botanical systematics over the period in question. The best general studies of systematic thought in the late eighteenth century remain those of Henri Daudin (1926a, 1926b). Nevertheless, there are substantial lacunae in our knowledge of almost every area of nineteenth-century systematics. Studies are needed on topics such as: the use of anatomy in botanical systematics in the nineteenth century; national differences in approaches to systematics; the debates about the relationships between the monocotyledons (grasses and their relatives), dicotyledons (oaks, sunflowers, and the rest of the flowering plants), and gymnosperms (pines and cycads); the connections between the large central herbaria (in countries like England, France, and Germany) and colonial and provincial botanists; the role of the concept of organization in defining research programs; and what it is that makes a systematist a professional.

We also know little about most of the people I shall mention and their interactions. In the last fifty years there have been detailed studies of the life and work of only three botanists other than Linnaeus—Michel Adanson, Robert Brown, and Asa Gray—who produced classifications in the whole of the eighteenth and nineteenth centuries. Some of the issues raised in this book might, for example, be clarified by biographies of the key characters; a detailed treatment of the life and work of Antoine-Laurent de Jussieu himself is a pressing need, but beyond the scope of this book. Little is known about the systematic philosophy of even so important a figure as J. D. Hooker—who was a confidant of Darwin, a part of the dynasty that made the Royal Botanic Gardens at Kew one of the major systematic centers of the nineteenth century, and a prolific botanical author. Although the development of zoological systematics is peripheral to my main argument, a little more biographical information is available on this topic, and a number of recent studies have proved very helpful in my own work. However, we still have a long way to go before we can write a general history of systematics that compares developments within different groups of organisms.

As I have followed the various threads of this study, I have relied quite heavily on the advice of numerous colleagues as to useful papers to read. Frans Stafleu and Richard Cowan's monumental *Taxonomic Literature* (1976–1988) has been an indispensable guide to the general literature of systematic botany. The knowledge I have gained of the German literature is, unfortunately, less comprehensive than I would have liked. For many of the people I mention there are very extensive archival holdings—letters, unpublished manuscripts, and the like—which are almost entirely untouched by historians of biology. I have drawn on these only sparingly, except in appendix 2 where I explore the relationship between Augustin-Pyramus de Candolle and Antoine-Laurent de Jussieu; for this, Candolle's letters to his father are of the greatest importance. My own archival researches have focused on George Bentham (who figures prominently in the later chapters). Active between 1826 and 1883, Bentham represents the third generation of botanists working on the natural system, and he is a central figure in the establishment of the natural system and of generic and familial limits as conventions in the latter part of the nineteenth century.

Of course, the history of systematics cannot be understood from the systematic literature alone, no matter how closely it may be studied. The development of systematics has been particularly vulnerable to what might be thought of as extra-systematic, indeed extra-scientific influences. And I have tried to take these into account, as best I can. I hope that what follows is only the beginning of a reevaluation of our understanding of the history and conceptual underpinnings of systematics.

Peter F. Stevens
Harvard University Herbaria

The Development of Biological Systematics

CHAPTER ONE

Introduction

Antoine-Laurent de Jussieu published his *Genera plantarum secundum ordines naturales disposita* in 1789 almost as the Bastille was falling. In this modest volume he provided descriptions of all the genera of plants known, and he placed the great majority in exactly one hundred families, several of which he was the first to recognize. These families, the groupings of genera that he recognized within them, and the groups into which the families themselves were placed were Jussieu's natural arrangement of plants—the outline of the so-called natural method—and have been the backbone of botanical classifications ever since. Jussieu wrote nothing on botany during the subsequent decade of the French Revolution, being involved first in the administration of hospitals and hospices in Paris, and from 1792 onward in the reorganization of the library and the Muséum nationale.[1] Early in the nineteenth century, however, he resumed publication and in a series of important papers that came out over a period of twenty years he described additional plant families, modified the limits of those he had earlier proposed, and described many new genera and species.

In the Introduction to the *Genera plantarum,* Jussieu made it clear that his natural method was both a natural series—an arrangement of plants that reflected natural relationships—as well as a method, the guidelines to be followed in constructing the series. The *Genera plantarum* was influential both because it was the first accepted treatment in which all plant taxa at and above the rank of genus were placed in a natural sequence and clearly described, and also because it included a clear outline of the principles by which the natural method was to be executed.[2] A. G. Morton suggested that Jussieu's work was of more general importance—indeed, that it was central to the development of biology as a whole:

> It fell to Antoine-Laurent de Jussieu to make natural classification a universally accepted concept and primary aim in botany. The principles of the method and an impressive example of the result of applying them to all known plants were published in the *Genera plantarum* in 1789, making that year as decisive a turning point in biology as it was in the social and political history of Europe.[3]

Jussieu and His Place in History

A revolution in biology is made to accompany the revolution in French society. Georges Cuvier made clear what he saw as the central importance of Jussieu's work when he ascribed to Jussieu the idea of subordinating characters in terms of their systematic importance. Cuvier wrote in 1845, "This [Jussieu's] work produced a real revolution in botany, since it was not until after its publication that one studied plants from the point of view of the relationships between them, and according to the totality of their organization."[4] (All translations are mine unless otherwise noted.) Some authors have suggested that it was Jussieu's great knowledge of plants rather than the originality of his basic concepts that made his work so important, and there has been extensive discussion about the relationship between the ideas of Jussieu and those of his uncle, Bernard de Jussieu, sous-demonstrateur de l'extérieur des plantes at the Jardin du Roi and member of the Académie Royale, with whom Antoine-Laurent lived from 1765 to 1777.[5] But historians and contemporary commentators alike are agreed as to the general importance of the ideas about natural classification—whatever their genesis—that were finally spelled out by the nephew.[6] Jussieu's work indeed influenced the young Georges Cuvier,[7] and I shall discuss the possible nature of this influence in chapter 4.

On the other hand Ernst Mayr, in his monumental *The Growth of Biological Thought*, barely mentioned Jussieu, preferring to credit the approach that he believed had been adopted by Michel Adanson, who had advocated recognizing relationships by using numerous features from all parts of the plant.[8] Mayr noted approvingly that "virtually unnoticed by the practitioners themselves [the makers of classifications in the early nineteenth century], the use of single key characters for the establishment of higher taxa was being replaced by the grouping of species (or other lower taxa) into higher taxa on the basis of character combinations. Upward classification was becoming a matter of course."[9]

Mayr also thought that the groups of organisms being recognized by naturalists in the nineteenth century were more or less discrete. However, such an upward or synthetic approach, that of forming groups progressively greater in extent,[10] was explicitly adopted by Antoine-Laurent de Jussieu in 1789 and was the basis of his natural method;[11] it was also favored by many other naturalists in the late eighteenth and early nineteenth centuries.[12] These naturalists all believed in continuity, that is, nature did not have discrete groups. Thus Jean-Baptiste de Lamarck[13] provided clear instructions about how to construct the natural series in the context of continuity; again, a synthetic approach was preferred. For both Jussieu and Lamarck groupings such as genera and families were part of a continuous series. Indeed, Ilse Jahn has suggested that in the later part of the eighteenth century, the period during which the natural method was developed in botany, all systems in which groupings were discrete were considered to be artificial.[14]

The continuity of nature in which Lamarck and Jussieu believed is one of the elements of the classical idea of plenitude. Arthur Lovejoy's 1936 book, *The Great Chain of Being*, is still required reading for anybody interested in the history of this idea, or rather this set of associated ideas. Lovejoy[15] isolated three principles within the comprehensive idea of plenitude. First, he described the idea of plenitude in the strict sense as the view that "the universe is a *plenum formarum* in which the range of conceivable diversity of *kinds* of living things is exhaustively exemplified."[16] Second, the associated *principle of continuity* implies that if there is a theoretically possible intermediate form between two existent species, that form must exist somewhere, and so *ad infinitum*. More specifically, forms in the world are connected by an infinitely graded series of nuances or intermediates. Finally, there is the idea of a *scala naturae*—the great chain of being itself—in which the natural world is arranged in a series from the simple to the complex, imperfect to perfect, hydra to human, and thence to god. These three ideas are dissociable; as we shall see, systematists who advocated the synthetic approach often believed in continuity, only sometimes in the idea of the chain of being, and still less frequently in plenitude. Although saying little about the role that ideas of continuity played in the work of late eighteenth and nineteenth-century naturalists, Henri Daudin[17] discussed in detail the often close relationship between ideas of continuity and natural classifications earlier in the eighteenth century.

It is with the vicissitudes of the idea of continuity itself that I shall be mostly concerned in the following pages. I shall emphasize the close relationship between the idea of continuity and the natural method—both theory and practice—in the work of Antoine-Laurent de Jussieu, and the consequence of this association for the subsequent development of the natural method, especially in botany. My focus will be on the higher taxa[18] in classifications: groupings of genera into families, families into classes, and so on. I will have less to say on the category of genus and will largely ignore species concepts, because at these higher levels it matters little whether species are discrete, or whether individuals alone are real. In other words, for my purposes it makes no difference if the water coming out of the tap, alias nature, is coming in a continuous stream, or if it is continually dripping.[19]

Despite the context in which ideas of the natural method developed, the idea that the forms assumed by living beings could be assembled into a simple, linear, continuous series was beginning to be superseded by the end of the eighteenth century.[20] Thus several well-known naturalists including Vitalinio Donati, Jean Hermann, and Pyotr Simon Pallas[21] suggested that relationships were web-, net-, or even tree-like, although it cannot be overemphasized that all these representations of nature still allow continuity. But the very idea of continuity came into question in the early years of the nineteenth century in the studies of Cuvier, Charles-François Brisseau de Mirbel, and Augustin-Pyramus de Candolle;[22] all three perceived discrete groups in nature separated by real gaps. Furthermore, the value of upward classification was also questioned by Cuvier and Candolle,[23] and, somewhat later, by naturalists such as Alfred Russel Wallace.[24] The method of Jussieu was thus transformed into that of Cuvier: out of a method that admitted no discrete groups in nature, that was based firmly on synthesis (with an analytical component apparently of secondary importance), and that largely used external features in grouping, came a method that admitted definite groups (at least after 1812), that was more analytical, and that was based on anatomical studies and the subordination of characters.

Our current understanding of the changes associated with the early development of the natural system in the period 1780–1850 can be outlined easily. Jussieu's work is generally acknowledged to be the most influential of early attempts to elucidate the structure of natural relationships. His approach was widely adopted by botanists and it also influenced zoologists, most notably Cuvier. By the end of the eighteenth century it had become commonplace to claim that analytical classifications, systems, in which groups were successively subdivided, were necessarily artificial; it was upward classification, method, or synthesis, the formation of successively larger groups, that was believed to lead to natural arrangements of organisms.[25] Naturalists' beliefs about the living world changed: relationships were no longer seen as being continuous, and groups were considered to be more or less discrete. The widespread use of comparative anatomy is linked to this change, and this discipline spread widely in zoology. Organisms were assigned to genera and families that were generally thought of as discrete and sharply bounded entities. Any continuity between them was by way of intermediates (at best a necessary evil), types of one sort or another, or (after 1859) extinct ancestors. Botanists have, however, placed less emphasis on the distinctness of groups than have zoologists. For most systematists working after 1859, the groupings of organisms in the natural system have been interpreted as being the result of evolution. The appropriate image for the post-Darwinian system is a much-branched tree, perhaps with some grafting between the branches.

Thus there is a paradox. To oversimplify grossly: Jussieu and many of his contemporaries believed in continuity; a number of their successors did not. Jussieu believed that relationships in the natural system could be recognized only by synthesis, whereas some of his successors did not. Yet it has been commonplace to continue to emphasize the great value of synthesis in systematic studies, although it is supposed to have been Jussieu's successors who promoted this idea.[26] In accounts of the history of systematics, if ideas of continuity are mentioned as surviving into the nineteenth century, this persistence is generally thought to be connected with the persistence of Lamarckian ideas.[27] Clearly, there are fundamental problems in our historical understanding of the development of the natural system. Yet we continue to talk about *the* natural system, as if everything that has appeared since the publication of the *Genera plantarum* is but a simple modification of Jussieu's arrangement. Thus Michel Guédès observed in 1982, "natural classification has been the aim of all taxonomists since the 18th century and before," and he clearly thought there was but one nature and one gradually improving natural system.[28] Guédès' point is true enough—but only if differing ideas of nature are ignored. To the extent that it is believed to be true it raises fundamental questions about the nature of the systematic discipline.

Yet much of the commentary in the rather sparse literature on developments in systematics in the nineteenth century either suggests that the history of systematics is an uninteresting problem or expresses uncertainty or even bewilderment as to what happened.[29] John C. Greene[30] admitted in 1971 that little was known of this period; for him it was John Ray, Joseph Pitton de Tournefort, and especially Carolus Linnaeus, who established the conceptual framework of systematics. Greene's idea of systematics involved simply the naming, classification, and description of the productions of the earth; the natural system seemed to show a rather uniform development. There has since been more discussion on the development of natural history, of which classification studies form a major part. The value of this discussion has, however, been compromised because of our lack of understanding of the development of the natural method. Thus Paul Farber in 1982 observed that natural history had as its focus the description and classification of natural objects, yet he thought that the history of nineteenth-century classification—and so, by implication, much of the history of natural history itself—remained to be written.[31] Only after we examine the theory and practice of the naturalists who made classifications in the late eighteenth and early nineteenth centuries can we discuss topics like the development of natural history in the nineteenth century, the reception of evolutionary ideas in systematics, the significance of the unchanging nature of systematic practice, and the position of systematics and systematists in science.

In this book I will challenge our current understanding of the history of systematics by examining the work of Antoine-Laurent de

Jussieu. The period between 1773 (when Jussieu read his first systematic memoir at the Académie) and 1859 (when Darwin's *Origin* was published) will be the focus. I do this because historians of systematics have not taken Jussieu's view of the living world adequately into account; indeed, the discipline has all but ignored it. Jussieu is usually credited with being the first to satisfactorily flesh out the botanical natural system, and systematists since have usually assumed that his system and theirs were fundamentally the same. There never has been a full analysis of Jussieu's *Genera plantarum*, or of any of his other work, to see if and how his classificatory practice relates to his general theory and the principle of continuity in particular. Such an analysis is carried out in the chapters that follow.

As I present my results, it will become clear that Jussieu's genera and families are not discrete and sharply bounded entities. He *expected* continuity between taxa and he believed that any gaps would eventually be obliterated by the discovery of living intermediates. Not only must intermediates exist; Jussieu deemed them essential for the very discovery of the natural order. The appropriate image of Jussieu's natural order is thus a continuous landscape, unbroken line, or net. Accordingly, synthesis (the gradual building up of groups) was the appropriate way to proceed, because only then could continuity be recognized. Anatomical observations were considered of little importance, largely because they were irrelevant, although I will also suggest that plants do not lend themselves readily to Cuvieran-style anatomy.

More generally, it is clear that the ontological status of Jussieu's taxa is very different from the modern viewpoint. Furthermore, the principles of classification that he outlined can be related directly to his view of the living world, one in which the principle of continuity was paramount. Jussieu believed in this principle to the end of his life; as he emphasized in the posthumously published Introduction to his *Genera plantarum* (which he had extensively rewritten), "*natura non facit saltus*."[32] Because his theoretical beliefs can be related to his view of the world, a study of his actual practice is of considerable interest.

Summary of Chapters

This book consists of two parts. In chapters 2 through 6 I establish how systematists of the late seventeenth and early eighteenth centuries saw the world and how they went about recognizing groups and relationships. Chapter 7 evaluates whether systematic practice changed over the period 1789–1859, and in it I summarize the previous chapters. In chapters 8 through 11, I look at how nature was understood by systematists of that era, and the relationships between systematics, other areas of science, and society. In these later chapters my topic becomes broader, and I move beyond botany and higher taxa to touch briefly on systematics in zoology and studies at the species level. I also discuss developments in systematics in the second half of the nineteenth century.

In chapter 2 I begin by looking at the Introduction to the *Flore françoise* of Jean-Baptiste de Lamarck (published in 1778, a decade before the *Genera plantarum* appeared) and at some of Lamarck's other early writings. Here we find an exceptionally clear analysis of the interrelated problems of how to construct the natural arrangement of plants if there are no gaps in nature, how to relate this arrangement to classifications, and how best to identify plants.

In chapter 3 my focus shifts to an extended treatment of Jussieu's theory and practice. The Introduction to the *Genera plantarum* of 1789 and three of Jussieu's important precursory papers[33] (for which the first English translations are provided in appendix 1[34]) take pride of place in my analysis of Jussieu's theory. I take examples of Jussieu's practice from the body of the *Genera plantarum* and from his later revisionary work. Jussieu's theory is shown to be consistent with a belief in continuity. His practice—reliance on synthesis in forming groups, arrangement of plant groups so that characters form a continuum, and modification of group limits depending on the numbers of included taxa—is also consistent with this belief.

In chapter 4 I turn to some developments in the understanding of natural relationships in the early nineteenth century which are connected with Jussieu's work, but which nevertheless differ substantially from it. The three people I discuss are Georges Cuvier, who began his zoological career in the 1790s, Charles-François Brisseau de Mirbel, and Augustin-Pyramus de Candolle. Mirbel and Candolle were both botanists and slightly younger than Cuvier; they became active in the first decade of the nineteenth century. I provide no more than a summary of the main elements of each of their approaches to systematic theory, but this is sufficient to show that all to a varying degree believed both in analysis (that is, in successively subdividing groups) and the existence of discrete groups in nature. Some of Candolle's letters to his father clarify the development of his ideas and also the personal and professional relationships between Candolle and Jussieu, but so as not to disturb the flow of the argument, they are placed in appendix 2.

In chapter 5 I briefly analyze the work of other botanists, in particular those from France and Great Britain, who were active in the first few decades of the nineteenth century. These include followers of Jussieu (mostly French), as well as other practitioners of the natural method (mostly British) who were nevertheless rather vague as to how nature could be represented. The chapter concludes with a discussion of the relevance of anatomy and paleontology to systematists of this era, the connection between characters and relationships, and the antithetical approaches of analysis and synthesis.

Chapter 6 entails a brief discussion of mostly French- and German-speaking botanists who used types in their systematic work, and whose classifications therefore were usually based on an assumption that groups are discrete. Typological thought tended to be associated with a vastly different conceptualization of nature than that held by those who presumed continuity and intermediates, but I show that typological thought did not provide a viable alternative to a continuity-based approach for botanists.

Having established that many botanists could be considered followers of Jussieu, in that they believed that nature was more or less continuous, in chapter 7 I look in more detail at the persistence of ideas of continuity in nineteenth-century systematic work. I suggest that this was informed by what I call "continuity-in-practice"—whatever the views of nature expressed by these systematists, their actual practice implied that there were no discrete groups.

Chapter 8 then broadens the discussion: I examine how systematists understood the living world. I argue that in the early nineteenth century, classifications often failed to provide any real understanding of nature; diagrams of relationships often implied a rather disordered natural world. Exceptions are some "ideal" systems, and the emphasis by some botanists on analogies and parallelisms also implied that there were some regularities in nature.

In chapter 9 I look at the relationship between systematics, natural history, and biology and the related issues of the status of natural history and systematics in science. I show that nineteenth-century systematics and natural history both retained important features of their common ancestry in eighteenth-century natural history, and that both, although more particularly the latter, had become almost marginalized in science by the middle of the century.

In chapter 10 I examine the stability of the natural system. I conclude that this stability owed to the disinterest of systematists in theory, how they learned their craft, and the need for the users of classifications to have stable names.

In chapter 11 I discuss briefly the general issue of continuity and change in systematics in the whole period from 1735 (when Linnaeus started publishing) to 1900, and how it relates to ideas of revolution in science or to changes in epistemes, that is, ways of seeing and understanding the world. In concluding, I suggest that the unconscious adoption of a distinctive theory of nature predisposed systematists to see continuity and reticulations in nature. Hopes that the names in plant classifications would be the basis of scientific discourse were compromised by new knowledge of the historical basis of life and dashed by the unchanging nature of systematic practice. Systematists were unknowingly caught between the discontinuities of the living world and the continuity implied by their practice; they were drawn to make changes dictated by the needs of their science, yet were restrained by tradition and the needs of the nonscientific users of classifications. Classifications could thus become stabilized only because of an uneasy truce struck between convention and botanists' ideas of relationship.

A Prologue: Groups, Systems, and Systematists

One of the themes of this book is the ambiguity of many of the key words and phrases used by systematists. The reader may have noted that in the previous passages I referred to botanists, naturalists, and systematists—yet these words are not strictly interchangeable. Similarly, "nature," "method," "system," "morphology," "organization," "analogy," and many other words have a variety of different meanings.

Likewise, the critical distinction between grouping (forming taxa) and ranking (placement of those taxa in a hierarchy) is rarely made explicit or even implicit in the historical literature.[35] The confusion so engendered further muddles interpretation of the answers to questions such as, Did Jussieu (or George Bentham, or Linnaeus) think that genera were real? Real as groups or as ranks—or as something else? Different answers to this question may mean the same thing: Yes may mean that genera are real as groups (but they may not be members of a rank with particular properties); No may mean that they are not members of a discrete rank (but they still could be members of discrete groups). Another set of distinctions drawn by George Gaylord Simpson is helpful here: "A group is nonarbitrary as to inclusion if all its members are continuous by an appropriate criterion, and nonarbitrary as to exclusion if it is discontinuous [discrete] from any other group by the same criterion."[36] By such criteria we will see that Jussieu's genera were arbitrary as to exclusion, i.e., their limits, and nonarbitrary as to inclusion, that is, their contents. Conversely, by Simpson's standard Bentham's genera were nonarbitrary as to exclusion but arbitrary as to inclusion, while Linnaeus's genera were nonarbitrary as to both exclusion and inclusion. Jussieu's genera were thus not real as groups; they were not discrete and they were not bounded by gaps in nature. Bentham's and Linnaeus's genera were real as groups, but only Linnaeus's category of genus was real as a rank. Nevertheless, it could be argued that all three had, in some way, captured a part of nature in their genera, which were thus "real."[37] Classifications were made up of groups. Groups and classifications represented nature, and by easy elision they become real. I return to this issue in chapter 7.

Words like "method," "system," and "systematics" are perhaps *the* key words used by the people whom I discuss here, and I must clear up some of the ambiguities surrounding their use.[38] First, as to the distinction between taxonomy and systematics, Simpson offered a much-quoted definition of systematics: "the scientific study of the kinds and diversity of organisms and of any and all relationships among them."[39] Classification was the grouping of organisms into the hierarchy of a classification; taxonomy was the theoretical study of classification.[40] For Frans Stafleu,[41] on the other hand, taxonomy was represented by keys, systematics by interpretive relationships. The connection between taxonomy, in the sense of classifications and keys, and systematics, in the sense of the relationships between organisms, is one of the themes of this book. Recently a different distinction has been drawn between systematization and classification, the former being an ordering according to element/system or part/whole relationships, the latter of categories based on common properties.[42] It will become clear that Lamarck (and also Jussieu and his followers) was a systematist in this latter sense, although Simpson would not qualify as a systematist.

There are other problems with the words "system" and "systematist." I will argue below that Jussieu's approach to organizing nature by the progressive building up of ever-larger groups (that is, synthesis, or what he called method) largely prevailed. With Jussieu and his followers a sharp distinction between artificial system and natural method appeared in the literature, with Linnaeus singled out as an arch maker of artificial systems. It is not irrelevant to note that Joseph Pitton de Tournefort, a Frenchman, produced what he called a method, while the botanical arrangement of Linnaeus, a Swede, followed his sexual system, avowedly more or less artificial, and that Linnaeus never fully explained the principles that informed those natural groupings he did propose.[43] However, in the middle part of the eighteenth century "method" and "system" were not clearly distinguished. Linnaeus did not sharply oppose the two; in the analytical summary of the contents of the *Classes plantarum* he implied that the classifications of all the authors he included there were methods, but in the body of the book they are referred to as being either methods (usually) or systems—except for the classification of Cesalpino, which is a method on one page and a system on the next.[44] In his earliest writings Jussieu himself was not clear on the distinction. His references to "méthode" and "méthodistes" in his paper on the Ranunculaceae read in 1773 were often negative,[45] and he did not distinguish sharply between system and method there. A year later he ended his paper on the new arrangement of plants in the Jardin Royal (the Jardin du Roi) with a dismissal of system—which he contrasted with the natural order.[46]

The later positive associations of "method" and negative connotations of "system" may represent a definite inversion of earlier meanings,[47] or at least the stabilization of previously fluctuating usages. It should not be forgotten that the word "system," in the sense of general theories about the organization of nature, had fallen into disrepute in philosophical circles in eighteenth-century France. Even in 1733 Bernard de Fontenelle, the powerful secretary of the Académie, noted in the preface to *L'Histoire de l'Académie des Sciences depuis 1666 jusqu'en 1699* that it was policy not to include such general systems in its publications.[48] The "esprit de système" in which general systems of nature (for example) were deduced from untestable first principles came under widespread attack.[49] Rather, "méthode," in which the problem was first carefully analyzed, the naturalist then proceeding by observation from the simple to more complex, was the only appropriate way to proceed.

Nevertheless, the definitions or redefinitions of "system" and "method" were not universally accepted, and in the early nineteenth century inverted definitions were also current.[50] Despite the opprobrium that became attached to artificial systems, and despite the desire of French-speaking botanists in particular to improve on the natural method, makers of classifications in general or followers of the natural method in particular were only infrequently called "méthodistes."[51] English-speaking authors of the nineteenth century, in particular, favored "systematist" over "methodist"; but since they often used inverted meanings of the words, their "systematist" was the same as "méthodiste" of their French colleagues.[52] Thus William Whewell referred to Georges Cuvier as "the great systematist" of fishes, while Cuvier himself called Linnaeus (with approval) a "méthodiste";[53] although they were using apparently different terms, Whewell and Cuvier were referring to basically the same thing. Ultimately the usage more common in English prevailed, as "systematist" did not have the stigma that might be attached to it in France. But there is an almost mundane reason why systematists should have triumphed over methodists: the word "methodist" was preempted, being widely used as the name for a Protestant sect.[54]

In the next few chapters, in particular, I shall avoid use of the words "taxonomist" and "systematist"—both because of their ambiguity and because it would be inappropriate to apply the former to a naturalist working before 1813 (when the term was coined by Augustin-Pyramus de Candolle). I shall call people who make classifications and systems simply naturalists, botanists, or zoologists, although these words are themselves heavily freighted with a variety of connotations.[55] Similarly, what came to be called the natural system I shall call *the natural arrangement* or *the natural series*. Unless otherwise defined, *synthesis* will refer to the progressive grouping of species into genera, genera into families, etc., *analysis* to the sequential subdivision of taxa.

CHAPTER TWO

Lamarck on Continuity and Classifications

Jean-Baptiste-Pierre-Antoine de Monet de Lamarck (1744–1829) made his initial reputation as a naturalist by his botanical writings. The three-volume *Flore françoise*, which came out in 1778 and was apparently written as the result of a wager, was his first publication in natural history. After leaving the army he had moved to Paris and became acquainted with the circle of botanists associated with the Jardin du Roi. There he met Bernard de Jussieu, the central figure at the Jardin. In the course of writing the *Flore*, Lamarck had attracted the patronage of Georges-Louis Leclerc, Comte de Buffon (1707–1788). The Abbé René-Just Haüy (1743–1822), something of a botanist himself, was closely associated with its writing, among other things making sure that the style was sufficiently elegant to suit Buffon's taste. The zoologist and anatomist Louis-Jean-Marie Daubenton (1716–1799) was also connected with the preface to the *Flore*, apparently being assigned by Buffon to oversee Lamarck's prentice work.[1]

Nature

In light of the philosophies of his associates, it is not surprising that ideas of continuity pervaded Lamarck's Introduction to his *Flora*. There he drew a connection between the existence of a continuous *scala naturae* and the assertion that the limits of higher taxa were as a consequence necessarily arbitrary.[2] He was convinced that the continuity of the natural order ensured that any one organism was an integral part of a seamless whole that made up nature.

> The goal of a natural order . . . is to link all our ideas, so that we can grasp all the commonalities by which beings hold together, the one to the other, and not to bring any object to our consideration without showing us at the same time everything that exists above and below, and so to train oneself in these larger panoramas which take in the whole sweep of the subject, and which are as it were the vision of a genius.[3]

He then proceeded to describe how to construct this natural order, and to show its immediate value for a naturalist: "having determined some plants as being the first of the order, one places immediately afterward that plant which appears, of all those known, to have most relationship ['rapport'] with it, and continues the same gradation of nuances until one comes to the plant that differs most from the first, and which for this reason will form as it were the last link in the chain."[4] Lamarck continues, "The order that is being discussed here, instead of being a confusing mass of names and ranks thrown together at random, will on the contrary form a whole subject to fixed rules, which, however, do not divide it, and do nothing except to determine the place which each species must occupy in the general series."[5] There was a simple, linear sequence and it was possible to assign plants to their correct place in the sequence—but not by using divisive or analytic methods.

Lamarck thought that it was important to outline the series as a whole, because what he called the factitious and arbitrary divisions of naturalists could not be used to judge either the laws to which living beings submitted or the true relationships which grouped them. In what he considered was a major metaphysical error, naturalists had confused ways of reconstructing the order of nature and those which helped only in assigning names to organisms.[6] Division was useful only when naming plants and was, Lamarck thought, an absolute stranger to the march of nature.[7] The "laws" of nature became apparent only when the undivided continuity of life could be perceived, or, better, it was the laws of nature that ensured the undivided continuity of life. There were no abrupt changes in this continuity; rather, there was a gradual transition between the first and the last members of the series.

Arrangements and Classifications Distinguished and Reconciled

The utility of the natural arrangement was clear: it was an *order*, and plants could be assigned their position in the overall order by the application of *rules*. Discussion of grouping and ranking is, however, conspicuously absent in Lamarck's *Flore*; indeed, such activities would be irrelevant in any description of the natural arrangement.

Lamarck identified three main problems in his endeavor.

1. *To determine the plant that should be placed first, and that would serve as the fixed point from which one departs and then arranges ["graduer"] the entire order, and arrives by a natural succession of similarities to the ultimate limit of the plant kingdom.*
2. *To establish the rules that should guide the Observer in the linking of species.*
3. *To find a way for orienting oneself in an order that admits no line of division.*[8]

The first problem was perhaps the least of Lamarck's worries. It was easier to decide which plant was least organized, the least complex, the least complete in its parts, than it was to judge which was the most "vivante," the most organized or the most perfect. Thus one should start with the simplest plant, but Lamarck noted that the order, once established, could readily be reversed.[9] In the first volume of the *Flore françoise* he began with simple organisms like mosses, fungi, and their ilk—all of which lack flowers. The first family mentioned in volume 2 is the Compositae, a family in which the flower heads consist of numerous small flowers surrounded by involucral bracts, while the flowers themselves have petals that are fused into tubular or straplike structures, in either case consisting of but a single piece. This seemingly counter intuitive arrangement of placing a family as complex as the composites at the beginning of volume 2 was perhaps, in part, because plants capable of movement (such as *Dionaea muscipula* and *Mimosa pudica*) were in the Polypetalae, and this family would therefore need to be placed later in the series, closer to animals.[10] Also, when looking at individual flowers, those of Compositae appear to be simple: the calyx is often modified as a pappus; the corolla is in one piece, hence appears to have only a single petal; there are only five stamens; and there is a single seed. He presumably thought that plants like Annonaceae, which he included in the Polypetalae, had more complete and complex flowers since they can be interpreted as having both sepals and petals (which are in threes), as well as numerous stamens and carpels; all parts of the flower are usually free from one another. The overall series was thus as follows: plants with no petals, then plants with one petal, and finally, plants with many petals. The sequence finished with the monocotyledons, in particular, the grasses. (Grasses present problems when placing plants in a series. They are obviously monocotyledonous, and many monocotyledons have separate sepals and petals (the two may be indistinguishable), but grasses lack these parts. However, other structures associated with the grass flower were interpreted as calycine by Lamarck—and later by Jussieu, as well.)

A few years later, in 1786, he reversed and in part reorganized the whole sequence. For the entry Classes in the *Encyclopédie méthodique* Lamarck began the botanical series with what he regarded to be the most complex plants: the Polypetalae. Although he did not begin with plants capable of movement, but with the Annonaceae, he continued to insist on the importance of sensitivity as indicating an approach of plants with this property to animals.[11] He also argued that his arrangement was based on the natural relationships of plants and that it showed "the gradation, either of the number or of the perfection, of the essential organs."[12] Lamarck continued the sequence with the Monopetalées, Composées, Incomplètes, and Unilobées, finishing with the Cryptogames.[13] However, it is unlikely that Lamarck's arrangement in flowering plants was linear in other than broad outline; as in the *Philosophie zoologique* later on, his idea of nature was certainly not one of a simple, linear progression.[14] But even in 1786 Lamarck suggested that only an order or sequence of *classes* could be established;[15] at lower levels of the taxonomic hierarchy organisms yet to be discovered caused too many gaps in the natural arrangement for any sequence to be discernible. This is in practice, if somewhat less in theory, very similar to the position he adopts in the *Philosophie* where the major groupings of animals were only doubtfully joined.

As for the second problem, that of establishing rules by which plants should be linked, Lamarck attempted to develop an objective measure of the degrees of similarity between the plants he was comparing. Two plants were placed closer to each other than to a third plant if they showed a greater similarity in a numerical comparison.[16] In this comparison, features of the flower and fruit in particular were used, and were assigned values in proportion to the frequency with which they occurred in plants—the greater the proportion of plants in which a feature was found, the greater the importance that feature must have for the life of the plant. It should be noted, however, that these values could be modified if there were differences in the expression of the feature being compared. Thus although *Nymphaea* and *Podophyllum* both had the same number of sepals ("folioles"), this similarity counted for only a similarity rating of 10, rather than 21 (the maximum number for this feature; sepals are very widespread in flowering plants, and hence implicitly important). The lower rating was because the sepals were persistent in the one and fell off easily in the other. *Phytolacca*, which Lamarck included in the three-way comparison to demonstrate how his method worked, entirely lacked a calyx and so was given a value of zero for this character.[17] It is important to remember that the features of the plant in which Lamarck was interested were largely those of external form.[18]

The final problem was that of orienting the naturalist in the seamless natural arrangement. Here it was necessary to use art in observing nature. Lamarck presented part of his natural arrangement in the form of a table in which genera are listed and families and higher groups are also indicated (figure 1). He included distinctive characters for the groupings ("saillies particulières"), and these groupings could be given convenient names by the naturalist. The characters in question were not divisive, separating discrete groups; rather, they simply highlighted those parts of the arrangement a naturalist might wish to emphasize,[19] serving to situate him in a restricted portion of the seamless natural order. Thus although the order as a whole (the arrangement) was that of nature, because it was the naturalist who decided which characters should be emphasized, groupings circumscribed using those (or any other) characters—the classification itself—must be alien to nature. Furthermore, Lamarck held that the characters included in the table were found only in those genera immediately opposite the place where those characters were inserted. Genera a little further away might have

ORDRE NATUREL.

SÉRIE GÉNÉRALE des genres rapprochés en raiſon de leurs rapports.	SAILLIES PARTICULIÈRES formées par certaines affinités remarquables.	RAPPORTS GÉNÉRAUX & éloignés, indiquant la perfection graduée des organes.
Agaricus T		Fructification abſolument inconnue & inſenſible.
Boletus.		
Fungus.		
Hydnum.	*Champignons.* Subſtance ſpongieuſe, lamellée ou poreuſe, & qui, ſous diverſes formes, s'étend en hauteur ou eſt très-ramaſſée.	
Phallus.		
Elvela.		
Clathrus.		
Peziza.		
Lycoperdon.		
Clavaria.		
Mucor.		
Byſſus.		
Conferva.		
Ulva.		
Tremella.	*Algues.* Subſtance aplatie, membraneuſe, & qui, ſous diverſes ramifications, s'étend en longueur, & produit des cupules floriformes.	
Fucus.		
Lichen.		
Targionia.		
Anthoceros.		
Riccia.		
Blaſia.		
Marchantia.		
Jungermannia.		
Buxbaunia.		
Hypnum.		Fructification ſenſible, mais indiſtincte ou peu connue.
Brium.		
Mnium.		
Polytrichum.	*Mouſſes.* Feuilles nombreuſes, & diſpoſées en gazon, ou embriquées autour des tiges qui produiſent des urnes anthériformes.	
Splachnum.		
Fontinalis.		
Porella.		
Phaſcum.		
Sphagnum.		
Lycopodium.		
Equiſetum.		
Iſoetes.		
Pilularia.		
Marſilea.		
Ophiogloſſum.		
Oſmunda.		
Onoclea.		
Pteris.	*Fougères.* Feuilles toutes radicales, roulées en croſſe avant leur développement, & chargées de pouſſière féminiforme.	
Aſplenium.		
Trichomanes.		
Blechnum.		
Hemionitis.		
Lonchitis.		
Adiantum.		
Acroſticum.		
Polypodium.		

FIGURE 1

The Lamarckian series. Genera are listed sequentially; groupings of genera are indicated by characters; no boundaries between these groupings are shown. Source: Lamarck 1778, 1: cxiv–cxv; top of p. cxv removed, and p. cxv placed immediately under p. cxiv.

only some of these characters, and there would be an insensible gradation to genera having a wholly different set of characters. The same was true of the characters listed in the third column, "rapports généraux," that were found in larger groupings of plants. The relationship between characters and groups in Lamarck's arrangement is shown in figure 2A; note that the limits of groups and those of the distributions of individual characters are not necessarily the same. Lamarck's solution would allow only some plants that he assigned to a particular family to be placed in that family if only a single character were used. As more characters were added, it might be possible to improve the chances of placing a plant accurately; the result, as can be seen in the figure, would be a polythetic group. Such a group lacks defining characters possessed by all and restricted to only its members. Rather, although some of the characters may be found in all group members, they are also found in other groups, and while other characters may indeed be unique, they are not found throughout the group.[20]

Not surprisingly, Lamarck considered higher taxa to be arbitrary affairs, that is, arbitrary as to exclusion but not to inclusion.[21] For example, he thought that distinct genera were not natural, and that the distinctness of large, isolated genera would probably disappear as further collecting led to the discovery of undescribed species.[22] Nature was not yet fully known. As more organisms were discovered the continuum would become more evident and the apparent discontinuities that delimited such groupings would disappear.

Lamarck also mentioned two other desirable features of genera, and by extension of groups at all hierarchical levels. First, he felt that genera should be neither too big nor too small.[23] If nature were subdivided, the naturalist could characterize the subdivisions; whereas if nature were left undivided, the naturalist would be able to say nothing about the entire group since the members at either end would have nothing in common. Thus when he described the genus lichen in his *Encyclopédie méthodique* he noted that the undivided genus was so large and variable that there was no feature in common to all of its included species—perhaps it would be better to subdivide it.[24]

Second, and clearly related to the first feature, Lamarck felt that genera should be readily distinguishable. He thought that the features used to distinguish genera could be artificial, and did not have to be chosen from among the primary characters used to construct the natural order.

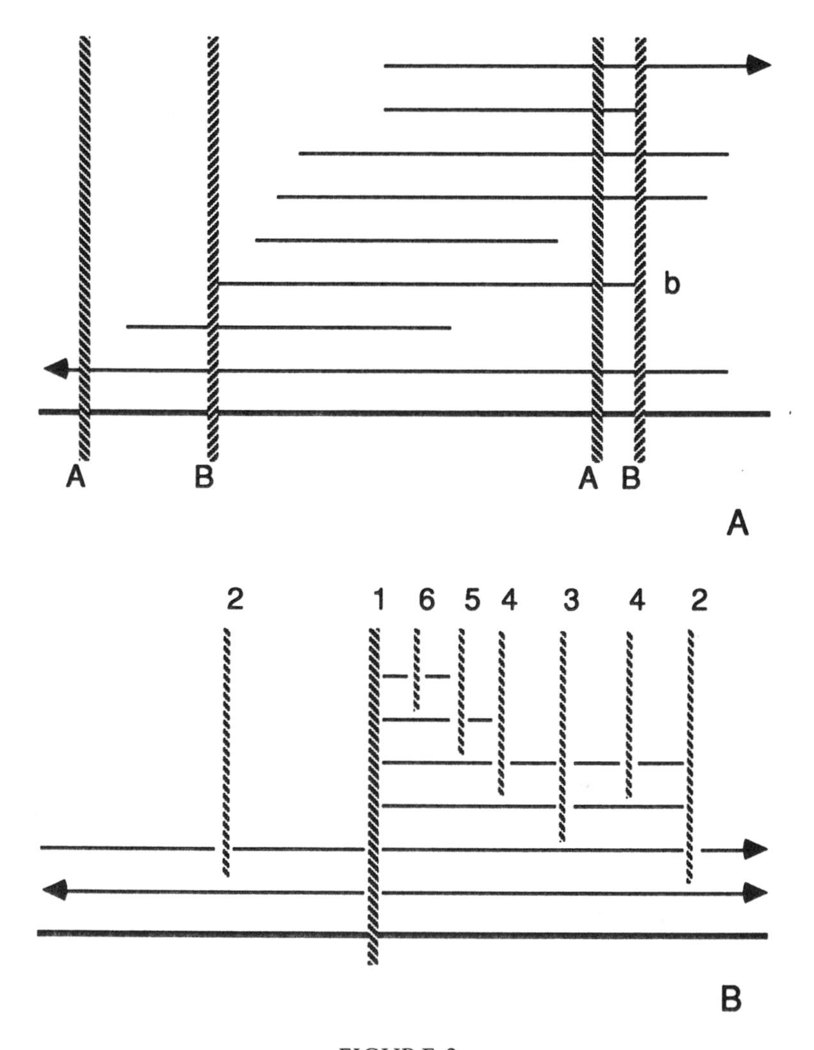

FIGURE 2

Relationships between characters and groups in a continuous series. Characters subordinated (A) or not subordinated (B). Thick, horizontal line represents the continuous series; thin horizontal lines, characters; arrows indicate characters occurring in other organisms adjacent to those in the diagram; vertical lines, taxon limits. 2A. *Continuity.* Vertical lines indicate family limits, with alternative circumscriptions of families indicated by A...A, B...B; in the latter case the family is incidentally monothetic by virtue of its possession of character b. 2B. *Hierarchy.* Characters progressively higher up the diagram are dependent on or subordinated to those below; thinner vertical lines indicate groupings of the hierarchy that are progressively lower in rank the shallower they are.

The latter characters might be inconspicuous, but secondary (often vegetative) characters were perfectly appropriate when the primary characters of the fructification were difficult to see and hence inadequate. Thus Jean-Louis-Marie Poiret (1755–1834), one of his collaborators in the *Encyclopédie*, separated *Trifolium* and *Melilotus* by features of this kind.[25]

Lamarck also described with great clarity the basic principles he used when constructing identification keys, analytic (artificial) methods that would allow the name of the plant to be determined in the most accurate and expeditious fashion possible.[26] Lamarck's recommendations for writing keys can hardly be improved; indeed, he discovered the practice of optimization, suggesting that symmetrical keys with leads of approximately equal length were more likely to lead the user to the correct identification than asymmetrical keys with leads of greatly different lengths. Even if there were 100,000 species, he realized that it would be possible to assign the correct name to any specimen by using only sixteen consecutive binary divisions in the key.[27] Lamarck insisted that neither method (grouping using many characters) nor system (grouping using few characters) was a substitute for analysis when it came to assigning names to plants.[28]

Thus distinctions between analysis and synthesis, and between identification, classification, and the arrangement of the natural order, were clear to Lamarck.[29] Division could only disturb relationships and destroy the natural order. Synthesis, on the other hand, resulted in groups joined by nuances and imperceptible changes—in short, the arrangement of nature, the natural order manifest. Taxonomic groups and classifications were made by the naturalist and were no part of nature, but they allowed the naturalist to talk about nature.[30] Although life as a whole was indivisible, for the immediate needs of the naturalist characters could be emphasized, and groups formed and named; and the name of the group to which an unknown organism belonged could be readily determined by an analytic process. It was by the process of synthesis, however, that the order of being—nature in its undivided entirety—was discovered.[31] Interestingly, although synthesis came to play an important role in classificatory methodology, Lamarck's work was eventually remembered mainly for the novel identification keys that were included in it.[32]

CHAPTER THREE

The Natural Method of Antoine-Laurent de Jussieu

Lamarck never provided a succinct natural arrangement of plants. In 1783 he began his mammoth *Encyclopédie méthodique* in which descriptions of all genera and families were included, but this was not completed for 35 years.[1] Moreover, he changed the sequence of his families, so he provided no "system" for arranging a cabinet or writing a flora. In 1764 Michel Adanson (1727–1806) had already outlined a natural order of plants and described all families and genera, the latter in tables of characters varying at a generic level within a family,[2] yet his work was largely disregarded. This was partly because of his idiosyncratic orthography and his rejection of Linnaean nomenclature. Linnaeus's binomials were already proving their value; Antoine-Laurent de Jussieu (1748–1836) himself introducing them into the Jardin du Roi in Paris, and Linnaeus's reform of generic nomenclature was also useful, even if some grumbled at the changes he had made. In contrast, Adanson seemed to reject weighting, simultaneously adopting as an ideal a classification based on all features of the plant; this was seen as being a decidedly impracticable approach to the study of nature. Moreover, as Lamarck observed,[3] Adanson's family descriptions were inordinately long and could not be shortened. Adanson's work seems to have been deliberately ignored by

Jussieu, and the methodology underlying it became the source of much controversy.[4]

At about this time it seemed to some naturalists that the production of an account of nature that included descriptions of families was impossible. Colin Milne (c. 1743–1815), writing in 1770, observed that many of the links in the great chain were unknown, with the result that many plants could not be placed in natural families and "devoid of uniform relations betwixt themselves, they cannot constitute new families; they remain, in some sense, solitary."[5] We shall see that the issue of such solitary genera exercised Jussieu a great deal, although only 137 out of a total of 1,754 genera he recognized in 1789 were not placed in families. The gaps in the chain of being also presented problems,[6] but even if nature in its entirety were known, description and delimitation might still be problematic. Sir J. E. Smith (1759–1828), a believer in continuity, observed that there needed to be a precise technical "limitation" of the characters of a group, yet groups could not be so limited.[7] This problem was apparently solved by Jussieu, although I will argue that his solution caused substantial problems for his successors.

Antoine-Laurent de Jussieu thus provided in his *Genera plantarum* of 1789 the first generally accepted natural ordering for a major group of organisms. Both here and in his later work, the value of features of the fruit and seed in classification are emphasized, stimulated in part by the important publications of Joseph Gaertner (1732–1791) and his son, Carl Friedrich von Gaertner (1772–1850), on the structure of the fruits and seeds of flowering plants.[8] As Cuvier[9] implied, the success of Jussieu's method was associated with its incorporation of Gaertner's careful studies.

Jussieu's active taxonomic life spanned half a century, albeit interrupted by the Revolution of 1789–1799 and subsequent upheavals.[10] Son of a pharmacist living in Lyons, Jussieu had three uncles (Antoine, Bernard, and Joseph) who were all members of the Académie. Bernard and Joseph were, in fact, well known botanists. Jussieu's botanical tutelage was under Bernard (1699–1777), with whom he lived in Paris from 1765 until Bernard's death; the two thought similarly about the natural method. Antoine-Laurent became a doctor of medicine in 1770, defending a thesis with the title "An oeconomiam animalem inter et vegetalem analogia?"[11] In 1773 his paper, "Examen de la famille des renoncules" was read at the Académie, and he was elected adjoint botaniste, although he was not quite 25 years old. (He replaced Adanson, who was promoted Associé.) His career was firmly established,[12] although he did not replace Louis Guilliame Lemonnier (1717–1799) as professor of botany at the Jardin du Roi in 1785—the young René Louiche Desfontaines (1750–1833), instead, was awarded that post.[13] Nevertheless, Jussieu was made a Professeur de botanique à la campagne when natural history in Paris was reorganized in 1793.[14] Elected second director of the Muséum in 1800, he left Paris in 1826, largely on account of his failing eyesight—a hereditary problem—and lived ten years with his family in the country before he died in 1836. Although the natural method spread relatively slowly outside France,[15] in part perhaps because of the political tumult throughout Europe, by Jussieu's death it had triumphed almost everywhere.

In this chapter I outline Jussieu's theory, emphasizing how central his belief in continuity was for that theory. I then turn to an analysis of his practice, focussing on three main points: the general sequence of groups in the *Genera plantarum*, the relationship between the size and circumscription of groups, and the prevalence of linking genera and families and the effect of such linkage on the distinctness of higher taxa. But a warning is necessary: if we are to understand Jussieu, we must divest ourselves of any preconceptions as to what we think his classification represents. In particular, we should not assume that his hierarchy is a hierarchy of classes like that of Linnaeus (or perhaps Augustin-Pyramus de Candolle), and we certainly should not assume that the numerous genera and families that Jussieu described were either discrete or had essences or types.[16] Furthermore, continuity can be as well represented by anastomosing lines as by the unbranched *scala naturae*, and it is likely that Jussieu believed in the former, particularly as he got older.[17] Thus we should not necessarily expect the discussion of relationships in the *Genera plantarum* to lend itself completely to an interpretation based on simple linearity, although we shall see that the extent to which this interpretation may be valid is remarkable.

Theory

The Shape of Nature

The idea of continuity pervades the writings of Antoine-Laurent de Jussieu. Henri Daudin[18] noted that the discovery of the continuous natural order was the goal of Antoine-Laurent's early work, and so it remained; this was a goal of most of his contemporaries, too.[19] Almost to the last Jussieu was refining the analogy by which the natural system was to be visualized, ensuring that it best conveyed the idea of connectedness and links between indistinctly separated groupings of organisms, rather than varying distances between clearly delimited groups.[20]

Jussieu thought that the arrangement of his uncle, Bernard, was imperfect in that the points of transitions between the "classes" were not demonstrated.[21] He nevertheless observed with approval that Bernard was trying to fill in the gaps between the natural groupings recognized.[22] That continuity was the touchstone of Antoine-Laurent's own view of nature was perfectly clear, at least to his early readers. One cannot do better on this score than to quote the summary of the *Genera plantarum* written on behalf of the Société Royale de Médécine by the physician Jean Noël Hallé (1754–1822).[23] At the same time, this summary also suggests conflict between the idea of continuity and that of groups, the latter being almost *ex definitio* discrete. This conflict pervades Jussieu's work, and has considerably obscured our understanding of it.

> M. de Jussieu's work has as its goal to join all known plants in an order which, never interrupting the analogies by which the different individuals of this kingdom appear to be drawn together, presents them in a sequence that is in such a way continuous by nuances and relationships, that this chain needs nothing to be complete, other than the addition of plants that naturalists have not yet discovered or observed: this he calls the natural method.
>
> This method, if one can imagine it finished, would not present major intersections [and] well separated divisions, as with the artificial methods; but each plant placed between its neighbors ["entre ses analogues"] would always find itself as if en famille, and one would be unable to tell where one particular series began or ended. Nevertheless, one would see in the whole ensemble the different modification of the same organization forming, by special analogies, the principal groups of almost [much] related species; these groups, placed one after another, would touch and join each other by the nuances of the last species, their centers always distinct but their peripheries always joined.
>
> The existence of these groups cannot be doubted; it is incontestably demonstrated by those resemblances that not only join several species of plants in different genera, but which further join, in a similar fashion, several different genera in a way so obvious that, even in artificial meth-

> ods, botanists have sometimes had to abandon even the principal characters of their classification so as not to separate genera that nature forces them to join. One sees some remarkable examples of this in Tournefort and Linnaeus.[24]

Hallé apparently perversely suggests that the reality of groups is evident because of their continuity with other groups and their resultant undefinability. These groups were but fragments of the natural order, an order which was continuous; the fragments were arbitrary as to their limits but not, however, as to the groups they included. At the same time, it seemed that nature was divisible and groups recognizable, and these groups contained taxa that represented modifications of the one organization[25] or plan. In Hallé's summary the word "nuance," the gradual change that was the hallmark of undivided nature, is opposed to "group" and "organization," words which suggest a very different world, while ambiguous words like "analogie" and "rapports" also abound.

Jussieu himself was clear that his method depended on the idea of continuity:

> The natural method, the truest and best aim of science, has been and still is hidden in part from Botanists because of a lack of knowledge of all the plants to be arranged, and of the principles of arrangement. First, a continuous series of beings cannot really be established unless all have first become known. Botanical journeys produce discoveries daily; but there are probably many more exotic beings remaining to be discovered than have already been found, [thus] an absolute knowledge of all existing beings is never ["nec unquam"] to be hoped for. But, even though all regions of the earth have not yet been explored, the rest [unknown areas] have nonetheless been noted by Geographers and carefully arranged in maps, each in its own location. Again, although numerous links of a chain may be missing or scattered at a distance, many are easily connected at different points in the chain, which are joined at various later times by new intermediate links, and gradually come to be separated by fewer intervening spaces. Therefore, the method ought to be worked out so that it imitates a chain or map, despite the lack of exotic genera.[26]

The nature whose arrangement Jussieu was trying to discern was continuous; the gaps that existed would be filled by plants that, although not yet discovered, existed somewhere on the surface of the globe.[27] This continuity was quite clear in those examples of the natural order that were already known. Jussieu's natural method was thus a reasonable goal for naturalists: the portions of continuous nature that naturalists could recognize allowed them to demonstrate the method by which the whole was to be constituted, and also to confirm its validity, while future discoveries would, at least in principle, provide the rest of the materials needed for this task.

> This long-desired arrangement, far superior to all others, alone truly uniform and simple, always in conformity with the laws of affinities, is the so-called *natural method*, which links all kinds of plants by an unbroken bond, and proceeds step by step from simple to composite, from the smallest to the largest in a continuous series, as though a chain whose links represent so many species and groups of species, or like a geographical map in which species, like districts, are distributed by territories and provinces and kingdoms.[28]

In visualizing the shape the natural order might take, Jussieu emphasized the impression of continuity that the use of the analogies of chains and geographical maps conveyed. Between the boundaries of the different administrative units of a country there were no physical gaps, and in one step one could move from one province to another; links of chains were firmly interconnected, and whether they represented species or groups of species, they followed each other in a regular succession. However, as Jussieu had noted, different plant families had their own facies, or characteristic appearance, yet at the same time they were united by genera.[29] They were both groups and part of a continuum, the same contradiction apparent in Hallé's summary. To develop the analogy Jussieu used, a distinctive dialect in one of the provinces might reflect some commonality that might fuel later attempts to gain independence; the links in the chain could be prised apart.

Detecting the Natural Order

Earlier, Jussieu had expressed his feelings about the search for the natural order as he grappled with the analogy Buffon had used when he characterized—or perhaps caricatured is a better word—the way in which naturalists, in particular Linnaeus and his followers, recognized relationships. For Buffon, the search by such naturalists for general and perfect methods that would allow them to use single characters when recognizing relationships in the natural system was like the search for the philosopher's stone—it was fated to fail.[30] But Jussieu did not see the matter in quite the same light.

> This most perfect method is that which is closest to the natural order, or, rather, it will be the natural order itself, of which all the artificial methods are nothing but very imperfect imitations. The route to be followed in finding it is different from that we have followed up to now; it is true that the advantages which must result from success are balanced by the difficulty of the enterprise: the natural order is like the philosopher's stone of the chemists. The impossibility of uniting, or even of being acquainted with, all the plants which must make up the general chain will always be an insurmountable obstacle and will leave gaps that are difficult to fill; but Nature has scattered the material intended for the construction of this order, she allows us to catch at least a glimpse of the principles on which it is based. Among the characters that plants provide, there are some essential, general and invariable, which, it appears, must serve as the basis of the order that we seek. They are not arbitrary, but based on observation, and are not to be obtained except by proceeding from the particular to the general.[31]

There are two issues involved: a synthesis that is the building up the whole from its smallest parts and the use of many characters. The two together would give the natural method or order, and Jussieu's philosopher's stone was at last within reach of the naturalist. The true student of nature investigated the whole plant, looking at all its characters, and so, omitting nothing and analyzing the relationships of characters in different plants, could attain a complete knowledge of the relationships of plants.[32] Like Buffon, Jussieu never wearied of saying that the use of single characters led to error, although we shall see that he had in mind both the problem of identification as well as that of establishing relationships. For the former, he observed that a drawback of "artificial methods" was that a complete flower was needed to place the plant correctly, but, by using his approach, even a plant known only from its vegetative parts could often be placed in the natural series.[33] He felt that families and other higher taxa had to be recognized by combinations of characters, because the secondary characters (primary characters were useful at higher levels in the hierarchy) that were used as diagnostic sometimes varied.[34] These combinations gave the taxa their diagnostic characters; such taxa would not be recognizable as entities by any analytic method that required all members of a taxon to have the same single feature. Even if natural families possessed invariant secondary characters, these would not be unique at the level of plants in their entirety, since these characters might also be found in a family or group of families with quite different primary characters and so in a completely different part of the natural series.

Jussieu made, at least in theory, an important distinction between differentiae, which were external features, and characters proper, which were internal features that indicated the organization of plants and hence their true nature.[35] The two were connected, however. In 1788 he observed that "characters or external signs" were intimately linked to differences in the properties of plants, and these properties could be related to the "elementary principles" that made up plants;[36] characters were not superficial or trivial differentiae. In the *Genera plantarum* he suggested that "adjunct and secondary exterior signs" of animals were likely to be associated with internal differences (that is, in the structure of their hearts) that divided them into major groups.[37] However, although he exhaustively listed the differentiae of plants,[38] he neither listed true characters nor demonstrated that the characters he used and on which he placed so much emphasis had a basis in the organization of plants. He also seems to have contradicted his statement on the inner groundedness of characters when he suggested in 1774 that anatomical studies were needed only of animals. Animals had a variety of functions, a full understanding of which obviously depended on knowledge of the internal structure of the organisms. In plants, however, the functions of all characters were known, and almost all characters were external.[39] Yet in 1774 Jussieu had noted that differences in external features of plants depended on differences in the embryo, while he argued against reliance on external characters for delimiting higher-level groups since those would yield characters useful only at a low level in the classification.[40] Indeed, gross anatomy was part of Jussieu's program for the understanding of natural relationships, even if cell-level studies were not, and dissection was necessary for the observations on fruits and seeds so important for this work.[41]

There is, however, ambiguity in Jussieu's rationale because it is not clear how Jussieu interpreted plant structure. In the *Genera plantarum* he described the plant body in Linnaean terms, noting, for example, that the alburnum (sapwood) and liber (probably phloem) developed on the interface of the cortex and the medulla, and that changes in the purity and amount of the sap caused many of the changes in the form of the growing plant. Like Linnaeus, Jussieu thought that the corolla was continuous with the liber (not with the cortical epidermis), that there was a connection between the doubling of flowers and the level of nutrition of the plant, and that the seed was the "compendium" of the future plant.[42] Linnaeus himself believed that classifications should be based on internal characters, yet at the same time he insisted that the important characters were those of the fructification that were readily visible. There was, however, no real contradiction for Linnaeus, since such characters actually represented the inner tissues of the organism, the adult structure that became external only after the metamorphosis of the vegetative plant. This metamorphosis was a kind of moulting process in which the outer layer of the plant was sloughed off revealing the important organs formed from the innermost tissue of the plant by interaction with the medulla.[43] To the extent that Jussieu agreed with Linnaeus, he would be able to have the argument both ways. He could insist on using internal characters, yet base his classification on largely external features of the fructification.[44]

But how were characters to be used in the method that allowed the naturalist to discern the shape of nature? Initially, Jussieu did not place emphasis on characters. He was more insistent on establishing the correct sequence of organisms. From this sequence, information about the relative value of characters would become apparent. Moreover, the natural method as the set of procedures to be followed when establishing relationships between organisms and the natural method as the natural order itself (the arrangement of organisms in nature) are intertwined in a complex fashion. The order, the arrangement of groups in a continuous series, was to be produced by the process of synthesis; this order would then indicate the way in which characters could be used when further building up this series and in evaluating the significance of the distributions of different characters in the series. There were two ways of understanding the method: either by proceeding from the species and forming ever larger fragments of the natural order (as he emphasized in 1774, proceeding from the particular to the general[45]) or by examining fragments of that natural order and uncovering the principles by which those fragments were held together.

From Simple to Composite

Jussieu was clear on how to go about forming the natural series: nature itself was arranged from simple to compound or composite, so the first way of attaining the method was a similar progress. From a uniform and concise series of observations and propositions, the principles of the method could safely be elucidated. Species were the starting point. The same law that forced all individuals that were alike into a single species also determined the union of similar species successively into genera, families, and even classes. The species itself should never be divided since the first and clearest law of nature ensured that all plants that were similar and connected by a continuous series of generations were placed in species.[46]

This argument was then extrapolated to the construction of groups of larger size,[47] although Jussieu did not specifically mention genealogical connections in his discussion of groupings above the level of species. Thus species which were alike, and so agreeing in a very large number of characters, were to be grouped into genera, and genera were to be similarly grouped into families. There were, however, no break-points in the natural series that could be used to circumscribe these groups, and their limits were largely decided by the botanist. Nature could not help here, and any breaks were, as Jussieu noted, "almost arbitrary":

> A genus is thence constituted, a congeries of similar species, of which nevertheless there is taught no accurate definition nor clear precept of construction, especially in the natural method. If there is any such, it differs very much from the systematic, it does not assign strict limits to genera, but through preserving the affinities connects all species by an imperceptible nexus, which it then for the sake of ordering the science joins in distinct genera, with almost arbitrary cut-off points, enriched with few or many species at one's pleasure, being called truly natural when they do not separate related species or disturb the entire series.[48]

Although Jussieu suggested that all characters should be used in assigning a plant to its place in the system, he cautioned that not all characters were equal in value. He also discussed this problem in his two early papers, on the Ranunculaceae (the buttercup family) and on the arrangement he had adopted in the Jardin du Roi. As with Linnaeus, characters of the fructification were most important and were to be privileged;[49] the most essential characters were those involved in the most important function of the plant, that is, its reproduction. Essential characters were to be taken from the embryo first, then from those organs that cooperated to produce it, the stamens and the stigma. These primary characters were the number of cotyledons, their position in the seed, and the relative position of attachment of the calyx, corolla, stamens, and pistils. These criteria would then constitute the three main groupings of plants and those of their subdivisions, the classes, and they were invariant. As Jussieu observed, "*Any character that varies in a particular case cannot have general value.*"[50]

It may be helpful to explain exactly what these primary characters are. Cotyledons, seedling leaves, vary in number. They are absent in ferns (although Jussieu did not understand how these plants reproduced). There is one cotyledon in *Allium* (onions), two in *Vicia* (beans), and several in *Pinus* (pines). The calyx, corolla, and stamens may be inserted separately below the ovary on the receptacle, the hypogynous condition (an example is *Ranunculus*, the buttercup—see also fig. 7). Calyx, corolla, and stamens may also be borne on top of the ovary, the epigynous condition (*Cichorium*, the chicory), or they may all join together and have a common point of insertion on the rest of the flower, which is usually below the ovary, the perigynous condition (*Rosa*, the rose).

Other features, however, might have to serve as markers for these essential characters if their observation is difficult because of their position in the flower or their minute size.[51] Such markers, nevertheless, would be of no essential value themselves. Detection of the primary characters might require markers if they were internal; in both plant and animal kingdoms, however, only characters of the third and fourth rank would be obtained strictly by examination of external features. These lower ranks, of course, would be distinguished by differentiae, not characters, if one is to use the distinction that Jussieu himself later drew.[52] Furthermore, I do not think that Jussieu regarded even the characters as essences of groups; I return to the issue of the relationship between characters and groups later in this chapter.

Species and genera were apparently grouped because of overall similarity in many characters used simultaneously. Yet in Jussieu's discussion of characters we find that these are in fact *ranked* in such a way that the presence or absence of one or a few characters examined in a hierarchical fashion could lead to the placement of a plant in its proper natural group.[53] Characters are treated as circumscribing not groups of progressively larger extent but of successively smaller size. The Cruciferae provide an example of this hierarchy of characters. The primary characters of the Cruciferae, hypogynous stamens and an embryo with two cotyledons, were constant throughout this family and many others, and were very high-level characters; the former character was a feature of a group of several classes,[54] the latter characterized the dicotyledons in their entirety.[55] The secondary characters (e.g., the polypetalous corolla and the lack of perisperm) were those of classes or families. The tertiary characters (e.g., the six tetradynamous stamens, two of which were slightly shorter than the other four) were largely family characters and sometimes varied. Quaternary characters, such as the shape of the fruit, were less constant and allowed the recognition of sections (groups of genera).[56] The result was that although the entire fructification was needed to place a plant in its right genus, at each hierarchical level only one or a few characters needed to be examined or to be included in the description.[57] Jussieu also discussed constant and variable characters at the generic level, using *Rosa* and *Valeriana* as examples and making similar points. He observed that some characters were frequently variable within genera, others were either stable or variable, while yet others were largely invariable.[58] Along similar lines, Jussieu thought that if some characters were dependent for their existence on the presence of another, the former characters, no matter how numerous, could only be equal in value to the latter.[59]

Thus the generic characters to which Jussieu so often referred[60] were the characters of a genus that, considered in an almost analytical fashion, allowed it to be assigned its place in the system as a whole, or the characters distinguishing that genus from the others in the family to which it belonged *plus* the characters that located the family in the higher groups to which it in turn belonged. Jussieu's early treatment of the limits of the Ranunculaceae followed similar lines; when considered hierarchically, the numerous characters that he described allowed him to form an aggregation of genera in a particular subgroup of the dicotyledons. Thus unlike an "Adansonian" approach to grouping where numerous characters of equal weight were all analyzed together,[61] characters were neither of equal weight nor were they all used together. This is the context in which Jussieu's assertions that "*one constant character becomes equal or superior to many inconstant ones*" and "*those plants ought to be grouped together which are related in a very large number of characters*"[62] are to be interpreted. In estimating affinities a character was not so much weighted and combined with others as assigned to its proper level or rank.[63] The botanist could decide on this ranking only if he had grouped plants because of their similarity in many characters. Unfortunately, we have seen that the rank of lower-level characters might differ in different groups, and it was only by what Jussieu disarmingly called "assiduous meditation on truly natural genera"[64] that characters could be assigned their right rank.

Variation inappropriate for the assigned rank of characters also occurred, and here, perhaps (although Jussieu is not clear), the use of more numerous characters of lower rank might be decisive in assigning a subordinate group to its correct superordinate group.[65] Yet in his discussion of the ranking of characters Jussieu gave the impression that it was absolute, and he was certainly so interpreted by some of his followers.[66] His criticism of Adanson's equal weighting of characters—that it was a mistake that no invariant and essential characters were allowed[67]—is consistent with this position. Nevertheless, variation of some characters was to be expected because aggregations of species and genera were a part of continuous nature; characters, especially the hierarchically less significant ones, were not necessarily constant in groups.[68] In the introduction to the *Genera plantarum*, Jussieu thus asserted that characters are to be considered hierarchically, and that they are constant within the groups they characterize. Although the possibility of variation is sometimes allowed, this is only at lower hierarchical levels.[69]

In outlining his first approach, that of synthesis, Jussieu mentioned several obviously natural groups, which he variously called families and orders, over whose circumscription and existence there could be no doubt. These were the Gramineae (grasses), Liliaceae (lilies), Compositae (daisies), Umbelliferae (carrots), Labiatae (mints), Cruciferae (cabbages), and Leguminosae (beans). Similarly natural genera were also mentioned.[70] These taxa, it must be remembered, were regarded by Jussieu as exemplars of the continuous natural order, not as discrete, bounded groups, nor even, other than incidentally, of Jussieu's own families and genera. That they represented part of the natural order, not discrete taxa, is strongly suggested by the fact that one of them, the Compositae, did not appear as a family in the main part of the *Genera plantarum*, but as an unnamed class made up of three families: Cichoraceae, Cinarocephalae, and Corymbiferae. The obviously natural families that Jussieu used as exemplars showed how plants were to be linked together; many characters were to be used, but strictly subordinated one to another. Within these families there were no major divisions, as befitted their status as exemplars of the continuous natural order. The great similarity between the genera included in the Labiatae, Umbelliferae, and Leguminosae was proof positive that they represented the true order of life. But this same similarity was an embarrassment and a disadvantage when it came to finding characters by which genera within the families could be recognized. Indeed, there was an inverse relationship between the distinctness of a family and that of the genera it contained.[71]

From Composite to Simple

The second way of arriving at the natural method was quite different. Here Jussieu used the same natural families mentioned above to test what he considered to be botanical laws.[72] He claimed that these laws, which really amount to a repetition of the assertion that the limits of characters and groups are identical, should be rejected if but a single character contradicted a family. There were some apparent exceptions, but Jussieu showed that further study of the character itself might explain them away, so confirming the ranking. Thus he was much concerned to show that the infra-familial variation of the character of cotyledon number, on which the primary division of plants was based, was only apparent and that cotyledon number was indeed absolute. The single cotyledon of *Ranunculus glacialis*, for example, was more accurately described as monophyllous rather than unilobed, and the apparently single cotyledon in fact was the result of the fusion of two cotyledons.[73] (Later in this chapter I will examine if Jussieu's belief in the constancy of his so-called laws was justified.)

It is in this context in particular that Jussieu suggested that the constancy of a character within and between families reflected its importance in the life of the plant.[74] As he had observed in 1774, the seed was functionally the most important part of the plant and was involved in the fundamental ("primitives") divisions of the natural order.[75] Bolstering his belief that he had chosen the right primary character, the number of cotyledons, by which to divide plants was the analogy he drew between the corculum or embryo of plants[76] and the heart of animals. Variation in the structure of the heart separated seven classes of animals, and the heart was the first living and moving structure evident in the embryo.[77] The heart was functionally essential for animal life, it showed the distinctively animalian feature of movement, and so it was not surprising that it was the first obvious structure to appear in the embryo. There were also seven major natural groupings of plants;[78] incubation and germination were similar phenomena;[79] and the heart and the plant embryo were similar structures.[80]

These functional considerations are advanced only in Jussieu's discussion of the embryo. They are barely mentioned in his subsequent treatment of the insertion of the stamens on the flower, and they are entirely ignored in his treatment of whether the corolla consisted of one or several pieces (although the latter omission is not unexpected since the condition of the corolla was a largely artificial feature used to increase the number of dicotyledonous classes). In any case, Jussieu's knowledge of the functions of plant parts was severely limited, despite his protestations to the contrary.[81]

Jussieu proceeded to elaborate on the characters by which plants could be successively subdivided into groups of ever smaller extent—much as he had outlined his first approach.[82] He thought that only the characters of the number of cotyledons, the presence or absence of a calyx, and the insertion of the stamens on the flower (whether on the calyx, the receptacle,[83] or the ovary) were suitable. Only seven groups were delimited by these characters. These were the Acotyledones, Monocotyledones, Dicotyledones (the distinction between which was based on cotyledon number) and three unnamed subgroupings of each of the two latter groups that were distinguished by differences in the insertion of the stamens (the first group was indivisible because it had no stamens). The three subgroups of the Dicotyledones were too large for convenience. To subdivide them Jussieu used the more artificial character of whether the corolla was absent or present, and, if present, whether it was monopetalous (a corolla that was undivided when it joined the rest of the flower, whether or not it was lobed—in most cases, we would rather loosely say today that the petals were all more or less fused) or whether it was polypetalous (consisting of several separate elements). In the rather lengthy discussion of this character, Jussieu showed that when a corolla was monopetalous the stamens were usually adnate (or joined) to the petals. He also implied that *because* there is this association, the character was not as artificial as it might otherwise seem; the insertion of the stamens was an important character, and some of this importance was transferred to the corolla. Jussieu thus at times discussed the insertion of the stamens and the insertion of the corolla as if they were interchangeable.[84]

In the *Genera plantarum* Jussieu seems to have been striving toward a numerical weighting system not radically dissimilar from that suggested by Lamarck a decade before. Near the end of the Introduction, Jussieu suggested that the weight of a character depended on its stability within the families he had recognized. Thus the structure of the corculum was sixty times more valuable than the leaf blade.[85] Unlike Lamarck, Jussieu evaluated the character in the context of its distribution within families, not in all plants. Furthermore, Jussieu apparently did not modify the value of a character according to its particular expression, as did Lamarck, but it is reasonable to suppose that his notion of character would include not simply the number of sepals, but their persistence (to use Lamarck's example).[86]

The Meaning of Affinity

Jussieu was interested in understanding the relationships between the different characters of the plant. I alluded to this when mentioning his use of nonessential differentiae to mark the existence of hard-to-see essential characters, the ambiguities found in his complicated discussion as to the importance of monopetaly and epipetaly, and the analytical procedure by which the natural method could be detected. Unfortunately, Jussieu's ideas on the subject of affinities are not at all clear. In his early paper on the Ranunculaceae he drew an analogy between botany and chemistry: the goal of the new science of botany was not memorization and naming, but understanding its "combinations and affinities, as in Chemistry, its problems as in Geometry."[87] "Affinité" here signifies an attraction or connection between different parts of the plant.

One of the most exciting areas of study in chemistry at that time was that leading to an understanding of the laws of affinity that regulated the combination of elements and the formation of compounds. That the characters used in the delimitation of major groupings of both plants and animals were similar suggested to Jussieu that some fundamental understanding of how organisms were built up or compounded was within his grasp; in both groups, it was the first-formed organ that was important.[88] Similarly, he hoped that botanical research would lead to an understanding of the connection between essential characters and nonessential differentiae that caused the latter to vary less than they might otherwise, and in animals at least, the comparable connection between internal characters and external, seemingly secondary, characters.[89] But the term "affinité" was also used when relationships between natural groups were being discussed.[90] Affinities here were what drew groups together, or attracted them; from an understanding of the interaction of characters, affinities of plants would become evident. The botanist would be able to discern the undivided continuum of life.

To take the argument further than Jussieu did, one can imagine an understanding of the affinity between characters providing a rational basis for their weighting; a subordination of characters would be firmly established (this is affinity in the first sense). The continuum that was life was a kind of individual, a unique, supraorganismic entity, manifest in the general similarity of characters between plants that were adjacent in the series (affinity in the second sense). Affinities (in the first sense) between characters meant that the shape of the continuum would be neither entirely haphazard nor featureless; certain combinations of characters were forbidden. If the affinities between characters that could coexist on a single plant were understood, a newly discovered or poorly known plant could be readily placed in its correct place in the sequence, and, perhaps more important, "organisation" or life itself could be understood. Jussieu did not clearly articulate this last goal, but we shall see that it was that of Georges Cuvier.

Indeed, as knowledge of plants improved, predictions could be made that were based on the affinities that existed between the different parts of plants. For example, if stamens were borne on the pistil in a particular flower, several other morphological facts could be predicted about that flower; similarly, monopetalous corollas nearly always bore stamens (epipetaly).[91] Exceptionally, even vegetative features might allow the prediction of floral form, as in the Rubiaceae.[92] Of course, the natural world was not always so tidy, and features such as the number of seeds per fruit seemed to be largely uncorrelated with natural groupings.[93] Correlations or affinities extended beyond the world of form, and not surprisingly—given Jussieu's ideas of the connection between the characters of the plant and its basic organization. Thus he suggested that there were correlations between the position a plant occupied in the natural arrangement and its pharmacological properties. For example, the detersive (cleansing, purging) properties of the genera placed last in the Umbelliferae also occurred in those Ranunculaceae which immediately followed.[94]

Discussion

It is clear that Jussieu was convinced, in theory at least, that nature was continuous. Even if all plants had not yet been discovered, nature had made the principles by which it operated evident in its natural groupings, those natural areas of continuity we could detect. The result of the application of his method would not be a classification in the sense of a hierarchy of groups each with its essential character; rather, it was an arrangement, an "ordinatio." But Jussieu emphasized that characters were needed when groups were recognized. He was of course fully aware how much the adoption of Linnaean nomenclature had helped botanical discussion by providing stable and unambiguous names for plants (he had introduced Linnaean names to the Jardin du Roi when he reorganized it in 1776), and it would be abundantly clear to him that what could not be named, and equally importantly identified and described, would effectively not exist for a naturalist. Hence come some of the ambiguities in Jussieu's work.

The emphasis on primary characters of the embryo would "simplify the task of botanists,"[95] but since some essential characters were hard to recognize, Jussieu linked them with more obvious differentiae that were not of themselves essential. Thus natural groups could both be based on essential characters *and* be recognizable. (One might remember in this context that Jussieu had observed with approval that his uncle was trying to ensure that natural groups had precisely this combination of features.[96]) Nevertheless, the problems caused by characters that were useful in grouping but were difficult to see is very evident in the *Genera plantarum*,[97] and they were exacerbated by Jussieu's insistence that the limits of groups and characters were the same. At the end of his book, Jussieu provided a rather sketchy key to the genera that he could not place in a family. These genera did not have the characters of the families he had described, nor did they have features linking them to each other, so even if a naturalist had a plant that could not be assigned to a family, there was still a possibility of finding out if it was known by using this key. Jussieu realized that it might still be difficult to identify plants, even those assigned to families, and he was apparently working on a key to all genera in the latter part of the 1780s. He promised that it would be published, but it never appeared.[98]

Although Jussieu tried to base the groups he recognized on what he called essential characters, I argue that these characters are not essences in an Aristotelian sense. There can hardly be an essence of what does not exist in nature, a subdivision of something that is whole only when undivided.[99] Furthermore, as James Edward Smith observed, the family (order) descriptions were notably vague: "So many exceptions encumber most of his explanations [definitions or characters of orders], that they mean in fact little or nothing."[100] Nevertheless, the descriptions of the families and other higher groupings in the *Genera plantarum* were one of the most important and original features of the book, and the characters that Jussieu repeatedly emphasized as being necessary when groups were subdivided were needed both for their characterization and their recognition.

Practice

Jussieu was not an exceptionally prolific writer by the standards of his day. His oeuvre is smaller than that of Linnaeus, Alphonse or Augustin-Pyramus de Candolle, John Lindley, and a number of botanists active in the nineteenth century. However, he was certainly productive, and although I focus on only one or two aspects of his practice, a complete study of his whole work from these few points of view would simply stupefy the reader. Hence I have been selective, but only in the sense of providing just enough examples to make my points.

It is commonplace when studying classificatory works to be faced with the problem of understanding the relationship between theory and practice. Thus we cannot be sure—indeed, it is unlikely—that Jussieu produced his arrangement entirely by synthesis—did he really look at *all* the species of plants to which he had access, and only then form them into groups of ever greater extent?[101] Classifications and descriptions tend to be accompanied by minimal discussion or evaluation of what was seen. Still less is there likely to be theoretical justification for the approach adopted. Nevertheless, in the *Genera plantarum* the extensive discussion in the Introduction outlining the theory of reconstructing the natural series is immediately followed by the natural series or arrangement Jussieu recognized. Hence the structure of his arrangement and the comments he made about the limits of groups (which of course may be justification after the fact) may be compared with his theory. From the descriptions Jussieu provided, the distributions of characters may be analyzed in more detail than he himself did, and we can see how these distributions relate to his theory. We can also look at Jussieu's later works and see how his theory and practice fared over time.[102]

In the ensuing discussion, I will focus on three points. The first concerns the general arrangement of major groups in the *Genera plantarum*, and how that relates to Jussieu's ideas on the overall shape of nature. The second point deals with the connections he drew between the size and circumscription of groups. Through it we can understand questions such as why the Compositae were used as an example of a natural family in the introduction to the *Genera plantarum*, yet were not recognized in the body of the work. The final point involves an exploration of the connection between the existence of linking groups and the distinctness of the larger groups they join.

The General Order of Nature

Despite the analogy Jussieu drew between nature and a geographical map, an analogy which might seem to preclude any progression in a subsequent linear arrangement of genera, the basic arrangement of plants in the *Genera plantarum* is decidedly linear, proceeding from small and simple to larger and more complex organisms (see figure 3). Most Acotyledones lack stamens, cotyledons, sepals and petals. The Monocotyledones have sepals, and also have a single cotyledon. Within the Dicotyledones, which of course have two cotyledons (or sometimes more, as we shall see), the first group, the Apetalae, are simpler or less complex ("composées") than the other groups because they, too, lack an organ, petals, found in the later plants of the series.[103] The Monopetalae are more complex, but there is only a single "petal" (i.e., the petals are all more or less fused); the stamens and the "petal" are usually intimately joined. In the last major group, the Polypetalae, the petals are separate from each other and from the stamens, so completing the whole series with plants in which there is the greatest distinction between the different organs making up the flower. The Diclines Irregulares, the last of the Dicotyledones, seem to fit in quite logically. Although petals are often absent, the sexes themselves are now separated, being found in different flowers, and the flowers of different sexes may even be found on different plants—the distinction between organs is becoming still more evident. The very last members of the Diclines Irregulares have more than two cotyledons.[104]

Jussieu recognized that there was no absolute distinction to be made between herbs and trees;[105] there was a continuum between the two. He was, however, not clear whether this was true for all plants, or whether there was simply a gradation within classes, as with animals. But even within animals, he saw a general progression from polyp to man, one of increasing perfection and, to a certain extent, of increasing size.[106] In general, Jussieu seems to have been looking for a rather simple progression, and he continued to use language compatible with the existence of a *scala naturae* in the posthumously published version of the Introduction to the *Genera plantarum*.[107]

Jussieu on occasion described the pattern of variation within a family as a degeneration toward the two peripheries,[108] and this would seem to contradict any idea of a general progression, or of continuity itself. It implies that there is something more perfect or complete that is degenerating, that there are groups, perhaps even types, but the language Jussieu used is ambiguous and does not appear elsewhere in his work.[109] It is unlikely that he meant degeneration in this sense, however, since he mentioned the word in the course of emphasizing that there were transitions between families. The degenerate genera were simply those that did not have the characters of the family to which they were assigned; as with Lamarckian groupings, peripheral members of families tended to have some of the characters of the next family in the sequence.

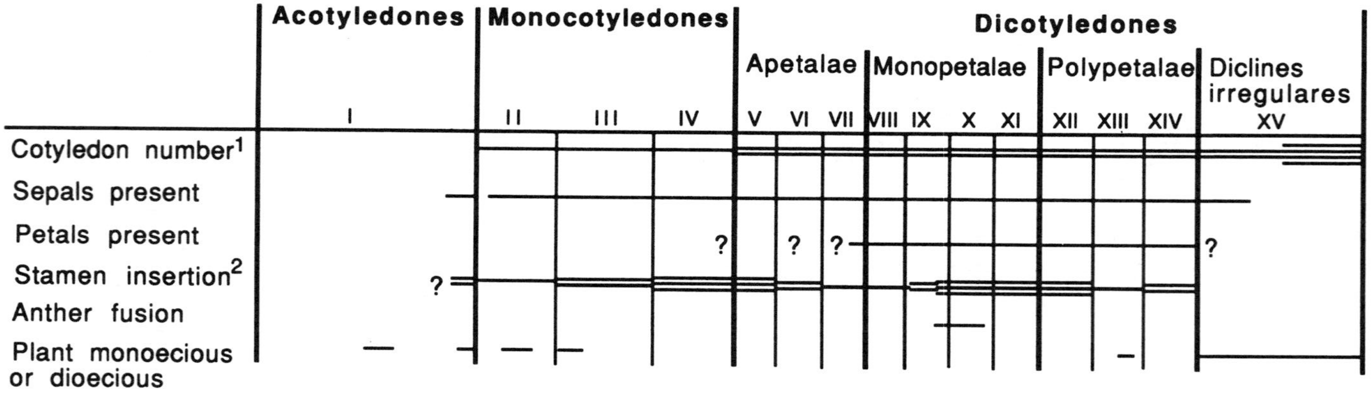

Acotyledones | Monocotyledones | Dicotyledones

Apetalae | Monopetalae | Polypetalae | Diclines irregulares

I | II | III | IV | V | VI | VII | VIII | IX | X | XI | XII | XIII | XIV | XV

Cotyledon number[1]

Sepals present

Petals present

Stamen insertion[2]

Anther fusion

Plant monoecious or dioecious

1. No line, none; — one; = two; ≡ more than two.

2. No line, not applicable; — hypogynous; = perigynous; ≡ epigynous.

FIGURE 3

Distribution of characters in Jussieu's general arrangement. Roman numerals, Apetalae, etc., and Acotyledones, etc., refer to classes; question marks denote uncertainties about the occurrence of a given character in a family. Information taken from Jussieu 1789.

Size and Circumscription of Groups

In any classification of a nature that is continuous there is an inescapable conflict—nature cannot be divided anywhere without doing violence to the system as a whole. Thus Jussieu observed in 1789 that *Valeriana* could be divided using characters from the fruits, but that even if it were divided the differences would be insufficient to separate the species in the natural series.[110] And as he had noted earlier, "It is immaterial if these genera [*Actaea* and *Podophyllum*] form a distinct section in the ranunculi or make part of a neighboring family; their relationships will always be the same; they will equally serve to establish an easy intercourse between two orders: this transition, which would be considered a defect in systems is a [mark of] perfection in the natural order."[111] Even at the level of the classes, he emphasized that there was no clean separation, since the extreme members of each class were joined to the next.[112]

Owing to the continuity in nature, the decision as to whether to maintain a group or to divide it depended on pragmatic considerations; groups had no ontological reality as groups. The main consideration to be observed when recognizing groups was that of size. Families ought to have more than a single genus in order "to generalize the character of the family."[113] This generalized family character was made up of characters found in all the genera of a family which then did not have to be repeated in individual generic descriptions; space was saved in the descriptions.[114] Jussieu's general aim was thus to generalize the characters of all higher taxa—that is, to base their descriptions on a summary of the characters of their subordinate taxa. The more related genera, the greater would be the pressure to recognize a family, and the more space that could be saved. Thus Jussieu noted that the last group of genera in his Apocineae were clearly different from the rest of the family, and if more genera were later found to belong to this group, they might form a family between the Apocineae and the adjacent family, the Sapotae.[115] (Note that in the time span covered by this book, the terminations of taxon names were not as formalized as they have since become. The names Apocineae, Apocinées, Apocinacées, and Apocynaceae all refer to the same family—but note that the first two names can also refer to a lower-level group, a tribe.) Similarly, he regarded the distinctive features of *Nyctago* as no exception, but "un mode d'organisation" that occurred in several other genera.[116]

The lower size limits of groups were clear—monotypic groups were of no use. Thus the small size of a group might make him hesitate to divide it,[117] and it is presumably because of considerations of size that Jussieu did not recognize monogeneric families in the *Genera plantarum*. Bigeneric families like the Typhae, Vites, and Cacti (the last included *Cactus* and *Ribes*, cacti and gooseberries) were either so discrete or the genera agreed in so many features that their small size did not militate against their recognition. Several other families had only three genera. Although some classes had only a single family (an example is the first class in the Polypetalae, which included only the Aristolochiae[118]) this was perhaps because they were in part artificial.

On the other hand, if groups were too big, they became unwieldy and could not be retained in one's memory. It is probable that the upper number Jussieu initially decided upon for both genera in families and families in the whole system was 100.[119] There are exactly 100 families in the *Genera plantarum*, and none of the families has more than 100 genera. The largest family, the Corymbiferae, has 99 genera, while the next largest, the Leguminosae, has 98. It is surely not irrelevant in this context that Jussieu divided what are now known as the Compositae—most of which had been recognized as a single family by Linnaeus.[120] To an uninitiated naturalist the Compositae might seem to be one of the most natural of all families, and, as we have seen, it was included as an example of a natural order in the Introduction to the *Genera plantarum*. However, the three families into which Jussieu subdivided it, the Cichoraceae, Cinarocephalae, and Corymbiferae, together have more than 100 genera.[121] More genera of these families were soon named, and when Jussieu reworked his *Genera plantarum* as he integrated Gaertner's observations on the structure of the seed with his own, he even suggested that the by-then unwieldy Corymbiferae would itself be best subdivided into four families based on *Eupatorium*, *Aster*, *Achillea*, and *Helianthus*, although he maintained his earlier circumscription of the family.[122]

Jussieu emphasized both that a group was to be subdivided when it got too big and that characters were needed for this.[123] Accordingly, he characterized the three families in which he placed genera from the Compositae. Thus the Cichoraceae had latex, and the flowers were ligulate and hermaphroditic; the other families lacked latex and had ligulate and tubular flowers. The Cinarocephalae had a pilose or plumose pappus (as did the Cichoraceae), and genera within it generally had a scaly receptacle (apart, for example, from the penultimate genus). The Corymbiferae quite often had a naked receptacle (the two other families did not), and those genera that had a scaly receptacle usually had naked seeds, or the pappus was paleaceous or dentate. Moreover, he had earlier observed that the Compositae in the broad sense did not have a single virtue common to all members, but that each subdivision had its own properties.[124] He had provided generalized characters for the three families and so they were perfectly formed according to his criteria.

Several additional genera of the Leguminosae, the bean family, were described after 1789, putting it over the magic number of 100. But perhaps unfortunately, in view of the subsequent protracted disagreement over whether it should consist of one family (the Leguminosae) or three (the Papilionaceae or Fabaceae, Caesalpiniaceae, and Mimosaceae), Jussieu never reworked it. The Rubiaceae, the coffee family, with 80 genera, were the third largest family in 1789, but Jussieu kept the family intact when he revised it in 1807, even though it then had over 110 genera. He noted, however, that many of these genera were small and might with advantage be combined so as to give a "greater extension to the generic character,"[125] i.e., the generic character would apply to more species. The Rubiaceae were also exceptional among plant families in that they could be readily recognized by vegetative features—they nearly always had opposite and entire leaves with prominent interpetiolar stipules. However, in view of Jussieu's dismemberment of the Compositae it is doubtful if this distinction would have kept the Rubiaceae from a similar fate if the family had become seriously oversized, although by 1820 it had 115 genera, as well as three more doubtfully included.[126]

Similarly, the presence of too many species in a genus provided grounds for its dismemberment into smaller genera, although here, too, Jussieu maintained that characters were needed for the subdivisions to be recognized. Linnaeus had included 24 species in the genus *Passiflora*, and it had accumulated about 80 species by the early 1800s. Jussieu observed, "This genus has enough species to be able to form a family by itself, and for that its subdivision into several genera is approved, if this division is well characterized and founded on very natural principles."[127] Of course, more than one genus would be needed if a family were to be recognized, yet a genus could become a family simply because more species were known, and not because of the greater variation shown by those species when compared to those known by earlier authors. Thus *Passiflora*, *Tacsonia*, *Murucuia*, and *Modecca*, the four genera Jussieu recognized in his Passifloraceae, all contained species known to Linnaeus. Similarly, there were many species in *Laurus*. In the case of *Laurus*, however, the characters available for its subdivision into separate genera were rather inconspicuous, but no others could be found. Jussieu didn't provide a full monograph, and the several genera that had already been described formed the basis of his Laurinées and allowed him to generalize its character.[128]

Thus the number of species in a genus was under the control of the naturalist, not of nature; in the Introduction to the *Genera plantarum*, having observed that genera were separated by "almost arbitrary" breaks, Jussieu went on to note that genera were "enriched with few or many species at one's leisure."[129] Almost twenty years later, although he emphasized that characters, differences in organization, were needed when genera were subdivided, the integrity of all large genera now seemed to be in danger. Jussieu mentioned *Melastoma*, *Carex*, *Erica*, *Hypericum*, *Salvia*, among others, as suitable candidates for division.[130]

At the very highest levels, too, the same principles operated. Classes that had large numbers of families should be subdivided if they were to be convenient to use.[131] However, here Jussieu ran into trouble. By using the character of how the stamens were inserted in the flower he was able to subdivide plants into seven classes. There were relatively few families of monocotyledons, and the three classes of monocots he recognized were based on important characters and were groups of reasonable size. There were, however, still too many families in the three classes of dicotyledons that he recognized using the same characters. Nevertheless, by using a less satisfactory secondary character, the nature of the corolla, the desired goal of "the proliferation of classes" in the dicotyledons, was achieved.[132] Initially, these classes were not named, perhaps because Jussieu was aware of their partly unnatural status, and only later[133] did he coin names for them. He recognized many informal groupings between the family and genus, and in his classification as a whole groups at all levels regularly contained very few (often about five) members.[134]

It is interesting that Jussieu did not recognize four major groups based on cotyledon number—plants might have none, one, two, or more than two cotyledons—despite the importance he assigned to this character in delimiting his main groupings of plants. He recognized that some of the Dicotyledones had more than two cotyledons, but noted that there were only a few genera involved, and in any case seedlings with two and those with many cotyledons were basically similar in their overall construction.[135]

Linking Genera and Families

Although characters were needed when groups were subdivided, this did not necessarily mean that natural groups could be characterized in an unambiguous fashion with characters unique to each group and found in all its members, despite Jussieu's protestations to the contrary in the Introduction to the *Genera plantarum*. He even implied as much himself in one of his last publications:

> We will note here that in following this natural rule [weighting of characters according to their constancy within genera] one often passes from one genus to another by insensible transitions, from the last species of one to the first species of the following, while artificial systems, which aim for well separated genera, do not make such genera except by separating those which have the closest affinity the one from the other.[136]

Genera reflecting affinity were gained only at the price of not being able to recognize them readily. Indeed, if one thinks about the likely consequences of subdividing continuous nature, Jussieu's familles, like Lamarck's, would be polythetic. The characters of a family or genus would be the characters of all its included members; as Jussieu noted, the general character of a genus was formed from all the particular characters of the species that made it up.[137] Characters constant within a family might also occur in adjacent families, or adjacent parts of adjacent families; other characters might be found in many, but not all, genera in the family (refer back to figure 2A in chapter 2).[138] Again, the result would be a catena-like distribution of characters, with the limits of characters and groups not being the same.

To what extent, therefore, do the distributions of characters in Jussieu's system support his claim that characters and groups were coterminous, or do his distributions conform more to the pattern to be expected if, as he often repeated, life was continuous? Unfortunately, one cannot assume that Jussieu's arrangement perfectly mirrored his idea of natural relationships, since he increased the number of dicotyledonous classes by using characters that he admitted were not of great importance. Nevertheless, he argued that it was only the arrangement of the classes or families, not their composition, that might be unnatural.[139] Thus he noted that although the Amaranthi and Caryophylleae were widely separated in the *Genera plantarum*, they should really be adjacent in a natural arrangement;[140] as families, however, they both represented a connected part of the natural series. I shall initially disregard this problem.

Even a brief perusal of the *Genera plantarum* shows that Jussieu described numerous individual genera or small groups of genera as being more or less intermediate between the family in which they were included and the next family in the sequence. Linking groups at these higher taxonomic levels remained a focus of his attention. Thus he found *Opercularia* of particular interest because it was intermediate between the Dipsaceae and Rubiaceae. Indeed, he thought that *Opercularia*, along with *Fedia* and *Valeriana*, might make a new order, although he did not describe it. The recognition of this order would both simplify the general character of the Dipsaceae (by removing aberrant genera) and form a "happy transition" (as he described it) between the Dipsaceae and Rubiaceae.[141] He remained hopeful that explorers would find plants intermediate between groups that were apparently distinct, so clarifying the order of nature; the botanist should try to increase the number of groups and thus to diminish the size of the gaps that still existed in nature.[142]

A particularly striking example of this expectation is provided by Jussieu's work on the Nyctaginaceae (Jussieu's spelling varies), the bougainvillea family. By 1803 he included ten genera in the family (there were only four in the *Genera plantarum*), and he knew of additional genera from South America that were as yet undescribed. It was also clear that the Nyctagines were not close to his Amaranthi, Plumbagines, and Plantagines, families with which he had earlier associated them, nor with the Dipsaceae, which also showed superficial similarities.[143] After an exhaustive discussion of the relationships of the family—which allowed him to reach no conclusions as to its relationships—he ended the article with a declaration of faith that such relationships, in the form of intermediate genera, would be found:

> From all these observations, it results that up to the present the Nyctaginaceae, of which the organization is very distinctive, do not have an affinity so far determined with other known families, that they can only be removed from them [other families] and occupy another place in the general series of plants. The naturalist occupied in the investigation of botanical relationships will draw moreover another conclusion from this. The obstacles that he encounters here in the precise determination of these relationships will confirm his ideas as to the existence of new orders of which no species has come to his knowledge. If nature, which establishes graduated affinities between all organized beings, has placed in distant parts[144] a family of plants so different to those that cover our [European] soil, there must equally be hidden in those same parts other families bordering it, and fitted to form the point of contact between this isolated order and the numerous series of European plants.[145]

Relationships Between Characters and Groups

As might be expected from the pervasiveness of intermediate groups in the natural series, especially those parts that were well understood, detailed analysis of the distribution of characters (I am now using the word loosely here—any feature of the plant) shows that the limits of the groups determined by Jussieu often do not coincide with the distribution of the features that nominally characterized them. Rather, the relationship between characters and groups is the same as Lamarck suggested they would be. A few examples must suffice because of the complexity of the variation patterns involved.

The first group of examples concerns variation at the level of the overall arrangement of the families. It will be remembered that Jussieu considered that an ordering of plants based on the number of cotyledons and on stamen insertion followed nature, but that a division based on the number of the petals—none, one, or several—was more helpful for grouping families into classes of convenient size. The distributions of all these characters are schematically outlined in figure 3, along with those of some other features. It can be seen that the characters often do not show the distributional behavior that would be expected if the groups they nominally delimit were distinct, or if such characters really were invariant in groups.

Jussieu arranged plants in a series that represented the increased complexity of the most important character, that of cotyledon number. Thus he began with the simply constructed Acotyledones. But since he thought that plants as a whole were not sharply set off from the rest of life, there would be connections between plants and animals. The first of the six families included in the Acotyledones is the Fungi, which he considered in part to be analogous with the animated zoophytes. The Fungi were very different from other plants in structure, in habit, and in how they flowered, and they formed a beginning and perhaps unfinished ("inchoantes") series of plants.[146] They were apparently intermediate between animals and plants, but they were also very close to some Algae, the next family in the series, although the Algae were poorly known. Jussieu described the genus *Lichen* as a group intermediate ("analogi") between the Algae and the third family, the Hepaticae, and so the series continued.[147]

After the Acotyledones came the Monocotyledones and then the Dicotyledones. The very last genera of the last class of the latter, the Dicotyledones Irregulares, were polycotyledonous; although they were not placed in their own major division, they were assigned a place in the series quite compatible with continuity. These genera were placed in the Coniferae verae, which was made up of *Juniperus*, *Thuya*, *Araucaria*, *Pinus*, and *Abies*.[148] Interestingly, however, Jussieu suggested that the relationships of at least some Coniferae were not simply with the adjacent Amentiferae. He toyed with the idea of a floral similarity between his Coniferae and the Aroideae, the first family of the Monocotyledones (both lack a corolla). He also saw possible linkage in the similarity in leaf type with the Musci (*Juniperus* in particular has leaves that are somewhat similar to those of mosses), or a linkage marked by vegetative and also floral similarity with *Equisetum* (*Equisetum* and the conifers—and *Casuarina*, which Jussieu included here—all have strobilate inflorescences and either scalelike or elongated leaves). *Equisetum* and the Musci were both Acotyledones, and to the extent that Jussieu thought that the Coniferae were really related to them, the natural arrangement was not simply continuous; it was circular.[149] In any event, the genera that ended Jussieu's whole series showed similarities to more than a single group, and the characters that nominally distinguished them also occurred in the groups adjacent to them in the series.

Possession or otherwise of a calyx was the next most important feature determining the natural arrangement, and a similar pattern is evident in the distribution of groups in which the flower has a calyx. All Acotyledones lacked a calyx—that is, all except the last family in the class, the Naiades, which is adjacent, of course, to the Monocotyledones.[150] The Monocotyledones and Dicotyledones alone were characterized by the presence of a calyx, although a calyx was described as being absent in the Aroideae and Cyperoideae, the first and third families in the first class of the Monocotyledones, and so adjacent, or almost so, to the Acotyledones. Not surprisingly, the calyx occurs in the very last class of plants, the Diclines Irregulares. But in the last family of this class, the Coniferae, only the first group of three genera was mentioned as having a calyx, although for the last genus of this group, *Taxus*, Jussieu described the calyx as being almost absent. The second group of genera, the Coniferae verae (the last in the entire sequence of plants), lacked a calyx entirely. Of course, if relationships were indeed in the form of a circle, this second group would be adjacent to the calyx-less Acotyledones.[151]

A third major character for these upper-level groupings was presence or absence of petals. Jussieu adopted a distinctive interpretation of sepals and petals: petals were those parts of the flower that were shed along with the stamens and were enclosed by the sepals; the sepals were, however, persistent.[152] Thus he thought that petals were largely lacking in the Monocotyledones because in many of its families all foliaceous parts enclosing the flower were similar in size, shape, color, and consistency. Also, they did not fall off after fertilization, but persisted on top of or below the fruit. Again, details of the position in the overall sequence of the families and genera that lack petals are intriguing. In my diagram of Jussieu's arrangement (figure 3), it can be seen that the first three classes in the Dicotyledones, which are nominally all petaloid, and the very last class of all, the Diclines Irregulares, lacked, or almost entirely lacked, petals. The Apetalae, made up of these first three classes, were adjacent to the Monocotyledones, nearly all of which lacked petals, and the Diclines Irregulares would be adjacent (in a circular arrangement) to the Acotyledones, also devoid of petals.

However, a yet more complex pattern of variation is evident when the family descriptions are carefully studied. Jussieu clearly described the last family of the third (and last) class of the Apetalae, the Plumbagines, as having petals. Similarly, the first family of the last class of the Dicotyledones as a whole (the Euphorbieae, of the Diclines Irregulares) was described variously as lacking petals, having a pseudopolypetalous corolla or inner divisions of the calyx that were petaloid, or perhaps really being polypetalous.[153] Even in the Monocotyledones, Jussieu described the last family, the Hydrocharides, as having a petaloid calyx; he was not even sure that it had a single cotyledon. In the antepenultimate family, the Cannae, he described the three outer parts of the calyx as mimicking an outer calyx (he also mentioned that in the internally positioned and very heterogeneous Junci the inner calyx might be petaloid). In general, taxa that are deviant in their calyx or corolla when compared to the descriptions of the groups in which they were included, were placed adjacent in the series as a whole to groups in which these "deviant" conditions were normal.

The next most important character hierarchically is that of the position of the ovary relative to the insertion of the other parts of the flower. Were the stamens, sepals, and petals (if present) inserted below the ovary (hypogynous), around it (perigynous), or on top of it (epigynous)?[154] In this case, Jussieu constructed a fairly regular sequence such that a class whose members are hypogynous is usually adjacent to one whose members are perigynous, as are classes with epigynous families; epigynous and hypogynous classes are usually not adjacent (figure 3).[155] Again, however, a more detailed analysis of the distribution of this set of characters shows a blurring of these boundaries. For example, the second class of the Monopetalae was described as being perigynous. The ovary of the last family, the Campanulaceae are, however, described as being inferior: "Stamens similarly inserted [high on the calyx], below the corolla, . . . Ovary inferior." The Campanulaceae abut avowedly epigynous (inferior) groups. This second class of the Monopetalae is very heterogeneous with respect to stamen insertion and ovary position. It includes the Rhododendra and Ericae, both of which have largely superior ovaries, the stamens usually being inserted only at the very base of the corolla and the sepals and petals arising at the base of the ovary (in most Monopetalae the stamens are inserted *on* the corolla). Interestingly, Jussieu placed a group of genera that all have an inferior ovary at the end of the Ericae. These are adjacent in the sequence to the Campanulaceae; all genera in this family have inferior ovaries.[156]

Proceeding one more step down the hierarchy of subdivision of classes, Jussieu divided the Monopetalae with epigynous flowers into two classes, both containing three families. The distinguishing feature he used was whether the anthers were connate (fused: the third class of the Monopetalae, containing the three families into which the Compositae had been subdivided) or distinct (the fourth class). The Campanulaceae, the last family of the second class and so adjacent to the third class, again show a distinctive pattern of variation. *Jasione* and *Lobelia*, the two genera placed at the end, have connate anthers, and are the only genera in the whole of the second class with this character. Furthermore, the Campanulaceae as a whole have latex, as has the first family of the third class, the Cichoraceae. *Jasione* and *Phyteuma*, respectively the last and antepenultimate genera of the Campanulaceae, also have an involucre surrounding the inflorescence, as have all members of the third class, as well as the Dipsaceae, the first family of the fourth class.[157] The last group of genera in the Corymbiferae, the last family in the third class and characterized, of course, by its connate anthers, is made up of *Ambrosia*, *Xanthium*, and *Nephelium*.[158] These genera have anthers that were described as "approximatis non coalitis," that is, basically free. They are adjacent to families in the fourth class, since in that class the anthers are always free. Similarly, Jussieu later placed the Olacaceae (his Fissilées) at the end of the group of Monopetalae with a hypogynous corolla because their corolla separated into several parts; they were next to the Symplocaceae, which was placed in a perigynous group, since that family also had an intermediate corolla when it came to the character of polypetaly versus monopetaly.[159]

We find characters showing the same overlapping pattern of distribution at lower taxonomic levels. Over a century ago, Édouard Bureau suggested that the "central" genera of Jussieu's Bignoniae family (i.e., those placed in the second of the three groupings recognized) were typical, and those placed in the first and third sections were more different the further away they were in the sequence from these central genera.[160] Indeed, genera at the ends of groups may not possess the characters of the group in which they are placed, but those of an adjacent group. Thus *Boerhaavia*, the last genus of the first (herbaceous) subgroup of the Nyctaginaceae, had some woody species, and established the transition to the second group, which contained genera with woody species.[161] A more comprehensive example is provided by four groups of genera linking the Guttiferae (the mangosteen and its relatives) and the Aurantia (oranges and lemons), both Monopetalae with hypogynous flowers.[162] In figure 4 the distribution of characters varying within and between these groups is shown. Jussieu described the first group of genera in the diagram as Guttiferae with a single style (the first group of genera in the Guttiferae lacked a style); the second are genera linking the Guttiferae and the Aurantia; the third group is the Aurantia spuria; and the fourth makes up the first part of the Aurantia vera. Again, the limits of groups, as indicated by horizontal bars of varying thickness between the genera, do not coincide well with discontinuities in the variation pattern of individual characters, indicated by horizontal bars (sometimes tentatively positioned) in the body of the illustration.

If these two intermediate groups, that is, the last group of the Guttiferae and the Aurantia spuria, were removed, there would be a sharper distinction between the Guttiferae and Aurantia; neither of these two intermediate groups of genera is associated with the Guttiferae today.[163] The taxonomic vicissitudes of the last group of genera in the Guttiferae are particularly interesting. This linking group proved to be unsatisfactory, the genera in it having the characters of other families not immediately adjacent to these two in the series. Jussieu transferred the first three genera in it to the Tiliaceae, while *Allophylus* was placed in the Sapindaceae.[164] This left a gap between the Guttiferae and the Aurantia. It was, however, soon filled, with *Marcgravia* and its poorly known relatives, *Antholoma* and *Norantea*, becoming the transitional genera.[165] It was only later and with evident reluctance that he based a family on these genera; the only genus he knew about in any detail was still *Marcgravia* itself (in fact, *Antholoma* is not related to *Marcgravia*, but belongs to the Elaeocarpaceae). To describe what was effectively a monotypic family would mean that its characters could not be generalized. He could only surmise what the general character of the Marcgraviaceae was, and it might well have to be changed when other genera that were well understood were included. However, he appears to have felt almost obliged to describe it, as Augustin-Pyramus de Candolle had mentioned the family by name and attributed it to him.[166] And elsewhere Jussieu gave the same impression of satisfaction when he moved a genus and as a result broke the series, but was then able to fill in the hole that he had made.[167]

	Leaves borne	Secretions/ Punctations	Sepal number	Stamen number	Stigma type	Fruit type	Seed number
Singana		?	3-5	many	"1"	fleshy capsule	many
Mesua	↑	latex	4	many	thickened	nut	1-4
Rheedia		resin	0	many	infundibular	berry	3
Calophyllum	opposite	latex	4	many	capitate	drupe	1
Vateria	alternate	resin	5	many	"1"	capsule	1
Elaeocarpus		?	4-5	16-20	"1"	?drupe	?1
Vatica	↓	—?—	5	15	"1"	?	?
Allophylus		?	4	8	"4"	?	?
Ximenia		neither	4	8	"1"	drupe	1
Heisteria		neither	5	10	sub 4-fid	drupe	1
Fissilia		neither	5-6	8	truncate or 3-angled	nut	1
Chalcas		punctate	5	10	capitate	berry	2
Bergera		↓	5	10	turbinate	↓	2
Murraya			5	10-12	capitate		1-2

FIGURE 4

Distribution of characters across the Guttiferae–Aurantia boundary. The broad bar in the list of genera marks family limits; narrower bars, the limits of groupings of genera within families; bars in the list of characters, possible discontinuities in those characters; question marks, no information provided; all genera in the direction of the arrow have that particular character. Information taken from Jussieu 1789.

Discussion

There appears to be a direct relationship between Jussieu's ideas of continuity in the living world and his circumscription, characterization, and arrangement of plant groups. Jussieu insisted that nature formed a continuum, and that the natural order was to be built up using species and then genera as the building blocks; organisms linking "groups" were an integral part of this synthetic procedure. His practice of linking the number of taxa at any one rank directly to the number of taxa at the next lowest rank is consistent with this position. Groups formed by the division of the natural continuum could be of any size because the ultimate reason for their recognition was convenience for memorization. Despite Jussieu's assertions, characters very frequently do not delineate groups, even when they are described as characterizing them, although he might well have expected this; transitions, as he remarked, were a mark of perfection in the natural order.[168]

Indeed, almost every feature that Jussieu emphasized as being important in delimiting major groupings in the natural series transgresses slightly the lines dividing groups. Groups and characters are not coterminous, despite his claims. Classes and orders were circumscribed as they were because Jussieu used additional features that he included in his family descriptions but that were not in the prefatory discussion. Thus the character of the number of seeds produced by the flower was apparently used to separate the Campanulaceae, members of which had several seeds, from the whole of the third class, which had only one; many-seeded fruits were general in the second class.[169] Similarly, the Rhododendra and Ericae, adjacent families in the series, could, Jussieu insisted, always be separated by their fruits; these dehisced along the lines of the septae in the Rhododendra and down the middle of the loculi in the Ericae.[170]

But Jussieu seems to have used another way of making his descriptions seem deceptively clear cut. Thus he might exclude from family descriptions genera he placed in the family, but only as "genera affinia." For example, *Drosera* and *Parnassia*, with five and four stigmas respectively, were placed at the end of the "Genera Capparidibus affinia"; the stigma of the Capparides was described as being "simplex."The Rhamnaceae were described as having a superior ovary, but five genera that have inferior ovaries were placed at the end of the family.[171] Jussieu's practice in this respect is that of Lamarck: the characters of a family simply refer to a limited group of genera in the continuous series, and Jussieu's descriptions did not always encompass all members of the family. The greater the segment of the series nominally encompassed by the description, the vaguer the description.[172] There is clearly a conflict between the family as represented by a description and the family as part of the natural series.

Scott Atran[173] has recently suggested that Jussieu may not have thought that his families were arbitrarily circumscribed, and the phenomenal continuity of his families may be compatible with a more typological notion of each family having a discrete organizational plan;

a morphologically polythetic family might be physiologically monothetic. However, all the evidence suggests that Jussieu did consider his family and generic limits to be basically arbitrary, and although the second suggestion is indeed possible, Jussieu does not seem to have adopted this position, and he certainly had no evidence on which he could base such a position.[174] Again there remain ambiguities. Jussieu mentioned groups that "dans l'origine" had a corolla, and then lost it.[175] This suggests groups and types, but also taxa in a part of the series that he had characterized by having a corolla, although these taxa did not. But the very emphasis he placed on characters would make it seem that there were groups, perhaps with types or essences; characters themselves were discrete, even if taxa were ultimately separated only by infinitely graded nuances.[176] Furthermore, the correlations and dependencies between characters he suggested prefigure the subordination of characters (to be discussed in the next chapter). Again groups appear (figure 2B).[177]

William Roscoe, a critic of Jussieu, commented in 1813 that Jussieu's *genera affinia* were a defect of the system, blurring the distinctions between groups, and that the characters of families were not at all evident in Jussieu's work.[178] But this comment seems to have been made with the assumption that Jussieu had been intending to describe discrete groups. For Jussieu, these *genera affinia* were in fact evidence of the success of the method, not its failure. The fact that families could not be characterized was a consequence of continuity. In general, the extent to which Jussieu was able to arrange genera so as to maintain continuity between the groups with overlap in the most important characters can only be described as remarkable. One can therefore suppose it was deliberate; indeed, there is a comparable gradation of variation in artificial characters as well. Thus he emphasized that the presence or absence of a corolla was not a good indicator of natural relationships, and its use led to the artificial separation of the Amaranthi and Caryophylleae in the *Genera plantarum*.[179] But this is one of the characters that exhibited a particularly striking pattern of overlapping variation (figure 3). Synthesis and the establishment of the continuous natural order was the touchstone of his method: in the 1870s, George Bentham recalled that Jussieu had told him that for a systematist to know his genera he had to know his species first.[180]

But to leave the matter here, even if only provisionally, would be to ignore other dimensions to the problem. There were heated arguments in Paris toward the latter part of the eighteenth century over what constituted good scientific practice,[181] and Jussieu's work cannot be discussed without at least outlining these arguments, although they are of considerable complexity. It is not certain that these arguments over scientific practice actually affect my portrayal of Jussieu's work and intent, but they may nevertheless contribute to a better understanding of Jussieu's theory, particularly the importance he placed on method and the relationship between his two ways of attaining the natural method. They may also affect more generally interpretation of the rhetoric of the natural method.

In the first place, one might wonder whether Jussieu's methodological prescriptions can be interpreted in the context of Cartesian science. In the *Discours de la methode*, published in 1637, René Descartes provided clear instructions for good scientific procedure. The Cartesian method entailed: (1) Recognizing that the only things that could be accepted as true must be clearly known to be such, and so evident to a mind cleared of all preconceptions that there would be no doubt about it; (2) Dividing a problem into as many parts as possible (here called analysis); (3) Proceeding from the simple to the more complex (synthesis); and (4) Making complete enumerations.[182] Obviously, the first requirement tends to be in the eye of the believer, and it can be argued that Jussieu accepted ideas of his predecessors that did not fulfill the first criterion;[183] that would not, however, have been Jussieu's opinion. The three other criteria do describe Jussieu's practice quite closely. The problem, that of understanding relationships, was decomposed into its smallest parts—the species—the relationships of which were then to be established. These species were first grouped into genera, and then into higher groups, and the whole was arranged so as to start with the simplest plants, ending with the most complex. The *Genera plantarum* would also satisfy Descartes's fourth criterion in that it was a complete enumeration of plants known at the time, although it was still manifestly incomplete. Furthermore, Descartes emphasized the importance of clear and distinct knowledge,[184] just as Jussieu's groups were clearly described by characters that had been carefully defined. To the extent that Jussieu's methodology was consciously Cartesian—and I can provide no direct evidence that it was—Cartesian science and the requirements for recognizing relationships in a seamless natural world seem compatible, even if the antithesis between synthesis and analysis as it was developed by Jussieu and his followers springs more from the immediate circumstances surrounding the development of the natural method, that is, the conflict with Linnaeus.

But Descartes's ideas had been modified by his successors, and in the eighteenth century both John Locke and Isaac Newton had staunch adherents in Paris;[185] their views of the world and of good scientific practice informed debate there. The Cartesian approach was infused with mathematics,[186] and it was hoped that mathematics would provide a language by which nature could be described unambiguously, a problem of great interest in the late eighteenth century. Precision and clarity in language was seen as being essential, if not for yielding truth directly, at least for progress in science and discussing its findings.[187] A philosophe who campaigned against systems and for such a mathematical language was Étienne Bonnot, Abbé de Condillac (1714–1780).[188] As C. C. Gillispie observed, the analytical approach favored by Condillac "dissipates that confusion [caused by unanalyzed experience] by finding the science its proper language, a systematic nomenclature chosen to identify the thing by the name, associate the idea with its object, and fasten the memory to nature."[189] Mathematics was particularly important here because as a language it was an instrument of analysis, and Condillac favored the analytic algebraic over the synthetic geometrical approach;[190] geometry involved deductions from general definitions. Condillac argued strongly against systems not based on facts and observations, and here Leibniz and his monads came in for particularly severe criticism.[191]

At the end of the eighteenth century approaches to nature based on geometry, system, synthesis, composition, instruction, demonstration, and deduction were opposed to those based on algebra, method, analysis, decomposition, discovery, invention, and induction.[192] In the *Genera plantarum* Jussieu shows himself to be aware of the distinction between synthesis and analysis and between system and method.[193] Shortly after its publication Georges Cuvier linked analysis and induction, proceeding from the part to the whole, and contrasted this approach with synthesis, proceeding from the whole to its parts.[194] But there is also a synthesis that involves the building up of a whole from its parts, an activity that could be carried out after Cartesian analysis, the decomposition of a complex whole into its unitary parts, or really it was simply the other half of this analysis; decomposition was followed by recomposition.[195] This approach, allowed by Condillac and the basis of "l'esprit systématique," could serve as the basis of analysis in another sense, the discovery of general principles from underlying concrete phenomena.[196]

Jussieu's prescriptions for the natural method can be interpreted in several ways, but they are not a simple response to the concerns just discussed. In the Introduction to the *Genera plantarum*, Jussieu addressed the problem of the proper language for nature by providing an extended series of definitions of plant parts.[197] Although Linnaeus[198] had noted that the highest method in natural science was the mathematical method, which proceeded from the simple to the complex, a mathematical language for natural history was possible only by analogy, with the species (or individual) being the basic unit of the natural historian, and synthesis the favored procedure. Certainly, in Jussieu's work there is no weighing, little measuring, and few numbers other than those pertaining to the different parts of the flower and of the cotyledons.[199] But in his overall arrangement of plants, in basing the natural method on species, and in arranging his overall series from simple to complex, Jussieu could be seen as following Condillac's prescriptions of forty years before.[200] Jussieu was promoting synthesis in the sense of building up the whole natural order from its parts, and at the same time he hoped that an analysis of that order would allow him to uncover general principles such as the relationships between different organs.

It is surely directly from continuity itself, rather than from mathematics, that the emphasis on continuity and distance made by Jussieu, as well as by other naturalists at this time who were ordering and naming nature, take their origin.[201] Given his repeated comments about the existence of intermediates and the general compatibility of both his theory and his practice with continuity, this view of nature has to be allowed a major role in the framing and execution of his method. Moreover, we shall see in chapter 5 that his championing of synthesis as an integral part of the natural method is similar to the position adopted by other proponents of continuity. However, the issue of the relationship between classification and language, which in a sense underlies everything else, is left to the epilogue.

CHAPTER FOUR

Jussieu and the Next Generation: Breaking with Continuity?

There was agreement by the first part of the nineteenth century that the main goal of naturalists, whether working on plants or on animals, was to find the natural order. France was the center of botanical developments in the late eighteenth century; important early works that popularized the natural method included those of Étienne Pierre Ventenat and Jean Henri Jaume Saint-Hilaire.[1] The first part of Joseph Gaertner's *De fructibus seminibus plantarum* came out in 1788. Gaertner, who worked at Calw, a small town in Baden-Württenburg, emphasized the value of characters of the fruit and seed in establishing relationships.[2] His observations supported and extended Jussieu's findings, and botanists throughout much of Europe adopted the natural method. In the English-speaking world Robert Brown's unfinished *Prodromus florae novae hollandiae* of 1810[3] was the single most important work popularizing the natural method. In general, however, both Sweden and England—the one the heir of the Linnaean tradition, the other (through James Smith) the purchaser of his memory—were rather slower than other parts of Europe to change from the artificial Linnaean system to the natural method.[4]

The general utilization of the principles of the natural method might seem to promise a period of development and, ultimately, stabilization of natural relationships. Accepted wisdom has it that evidence accumulated showing that groups were discrete, and extinction also became accepted. Yet botanists in particular found it difficult to settle on an agreed system for their higher groupings. No fewer than twenty-four natural systems were produced during Robert Brown's lifetime (1773–1858),[5] and they continued to be produced throughout the century; hardly surprisingly, many botanists were skeptical of the relationships that were suggested in them. The analogies used when discussing systems in general suggested reticulating relationships in two or sometimes even three dimensions,[6] seemingly reflecting this fundamental skepticism; groups might show a diversity of relationships even within the one system. Moreover, those who disavowed the existence of a general *scala naturae* did not at the same time insist that groups were discrete, and continued to discuss relationships in terms appropriate if there were such a *scala*, at least locally. The relative highness and lowness of groups remained of central importance.

To aid an understanding of how this confusing situation arose, I first turn to an investigation of the development of the natural method in both botany and zoology in the early years of the nineteenth century. Again, I emphasize both concepts and their application. This should lead to an appreciation of the extent to which essential elements of Jussieu's theory and practice persisted and how the *Genera plantarum* relates to subsequent developments of the natural system. I set the scene in this chapter by discussing the work of three prominent naturalists who were all active in the first part of the nineteenth century: Georges Cuvier, Augustin-Pyramus de Candolle, and Charles-François Brisseau de Mirbel. Cuvier is included both to see how his work relates to its Jussiaean base and also to clarify the nature and significance of the more purely botanical developments; I do not discuss his later work in particular in more than passing. Although all three mention Jussieu with great respect, in no sense are they slavish adherents of his method, and they can be considered his followers only insofar that all were interested in "natural" relationships. Cuvier and Candolle are perhaps the greatest zoological and botanical systematists of the nineteenth century. However, I shall argue that their approaches—which have much in common—do not come to dominate botanical systematics in that century.

Before proceeding further, a clarification of what is meant by anatomy is in order, since the value of comparative anatomy in the naturalist's endeavors was an important issue for all three authors. As E. H. Merriam observed,[7] Cuvier, a champion of the value of comparative anatomy, observed only gross anatomy. He dissected organisms, perhaps using a lens to clarify what the scalpel disclosed, and he was primarily interested in bones, muscles, and their connections, as well as viscera in general and the courses of nerves and blood vessels. The comparable botanical activity involved dissection of the flower, with observations being made on placentation and the nature of the ovule—was it straight, inverted 180°, or curved in a more complex fashion? Detailed studies were also made of the anatomy of the seed. In both cases a lens was often used.[8] However, what became comparative anatomy in plants was a rather different operation in that the use of a mounted lens or microscope was essential. It involved the cutting of sections and the observation of cell type and tissue arrangement, more like animal histology. Nevertheless, the observations of cells and tissues by botanists and of muscle, nerve, and bone by zoologists have both been called comparative anatomy.

Georges Cuvier

Léopold-Chrétien-Frédéric-Dagobert (Georges) Cuvier (1769–1832) was arguably the most prominent zoologist in Europe when he died. He was also a very influential figure in both scientific and secular politics as long-time secretary to the Institut and Peer of France.[9] When he moved to Paris in 1795, he collaborated closely with Étienne Geoffroy Saint-Hilaire (1732–1844), who later came to argue strongly for the idea that all animals were fundamentally similar, an idea that Cuvier equally strongly opposed.[10] However, some authors have noted that when Cuvier was young, he was less hostile to such ideas.[11] Indeed, he was early much influenced by Jussieu, although the nature of this influence has not been discussed in any detail. Even before Cuvier met Geoffroy Saint-Hilaire, his work shows many of the features of a believer in continuity, and it is in this context that Jussieu's ideas become of great importance.[12] Cuvier later moved away from this position, and he became renowned for his studies on comparative anatomy (especially that of vertebrate animals, both living and fossil), his championing of the subordination of characters,[13] and his assertion that there were four quite distinct major groupings of animals.

When Cuvier moved to France from Württemberg in 1788, he developed his anatomical studies, building on the approach adumbrated by the great anatomist and physiologist Félix Vicq d'Azyr (1748–1794). Comparative anatomy came to be the most important tool that he used in his studies of animal life. Thus in 1795 he rejected the value of mollusc shells in classification—they might be the basis of a special classification, but the "grand system" would entail dissections of the organisms inhabiting the shells.[14] Yet initially Cuvier (and also Geoffroy Saint-Hilaire)[15] implied that the characters used in natural classifications could also be used for the identification of the organisms classified. In a joint paper on the classification of mammals, Cuvier and Geoffroy Saint-Hilaire wanted to combine the advantages of the natural order with that of dichotomous division. They suggested that classification should be based on organs of touch, rather than nutrition, since the former were external and had other organs dependent on them. The proposed method thus provided both numerous characters for the subdivision of the group and features that would allow taxa to be recognized. Characters involved in nutrition were usually not so easy to see. Like Jussieu, Cuvier and Geoffroy Saint-Hilaire took generally accepted families and found the characters that were constant within them.[16]

One of the main elements of Cuvier's approach to the study of animals was his proposal that there was a functional correlation between parts: "*General natural history* . . . determines the laws of *coexistence* of its [the animal's] features ['propriétés']; because each of these features implies or necessarily excludes a certain number of others."[17] This was a proposal quite compatible with Jussieu's ideas of affinity between or interdependence of characters. These laws were reflected in the organization of animals, which in turn was connected with their inner nature; organization was also in harmony with external nature.[18] Characters connected with the very existence of organisms were most important; those concerned with the relationships between one organism and another were less so.[19] This physiological-functional-ecological hierarchy was linked to a parallel hierarchy of relative invariance of characters in taxonomic groups. In the principle of the subordination of characters, the functional interrelationships of the organism were linked to the taxonomic hierarchy; there was a subordination of characters, constant characters were more important and (here Cuvier referred to

Jussieu) lower-rank, subordinate characters were less constant.[20] This relationship was integral to his extensive comparative researches. The relative importance of organs could be fixed, and once established, it could be extended, since it would be the same in different organisms;[21] the "calcul" of characters first achieved by botanists could be extended to the animal kingdom.[22]

Cuvier contrasted two approaches for discovering the natural arrangement:

> One can arrive at this goal by two different means, which can serve as the proof and verification of each other: the first, which every one uses naturally, is to go from the observation of species, to grouping them in genera, and their collection into a higher grouping, according to the totality of their attributes; the second, which most modern naturalists have used, is to fix in advance certain bases of division, according to which beings are arranged as one observes them.
>
> The first means cannot be wrong; but it is applicable only to a being of which one had perfect knowledge. The second is more generally used, but it can be in error. When the bases of division which have been adopted do not contradict the groupings which are the result of observation, the two means are in agreement, and one can be certain that the method is good.[23]

The two procedures ("moyens," means), the synthetic and analytic, are equivalent to those of Jussieu. However, Cuvier's emphasis differs from that of Jussieu, and it was the development of the second that was initially Cuvier's goal, and ideally, at least, his method was analytical rather than synthetic.[24]

Cuvier is generally supposed to have rejected the idea of a *scala naturae* and, implicitly, the idea of continuity, but the two ideas can be dissociated.[25] The work of Jussieu was the young Cuvier's guide to botany in particular and to the method by which relationships in general should be established. Moreover, as Outram has remarked, Jussieu became Cuvier's "friend and patron."[26] In this context, the extent to which ideas of continuity are evident in Cuvier's early work in particular has been overlooked. If we examine these early writings carefully, we find numerous comments that suggest that he believed that there was some sort of a continuum in nature. He noted in 1790 that "classes, orders [families], genera are simple abstractions made by man, and nothing comparable exists in nature";[27] species, however, were another matter. This is very similar to Jussieu's approach to the nature of higher taxa. Of course, such a statement is silent about what *did* exist in nature, but there is little evidence that Cuvier at this early stage of his career thought that there were discrete taxa, at least, discrete taxa above the rank of species. Thus he wrote to Christian Heinrich Pfaff fourteen months later in October 1791, "I have studied him [Jussieu's *Genera plantarum*] for a year, and I cannot admire enough the sense with which the plants are arranged there, their analyses indicated, and the very small characters by which different families merge one into another ['se perdent les unes dans les autres'] indicated with care." And still later, "His merit is not really in very detailed descriptions, or in a large number of species being described, which is often only a sign of a lack of criticism, but it is in the philosophical manner of seeing things, and finding the delicate threads by which plants may be held together, making the whole a painting."[28] One of the impressions that the young Cuvier gained from Jussieu's work was not that groups could be placed in different ranks largely on the whim of the naturalist, but that discrete groups did not exist at all. Clearly, Jussieu's work provided a satisfying "toute ensemble" for Cuvier, and it was Jussieu's subtle evaluation of connections that he found especially satisfactory.

Cuvier indeed specifically rejected the notion of a *single* chain.[29] As he wrote rather later, "What laws would have unnecessarily constrained the creator to produce useless forms, merely to fill voids in a chain, which is only a speculation of the mind?"[30] He suggested almost presciently that a "grande chaine" should ultimately be one of "generation" and should be based on all features of the organism, both external and anatomical. It was clear to him by early 1791 that an unbranched chain could be produced only if a naturalist looked at but a single aspect of an animal;[31] different features of the organism "degraded" (sic) at different rates and would produce different chains.[32] However, although he rejected an unbranched chain, he was otherwise silent about how he saw nature. He seems to have thought in terms of a branching continuity, but whether this pattern initially extended throughout the animal kingdom, or was restricted to smaller groups, is unclear.

When in Normandy Cuvier had studied sowbugs and their relations, and he published his findings in the short-lived *Journal d'Histoire Naturelle*. He began a paper that appeared in 1792 with a direct statement that relationships between these arthropods and their relatives were continuous; nature did not make leaps.

> The water sowbugs with four antennae (*Aselles* of Geoffroy) are the nuance between the sowbugs and the little "squilles" (*Gammarus* Fabr.) which themselves join the *Aselles* to the "écrevisses" with a long tail (*Astacus* Fabr.) by an almost continuous chain.
>
> Here, as elsewhere, nature never makes a jump, and there is a sowbug similar to all the others that approaches the *Aselles* by a sort of rudiment which represents their second antennae.[33]

And in the body of the paper he discussed the relationships between the two genera he was describing as follows:

> One sees then the *On[iscus] armadillo* and *globator* form even within this genus a distinct family [sic], characterized by the rounded tail, and by the ability to roll up which is common to it and the following genus. One can note yet another link which unites the genus of the sowbugs with that of the pillbugs. It is the double edge of the thorax of *Oniscus globator*, like that of my two *Armadillo*.[34]

Jussieu would have used very similar language, and Geoffroy Saint-Hilaire adopted a similar position. The latter clarified his ideas on relationships in a monograph of the lemur:

> If we particularly consider a class of animals, it is there above all that its plan will become evident to us; we will find that the different forms under which [nature] was pleased to make each species exist are all derived the one from the others; it suffices for her to change some of the proportions of the organs to make them fit for such functions, or to extend or restrain their uses. . . .
>
> Thus the forms in each class of animals, however varied, all result in the end from differences in proportions in organs common to all. Nature refuses to employ new ones. Thus all the most essential differences that affect each family within the same class come only from another arrangement, from another complication, in short, from a modification of these same organs.[35]

This almost combinatorial approach to variation might allow groupings within each class, but Geoffroy Saint-Hilaire's "plan" came to encompass all animals. Comments that Cuvier made suggest that he, too, thought that there could be gradual transitions at least between animals with the same basic plan. He and Geoffroy Saint-Hilaire noted in their monograph on the orang-outang that travelers often did not recognize the different species of monkey, although these differences were evident to naturalists who could make more careful observations. These species were degradations of other species, hybrids perhaps, that were separated by very subtle differences—"nuances peu variées," "changemens légers"—the language of continuity.[36] As he observed a decade later,

> [T]hese smooth and invisible nuances are observed only as long as one remains within the same combination of principal organs, as long as the major central provinces remain the same. . . . but as soon as one passes to those which have other principal combinations, there is no more resemblance in anything, and one cannot mistake the interval or marked leap.[37]

The language Cuvier used in his discussion of how relationships were ordered not only suggested continuity but a linear continuity, and even the chain of beings figures quite prominently.[38] As he remarked when commenting on the anatomy of the jellyfish, *Medusa*, he observed "in the chain of beings the heart disappeared sooner than the stomach."[39] In general, his language is very similar to that used by Jussieu and Lamarck when they described the patterns of relationships they observed in nature—which was, for them, continuous.[40] The language is appropriate for the synthetic approach; and although we have seen that this approach was less favored by Cuvier, it might well be carried out at the lower levels of the hierarchy where, even if the perfect knowledge he claimed was needed for it to be prosecuted successfully was lacking, direct relationships would seem more evident. Most of the examples I have been using come from these lower levels, although an exception, a comment on the relationships of the tarsier, *Didelphis*, is in the same spirit: "This genus could be considered as the link uniting quadrumana to Chiroptera or bats."[41] But it should not be forgotten that for purely didactic reasons it is often simplest to discuss relationships as if they were continuous, a point to which I will return in chapter 7.

Yet Cuvier's work on fossils had already made it clear that at least some fossil organisms had no parallel among living beings—there was extinction.[42] He does not seem to have used the fact of extinction to justify a belief in the existence of distinct groups, and he was actively working on fossil animals at the same time that he discussed relationships between extant organisms as if they were largely continuous.[43] Nevertheless, as his studies advanced, it became apparent to him that the issue of discontinuities was not as he had initially seen it. More or less distinct groups did exist, and although further discoveries might make their distinctness less obvious, we find him taking an explicitly anti-Jussiaean position in 1804 when he suggested that *any* distinct form should be described.

> Every time that one finds in organized beings some form that does not allow itself to be at all exactly compared with those [forms] of families or natural groups already known, one can presume that one has discovered the first collection, the first indication of some group, of some new family; it seems that nature has been too fecund to create any principal form without adorning it with all the necessary details of which it is capable. . . .
>
> They [naturalists] are not always bold enough to establish orders when they have only a single genus to place; and we see that one of the most celebrated among us [Jussieu] preferred rather to place in sequence at the end of his work genera which cannot be placed in any of his families, rather than make for each its own family.
>
> It seems to us that to give a complete idea of the series of being, each of them should occupy its place, whether it is isolated or whether numerous beings surround it or link it to the rest of the system.
>
> Besides, those apparent gaps are often only because we do not yet know all beings, and it would be good to mark them [by describing a separate genus or family], if only to make observers more assiduous in finding them, because several examples show that we have to wait too long in this respect.[44]

Animal life did not form an indivisible series, and Cuvier's early ideas of the animal hierarchy as they were reflected in his classifications seemed to him incorrect. In 1812 he claimed that all animals could be placed in one of four major and near-equivalent divisions, "les quâtre embranchements." Each division could be characterized by the possession of positive features, not simply by the absence of a feature. The old Invertebrata were simply a privative group, animals that lacked backbones, and so had to be subdivided and the subdivisions characterized by positive features. He emphasized that there were no connections, no nuances, between these divisions, which were isolated both from each other and from the rest of the living world. This separated his conception of the world from that of most of his predecessors and many of his contemporaries. However, he continued to describe variation at lower levels within major divisions as if it were variation on a theme, organisms being distinguished by "dégradations" of the characteristic features of the "embranchement."[45] Genera and families might still be very hard to separate both in theory and in practice. At the same time, however, he subdivided each of his four embranchements into four more groups;[46] by implication these subdivisions were also discrete, although he did not say so. Here the dependencies and relationships between characters that became apparent through his understanding of the subordination of characters would imply that only certain combinations of characters were possible, and that some characters were more important than others.[47] So long as characters were sharply distinguished—and there never seemed to be any doubt about this—discrete groups would result (figure 2B).[48] Lamarck's "pretended subordination of characters"[49] was real for Cuvier; although it might rupture Lamarck's continuum, it produced discontinuity in relationships and discrete groups for Cuvier.

Charles-François Brisseau de Mirbel

The main interest of Charles-François Brisseau de Mirbel (1776–1854) was not in cataloging plants and establishing their relationships; he largely made his reputation with the physiological-anatomical studies summarized in his *Exposition de la théorie de l'organisation végétale* that appeared in 1809.[50] But early in his career he also wrote some general treatments of plants. With the clergyman Nicolas Jolyclerc (1746–1817) he produced the *Histoire naturelle, générale et particulière, des plantes* (1802–1806), and, with Lamarck, the *Histoire naturelle des végétaux, classés par familles* (this came out in 1803).[51] Both of these multi-volume works were part of the Suites à Buffon accompanying the octavo edition of Buffon's unfinished *Histoire naturelle.*

Mirbel was the successful candidate at the last election to the Institut that Augustin-Pyramus de Candolle contested (in 1808, see appendix 2). After 1810 Mirbel worked mostly on physiological, developmental, and organographic problems—as a glance at his biography shows—and it is for such work that he is remembered. In an obituary of Mirbel, Anselme Payen (1795–1871) noted of Mirbel's findings on the development of the ovule and pollen that not only did they "directly interest anatomy and vegetable physiology, but further they furnish philosophical botany with characters the more important in that they often give them the sanction of physiology." Jean-Baptiste Payer (1818–1860) claimed that Mirbel's work had stimulated "the completely new and completely French science of plant development," although in this latter remark there is more than a touch of hyperbole and self-justification, since Payer himself was a well-known developmental morphologist.[52]

Mirbel early adopted an approach to the search for relationships and the shape of the natural arrangement that was distinctively different from that of Jussieu.[53] Two papers on the Labiatae published soon after his election to the Institut are particularly relevant; here his position on classifications and method is expressed clearly.[54] Nevertheless, he noted that the Jussieus, uncle and nephew both, were justly called legislators of botany, and he also quoted Cuvier's estimation that Antoine-Laurent de Jussieu's work in the observational sciences was comparable to that of Lavoisier in the experimental sciences.[55] Yet the approach to observation, classification, and system formation that he advocated shows major departures from Jussieu's prescriptions, and it is much more similar to the position of Candolle and in particular to that of Cuvier.

When he was young, Mirbel carried out extensive anatomical studies. He recognized that anatomy was a poorly understood discipline. Other sciences were developing fast, but botanists did not know the first thing about anatomy, this discipline having advanced little since the days of Anton van Leeuwenhoek (1632–1723), Marcello Malpighi (1627–1694), and Nehemiah Grew (1641–1712) over a century before.[56] He described how he went about his anatomical work. He had spent six months carefully looking at all aspects of the anatomy of elder, *Sambucus*, using four or five different microscopes; he then extended his studies to other plants. In his use of the microscope, he descended to the level of histology, rather than staying at the level of Jussieu's gross anatomy.

But not only did he look at the organisms in a different way from that of Jussieu; he also saw relationships in a distinctly different way.

> One should not, however, think that groups can be formed without order and without law. If all beings are related by reciprocal relationships, it is nevertheless evident that these links are not equally strong throughout; and to impress this truth by a striking example, it will suffice for me to say that the relationships that exist between quadrupeds are more numerous than those that unite quadrupeds and birds, and all animals have more relationships with each other than with minerals. It is these gradations ["nuances"] which are important to understand, not only in the great divisions where they are easy to grasp, but even in the smallest details, where they often escape the most penetrating mind.[57]

Mirbel clearly, yet subtly, switched Jussieu's emphasis on the largely featureless continuum between all organisms to the different degrees of relationship that existed between groups of organisms. The study of "nuances" was still of the greatest importance, but these nuances suggested an entirely different view of life. One can imagine the botanical universe of Jussieu consisting of a continuum subdivided by lines (figure 5A; this would be consistent with Linnaeus's map analogy), or of areas in which species or genera existed in equal density, the places where adjacent bundles of sticks touched, or where there were no species or genera at all, the interstices between these bundles (figure 5B and 5C: Jussieu's later analogy). The nuances mentioned so frequently by Jussieu established the continuum. In Mirbel's universe, groups of organisms differed in how far apart they were, and it was the nuances in these differences that were of interest (figure 5D). He later suggested that all families were separated by wide gaps, any links between them being at most tenuous, so groupings between the family and the four main divisions of plants could not be recognized.[58] Other points of difference between Mirbel and Jussieu are that Mirbel rejected the close analogy Jussieu drew between eggs and seeds, since plants (unlike animals) took up inorganic substances, and he even questioned Jussieu's interpretation of the nature of the calyx.[59]

In his work on the Labiatae, Mirbel stated that his goal was to understand the relationships between characters rather than groups. System and method were of course to be separated: "But if, neglecting the true end of studies of natural relationships, we persist in looking for the advantages of a systematic arrangement, we will deface the work of Nature, and at the same time we will fall into the most defective of all the systems."[60] Mirbel was promoting the value of anatom-

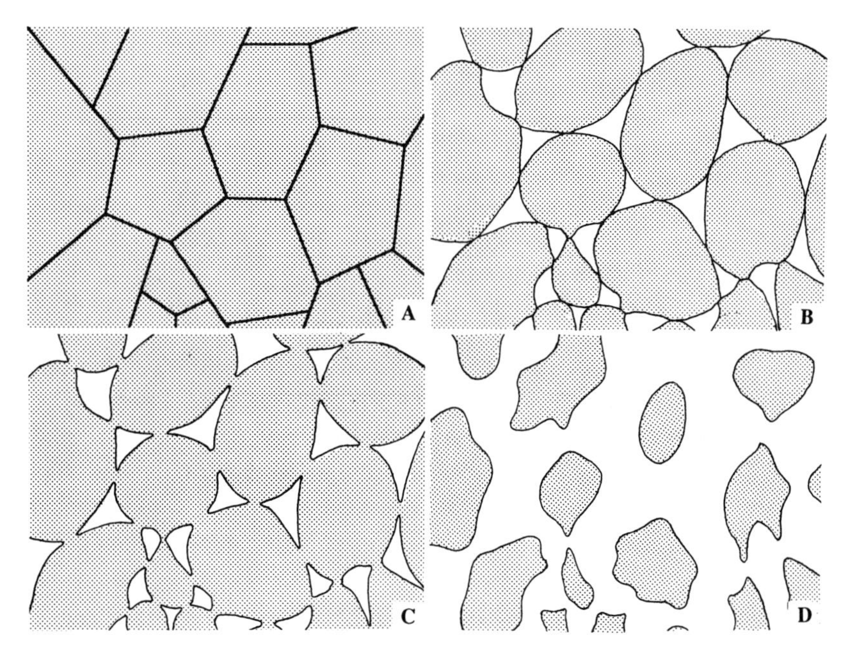

FIGURE 5

From continuity to groups. A is the Linnaean map (limits of groups arbitrary); B, the Jussiaean fasciculus (groups discrete, but adjacent); C, the Jussiaean fasciculus, an alternative interpretation (groups not discrete, but directly linked by intermediates); D, the Candollean archipelago (groups discrete, but separate). Lines indicate limits of groups; shaded area is morphospace known (or believed) to be occupied by organisms.

ical studies in the study of natural relationships, and any systematic arrangement would be based on external characters. It was rather in the necessary linkage ("enchaînement necessaire") of characters that their real importance was to be found, and this was far from being the same in different groups. This importance, or value, was rational, not arbitrary, and was to be based on the general system of organization of the plant—that is, how the plant was put together.[61] It depended less on the constancy of characters than on the necessity for their coexistence.[62] Thus the most important characters could not be changed without affecting many others, while less fundamental changes represented only slight modifications of organs that did not affect their symmetry.[63]

The use of analogy in such circumstances Mirbel thought was a double-edged weapon:

> In truth, experience shows us that when organs are in [a relationship of] mutual dependence, they are constantly associated, and we think we can surmise, by virtue of analogy, that constant association may indicate mutual dependence, even when this dependence escapes us, but that is clearly no more than a simple conjecture, and one should not establish the theory of natural families on facts so vague.[64]

Arguments of the kind *post ergo propter* were not permissible. But this objectionable kind of argument had been used elsewhere. "Familles en groupes" had features in common, so naturalists expected to find such features in families equally natural (although in a different way) that had been formed by "enchaînement." However, such families were united only by hypothetical characters taken from obscure details of the organization of the plant. Exceptions could easily be argued away, and Mirbel observed that since the characters were so small and equivocal, it was easy to turn truth into illusion. As an example he gave that of a species in the Primulaceae that was not hypogynous like the rest. Some naturalists resolved the issue not by direct observation, which would have showed that the plant was perigynous, but by dangerous analogy—other plants in the family were hypogynous, so this one must really be so also.[65] (Yet Mirbel himself apparently followed this style of argument when describing all the Labiatae as having four stamens, while at the same time recognizing that a number of genera had only two. Here he implicitly recognized a type which had the structure most commonly found in the family.[66])

Mirbel thought that reproductive characters were less important than they were supposed to be; after a period of neglect, too much attention had been paid to such characters to the detriment of the study of other features of the plant.[67] It was on similarities of organization, the whole structure of the organism, that families should be based, not on such minute and inaccurately observed features such as the insertion of stamens or the presence of perisperm.[68] The former character was of course of prime importance for Jussieu when he delimited classes in the *Genera plantarum*, while the latter he used especially in the subsequent revisions of families he had first recognized there. Mirbel again suggested that one should look less at the linkages between families than at the numerous external and visible characters by which nature separated them.[69] This is in contrast with the approach taken by the two Jussieus, who made linkages using minute characters. Although Mirbel is ostensibly criticizing Jussiaeans, not the Jussieus, he is implicitly disagreeing with Antoine-Laurent de Jussieu.

Mirbel emphasized that anatomical and physiological characters needed to be incorporated into the natural system, and he repeatedly invoked the name of Theophrastus as justification for this position;[70] anatomy and systematics should be integrated. Natural history was not simply the naming of plants, using a few conventional features. It should include physiognomy, the internal organization (anatomy) of the plant, and how the plant grew.[71]

In Mirbel's study of the Labiatae these principles were put into practice. He looked at the epidermis, the structure of the hairs, and petiole, stem, and even some floral anatomy. He found to his satisfaction that there was indeed substantial agreement between suites of anatomical characters and vegetative and reproductive characters—stem anatomy, its shape, the position of the leaves and those of the flowers were all linked. This demonstrated a fundamental axiom: "*the more the mutual dependence of characters is marked, the more they contribute to the aggregation of species and the formation of natural groups*."[72] He was able to show a "curious" agreement between the external form of the petiole, its anatomy, and its physiology, and he suggested that the ability of the apparently feeble petioles to support the lamina in fact depended on a mechanically appropriate disposition of the vascular tissue. It was the search for such an understanding of the coexistence of characters that remained one of Mirbel's main goals, as in his *Élémens de physiologie végétale et de botanique* of 1815. Alexandre-Henri Gabriel de Cassini (1781–1832), in reviewing the book, noted however that Mirbel had not demonstrated that there was any functional necessity for this coexistence, but merely that characters occurred together.[73] In effect, Cassini suggested that Mirbel had been hoist by his own petard.

It was Mirbel's hope that an understanding of the linkage between characters would allow him to address one of the central problems in botany. How could the characters that were responsible for the aggregation of species "en famille"—clearly recognizable groups whose genera were often more or less indistinguishable—be recognized? And was it possible to determine in advance how far these characters could vary in species yet to be discovered, without the "type" (sic) of the family being completely obliterated?[74] Mirbel frequently commented about variation that he considered to be compatible with the real characters of the Labiatae. Thus he felt that a labiate with four separate styles was not an impossibility since there were four separate strands of conducting tissue in the style. There also might be more than one embryo in each nutlet, or the fruitlets might even be capsular. However, in any plant correctly assigned to the Labiatae, the styles would never be borne on the pericarp, and the pericarps themselves would always be distinct.[75]

But the discovery of this type might depend on the kind of family being studied. Mirbel drew a formal distinction between two kinds of family. All the members of "familles en groupes" were very similar, agreeing in both vegetative and reproductive features, and it was easy to gain an idea of "the abstract image" of such families. On the other hand, very little, if anything, could be said about all the members of the other main kind of family, "familles par enchaînement." As there was little in common between the genera at the ends of the chains of relationships that made up such families (individual genera being connected by nuances), such relationships were very subtle and difficult to grasp; their recognition was the test of a systematist's acumen.[76] Later he recognized families and genera of a third kind, using examples mostly taken from the Labiatae. In addition to "genres par enchaînement," such as *Melissa* and *Thymus*, and "genres par groupes," genera like *Scutellaria*, as well as *Dianthus* and *Rosa*, there were "genres systématiques," like *Salvia*. In this last kind of genus, the species were linked only by a single character. Mirbel emphasized the main difference between grouping and chaining when he noted that groups made by the former process were monotypic, while those made by the latter were polytypic.[77] The recognition of "genres par groupes" was independent of any system, and their "metaphysical existence" was as evident to the naturalist as the material existence of individuals. On the other hand, "genres systématiques" or "familles systématiques" were likely to be artificial, and here he mentioned as an example the three families into which Jussieu divided the Compositae.[78]

Although Mirbel thought that botany should be "solid, profound, and vast," not necessarily simple, the profundity in which he was interested had nothing directly to do with classifications and relationships.

Normally, natural groups were more or less self-evident. However, experience might show that a microscope was needed to establish natural relationships because the most constant characters, those most useful in explaining physiological phenomena (this really gives the game away), might only then be visible.[79] In the search for natural relationships it was synthesis, not analysis, that was the preferred approach, as it was implicitly the former that led to an understanding of natural relationships, although he felt that ideally the two should be combined.[80]

Augustin-Pyramus de Candolle

Augustin-Pyramus de Candolle (1778–1841) was, along with Robert Brown, the most prominent botanical systematist of the first half of the nineteenth century.[81] Unlike Brown, he published a great deal, writing important studies on the theory and practice of classification, plant geography, morphology, and physiology. He also edited "the great Prodromus," which he began in 1827, and he himself contributed very largely to the endeavor.[82] Like Cuvier, Candolle was also active politically and in philanthropy.

Candolle was explicit in his theoretical writings both that relationships between taxa could not be represented by a linear *scala* and that relationships were not continuous; taxa were discrete entities. In arguing against continuity, he observed that no example that purported to show continuity stood up to close examination: the ornithorhynchus was not intermediate between mammals and reptiles, nor the whale between mammals and fish. He also wondered why all the plants discovered since the time of Linnaeus had not filled in the gaps between groups.[83] His preferred analogy when discussing relationships, that between nature and a map of a country, suggested to him discontinuity rather than continuity[84] (see figure 5D). Differences in the closeness of affinity between groups could be compared with the distances between centers of population on this map,[85] the areas in between being devoid of inhabitants. He thought that genera as yet unplaced in the natural system could also be likened to islands at varying distances from the continent. Although in the later part of his life Candolle produced diagrams that implied that relationships within families in particular were more or less circular, with adjacent families touching, there is no indication that he thought that there was real continuity between groups, even between taxa below the level of family.[86] Furthermore, he thought that botanists interested in general ideas should concentrate on establishing relationships at the higher levels of the hierarchy, so forming groupings between class and family.[87]

Not only did Candolle believe that variation was discontinuous; he believed that only through discovering the precise relationships of that hierarchically arranged discontinuity would an understanding of nature be obtained. The discreteness and reality of taxa was such that he repeatedly emphasized that the numbers of individuals in species, species in genera, or genera in families, did not affect the rank of these higher taxa.[88] Thus Candolle clearly explicated the relationships between classification, natural system, and laws when the natural system was not a continuous scala or web of relationships.[89]

Candolle was stating a theoretical position very different from that of Jussieu, as is evident in his three papers on the Compositae that appeared between 1810 and 1812.[90] In the first of these papers, Candolle outlined a distinctive approach. He emphasized that the number of species in a group should have no effect on the rank of that group—a principle, he says, that naturalists had already recognized in that it was then acceptable for both families and genera to be of very different sizes; the cryptogamist Dominique Sébastien Léman (1781–1829), in a review of the paper, singled out this point for notice.[91] Candolle's paper was about the Compositae, not the three families into which Jussieu had divided it, and Candolle also proceeded to suggest that *Equisetum*, *Globularia*, *Dillenia*, *Begonia*, and other groups properly belonged to monotypic families; all except the Begoniaceae were listed as such in his *Théorie élémentaire* that appeared shortly afterward. But we have seen that it was precisely such families that Jussieu would *not* recognize; he had placed *Begonia* in his "plantae incertae sedis" at the end of the *Genera plantarum*, while the other three genera were included in the Filices, Lysimachieae, and Magnolieae respectively, although in no case were they "typical" members of these families.[92] For Candolle, genera *incertae sedis* were either incompletely known representatives of known families or they represented separate, albeit small families; in neither case were they likely to link existing families.

However, Candolle at times adopted a contradictory position that came perilously close to subverting these principles of relationship. Thus he suggested that genera that were very different from others in a family and within which subgroupings of species could be recognized should not be subdivided, although if they were later placed in separate families they could then be subdivided.[93] This would seem to be a reasonable position, as each family could have different rules governing its subdivision. However, contradicting this interpretation is his tendency to equate the systematic value of characters *between* different families, albeit with some hesitation,[94] which implies that moving followed by subdivision would be appropriate only if the new family were removed to a different circle of affinity, or if relationships within the new family had previously been misunderstood.

Candolle's work on the Cactaceae may incline to the latter position. Scott Atran recently observed that when Candolle recognized a monogeneric Cactaceae in the *Flore française*, at the same time remarking that several genera, based on differences of habit, might have to be recognized, he was making an argument (based on common sense) that there were monotypic families, as well as articulating a theoretical point as to how to handle such atypical families. However, Candolle made no such comments about the Grossulariaceae, the family, also monotypic, which he had separated from Jussieu's ditypic, but manifestly and impermissibly heterogeneous, Cactaceae.[95] He thought that, in general, cases where tradition or the size of a group might influence either grouping or ranking were always questionable.[96] When he returned to the Cactaceae over twenty years later, however, he observed that it was "instinct" not to divide a genus when it was mixed with other genera because one then could see similarities. But when the genus was put by itself, subdivision would become natural, as one was then inclined to see differences within it.[97] Groupings within groups only became evident when the latter were separated. Candolle would not have been able to make such an argument for all the monogeneric families he recognized.

Candolle described three stages in the development of the natural arrangement: "tâtonnement" (blind groping), general comparison, and the subordination of characters. He suggested that it was the latter to which naturalists were aspiring.[98] Candolle adopted a more rigid approach to the subordination of characters than did Jussieu, and in Candolle's paper on the Compositae, the equal valuation of characters figured prominently. He compared families to see what sort of character varied within them and to try to make their included variation generally equivalent. Thus although the corolla was divided on one side in the Cichoriacées (one of the three subgroups into which he divided the Compositae), this did not mean it should be removed from other Compositae; *Lobelia* was not separated from the Campanulaceae, nor *Teucrium* from the Labiatae, although each differed from the rest of their families in which they were properly included in the same way that the Cichoriacées differed from the other Compositae. This interfamily subordination of characters contributed to Candolle's adoption of extended limits for the Compositae; the Compositae *sensu lato* were comparable to other families in their internal variation pattern. He supported his argument by showing that there were no other real differences between the three groups he recognized within the Compositae, and that there were analogous variations within them.[99]

Thus it was very largely the importance of the subordinated characters of a group that would determine its rank.[100] As with Cuvier, this subordination of characters was in theory not simply that apparent on inspection of the distribution of characters in a taxonomic hierarchy, although the two were in fact intimately connected. Candolle proposed that there was a connection between the taxonomic importance of a character and its function and/or anatomical basis—the functional approach, he thought, being rather more *a priori*.[101] He distinguished between two classes of characters, those concerned with the conservation of the individual and those involved in the conservation of the species. A classification based on either class of character would be the same.[102]

Understanding relationships was not simply a matter of comparing characters, even when they were subordinated—or perhaps especially when they were subordinated. As Candolle had earlier remarked in the course of his extended discussion on how to interpret floral morphology, "it is necessary to establish a good classification, to restore, by all the ways that observation and experience can provide, all irregular plants to their primitive and regular types, even though these types may be rarely met with, sometimes even ideal."[103] And later: "The entire theory of natural classification clearly rests on the intimate knowledge of organs and their modifications."[104]

But how was this to be done? The botanist had to study numerous aspects of plant parts, for example their number, position, size, shape, and function. Also important were the way in which they were attached one to another, whether they were fused or free,[105] and (added in the second edition of the *Théorie élémentaire*) whether there were intermediate structures formed by the degeneration of organs. The search for the basic symmetry of families or other higher taxa was perhaps the main goal for the taxonomist interested in the natural system and an understanding of nature. This symmetry can be loosely equated with "type," and it represents an abstraction of the family, not the totality of form in the family's constituent genera; it was its "general system of organization."[106] Moreover, extended descriptions of families made establishment of relationships difficult. But with a simplified description, that of the basic symmetry of a group, this task would be much easier.[107] In line with this approach, Candolle thought that if an organism was considered as an association of originally distinct organs—by an analogy with crystallography it would be an assemblage of different "molécules primitives"—how these basically distinct plant parts joined would become clear;[108] an understanding of the laws of plant construction was within reach.

Subordination of characters might have a role in the formation of groups, but perhaps less in the understanding of what a group was. Candolle was clear:

> The number of truly distinct organs was found to be greatly reduced when their nature needed to be analyzed; several of them to which an important function has been attributed were seen to be nothing but simple modifications the one of the other; one and the same organ under diverse appearances could be recognized, and as a result a thorough comparative anatomy practiced. Of course, comparisons deduced from beings that are too different must be distrusted, and not mentioned except with doubt and circumspection; but, for those few arguable examples that are mentioned with hesitation, how many which are not contested again have been obtained by this method of the association of organs.
>
> In particular, the whole of that numerous class of facts, known under the name of monstrosities, which were impossible to understand under the old system, and which one pretended to overlook so as to avoid studying them, the whole of this class, I repeat, took on a clarity and a new interest, once they were looked at from their real point of view, that is as clues for the recognition of the normal or basic ["primitive"] symmetry of beings. These monstrosities are, so to speak, the experiments that nature made for the benefit of the observer: there we see what organs are, when they are not joined together; there we recognize what they really are, when an accidental cause has not prevented them from enlarging. And in starting thus from the opinion that original nature is symmetry, that irregularity is the result of various factors which alter this symmetry, we understand that monstrosities are the result of certain changes in these factors, and that they can, as a result, sometimes make us understand the

cause of the change when its action has been increased or freed of any complication; sometimes show the symmetrical condition when the factors that change it have been either weakened or destroyed.

The whole theory of natural classification obviously depends on the detailed knowledge of organs and their modifications. The arrangement of plants in natural orders implies, according to me, that one day one will be able to establish the characters of these orders upon that which forms the basis of its symmetry, and to attribute the varied forms of species or genera to the action of factors that tend to alter the basic symmetry. Thus each family of plants, like each class of crystals, can be represented by a regular condition, sometimes actually visible, sometimes comprehensible by the mind; it is this that I call its *type*: fusions, abortions, degenerations, or multiplications, either separately or in combination, change this basic type, so as to give rise to the normal characters of the beings that make up [the families]. These modifications are constant within certain limits, just like the secondary forms of crystals. But each genus, each species is, by its very nature, more or less subject to each of the factors that determines them; because plants that have the same type are not as identical as crystals that have similar basic molecules. If botany is far behind mineralogy in this respect, this results, on the one hand, from the far greater number of forms and causes of action; and on the other, from the fact that all these cases are subject to a particular force (the vital force) of which the laws are far more obscure and more difficult to study than those of affinity and attraction.[109]

In this passage, Candolle integrated his approach to comparative morphology with the recognition of groups. Plant groups were not some totality of the variation shown by their included members, but each group had its own basic symmetry which was a subset of this variation and which was also more or less regular—and by this he meant that the type of a family could be represented by a more or less radially symmetrical flower. Candolle thought that the basic symmetry of a group was easily disturbed by natural forces, but he felt that he would be able to understand these forces by studying their results. He thus emphasized the value of terata—monstrosities, abnormalities—in unmasking the basic symmetries of plants, as they provided evidence of connections between plant forms.[110] Once these forces and connections were understood, then the details of the shape of nature, "the varied forms of species or genera," would be comprehensible. Candolle thus shared the goals of Cuvier and Mirbel, although he placed less emphasis on function.

Yet in Candolle's monographic work there is little mention of the notion of symmetry in the sense of the type of a group; the word was used in a more literal sense to refer to radially symmetrical flowers and, rarely, to describe parallel variation in different groups.[111] He also drew largely on characters taken from external form when delimiting groups, and he was ambiguous when he discussed the use of cell- and tissue-level anatomical observations. He was usually, but not always, supportive of the use of such data, and was himself certainly no stranger to the use of a microscope.[112] In his work on the Compositae he allowed that anatomy should play a part in furthering understanding of the limits and distinctness of taxa.[113] Although anatomy here is more the careful dissection of the flower and fruit (gross anatomy rather than structure evident at the level of tissues or cells), the use of the term anatomy, even in this sense, is somewhat unexpected. In the *Théorie élémentaire* he suggested that if one examined the anatomical features of organisms that were supposed to form links between groups, one would find that these links were superficial only. Unfortunately, the only examples he could give here were of animals like bats, whales, and the platypus.[114] He thought that anatomical characters would be most useful at about the level of family.[115] But even where anatomy might allow one to understand why characters that were normally relatively unimportant and variable were constant through whole families, like the opposite, entire leaves of the Rubiaceae, it was constancy rather than "theory" (sic: anatomy) that determined the value of a character.[116] Cell- and tissue-level anatomical studies did not figure in his monographic work; theory and practice were not combined.[117] Indeed, in his *Organographie végétale* he was somewhat equivocal about the value of such anatomical characters: they were often under the subordination of external characters, microscopic observations were not always easy to make, and there was much disagreement over fundamental matters.[118]

The general distribution of a character determined its importance. Thus in the introductory material that the young Candolle contributed to the second edition of the *Flore française*, he noted that the importance of characters might be determined from the importance of the function of that organ, or from the number of plants with that character[119]—potentially quite different procedures. He thought that genera would be placed in the wrong family if these principles were not followed, and this would then mask or disturb the basic symmetry of the family. He also felt that similar problems would also be caused by artificially reducing the number of taxonomic groups: families with very heterogeneous pharmacological properties might result if elements with very different "organizations" were included in them.[120] His later development of a more articulated theory of organogenesis and comparative morphology did not really change the uneasy relationship between taxonomic pattern and theoretical principles that often could not be applied in particular cases.

There is an even deeper tension that pervades Candolle's work and that is particularly evident in the structure of the first part of the *Théorie élémentaire*. There Candolle deals with the theory of classifications, or plant taxonomy. Of the three books that make up the first part of this opus, book one deals with the various ways classifications had been made in the past, and artificial and natural classifications are discussed. Book two is an extended treatment of the theory of natural classification, that is, the various ways in which organs might be modified, and how the taxonomist was to evaluate these modifications. Book three deals with the different degrees of association that exist between plants, how closely or distantly they might be related.

Here Candolle observed by way of introduction that there were two different ways that might be followed when grouping plants according to the principles that he had just discussed in book two. One of these ways was "le marche d'invention," or synthesis. The taxonomist started with individuals and made ever more inclusive groupings. But it was also possible in the course of the verification and exposition of ideas to start with general principles and establish *a priori* general classes, and then to subdivide them in the same way, so finishing with individuals.[121] The first approach was perhaps all that was practicable, but one could at least see both the possibility and the utility of the second. He said he would return to the second approach after discussing the various levels at which plants were grouped.[122] And when he finally did return to this second approach, some forty pages later, he began by discussing the shape of nature, posing the analogy between nature and a geographical map. However, he could not provide even an outline of such a map because relationships between families were so poorly known.[123] To fix the divisions between plants in a positive fashion and hence enable such a map to be produced was a goal to which "classificateurs" should aspire. Nevertheless, he had drawn a plan for the botanical garden at Geneva that was based on his idea of a map of relationships,[124] and what is very prob-

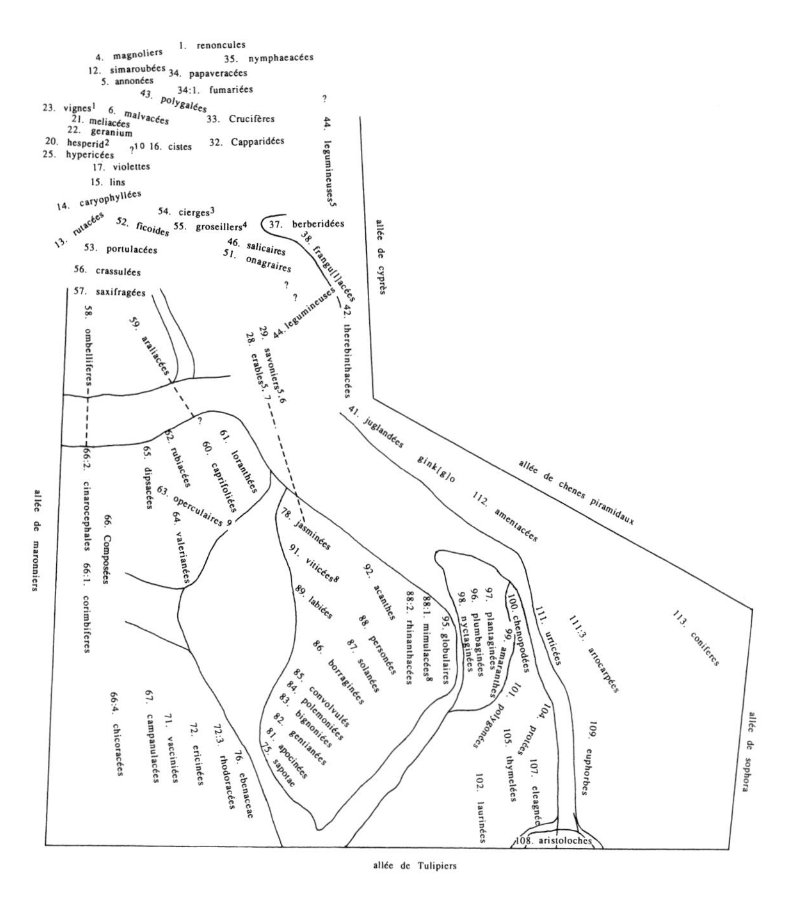

FIGURE 6

Candolle's "landscape of nature"—a botanical garden. Redrawn from copy at the Jardin botaniques, Genève; spelling unchanged (except where indicated), but some family names inverted for ease of reading. Numbers in front of families have been added, and are those given in Augustin-Pyramus de Candolle's *Théorie élémentaire* (1813b: 213–218). Continuous lines represent paths or lawns; dashed lines, meaning uncertain, but perhaps indicating some kind of relationship; ? = word illegible; 1. = Vitaceae (Candolle's Sarmentacées); 2. = Hespéridées, now usually included in the Rutaceae, number 13 in Candolle's list; 3. = Cactaceae (Nopalées); 4. = Grossulariaceae; 5. family name crossed out; 6. = Sapindaceae; 7. = Aceraceae; 8. is now usually included in the Rubiaceae; 9. = Verbenaceae (Pyrénacées); 10 = Tiliacées (perhaps; the copy is almost illegible). Annotation by Alphonse de Candolle on the plan reads "Planting plan for a botanic garden made by Aug. Pyr. de Candolle, probably at the end of his stay at Montpellier." I am grateful to Hervé Burdet for his help in deciphering some of the words.

ably this map is reproduced here (figure 6). In general, families that are adjacent in the sequence he gave in the *Théorie élémentaire* are planted close to each other. It was by such a map that beginners in botany should come to understand the natural method,[125] and he felt he had provided examples of the approach to be adopted in his subsequent treatments of families like the Melastomataceae and Crassulaceae.[126]

This second and more analytical way of looking at plants, rather similar to that of Cuvier (even down to the use of positive characters when describing groups), thus seemed more a goal than an accomplishment in 1813. Candolle returned to the same issue of analysis and synthesis in his *Organographie végétale*, interestingly subtitled "to serve as a continuation and development of the basic theory of botany and in the introduction to plant physiology and in the description of families." He now inclined toward synthesis, preferring what he called "deductions" made from established facts to those made from premature general laws established *a priori*. The latter, he thought, often concerned interrelationships among organs. But some kind of generalization was needed, so at the other extreme, practitioners who were interested only in isolated facts, simple describers, also were failing to advance the science.[127]

Discussion

Candolle suggested in the *Théorie élémentaire* that zoological classification was easier and more certain than was botanical classification.[128] But a year before this book was published, although probably some years after the bulk of it had been written,[129] Cuvier had observed that the use of anatomy in zoology had reversed the relationship between botany and zoology that had obtained toward the end of the eighteenth century. Then botanists emphasized function when establishing classifications, but zoologists used simple constancy of characters. Now botanists, who were still largely focused on the outside of the organism, used constancy, but zoologists could select important or influential characters because, largely thanks to anatomical studies, they now understood function.[130]

Candolle was proposing an integration of a rather different sort of anatomy with the study of groupings and relationships, and he persisted in calling for an integration of botany with "physique végétale" or "botanique organique." Botany, the study of plants distinct from one another, included taxonomy (the theory of classification), glossology (essentially terminology), onomatology (the laws of nomenclature) and phytography. "Physique végétale," the study of plants as organized and living beings, was made up of organography (of which anatomy was part), physiology (the functions of these organs), and pathology.[131] This division of botany in the broad sense reveals the problem Candolle faced. Areas that he thought were all relevant to the study of groupings and relationships had become the sole concerns of different botanists (an issue to be taken up in chapter 9). Candolle thought that botany *sensu stricto*, anatomy, and "physiologie végétale" should form a single science. But for "this latest revolution of the science" to be consolidated, a work was needed in which the principles of the natural method were exposed and the natural groups laid out.[132] Unfortunately, Candolle never provided a new natural series of families,[133] and his new way of looking at plants never really developed. Moreover, principle and practice were not always in accord—although this is perhaps not surprising, given the size of his oeuvre.

Overall, Cuvier, Candolle, and Mirbel all differ significantly from Jussieu in their approach to classification and the natural system. Most importantly, perhaps, all saw at least some gaps in nature; there was no continuity, no nuances linking each organism directly to at least two others in such a fashion that it would be impossible to say where any line dividing the continuum was to be drawn. Although Cuvier was not clear that all taxa, not simply his four embranchements, were sharply discontinuous, the issue of the discontinuity between these embranchements was a major element in the famous debates between Étienne Geoffroy Saint-Hilaire and himself in 1830.[134] It was only Cuvier, however, who made much of the fact that organisms had become extinct in the course of the earth's history, yet it is not clear that this bore on the issue of whether or not there were groups; the subordination of characters would explain the existence of groups (see figure 2B in chapter 2). Candolle rather more prosaically showed that all the collections made since the time of Linnaeus had not filled in the gaps between distinct families. He thereby specifically contradicted Jussieu's prediction that new collections would lead to the obliteration of such gaps. Both Mirbel and Candolle ignored some of Jussieu's family limits; neither followed him in his subdivision of the Compositae, and Candolle in particular was quite willing to recognize monogeneric families.[135] Perhaps indicative of the deep differences between Candolle and Jussieu is the fact that the word "classification" is common in the *Théorie élémentaire*, while the Introduction to Jussieu's *Genera plantarum* can be read almost without needing to invoke the idea of classification at all.[136]

Mirbel and, in particular, Candolle, also emphasized types, or the primitive or basic symmetry of groups, more so than did Cuvier and much more so than did Jussieu. Both Cuvier and Mirbel attempted to understand the hierarchical dependencies of characters, their subordination, in terms of their anatomy, physiology, and function. If these were understood, then the variation in plant groups would be understood and could be related to a single form, that of the type. Candolle's attempts to find the symmetry of plant groups depended more directly on his comparative morphological studies, but again, the goal was to find rules governing organic change.

Jussieu had early suggested that zoologists would have to look at anatomy if they were to make any real progress in establishing the natural order, since there were many different functions in animals, and the organs associated with them were internal.[137] Thus the comparative anatomical studies at which Cuvier was so conspicuously successful were in line with Jussieu's prescriptions, as Cuvier himself recognized. However, although both Candolle and Mirbel were interested in the role anatomy might play in systematic studies in botany, in neither case did their interest amount to much in the way of results, and, as we have seen, the anatomy in which they were interested was rather different from that studied by Cuvier.

The four authors had sometimes rather subtly differing ideas as to how characters that suggested relationships were to be used. Jussieu developed a hierarchy of characters that was a relative, not a general one. The insertion of stamens in the flower allowed the placement of families within both the monocotyledons and the dicotyledons separately, and in this case was of equal value in both groups. But at the familial and generic levels, in particular, the same character was not necessarily of equal importance. Candolle may have preferred a more generalized subordination, or hierarchy, of characters—characters might have a similar role to play in delimiting groups in different parts of the system—and Cuvier also initially inclined to this position. Mirbel, however, argued strongly for a relative rather than absolute hierarchy of characters, and in this respect at least his position was similar to that of Jussieu.

There thus seem to be two rather distinct positions with respect to the shape of nature and the basic approach adopted in group formation: that of Jussieu on the one hand, and that of Mirbel and Candolle (and also Cuvier) on the other. Yet the substantial differences between the two positions appear never to have been spelled out—neither by the participants in the debate, nor by their followers, nor (later) even by historians.[138] Indeed, Fuhlrott in his comparison of Jussieu and Candolle thought that the meaning and significance of the natural system in the work of the two was essentially the same—and this after having discussed at some length Candolle's assertion that there were discontinuities in nature.[139] Jussieu never indicated that groups were discrete, and to the end of his life he was looking for the linkages between groups. Cuvier, Mirbel, and Candolle all saw gaps in nature, and both Cuvier and Candolle in particular toyed with the idea that an analytical, rather than synthetic approach to the study of nature was to be preferred. Mirbel and Candolle disagreed with Jussieu on many other matters, Mirbel even rejecting some of the most important characters on which the Jussiaean arrangement was founded. Nevertheless, many aspects of the approaches adopted by Cuvier, Mirbel, and Candolle in their attempts to understand the relationship between characters, groups, and organisms can be seen in outline in the *Genera plantarum*; Jussieu himself, however, certainly had not taken the ideas very far.

Despite their acknowledged eminence, the work of these three authors cannot be considered as representative of the theory and practice of the natural method as it developed in the first part of the nineteenth century—even Candolle, we shall see, in no way represents botanists. Before attempting a summary, it is essential to look at the work of other French naturalists, and also that of some of their English and German colleagues; this task will occupy the next chapter. I will also examine typological thought in botany during the first half of the nineteenth century, and this will be the subject of chapter 6. In chapter 7 I spell out the implications of the preceding chapters in terms of systematic theory and practice. These implications are in turn the subject of the rest of the book. I move beyond classifications and their construction to consider issues such as what relationships signified as well as the status of the discipline of systematics in the nineteenth century.

CHAPTER FIVE

Detection of the Natural Order: Growing Confusion

It is generally assumed that few naturalists in the early nineteenth century were overt believers in continuity and that still fewer were believers in an unbranched *scala naturae*. It is also assumed that as the nineteenth century wore on such ideas became even less common. But then, why did Henri-Marie Ducrotay de Blainville (1770–1850), Cuvier's successor at the Muséum nationale d'Histoire naturelle, expect to find fossil intermediates between extant groups? Such an expectation would seem to presume that linear continuity would be established when fossil animals were added to the living.[1] Consider, too, that the *Phycologia generalis* of Friedrich Traugott Kützing (1807–1893) came under sharp attack by Carl Nägeli (1817–1891) for its "systematic abolition [of] any absolute difference"—Kützing made no sharp distinctions between species, genera, or even the terms used to describe the organisms on which was working.[2] A form of plenitude (as much as continuity) is incorporated into William Sharp Macleay's controversial quinarian system of 1819 and also into other systems in which all groups could be divided into the same number of subdivisions. More generally, a presumption of plenitude is inherent in idealistic systems where analogies as well as affinities were used to establish relationships between groups.[3]

But people like Blainville and Kützing were surely exceptions, and numerical systems, although much talked about in the first half of the nineteenth century, were ultimately considered of peripheral interest by most naturalists. More to the point, Augustin-Pyramus de Candolle and Mirbel claimed that there were gaps in nature. A similar view of nature is implicit in an observation made by Adolphe Théodore Brongniart in 1868. He noted that since Robert Brown's work on the plants of Australia almost half a century before there had been an interest in recognizing groupings of plants above the level of family. Brongniart himself had discounted attempts to fit all plants into a simple series.[4]

I turn now to consider the work of several of the more prominent naturalists in the period 1789–1859 (following the publication of Jussieu's *Genera plantarum* and up to about the time of Darwin's *Origin of Species*) who can be considered proponents of the natural method. What did they believe the shape of nature might be? And how did their views compare with those of their illustrious predecessor, Antoine-Laurent de Jussieu?

It is often difficult to obtain more than a tantalizing indication of such ideas, as the focus throughout this time period was resolutely on revisions, descriptions of new taxa, and flora writing. The natural method or system was simply assumed to underlie these works. Furthermore, it will be clear that the rejection of the *scala naturae*, although almost obligatory in literature of the early nineteenth century, does not necessarily bear on the issue of whether or not there were discrete groups in nature. Nevertheless, this chapter will show that a sometimes modified belief in continuity remained widespread.

The German-Speaking World

German-speaking naturalists as a group did not articulate a distinctive or even a clear view of the shape of nature. Some naturalists active in the late eighteenth century—examples are Gottfried Reinhold Treviranus (1776–1837) and Heinrich Friedrich Link (1767–1851)—saw nature as being basically continuous.[5] Carl Ludwig Willdenow (1765–1812), a major figure in botany at the turn of the century, conceded in his much-reprinted *Grundriss der Kräuterkunde* that there was no simple chain of being, but that relationships formed a network: perhaps in a hundred years we would understand the natural system.[6] He does not give much further indication of what his complex classifications represented. The Austrian botanist Baron Nicolaus Joseph von Jacquin (1727–1817) discussed his own ideas in rather more detail. He thought that genera and families—that is, all groupings—were all arbitrary; the only natural genus was monotypic. Nature was continuous, although not in any simple fashion, being more a branched chain; this continuity was evident at least above the level of species, perhaps below. He observed, "Nature shows us species, between which our imagination believes we can discover distinctions, which were never in nature."[7] Franz Peter Cassel (1784–1821), who became a professor of botany in Gent, Belgium, did not articulate such a clear vision of nature. The most natural genera and families were those whose included members were very similar, and some genera, at least, were distinguished from all others. Yet plants as a whole formed an ascending series.[8]

If these German-speaking naturalists had attempted to understand the ideas of Candolle and Jussieu by reading commentators and translators speaking their own language, they still would not have found a clear idea of nature. Candolle's views were distinctly changed in the German translation—better, adaptation—of the *Théorie élémentaire de la botanique*, which was made by Kurt Polycarp Sprengel (1766–1833), professor of botany at Halle (and of course in *its* translation into English by Jameson). Here Sprengel suggested that there was indeed no simple series and that some groups were quite distinct, yet that the links in the chain of nature, all connected, and with no proper separation of the parts evident, were being daily more and more confirmed by observation. There was no uninterrupted progression "from inferior to higher degrees of perfection" because the same forms were repeated in different families; yet the example of method that was provided showed plants being arranged in exactly such an uninterrupted progression.[9] Similarly, Friedrich Siegmund Voigt (1781–1850), professor at Jena, described the most recent improvement of what he called the Jussiaean system by emphasizing only the characters being used. He mentioned Étienne Pierre Ventenat's weightings of characters, but there is nothing in Voigt's text about continuity.[10] Carl Fuhlrott (1804–1877) compared Jussieu's and Candolle's approaches in detail, but without mentioning their different ideas about the shape of nature.[11]

But by the 1820s many German-speaking botanists were becoming interested in types, an issue to be discussed in the next chapter. Although types and continuity are not absolutely antithetical, typological thought would tend to encourage ideas that groups were discrete, and comments to this effect appear in the literature at this time.[12] Furthermore, morphology, microscopic studies, and Goethe's ideas of metamorphosis were widely discussed in Germany, and the emphasis switched from an understanding of nature through classifications to an attempt to discover nature's laws and regularities through morphological and physiological studies. (This subject will be revisited in chapter 9.)

French Continuity

As Jean-Henri Jaume Saint-Hilaire (1772–1845) rightly observed in 1805,[13] it was largely French workers who were developing the natural method—that is, who were perfecting ideas of natural relationships in plants, refining the circumscriptions of natural taxa and clarifying their relationships. Problems owing to their presumption that the organic world was continuous are particularly evident in their work.

I have mostly ignored Michel Adanson (1727–1806) so far, mainly because (as I mentioned in chapter 3) he seems to have had little direct influence on the early development of the natural system. However, his method and his general arrangement of families have several of the features associated with continuity. Adanson's general approach was synthetic: describe one species in detail, and then add only differences when describing the next.[14] Adanson observed that although there might be continuity in God's eyes, this continuity was not evident to us ("et cela suffit"[15]), so one should go ahead and recognize groups. This first species described might be a randomly chosen (or particularly well-known) species in a continuous series just as much as the type of a distinct group. Indeed, from the relationships he mentions his arrangement as a whole appears to form a basically linear series.[16] That Adanson thought of nature as forming some sort of series is also suggested by his comment that when it came to understanding a family, it was sufficient to know all the parts of one, two, or three genera—perhaps one from the middle and two from either extremity. Hardly surprisingly, this polarization of variation also suggested a blurring of the gaps between groups. For plants that were intermediate between families, Adanson promised to develop a method that would allow such plants to be placed unequivocally,[17] but he never followed through. He also made detailed predictions as to how many organisms remained undiscovered;[18] these predictions also suggest that nature as a whole and what groups it contained were in his view basically linear, or somehow known, with Adanson extrapolating from his knowledge of collectors and collections.

It is to be expected that in the work of Jussieu's apologists, such as Jaume Saint-Hilaire and especially Étienne Pierre Ventenat (1757–1808), librarian of the Panthéon, more obvious ideas of continuity would be clear; even if there were gaps in nature now, there would be continuity when all the plants of the world were known. Jaume Saint-Hilaire noted that species were created by nature, but genera were not. Genera were simply divisions established to make the study of the science easier; the same was true of higher taxa. When he discussed the subdivision of "natural" Linnaean genera, he suggested that they could often be subdivided by characters of flower and fruit. The reader might get the impression that these subdivisions were clearly separable,[19] however, this would be a forced reading. Rather, Jaume provided detailed notes on the "enchaînement" of orders in each class, and also on the intergradations, degradations, and linkages that occurred between orders. The nettle order (the Urticaceae) provided a particularly good example: "Furthermore, many orders are terminated by genera analogous to plants which they contain and which seem to be scattered here and there by nature to fill the gaps and to form intermediary nuances."[20] Thus the receptacle was closed in the fig, open in *Ambora*, everted in *Dorstenia*, and turned upside down in *Perebea*; it became central in *Artocarpus* and the mulberry, and elongated in the "spherical nettle"; finally, nettles with panicles gradually approached the shape of the inflorescence in the Amentacées, the next order in the sequence.[21] Indeed, Jaume Saint-Hilaire thought that the establishment of a Lamarckian series of plants, the two extremes of which would be diametrically different plants, should be the goal for modern botanists. On the other hand, Jussieu had (as Jaume put it) "fixed the characters" of families to facilitate the understanding of natural relationships.[22] Unfortunately, the significance of this latter remark is unclear: were fixed characters needed, not to help in the formation of a Lamarckian series but to characterize groupings in it that would help the reader fix parts of such a series in the memory? or were there discrete and characterizable groupings in nature?

Étienne Pierre Ventenat's four-volume *Tableau du règne végétal* also follows Jussieu, as its subtitle, "Sélon la méthode de Jussieu," indicates.[23] Ventenat thought that relationships were best visualized as a map,[24] with the goal of botanists being to find the correct latitude and longitude of unknown plants[25] and so better to fix the similarities with countries (that is, organisms) already known. Like Jaume, Ventenat thought that relationships were likely to be continuous and that at least higher taxa were arbitrary; higher taxa were certainly not the work of nature, as species themselves were. Genera like *Rosa* and *Ranunculus* might indeed be discrete, but others were not. If nature formed an indivisible whole, naturalists would still subdivide it, but Ventenat did not commit himself fully as to what he thought the shape of nature was.[26] However, he earlier suggested arranging all plants in an insensible gradation from the most simple to the most complex,[27] and he discussed character associations of groups within the natural series in terms similar to those used by Jussieu and Adanson.[28] Elsewhere, Ventenat singled out for comment newly discovered genera like *Cobaea*; this "brings together the family of the Polemoniaceae with that of the Bignoniaceae, by establishing a so as to say perfectly natural passage between the two orders."[29]

Achille Richard (1794–1852), who was elected to the Académie in 1834, and Adrien-Henri-Laurent de Jussieu (1797–1853)—the only son of Antoine-Laurent de Jussieu, Professor of Botany at the Muséum d'Histoire naturelle, and the third generation of the family in the Académie—espoused the same general philosophy of nature and approach to the detection of natural relationships.[30] Richard wrote a long article on "Méthode" in the *Dictionnaire Classique d'Histoire Naturelle*. There he first suggested that nature was composed of groups. These were evident because nature had made the members of natural families resemble each other in their organization (by this he probably meant, in their important characters) to a greater degree than did individuals taken at random from as many families.[31] But then his position becomes less clear. Richard thought that nature had modified the organization of each plant to yield a series of almost invisible transitions between simple and more complex organisms.[32] The establishment of continuity seems to have been the ultimate goal of Richard's ideal botanist. Members of natural families necessarily showed gradual transitions among themselves, anything else being counter to the general harmony of nature. He suggested that the analogy Linnaeus used for nature, that of a map, was perfectly suitable.[33] He also observed that only individuals were real, all other groupings being man-made and needed only because the naturalist could not grasp both the extent and the details of the variation in nature.[34] All in all, Richard held to a thoroughly Jussiaean view of the world.

Adrien de Jussieu's views are clearly expressed in his much reprinted and translated textbook on botany in the *Cours élémentaire d'histoire naturelle*, which first appeared in 1843.[35] He suggested that the organic kingdom as a whole was an indivisible unity; there were no clear lines of demarcation because nature did not make leaps. There were not even rules without exceptions.[36] Relationships between groups were very definitely reticulating; it was impossible to form a linear series because emphasizing a relationship in one direction necessarily entailed simultaneously breaking relationships in another.[37] Some confusion of relationships was to be expected within the most natural kind of families, "familles par groupe," since there the genera were closely related, although in no apparent order. But within less natural families, "familles par enchaînement," relationships might be more clear-cut.[38] He saw his father's natural method as being an indispensable preliminary to the study of plants; the families Antoine-Laurent had recognized still conserved relationships when subdivided, as only rank was being changed, not relationships—the process being rather like the subdivision of a province into several départements.[39]

However, Adrien saw many species and some genera as being natural units ("unités naturelles"), and he applauded work that was leading to the recognition of groupings above the family level.[40] He showed how easy it was to change the limits of genera without making them any more or any less natural. In working through an example he provided, he seemed to state clearly that some species are more closely related than others, not because they are separated by fewer intermediate species, but because in nature gaps between species are of different sizes.[41] Given that rank could be changed so readily, it is not surprising that he felt that groups at the same rank were not necessarily equivalent, yet he suggested that one could still produce a general system, with groups "legitimized by general assent, and their characters firmly settled."[42]

Rank was not all that would have to be legitimized, since the very circumscription of groups presented problems. This becomes evident in

Adrien's revision of a large and difficult family, the Euphorbiaceae.[43] He felt that genera should be arranged according to the natural series, with some genera establishing the passage from one section to another. Thus the six groupings he obtained in the Euphorbiaceae were distinct when their centers were compared, but they were linked via their peripheral genera. Groups in the Rutaceae were similarly indistinct (see figure 10 in chapter 7).[44] And even when Adrien (rather unusually for his time) compared natural relationships with a tree in his *Cours élémentaire*, his purpose was to promote the idea that there was a general series—despite realizing, along with Richard, that there was no simple series. Moreover, Adrien felt that there could also be grafting between branches.[45] In an article that appeared in 1848, only four years before his death, he returned to the ideas he discussed in his treatment of the Euphorbiaceae over twenty years before: the characters of groups were clear only in their center, but the peripheries of the group intergraded with other groups.[46] He felt that "the expert genealogist" would know how to attach a hitherto unknown species to the common stock.[47] It is difficult to distinguish this position from that of Lamarck in the *Flore françoise* of 1778 or from that of William Whewell and Jules Émile Planchon (shorn of mention of types), which will be described in the next chapter.

Overall, Adrien de Jussieu had not articulated a view of nature that was distinctly different from that of his father. The two also looked at plants in a similar fashion. Adolphe Théodore Brongniart wrote in an obituary that Adrien's work was exemplary: he had converted descriptive botany from a science of generic and specific distinctions into a veritable anatomical and often physiological study—in Brongniart's view, the only real basis of the natural method.[48] However, in the paper Brongniart cited as an exemplar, Adrien's study of the Malpighiaceae that appeared in 1843,[49] anatomy occupies a peripheral position. Furthermore, the illustrations there convey little information that could not have been obtained seventy years before, although details of the seed and embryo are meticulously documented and floral diagrams are provided (figure 7).[50]

British Ambivalence

Turning to the natural method as it developed in Great Britain, five people are particularly important in the period 1810–1859: John Lindley, George Walker Arnott, William Jackson Hooker, George Bentham, and especially Robert Brown—undoubtedly the greatest botanist of them all. Unfortunately, little is known about the philosophy of anyone in this group. Brown was notably reticent,[51] while Hooker, more prolific, seems to have been disinclined to discuss such matters and wrote little on the natural method.[52] Bentham did describe his theory of classification; however, much of his more theoretical work was never published. Although Lindley, a prolific author, explained his philosophy in his numerous introductions to or treatments of the natural system, neither his nor Bentham's systematic theory has been studied in any detail.[53] Arnott wrote an article on "Botany" in the seventh edition of the *Encyclopaedia Britannica* and there discussed the natural method in detail; this article was often cited by his contemporaries.[54]

In so far as a vision of nature had been articulated by their English predecessors, it was a form of continuity. Thus James Edward Smith (1759–1828), founder and long-time president of the Linnean Society, saw nature as "an infinitely complex chain, every link of which is essential."[55] Overall, his vision of nature was similar to that of Antoine-Laurent de Jussieu as expressed in 1789, and the British proponents of the natural method or system either implicitly agreed with Jussieu or were generally uncertain as to the shape of nature. They certainly did not emphasize the discreteness of groups.

Robert Brown (1773–1858), "Jupiter botanicus" to commentators surveying the botanical scene, is both a particularly critical and particularly frustrating figure. If any one publication was central to the introduction of the natural method in England, it was Brown's unfinished *Prodromus flora novae-hollandiae*.[56] He also wrote a number of important papers[57] in which he used a parenthetical and cautious style to make acute remarks about the limits and relationships of genera and families. Brown was an excellent observer, quite at home with a microscope, although no comparative anatomist. Yet we know little about how he saw nature and classifications. In the *Prodromus* he made it clear that the natural order was more reticulate than linear, and so there was no possibility of establishing a general series.[58] But if there was no general linear series, there might also be no obvious groups. This is apparent in his exchange of letters with James Smith in the early part of 1810 over Smith's wish to name *Brunonia* after Brown—an exchange that provides a tantalizing glimpse of Brown's ideas as to what taxonomic groups

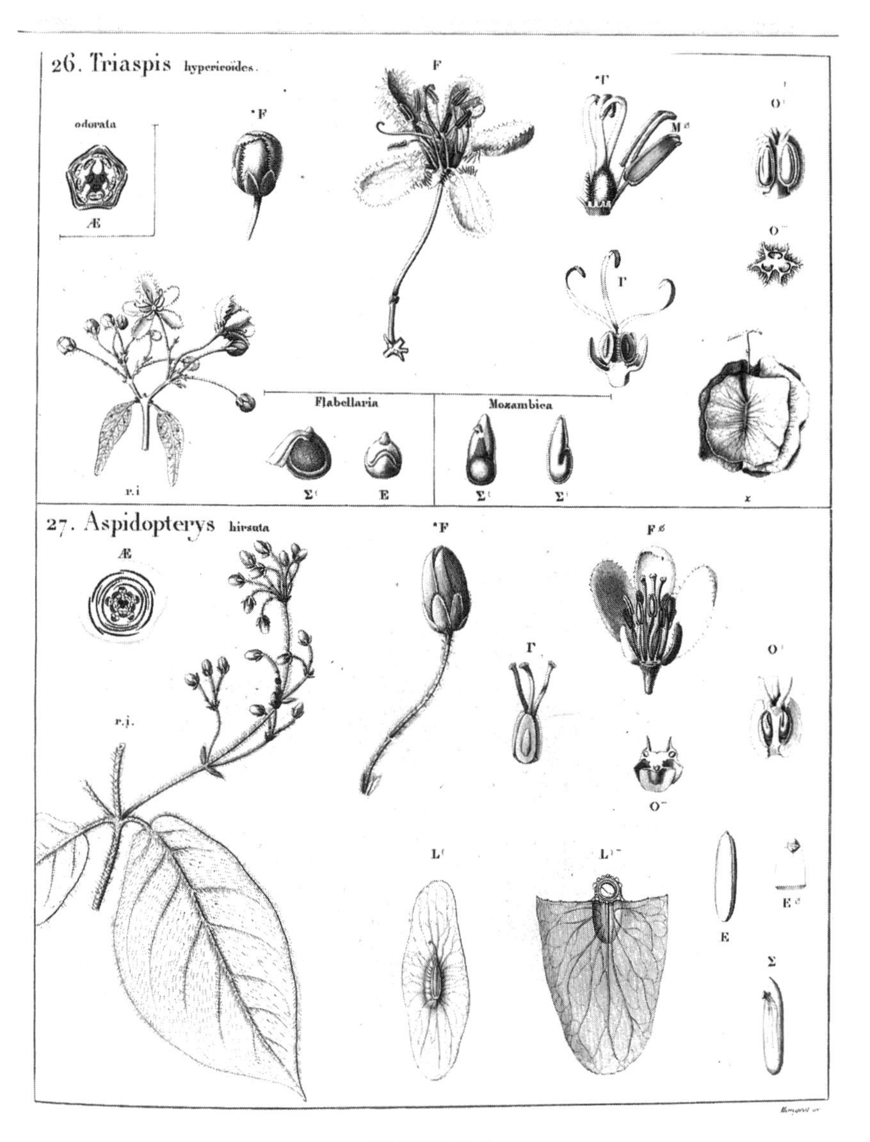

FIGURE 7

Information used in establishing relationships in 1843. Illustrations of members of the Malpighiaceae. The floral diagrams (Æ at top left of each illustration) are colored in the original. The lettering of the diagrams follows a convention that Adrien de Jussieu used, whereby a particular letter always refers to the same part of the flower examined in the same way. Source: Adrien de Jussieu 1843a: pl. 18.

might represent. Brown thought that "upon the whole there is surely no such thing as *Natural Orders* in nature and my scepticism would lead me a step further, that Brunonia be odd in any order." He declared that *Brunonia* "itself broke down the barrier between Syngenesiae [Compositae] and Campanulaceae, Dipsaceae and Globularieae."[59] Thus a paradox: families, which Brown called natural orders, were only constructs of the botanist's imagination. But Brown's silence as to whether nature referred to groups, ranks, or something else, makes it difficult to interpret his position. It was, nevertheless, probably the very existence of discrete groups that was the matter at issue.

That Brown did think nature was continuous is made more likely when we find that he justified his separation of the Apocineae and the Asclepiadeae on grounds very similar to those used by Antoine-Laurent de Jussieu when Jussieu split large groups within which relationships were continuous. Brown observed that plants that were being discovered made existing groups large and difficult to circumscribe, although initially these groups had been quite distinct. Examples were the Jussiaean Rubiaceae and Apocineae; if the two were combined the resultant group would be too big, and any distinguishing characters, if they could be obtained, would be very vague and "clogged" by numerous exceptions.[60] However, the Apocineae could be subdivided by a single character, the nature of the androecium, that was both obvious and important and "preserves the natural series unbroken"; it had the additional advantage of dividing the group into two rather equal parts.[61] Similarly, the large size of the Ericeae was one of the factors that induced him to describe the Epacrideae as a related, but separate family.[62] Brown was perfectly aware that taxa at the same rank were not necessarily equivalent, in large part for reasons associated with the classificatory history of different groups.[63]

Brown disliked typological thought, whether of groups or of characters—although, as in Jussieu's work, the features listed as characterizing the natural orders might not include those found in some of the constituent genera.[64] An apparent exception occurs in his work on the Gesneriaceae of Java.[65] Here he suggested that the ovaries of many flowering plants could be considered as modifications of several fused legumen-type modified leaves. Although he used the term "confluence" to describe the nature of many ovaries, he observed that there was no evidence of this; ovaries might be modified leaves, but the nature of such a modification was unclear. Nevertheless, having come up with this abstraction, he then used it to understand anomalous ovaries of genera like *Hydnora* and *Rafflesia*. Throughout his work, it is clear that determining what a structure "was" might "in some degree lead to correct notions of the affinities of plants,"[66] and that an enquiry into "the real relation which . . . a structure bears to the more ordinary states"[67] was integral to his success in the detection and understanding of relationships. Brown was a consummate observer, and it was against the background of his detailed and extensive morphological knowledge that he made and compared his cautiously made abstractions. From this point of view he is more Candollean than Jussiaean.

George Bentham (1800–1884) is considered to be one of the greatest nineteenth-century monographers and flora-writers, largely because of his treatments for Candolle's *Prodromus*, his important studies on the Leguminosae and Labiatae, his seven-volume *Flora australiensis*,[68] and his collaboration with J. D. Hooker (1817–1911) on the *Genera plantarum*[69] (Bentham wrote the larger part). Although he had no formal schooling and never occupied a paid position (he was moderately well off), his reputation as a general systematist is second to none in the nineteenth century. If we focus on his work before 1860, we find several indications that in his general approach and even in his worldview he was Jussiaean.[70] His first work at all connected with botany is his translation—really a considerably modified new edition—of the "Essay on Nomenclature and Classification," an appendix to a book on education, *Chrestomathia*, written by his uncle, Jeremy Bentham, the utilitarian philosopher. There George Bentham expressed concerns about the effect that the discovery of intermediate organisms, such as a vertebrate with blood vessels but no nerves, would have on a key that he drew up; this key had at least some of the features of a synoptic key, i.e., it conserved natural divisions. He thought that our ignorance of so many beings, by implication as yet undiscovered, would be even more damaging to finer subdivisions of the key.[71] Intermediates blurred the distinctions between natural groups.

In his classificatory work Bentham early took the position that the limits of genera in the Labiatae were likely to be more or less arbitrary. As one of Jussieu's unquestionably natural families, its limits were clear, but those of its genera were less so. Bentham wrote in the introduction to his classic monograph of the family,

> I have always considered, that, although certainly we do observe certain groups of plants more nearly allied to each other than to any others, and distinguished from their several affinities by a sort of hiatus, yet even now, many as are the species which remain to be discovered, it is often very difficult to say between which two species the hiatus is most marked, and the more plants we discover the more are these hiatuses filled up. A genus, therefore, has seldom any real existence in nature as a positively determined group, and must rather be considered as a mere contrivance for assisting us in comparing and studying the enormous multitude of species, which, without arrangement, our minds could not embrace.
>
> But although our genera be not in nature, yet the nearer we follow what is in nature in grouping our plants the more useful is our labour.[72]

In general, all groups were equally natural, although the more sharply defined (positively determined) a group was, the more natural.[73] But Bentham seems to have had a general expectation that there would be intermediates between groups. In an unsigned review of a paper by Alphonse de Candolle on the Apocynaceae, he questioned the assertion made by Candolle[74] that families were strictly limited and could be likened to islands.

> We botanists cannot be so mathematically exact as geographers, and where an isthmus is very narrow, we must class the peninsula with the island. How often does it happen that two large orders, say of five hundred to two thousand or three thousand species, totally distinct from each other in all these species by a series of constant characters, are yet connected by some small isolated genus of a dozen, half a dozen, nay a single species in which these characters are so inconstant, uncertain or variously combined as to leave no room for the strait, through which we ought to navigate between the two islands.[75]

Bentham, in one of his rare uses of an analogy for the natural system, took the variation in a family to be that shown by its included genera. He compared relationships in plants to a wooded landscape:[76] the denser copses represented families like the Compositae, the less densely wooded areas encompassed pairs of families such as the Malvaceae and Tiliaceae or the Melastomataceae and Myrtaceae. But it is unclear to what extent Bentham saw real continuity or simply groups with their edges blurred. It is perhaps more likely that the latter was usually the case except, of course, within families like the Labiatae.[77] And the way in which these quasi-continuous relationships were recognized was by cautious synthesis or building up of groups: "And let it not be forgotten, that although the analytical process carried to the uttermost is necessary for ascertaining the facts upon which botanical science is based [i.e., in identification], it is a judicious synthesis alone which can enable the human mind to take anything like a comprehensive view of those facts, to deduce from them the principles of the science, or to communicate to others either facts or principles."[78]

Bentham had a distinctive attitude about the data that a botanist could use in establishing relationships. Characters useful for "descriptive and systematic botany" were to be taken from the outside of the organism—from a study of the forms of organisms and their general structure.[79] But cell-level anatomy in particular was no part of this because it was so difficult to interpret reliably.[80] Typological thought in particular and speculative work in general had no place for him in his work, and theory should not intrude into a naturalist's endeavor at any level, including the naming of organs. Although both John Lindley and Asa Gray at different times[81] questioned Bentham's narrow approach to characters and observation—unquestionably much more limited than that of Robert Brown or Augustin-Pyramus de Candolle—Bentham's approach was quite common among English-speaking naturalists.

John Lindley (1799–1865) was secretary to the Horticultural Society, editor of the *Gardener's Chronicle and Agricultural Gazette*, the first professor of botany at the University College of London, and a tireless popularizer of the natural system. In several books written for semi-popular audiences, he articulated the principles of the natural method, at the same time being frank as to what he saw as its problems. Lindley was unclear as to the discreteness of groups in nature.[82] He believed that relationships between groups were reticulating, or, as he expressed it, like rays coming from the center of a sphere.[83] Yet although relationships were basically continuous and reticulating, he felt it indisputable that something indeed allowed a group to be recognized, at least when compared with groups that showed the closest affinity, that is, those similar in all essential characters.[84] Higher groups from genera to classes had no real existence in nature, or, as he put it, "groups are nothing more than an expression of tendencies [nixus] on the part of the plants they comprehend, to assume a particular mode of development."[85] Lindley had developed this point in his 1830 *Introduction to the Natural System of Botany*:

> The principle upon which I understand the natural system of botany to be founded is, that the affinities of plants may be determined by a consideration of all the points of resemblance between their various parts, properties, and qualities; and that hence an arrangement may be deduced in which those species will be placed next each other which have the greatest degree of relationship; and that consequently the quality or structure of an imperfectly known plant may be determined by those of another which is well known. Hence arises its superiority over arbitrary or artificial systems, such as that of Linnaeus, in which there is no combination of ideas, but which are mere collections of isolated facts, not having any distinct relation to each other.
>
> This is the only intelligible meaning that can be attached to the term Natural System, of which Nature herself, who creates species only, knows nothing. It is absurd to think that our genera, orders, classes, and the like, are more than mere contrivances to facilitate the arrangement of our ideas with regard to species. A genus, order, or class, is therefore called natural, not because it exists in Nature, but because it comprehends species naturally resembling each other more than they resemble anything else.[86]

The natural system might be superior to the artificial, but what it represented was unclear: what really exists in Nature? God had created species, but He had not obviously imposed a particular structure on creation in its entirety. Lindley did not adopt the position of Louis Agassiz (1807–1873), his near-contemporary, who insisted that the ranks of the hierarchy used in the classification of life were "instituted by the Divine intelligence as the categories of his mode of thinking."[87] For Agassiz, the Divine intelligence had imposed a class hierarchy on life and then exerted control on the creatures of the creation down to details of their distribution.[88] For Lindley, groups could not be placed in a particular rank within an ordered system, but he could still believe that there were groups. He noted approvingly, "the axiom of [Elias Magnus] Fries, that each division or sphere indicates one single idea, and hence the character of each division [classes to families] is best expressed by some simple notion,"[89] and elsewhere he returned to the idea that major groups could be satisfactorily indicated by one or two characters.

But even if the character of a group can be expressed by "some simple notion," this may not bear on how the group is used when establishing relationships, nor even on whether the group exists in reality. Indeed, when it came to actually forming groups, Lindley found that there were no rules that could be applied. It was a matter of considerable importance for him that the rank of a character was connected with its physiological importance.[90] Nevertheless, ideas of the relative physiological importance of characters were largely irrelevant because little was known about such matters in flowering plants and what little was known suggested little variation. Ultimately, everything was left to the skill of the individual botanist,[91] but even then the existence of discrete groups seemed problematic:

> When you shall have made yourself acquainted with all the principal forms under which Nature presents herself, and shall have studied the various links by which one kind of matter is connected with another, you will probably arrive at the conclusion, that there is no such thing as a definition in Natural History; and hence you will come to doubt: so insensible, and at the same time so complete, will you find the gradations between *men* and *trees*, between yourselves and those gigantic creatures of which the timbers of this building may be called the bones,—hence I say you may come to doubt whether any one can truly define the boundaries within which the two kingdoms, that these bodies represent, are mutually confined.[92]

These last comments were made in an address validating and justifying the science of botany that Lindley made during the celebrations surrounding the institution of the University of London in April 1829. To Lindley, the principles of botany were still unsettled, and its purposes, except as an accomplishment, were far from being understood; this was in manifest contrast to the situation in zoology.[93] All that even a perfect natural system in botany could achieve was to make it easy to find the name of a given species and to indicate the mutual affinities of species so that one could use the structure and properties, presumably including the medicinal properties, of a known plant to judge by analogy (induce) those of an unknown.[94]

George Arnott Walker Arnott (1799–1868) was regius professor of botany in Glasgow from 1845 to 1868. As previously noted, his article "Botany" in the seventh edition of *The Encyclopaedia Britannica* was considered to be an important statement of the natural method for English-speaking botanists. Arnott was clear about the overall shape of nature. There was a continuum: zoophytes linked animals and plants,[95] while within the latter "from the highest organized plant to the lowest, all form a chain, in which a link broken or lost disconnects the whole, and to which the recent additions of new links, in the shape of new species, has tended much to the increase of our knowledge."[96] As we might expect that links remained to be discovered between isolated genera and the rest of nature, then if nature were laid out in front of us, orders, families, and genera would be found to exist only as our own conceptions.[97] Nature presented to our eyes only individuals, and Arnott considered species to contain individual members resembling each other and also reproducing themselves; a genus had the same relationship to a species as a species had to an individual. Higher groups should be formed in accordance with Nature's laws, that is, by analogy with the way in which species were formed.[98]

Arnott was interested in the study of organization and in the functions of plant parts.[99] He quoted Gilbert White of Selbourne to the effect that we should study the laws of vegetation and study nature philosophically. White himself had advised botanists not to become immersed in the study of the minute distinctions between species of obscure genera, but to look at useful plants, and "graft the gardener, the planter, the husbandman on to the phytologist."[100] There was no clear indication of the shape of nature there, either.

Arnott, Bentham, and Lindley in different ways raised the issue of communication between the makers and users of classification. However, the success of such communication would depend on the groupings and concepts that underlay plant names. Here the reader was thrown back on "nature," and who was using the names and for what purpose (issues to be explored in chapter 10). Arnott and W. J. Hooker in 1855 quoted verse in their introduction to the seventh edition of the *British Flora*:

> Each [plant] . . . holds a rank
> Important in the plan of Him who fram'd
> This scale of beings; holds a rank, which
> Lost would break the chain, and leave behind a gap
> Which Nature's self would rue.[101]

Although the *British Flora* was written for nonexperts, the message is clear and returns us to the idea that plant names represented parts cut out from a seamless nature. In 1821 Adrian Hardy Haworth had suggested that there might actually be groups in nature in his work on the Saxifragaceae: generic groups were plants "constructed as it were on one model, and so greatly resembling each other in habit, size, and appearance, that a Geologist might say, they are evidently 'all of the same *formation*'."[102] Such an endorsement was unusual, and British botanists by and large believed in a modified form of continuity.

Fossils and Groups

Fossils and the role they played in a naturalist's understanding of patterns and groupings in nature have thus far barely been mentioned. Fossil plants in general were poorly understood at the beginning of the nineteenth century.[103] Furthermore, as Janet Browne[104] stressed in her discussion of ideas of earth history, fossils were usually thrown in as supplementary data for hypotheses that had already been conceived and which had other evidence—in her case, geological—in their favor. In the late eighteenth and early nineteenth centuries numerous remarkable fossils were being discovered—mostly of animals but also some of plants (predominantly conifers and "ferns"). It was becoming clear that these curiosities were actually the remains or impressions of extinct organisms.[105] But in general, fossils were invoked neither to explain the existence of discrete groups of extant organisms nor to provide linkages between such groups; rather, fossils were directly interpreted in terms of living organisms.[106]

Alphonse de Candolle (1806–1893) may be something of an exception in this respect.[107] By 1820 at the latest, his father believed in extinction and was consulting with Adolphe Brongniart about fossils.[108] However, Augustin-Pyramus's own belief that there were groups in nature[109] depended on his interpretation of the relationships shown by extant organisms. Robert Brown was much interested in fossils, becoming an authority in the field. Unfortunately, his descriptions and memoranda have disappeared,[110] and he published only one short paper on fossils rather late in his life,[111] so we cannot say anything about how his work on paleontology may have influenced his work on extant organisms.

John Lindley also had an extensive knowledge of fossil plants. With the geologist William Hutton (1789–1860), he wrote a three-volume treatise entitled *The Fossil Flora of Great Britain*. Although Lindley and Hutton noted there had been changes in the surface of the globe and that there was a sequential basis to the animal fossil record, they thought evidence of change was rather inconspicuous in the plant record. In coal formations were regularly found Cactaceae or Euphorbiaceae, palms and other monocotyledons, and large coniferous trees resembling what were then called Lycopodiaceae.[112] Hardly surprisingly, Lindley and Hutton questioned the theory of progressive development;[113] although the simplest organisms might be absent from the coal measures, members of the Musaceae and Marantaceae (monocotyledonous flowering plants) could sometimes be found there. These were hardly plants with an imperfect organization. *Sigillaria* and *Stigmaria*, they thought, were really to be assigned to the Cactaceae, Musaceae, or Asclepiadaceae. Although Lindley seems to have allowed that extinctions might have occurred, there is no evidence that the fossil record made him think that extant organisms comprised distinct groups. Fossils are barely mentioned in his later discussions on the natural system.

The eminent French paleobotanist Adolphe Théodore Brongniart (1801–1876)[114] was also inclined to assign fossil plants to existing families, although he conceded that there was progression in the fossil record; the vegetable kingdom had gradually become more perfect. Fossils were either of vascular cryptogams or dicotyledonous gymnospermous phanerogams, so although the major divisions of the plant kingdom always had been represented, their proportions had changed.

The belief that gymnosperms should be included in the dicotyledons, or at least in the angiosperms, and the belief that gymnosperms were highly organized (both ideas were common until the latter part of the nineteenth century) confused the interpretation of the fossil record.[115] Other fossils also caused serious problems. The genus *Noeggerathia* in particular provided evidence for the antiquity of the monocotyledons; for example, *N. flabellata*, an uncommon fossil from the Carboniferous, was described by Lindley and Hutton as a palm.[116] *Noeggerathia flabellata* is now placed in the genus *Psygmophyllum*, and although its affinities are still uncertain, it is definitely no palm.

Some naturalists—although they may have been in the minority[117]—confidently expected to find living specimens of plants that were then known only as fossils. They tended to believe that the taxa represented in fossils were once widespread over a warmer earth and thus still likely to be alive in the tropics or in the southern hemisphere. This expectation fueled exploration.[118] Carl Friedrich Philipp von Martius (1794–1868) equated tropical plants he had seen in his travels in South America with fossil plants;[119] he found a tree fern, *Polypodium corcovadense*, in Brazil that he placed in the same genus as the fossils *Lepidodendron* and *Palmacites*. Similarly, he compared the remarkable genus *Lychnophora*, a member of the Compositae which he discovered and described, with the fossils *Lepidodendron dichotomum* and *L. laricinum*—both of which he transferred to the fossil genus, *Lychnophorites*.[120] The stout stems of *Lychnophora* with their prominent scars looked like stems of fossil *Lepidodendron*, a similarity enhanced by the linear or straplike leaves that the two have in common; *Lepidodendron* is a genus of flowerless plants not at all close to the Compositae.

Only rarely was the absence of fossils to explain problems with the natural system. Alphonse de Candolle observed that some naturalists hoped that fossils would fill in the gaps between classes, so giving a complete symmetry to the table of affinities which was currently irregular; but their hopes seemed unrealistic to him.[121] And Johann Friedrich Klotzsch (1805–1860)[122] felt that fossils would reveal the relationships of the Begoniaceae, a family quite distinct from the other dicotyledonous "Klassen-Typen" and whose relationships were unclear. Statements such as these imply that the naturalists making them thought that there were gaps in the order of nature, yet there was no consensus over the role fossils might play in helping a naturalist understand nature. Furthermore, even though Pierre Turpin suggested in 1820 that the loss of species might cause gaps in the order of nature, he basically believed in continuity of the here and now.[123]

Comparative Anatomy and the Natural System

We have seen that Cuvier and Mirbel in particular found anatomy important in their work. However, I also suggested that botanical comparative anatomy and zoological comparative anatomy were rather different disciplines (although this distinction has rarely been made by others). In what follows, *comparative anatomy* refers to cell- and tissue-level studies that depended on a microscope, rather than on simple dissection and the use of at most a hand-held lens. Four aspects to early discourses on the use of data provided by comparative anatomy in the evaluation of relationships will be examined in this section.

First, pre-Jussiaean botanists were seen as paying most attention to the outside of the organism.[124] Jussieu himself suggested why anatomical data were not likely to be of any use in botany—botanists had come to a good understanding of the functions of organs without their aid. Although Jussieu made a connection between the study of anatomy (what kind was unspecified, but he was probably thinking more of gross anatomy) and an understanding of function, as far as he was concerned the functions of organs, on which the importance of characters depended, were already well understood. Anatomy would add nothing to this knowledge, although he thought that the situation was quite different in many animal groups.[125] Despite his belief that external form was dependent on internal anatomy, Jussieu did not expect that gross anatomy would do other than confirm groupings based on external form. His generally rather negative comments about anatomy, or even the use of a lens, represented a common position at the time.[126]

Although some botanists took an opposing position,[127] there was certainly no general expectation that comparative anatomy would disclose groupings different from those that would be suggested by the use of external form alone. The Swedish mycologist Elias Magnus Fries (1794–1878) stressed the importance of "hidden" characters, but he linked them to "habit," a feature which however could unfortunately not be expressed in words.[128] Augustin-Pyramus de Candolle[129] thought that anatomy would confirm the value of leaf type and insertion—characters that were usually not very important—in the recognition of the Rubiaceae. As far as I know, nobody checked to see if there was a correlation between vegetative and anatomical characters in the Rubiaceae, and a knowledge of anatomy had nothing to do with the initial decision that the vegetative characters were important.

Adrien de Jussieu[130] advanced a broader argument that effectively devalued anatomical studies as a comparative tool in classification. He thought that experience showed in general that natural bodies which resembled each other on the outside resembled each other internally. Botanists had to find characters that necessarily coexisted, but at the same time they should focus on external characters that guaranteed real and intimate relationships. This coexistence is a functional interdependence of characters similar to that discussed earlier by Mirbel and Cuvier, but Adrien did not invoke anatomy in the search for such characters. Indeed, if anatomical characters necessarily co-varied with external characters, the only reason for looking at anatomy would be to discover external characters that did not co-vary with internal characters and then to place less importance on them than if there were covariance. In any event, Adrien had reversed the direction of relationship between the inside and the outside of the plant that his father had thought existed.

The assumption that anatomy would not tell systematists anything really new persisted for about a century after the publication of Jussieu's *Genera plantarum*. In an 1879 letter, George Bentham congratulated Casimir de Candolle (1836–1918, the third botanical generation of the Candolles) on his recent paper on the comparative anatomy of leaves;[131] Candolle's findings were less exactly in agreement with natural generic differences than Bentham's *a priori* ideas had led him to expect.[132]

A second aspect of the role of anatomy was that there were few examples of its use, and these seemed to support groups the existence of which had already been suggested by the examination of external morphology. In 1798 René Louiche Desfontaines (1750–1833), in his monograph on the structure of the stem in monocotyledonous and dicotyledonous plants, showed how anatomy could help one choose between alternative ideas of relationships. The stem anatomy of the Aristolochiaceae was consistent with the position Jussieu had assigned this family—not with that implied by other relationships that had been suggested;[133] they were dicotyledonous plants, not monocotyledons. But this anatomical evidence was interpreted as simply confirming Jussieu's judgment, rather than adding any fundamentally new element to the debate. Desfontaines suggested that the inside of the plant reflected its exterior form, and François Dagognet rightly observed that Desfontaines had satisfactorily linked the two.[134]

Mirbel's anatomical work on the Labiatae disclosed little of interest to the naturalist making classifications.[135] But some studies did demonstrate the value that comparative anatomy might have in understanding relationships. For example, Brongniart[136] showed that cycads were anatomically quite like conifers—with which they agreed in the morphology of their reproductive structures and their embryo—rather than palms, with which they shared only a general similarity in appearance. In general, Brongniart was a strong supporter of the use of anatomical data in systematics, and it seemed to him that "the whole world will admit that anatomical characters, those relating to the basic organization ['l'organisation intime'] are more valuable than external shape."[137] But as happened so often in such discussions, the only example he provided of its use was the distinction between monocotyledons and dicotyledons.

A third aspect of botanists' regard for the use of comparative anatomy is evident in the controversy over both the interpretation of what could be seen under a microscope, i.e., the anatomical data themselves, and the interpretation of anatomical data that were themselves not in question. Even in the middle of the nineteenth century some British scientists were unwilling to use the microscope because they distrusted what could be seen.[138] Even Augustin-Pyramus de Candolle was not sanguine about the use of anatomical data in botany, observing that "the fine ['délicate'] anatomy of plants is still, over the most fundamental points, in an uncertainty [that is] hopeless for the friends of truth." He went on to say,"the contradictions of observers on this subject are such, that it is not rare that several people looking together with the same microscope at the same fragment, see there or think to see different forms."[139] But the situation was not impossible. Although there were major advances in microscopy in the nineteenth century, the great anatomist Hugo von Mohl's remarkable studies in 1831 on the anatomy of palms showed what could be achieved by a careful investigator even early in the century (figure 8).[140] Nevertheless, in the early days of microscopy, it was not clear what drawings actually represented: were they individual sections or composites?[141] To skeptical observers like George Bentham the interpretation of a microscopical preparation was likely to be subjective, "theory" was often involved,[142] and the sampling was nearly always bad.[143]

Other controversies involved the interpretation of microscopical anatomy; the general pattern of what was seen was not, however, at issue. Thus the distinction between Exogens (gymnosperms and dicotyledons) and Endogens (monocotyledons)—terms invented by Augustin-Pyramus de Candolle[144]—became a matter of some disagreement. It was generally agreed that there were two main groups of flowering plants, the monocotyledons and dicotyledons, and that this

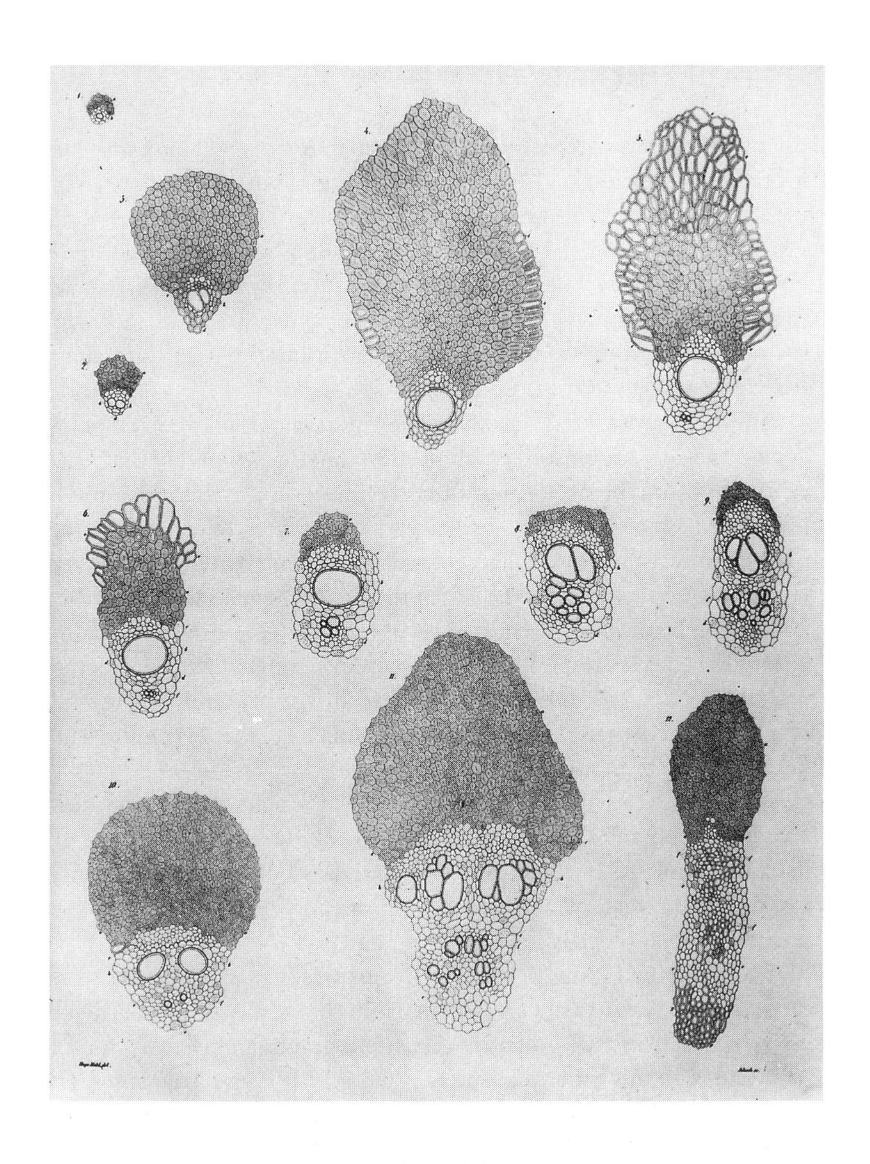

FIGURE 8

Anatomical studies in 1831. The drawings are taken from the vegetative parts of palms. Items 1–11, transverse sections of vascular bundles in the stem; item 12, bundle as it exits to the frond. 1–9, *Kunthia montana*; 10–12, *Leopoldinia pulchra.* Source: Mohl 1831: pl. E.

division was supported by both external form and anatomy. Desfontaines postulated in 1798 that a palm stem, with its scattered bundles in the center and more numerous bundles in the periphery, must grow from the center, while the dicotyledonous stem, with bast on the outside, grew in thickness at the periphery.[145] Although it became clear that this was an incorrect interpretation, it persisted, and in 1839 Mirbel was sent by the Académie to Africa to study the date palm and to resolve (it was hoped) the problem.[146]

Another controversy involved the systematic placement of the Nymphaeaceae. Augustin-Pyramus de Candolle[147] noted that those who supported the position of the Nymphaeaceae in the monocotyledons emphasized that characters taken from fruit and those from internal anatomy were not in agreement. Placing more reliance on fruit characters, he included the family in the dicotyledons, but he could not command a consensus. Auguste Trécul (1818–1896)[148] likewise pointed out that the correspondence between internal and external features was not without exception, yet he claimed that his careful anatomical studies showed that *Nuphar lutea*, a member of the Nymphaeaceae, belonged to the monocotyledons.

Fourth and finally, Achille Richard[149] thought that comparative anatomy in botany simply would not provide the diversity of data that it had in zoology. The anatomy of plants was both simpler and more uniform than that of animals, a position also taken by some other botanists and zoologists.[150]

The combined result of these problems was that anatomy was so little a part of a systematist's armamentarium that it was often not included in the study of true botany, that is, monographs and classifications. (I shall return to this issue in chapters 7 and 9.) Although Augustin-Pyramus de Candolle was no anatomist, he was largely sympathetic to the use of comparative anatomy as it related to the formation of groups.[151] Mirbel noted the desirability of integrating systematics and anatomy, but he never developed a school of comparative anatomy; his interests turned elsewhere.[152] In 1809 Mirbel did call for the development of a body of data on the comparative anatomy of the flower, as he thought that the vessels were obviously involved in reproduction and in the nutrition of the seed; one could expect functional, anatomical and systematic interrelationships. Vegetative anatomy, on the other hand, seemed to him to be of less interest.[153] He later downplayed the use of anatomy as a general aid in classification, and he did not recommend even the use of a hand lens in a naturalist's studies since groups were normally self-evident.[154] But more than a decade earlier, Desfontaines had called for studies of comparative vegetative anatomy,[155] and fifteen years later the prospectus for a new journal, the *Annales des Sciences Naturelles*, singled out for attention the absence of a comparative anatomy of plants. Furthermore, despite Jussieu's assertions, anatomical and functional studies were not linked in botany.[156] One underlying problem that we have already encountered was, as André-Marie-Constant Duméril (1774–1860)[157] and others noted, that plants had fewer organs and less localized functions than did animals, and anatomical observations had been of less use in plants than in animals. Animals, with their numerous internal organs and localized functions, were ideal subjects for comparative (gross) anatomy, as Candolle also recognized.[158] Comparative studies of cell-level anatomy in animals, histology, were not a part of the major changes in ideas of relationships there.

Even in the latter part of the nineteenth century neither Bentham nor his younger colleague Joseph Dalton Hooker (1817–1911, second director of Kew Gardens)[159] saw any use for comparative anatomy, although Hooker allowed that anatomy might be valuable in morphological studies (Bentham would have been less sanguine of this prospect). Nevertheless, Hooker suggested that it was because physiology was relatively unimportant in plants compared to animals that there could be no linkage of characters with function, and so anatomy, development, and morphology were of particular importance. He considered that the knife and the microscope were indispensable tools for the botanist in such situations.[160] But Hooker distinguished a comparative anatomy at the cell and tissue level of the vegetative parts of the plant (which was of little use to him in classification) from an anatomy pressed into the service of comparative morphology at the tissue and organ level (which did allow him to understand relationships). As Joseph Duval-Jouve (1810–1883) remarked, a sound knowledge of the structure, development, and function of organs was integral to establishing relationships; without such knowledge, organs could not be compared.[161]

In the first half of the nineteenth century, studies utilizing the microscope were overwhelmingly in subjects like organography, development, physiology, and cryptogamic botany.[162] Robert Brown's important work, which often entailed the use of a microscope, addressed just such issues concerning the nature of plant organs. Nevertheless, Brown was exceptional, and an anonymous author (actually, David Brewster) in the *Edinburgh Review* exhorted British botanists to take up anatomy and physiology, holding up Brown as an example of what could be done along these lines.[163] Such exhortations were to no avail, and Mirbel, Brewster, and others encouraged the use of anatomy outside the discipline of taxonomy itself. It is hardly surprising that Brongniart, in an obituary of Mirbel, observed that anatomy had to retrace its steps, and that comparative anatomy had to develop; it was even needed to confirm Payer's developmental work.[164] But there was little immediate change.

Discussion

To focus the discussion, I will outline the positions of two people, both of whom were largely interested in animals and are in some ways peripheral to our story, but who articulated the detection of the natural order in the context of very different ideas of nature. In 1764 the Genevan philosopher Charles Bonnet (1720–1793) stated with great clarity the position toward grouping and classification to be adopted if nature is continuous. Almost a century later the English naturalist Alfred Russel Wallace (1823–1913) advised what should be done in the relatively uncommon cases where there was continuity, in contrast with the commoner situation of discontinuity.

Bonnet contended that synthesis, Jussieu's first method of apprehending the natural arrangement, worked only if there were no gaps in nature. His overall view of nature was quite similar to that of Lamarck, and he drew out a series like that shown in Lamarck's table (see figure 1 in chapter 2), but without characters.[165] Bonnet was a vigorous champion of the principle of plenitude, of continuity, and of the *scala naturae*, although he suggested that there might be some branching of the chain.[166] He thought that if nothing was separated in nature, then human classifications, which implied discrete groups, could not be congruent with nature.[167]

As Bonnet observed, if intermediates were absent, there could be no connection between one being and another,[168] and continuity could not be established; fortunately, the intermediates were there. His instructions on how to construct this natural order read like a summary of Jussieu's introduction to the *Genera plantarum*: synthesis would be the *modus operandi*; intermediates were important; the sequence should begin with the least perfect (simplest) things. Bonnet deliberately disregarded anatomy,[169] and more general characters distinguishing groups were mentioned only in the context of a more analytic way of looking at organisms.

> The Chain of Nature is thus constructed by going from the units that make things up to composite things, from the less perfect to the more perfect.
>
> But in considering it thus, and from a very general point of view, one never forgets that our way of understanding is not the normal way.
>
> We will glance only at the outside of Beings: we survey only the outer surface: the contemplator of Nature restricts himself to contemplation, and he does not undertake dissection. Perhaps we will pay a little more attention to less known or more overlooked species.[170]

And in his discussion of intermediate species, Bonnet noted,

> There are no gaps at all in nature; everything in it is graded, nuanced . . . . Among the characters which distinguish Beings, we find more or less general ones. From thence our distributions into Classes, into Genera, into Species.
>
> These Distributions will not divide. There are always between two Classes or two neighboring Genera *intermediate* Productions, which seem not to belong more to the one than to the other, and join them.[171]

Wallace spelled out the issue from his very different point of view. He felt that the analytical approach was inappropriate if there were no gaps between groups, because then its use led to endless disagreements over relationships. It was only when there were no gaps (for instance, in groups like sparrows, with their numerous and very similar species) that the principle of synthesis or "agglutination" should be applied.

> But in attempting to carry out this system [separation of groups from the general mass] in a further division of the Passeres, no such satisfactory and generally accepted results have been produced. No systematist has been satisfied with the arrangements of his predecessors, and, after an endless variety of divisions and subdivisions, we are as far off from any generally accepted system of arrangement as ever.

> The reason of this we conceive to be, that we have to deal with a mass of species in which the series is so nearly complete, that there are no more of those great chasms separating considerable portions from each other, and that the affinities are so intricate and minutely varied, and so cut up as it were by minor gaps between genera and families, that any attempt to form great and well-marked subdivisions must fail, for the simple reason that such are not marked out by nature. In such a case, an *arrangement* may be possible, but a *classification* may not be so. We must therefore give up altogether the principle of *division*, and employ that of *agglutination* or juxtaposition.[172]

Wallace claimed that it was an article of zoological faith that there generally were gaps between groups caused by extinction in earlier epochs; the "principle of division" allowed such groups to be recognized.[173] From this perspective, the more analytic approach would usually be the appropriate tool for zoologists, and presumably also for those botanists who believed in the existence of discrete groups in nature. Synthesis or agglutination was appropriate only if there were no gaps in nature.

Antoine-Laurent de Jussieu thought that any classification of continuous nature was man-made. In this he was more or less echoed by his contemporaries Adanson, Lamarck, and Buffon. Buffon noted in the important "Premier discours" of his *Histoire naturelle, générale et particulière*[174] that he saw only disorder in nature—the Creator had thrown together a world of beings, some related and some not, in an infinity of combinations, some of which were harmonious and some not, and had subjected them to a perpetuity of destructions and renewals. All that the naturalist could do was to perceive particular effects—in the case of organisms, their form, etc.—compare them and combine them and then produce an order that had more to do with our own nature than with nature herself. Nature proceeded by imperceptible nuances which joined species and often also genera.[175] Lamarck stressed that it was the naturalist, not nature, who chose the characters in the linear series around which groups could be formed. Adanson was apparently less convinced, observing that although there might be continuity in God's eyes, we could go ahead and recognize groups, since this continuity was not evident to us.[176] Nevertheless, I suggested earlier in this chapter that even for Adanson the shape of nature was basically linear. Synthesis was seen as being an appropriate procedure to follow if nature was continuous.

This situation did not perceptibly change in the next seventy years. Although Mirbel and in particular Augustin-Pyramus de Candolle emphasized that nature was not continuous, and Candolle that the size of gaps between taxa determined their rank, the issues were very much less clear cut for many botanists in the middle of the nineteenth century. There are substantial ambiguities in the positions of George Bentham and Alphonse de Candolle. For Bentham, groups were more or less natural; but he never stated that they were, or at least ought to be in theory, sharply circumscribed. In practice groups were sufficiently recognizable, but there were nevertheless intermediates; taxa at the same rank were certainly not equivalent.[177] For Candolle, sharply circumscribed groups existed in theory; but when it came to attempting to define such a natural group, he was reduced to talking rather vaguely about tendencies, and hoping that somebody would eventually find the distinct characters he thought groups in nature should have.[178] Lindley seems to occupy a more Benthamian position, but he also adopted a Friesian idea of groups that led him to suggest that they existed not so much as their constituent members but as some central idea or type.[179] Robert Brown used his types to compare organs in members of different groups, rather than represent the quiddity of those groups; separate groups might well be closely connected, if not continuous, and rank did not necessarily mean anything.[180] Adrien de Jussieu showed how rank and circumscription of groups could be changed without affecting relationships, but his reader is left in the dark as to whether he thought distinct groups were of common occurrence. And another French botanist, Auguste-François Saint-Hilaire, claimed in 1840 that groups existed independently of our arbitrary classifications, yet he went on to say that we should not subdivide groups because future discoveries would tend to fill in the gaps between the subdivisions.[181] The other mostly French and English authors mentioned in this chapter similarly vacillated over the shape of nature and the nature of groups. Of course, the ambiguity may be heightened for us because the distinction between grouping and ranking that is so important in current discussions was then not commonly made, but we are given the impression that groups were arbitrary as to exclusion.

Weighting Individual Characters

That groups would tend to have, at the least, indistinct boundaries will be clear when we look at how naturalists decided whether or not a particular character was of any value in indicating relationships. There was much argument over how rigidly characters were to be ranked for this critically important function in systematics. Although Mirbel was at pains to convey that it was Jussieu's recent apologists such as Ventenat who had applied the doctrine of subordination of characters in too rigid a fashion—students were typically inclined to take the ideas of their master to extremes—it is easy to see how attempts to impose such a hierarchy developed.[182]

Ventenat[183] had provided a table in which various expressions of twelve floral characters were assigned absolute weights; variation in these characters would be of the same systematic importance wherever it occurred. Augustin-Pyramus de Candolle could, however, have been accused of the same failing as Ventenat, as we have seen that he was inclined to think that characters in different groups might be of similar importance. Alphonse de Candolle devoted considerable effort[184] to developing a hierarchy of characters ranked by their value. Taxa at different levels of the taxonomic hierarchy could, he thought, be characterized by different kinds of characters: genera could be distinguished by differences in the numbers of the parts of the flower, less important so-called fusions of these parts, the number and shape of the seeds, the relative position of the leaves, and so on; families differed in characters such as the symmetry of the flower, fusions of important floral organs, and the development of the ovule.

As more became known about plants, the rigidity of Jussieu's weighting—let alone that practiced by Ventenat—was largely discarded. Augustin-Pyramus de Candolle saw the subordination of characters, on which weighting depended, as the third, and yet-to-be-attained highest, stage in the development of natural classifications.[185] Yet there was little progress in actually developing ideas of subordination or weighting, and from 1800 onward comments were frequent about the impossibility of working out absolute principles when deciding how important a particular character might be in classification. As the ornithologist Hugh Strickland put it, "The importance of the same character manifestly varies in different departments of nature, and must therefore be estimated by moral rather than demonstrative evidence."[186] By this he meant that adequate and credible evidence for the value of a character could be provided, but not evidence deduced from axioms.[187] Similarly, John Lindley noted that "Affinities of plants . . . depend on more or less intricate combinations [of characters], the power of judging of which is the same test of a skillful botanist, as the appreciation of symptoms is that of a physician."[188] It is in this context that John Stevens Henslow observed that not only were there no precise laws (as there were in crystallography, allowing a naturalist to set the limits of species) but there were no criteria for establishing the relative values of genera, tribes, orders, and so on.[189] Although Antoine-Laurent de Jussieu thought that there was no exact geometry of nature, there could, he felt, still be a "calcul" of characters.[190] Under such circumstances large collections were seen to be at a premium, as sound taxonomic judgment could result only from the prolonged study of many plants.[191]

The establishment of principles that would allow a proper subordination of characters remained a desideratum, and Asa Gray, commenting on J. D. Hooker's work on the Balanophoraceae that was published in 1858, sadly observed that there were "no fixed and philosophical principles for the subordination of characters and the study of affinities in plants."[192] He hoped that Hooker would provide them. Hooker did not, and the selection of characters that were believed to provide evidence of relationships, and their weighting, remained problematical throughout the century.[193]

Three more specific approaches to weighting were sometimes adopted, although none commanded general assent. One sanctioned the use of *a priori* criteria for weighting. Another found value in *a posteriori* criteria for weighting. A third called for equal weighting of many characters.

No general *a priori* weighting scheme was developed; indeed, the most notorious systems were those in which individual characters were used as the basis of a new arrangement of plant families. At the end of the nineteenth century, Franz Buchenau (1831–1906) used a quotation from the physiologist Émil Heinrich Dubois-Reymond (1818–1896) as an epigraph for his monograph of the Juncaceae: "Absolute characters are the philosopher's stone in systematics."[194] (There is more than just a hint of Buffon in this quotation.) More common was the application of local weighting schemes. Thus the Swiss botanist Johannes Müller Argoviensis (1828–1896) sought a (locally) absolute weighting of characters for his work on the Euphorbiaceae, and he later cast the argument between himself and Henri Baillon (1827–1895) over how characters should be used in terms of Jussiaean weighting versus Adansonian nonweighting. "*Dr Baillon is Adansonian, he counts differences, he more or less rejects the subordination of characters*; I on the contrary belong to the systematic persuasion of Jussieu, R. Brown, de Candolle, etc., I place characters in the balance, *I evaluate differences*, and the subordination of characters is for me a basic and absolute necessity."[195] Of course, it is unlikely that Baillon consistently gave characters equal weight, but Müller was not alone in his attempt to weight characters using *a priori* criteria. Lindley divided *Nolana* using characters that separated genera in the related Convolvulaceae, and Adolphe Brongniart used largely similar generic characters throughout the Rhamnaceae.[196]

The *a posteriori* approach is exemplified by the distinctions that were being drawn between analogy and affinity (homology), and by how characters that expressed affinities were to be used to establish groups. For Augustin-Pyramus de Candolle, the distribution of characters that showed affinity might be affected by the principle of the subordination of characters, but such characters also showed a distinctive pattern of distribution—analogies linked different taxa independently, while affinities linked a similar set of taxa.[197] Along similar lines, Jules Émile Planchon recommended that naturalists should group plants, not by characters of detail of which analogies were found in all classes, but by "the *consensus* of *facies*, properties ['propriétés'], and of organization which summarizes the constitution of groups and almost always betrays their natural tendencies."[198] Such remarks are not often encountered in the literature and were made almost in passing. Clear guidelines for the establishment of *a posteriori* weighting schemes were not developed.

As to the third approach, attempts to determine relationships simply by counting the numbers of characters groups had in common (an approach similar to that advocated in 1764 by Michel Adanson) were very difficult to implement. This approach had to wait for accessible computer technology, which developed in the middle of the twentieth century. Nevertheless, authors like Baillon tried to determine groups and relationships in a more Adansonian fashion than did other botanists.

In general, statements about the relative importance of characters remained questionable. Systems based on individual characters that were heavily weighted were questioned, and no "decisive affinity, or analogy," whether taken from the fructification or from elsewhere in the plant, was established. It seemed that each and every character that was apparently important in some situation would be judged as having little to offer in another. Not only did all characters vary in their value depending on the group being examined, it sometimes seemed, as to Sir J. E. Smith in 1822, that no tangible character of an organism had any importance at all.

> What has just been remarked, of the inconstancy of number in the seeds of particular plants, and of its great diversity in species or genera nearly akin, may possibly diminish the apparent absurdity of considering the great differences between the fruit of *Begonia* and *Polygonum* or *Rumex*, and between that of some *Campanulaceae* and the *Compositae*, as a matter of but secondary importance, and may reconcile us to the opinion that such differences should give way, in both cases, to strong points of agreement. Even the great distinction between the inferior germen of *Begonia*, and the superior one of the Order of *Polygoneae*, Juss. 28,[199] is invalidated by the above instance of *Vaccinium*; and the coincidence of habit is so remarkable, that I cannot but confess myself very anxious to ascertain a decisive affinity, or analogy, in the fructification, lest the great fundamental principle of all sound botanical classification should, in any degree, be undermined.[200]

Almost thirty years later, Adolphe Brongniart echoed Smith's concerns; there were arguments over the details of subordination, and even if there were ever cases where a relationship suggested by one character could never be out-voted by relationships suggested by other characters.[201] Yet despite such problems, the lack of consensus on ranking, and the difficulties practitioners encountered in attempting to abide by the competing visions of nature, taxonomists continued to recognize genera, families, and higher-level groupings of plants.

Analysis Versus Synthesis

In chapter 3 I suggested that Antoine-Laurent de Jussieu linked the existence of character correlations with an analytical hierarchy of characters. The most important characters, those with the widest distribution in the arrangement, were discussed first, so proceeding from the top downward.[202] Jussieu himself sometimes discussed such correlations, for which he used the term "affinité," in functional terms. It is exactly these functional relationships and dependencies that so interested Cuvier and, to a somewhat lesser extent, Mirbel and Candolle; in their works we find suggestions that a more analytic way of forming groups might be preferable.

Jussieu's own approach to analysis was ambiguous. He used analysis when subdividing his classes, but this subdivision, as he himself admitted and later commentators have also observed, was largely for the sake of convenience and had nothing necessarily to do with nature.[203] But he did have in mind an analytic hierarchy of characters that was connected with their importance in the life of the organism.[204] As he later observed, it was important to know the functions of organs to determine their importance; the characters of natural groups were those that could coexist within a single functional individual.[205] Yet it was nominally the more synthetic, inductive and "intuitive assessment of overall similarity along traditional lines" that led to the grouping of species into genera, and genera into families.[206] But even there, Jussieu's discussion of the importance of characters for establishing relationships has a very hierarchical, if not analytical, cast. Thus a reader of Jussieu might be forgiven for thinking that characters in general were to be rigidly subordinated, albeit only in the local context, or that they were to be assigned fixed importance values across all plants. Jussieu's writings may also mislead one into thinking that the analytical approach was based on an understanding of the function of plant structure, or even that it was largely artificial. These are all very different ideas from those that provide the foundation for Jussieu's synthetic natural method.

Similar imprecision is evident in Augustin-Pyramus de Candolle's work. He emphasized, "the value of a character is in the compound ratio of the importance of the organ and the point of view from which it is considered."[207] The "point of view" that was most important to Candolle was the presence or absence of a character, its position and number was next, then its appearance, and finally its function and properties. The last was dependent on the first three; but, rather confusingly, the importance of an organ was also directly connected to the role it played in the life of the plant.[208] In this latter context a hierarchy of values could be established within two classes of functions, reproduction and nutrition, themselves incommensurable. But when Candolle questioned the synthetic approach, and discussed his more analytical approach to classification and system, he did not suggest the nature of the "general principles" by which plants should be divided.[209] Furthermore, as Asa Gray observed[210] in his review of J. D. Hooker and Thomas Thomson's *Flora indica*, "physiology" was more important in zoological than in botanical classification; there was no close and obvious connection between structure and function within flowering plants. In botany in particular the lack of functional knowledge meant that the synthetic approach as represented by some sort of study of the general distribution of characters necessarily preponderated.[211] At the same time all characters were not treated equally, even although no subordination of characters based on functional relationships was possible.

As was often noted by contemporary commentators and also subsequently,[212] the relationship between the functional and taxonomic hierarchies tended to be circular, or at the very least the synthetic and analytic approaches were intimately entwined. The analytical principles of the subordination of characters came from the distribution of characters in taxonomic groups, whether formed by synthesis or analysis; it was no wonder that the principles drawn from the one supported the other. As Jussieu observed, "these [principles of the natural method] are inherent in plants, and are readily apparent to the observer, and can be explained by a dual procedure such that the one functions in turn, so as to speak, as the confirmation of the other."[213] And Cuvier: "Here I have proceeded in part by ascending from the inferior to the superior divisions by the way of bringing things together and comparing them; in part also by descending from the superior to the inferior divisions by the principle of the subordination of characters; comparing carefully the results of the two methods, verifying the one by means of the other and taking pains to always establish the correspondence of the external and internal forms, which are equally integral parts of the essence of every animal."[214]

The general procedure is identical, except that Cuvier paid more attention to gross anatomy and function than did Jussieu. Jussieu's principles by which characters might be subordinated are, however, evident not only in his discussion of the function of characters but also in his distinction between characters and differentiae, and internal and external features. Other authors also connected an analytic, top-down study of characters with a functional weighting. John Stevens Henslow noted that systematists were searching for laws governing the natural system, and he clearly distinguished between the analytic and synthetic approaches. It was in the former, he thought, that the "principles by which the classification was effected"[215] were explained. In the context of my discussion here, Henslow was right. The analytical approach was connected with the justification of the natural arrangement, an explanation of why the pattern of characters was the way it was. This approach might provide an understanding of the existence of groups, or of the arrangement as a whole, by explaining the relationship between structure and function, the functional dependencies of organs, and the hierarchy of groups and characters. It was this explanatory power that was of great interest to Jussieu's immediate successors. Linnaeus may have had this in mind when he noted that the teacher descended from the general to the specific, but the author of "systems" proceeded in the opposite direction.[216]

It can be argued that the analytic or "Aristotelian" approach of Jussieu, Cuvier, and others differs (at least partly) from that of Linnaeus, notably from Linnaeus's largely stamen-based sexual system. There Linnaeus constructed higher groups by using a set of characters that he thought were important, but which did not yield the natural groupings he wanted to recognize in a truly natural system. Nevertheless, the characters of groups at different levels of such a Linnaean hierarchy would be at one with nature if they represented essences and were part of a system of analyzed entities.[217] Although Linnaeus was impressed with the functional importance of the stamens, he could not assert that they represented the essence of plants; the number of stamens was on the whole, however, an easy character to see.

There is another way of looking at the analytic and synthetic approaches that is suggested by Linnaeus's remark and which is implicit in Jussieu's discussion. Perhaps naturalists tended to use an analytic approach when subdividing nature into groups of manageable size, and then use a more synthetic approach in more detailed studies of these groups.[218] This raises the possibility that there may be a methodological dichotomy in group formation at about the level of the genus,[219] which is clearly an element in the whole argument, and it also leads to issues of human cognition that I shall discuss in chapter 7. Nineteenth-century naturalists did tend to work more intensively on smaller groups than did their forebears; the size of the groups might have made it possible to effect synthesis, perfect knowledge.[220] As knowledge of plants increased, groupings of families were also being recognized in the 1820s and 1830s,[221] perhaps also by a modified synthetic procedure. But by having to deal with many characters and taxa at these higher levels, problems in the weighting of characters would become particularly acute. It is in part for this reason that a century later Thomas Archibald Sprague felt that systematists still had so much to accomplish.[222]

Some contemporary commentators on the relative values of the synthetic and analytic approaches took a different tack, suggesting a link between this problem and different views of nature. Synthesis was seen by them as a flawed procedure rather than an integral part of the process of recognizing relationships. Some authors distrusted synthesis simply because it led to superficiality, and they hence preferred a more analytical approach when forming groups.[223] It seemed a positive recommendation that the analytic approach entailed the use of relatively few characters, at higher levels of the classification, at least;[224] these presumably could be studied more carefully than would be possible through attempts to synthesize relationships based on overall similarity.

But it was Étienne Pierre Ventenat, one of Jussieu's staunchest supporters, who perhaps inadvertently suggested a connection between Jussieu's view of the world and his methodological recommendations. Ventenat thought that because there were still major gaps in the order of nature, a more analytical and hierarchical approach to grouping and evaluation of characters than Jussieu had proposed would be useful.[225] Of course these gaps existed only because the species that occupied them had not yet been collected; when all plants were known there would be no gaps, and thus no place for the analytic approach. On the other hand, a naturalist who believed that there really were gaps in nature might well see his approach justified by Ventenat's expedient and temporary solution to detecting relationships.[226] Both Bonnet and Wallace, whom I mentioned earlier as illustrating diametrically different approaches to the ordering of nature, indeed linked the kind of patterns that existed in nature with how naturalists went about their work.

The need to identify plants placed an important constraint on the resolution of the debate about analysis and synthesis. Jussieu realized that naturalists needed to be able to characterize and recognize groups at all hierarchical levels. He believed that his groups had distinctive characters, since even if the essential characters on which a family (for example) was based were hard to recognize, it might well be possible to find conspicuous differentiae that, although not essential in themselves, were correlated with essential features. Similarly, one of the arguments Adanson advanced to defend his use of characters from all parts of the organism in his family descriptions was that a plant at any stage of its growth would be recognizable as belonging to a particular family.[227]

In chapter 2 I pointed out that Lamarck separated the process of recognizing relationships, and even forming groups, from that of identifying organisms. The former was synthetic, the latter analytic. Colin Milne[228] nicely captured the distinction between the two approaches when he compared them to two kinds of dictionaries; one arranged alphabetically (the artificial "method") would be convenient to use, the other would be rational and less useful to learners, having all derivative words placed under their primitives. Unlike Lamarck, Jussieu had not fully separated the operations involved in recognizing a plant from those involved in constructing the system in which that plant was to be placed.[229] And despite his promises, he never provided a key to the genera or even the families that he described. Nevertheless, following Lamarck's lead in the *Flore françoise*, strictly artificial keys soon came into use, for example in François Joseph Lestiboudois's *Botanographie belgique*, which appeared in 1781.[230] The distinctive characters Jussieu insisted that his groups needed only partly facilitated their recognition and naming. But the distinction between artificial keys, based on characters that were obvious to the eye, and natural systems, which might use characters that were decidedly inconvenient when it came to the problem of simply naming plants, was not always understood or accepted.[231] The young Asa Gray[232] thought that with an understanding of the morphological laws of nature a natural system would come to have readily characterized (or at least more readily understood) groups. Less orthodox workers such as Constantin Rafinesque explicitly considered the possibility that nature and keys could be combined.[233] Confusion persisted, and considerably later George Bentham[234] felt the need to exhort that identification and the recognition of relationships were very different operations. Bentham now could provide an explanation for this difference: characters used in recognizing relationships were essential, but nonadaptive and sometimes inconspicuous, while those helpful when naming plants were nonessential, adaptive, but often conspicuous.[235]

An exchange in 1829 shows how confused the arguments were becoming, at the same time suggesting that botanists and zoologists might have somewhat different views of nature. In 1827 James Ebenezer Bicheno (1785–1851), then secretary of the Linnean Society, published a paper "On systems and methods in natural history" in which he reviewed the differences between artificial and natural methods of classification. Bicheno's paper was reprinted in the *Philosophical Magazine*,[236] where it in turn drew a very negative review from William Sharp Macleay (1792–1865), and also comments from John Fleming (1785–1857) in the Tory *Quarterly Review*. Macleay, then in Cuba, in turn annihilated Fleming in a series of articles in the *Philosophical Magazine*. Bicheno, a botanist, was not pro-Jussiaean and preferred to maintain Linnaean groupings;[237] he nevertheless described the natural method in Jussiaean terms—which Macleay professed to find incomprehensible. Bicheno contrasted an ascending series that led to natural systems, which he called synthesis, with descending series, analysis, that led to artificial systems. He also compared Robert Brown and Augustin-Pyramus de Candolle, noting that Brown was chiefly involved in synthesis, whereas Candolle was making fresh combinations, that is, being more analytical.[238] Macleay professed to be surprised by Bicheno's assessment; taking a more Cuvieran position, he proposed that both synthesis and analysis were involved in systematics.[239] Macleay claimed that Bicheno's distinction "could have only been maintained . . . from love of paradox,"[240] although Bicheno was taking a very ordinary position for a botanist. Bicheno was not clear whether higher taxa were discrete, certainly when the process of synthesis was being emphasized,[241] while Macleay suggested that taxa at higher taxonomic levels were if anything more discrete, more comprehensive, as he put it, than those at lower levels.[242] Fleming seems to have picked up on Bicheno's ambivalence, and spent most of his review arguing that nature did not form a continuous series.[243]

Both Fleming and Macleay took groups to be more discrete than did Bicheno. Although Bicheno did not see the synthetic method as the one way by which natural relationships could be recognized, at about the same time, the amateur botanist Henry Thomas Colebrooke (1765–1837) thought that the shape of nature could be better described using nonhierarchical analogies rather than using Macleay's very hierarchical quinarian system.[244] Ideas of continuity are more evident in the botanical work of Adrian Hardy Haworth (1768–1833) than in his zoological studies.[245] In general, zoologists seemed happier with groupings and hierarchy, the botanists with continuity—although these were not the only issues.[246] But there was a further dimension to the problem. In 1823 William Whewell suggested that plants *could* be separated into distinct genera and families;[247] the limits of families seemed to be problematic only because they could not be defined. But Whewell claimed that natural historians had nevertheless arrived at "unanimity and certainty, with respect to the greatest part of their subdivisions."[248] Just how uncertain botanists were will become clear in chapter 8, and how they achieved what classificatory consensus there was is the subject of chapter 10.

CHAPTER SIX

Types, Groups, and Relationships

The botanists introduced in the previous chapter did not set out clear prescriptions for detecting natural relationships. In this chapter I explore the ways in which some botanists used typological thought to limit the extent of the variation they saw by reducing the unruliness of nature to invariant types. Instead of a group being represented by the sum total of the variation of its included genera and species, it was simplified and reduced to a type. Floral characters, in particular, were emphasized. Relationships were no longer diffuse, but focused on a central point. Augustin-Pyramus de Candolle's ideas of symmetry play a prominent role here, but at least elements of typological thought are evident in the work of many botanists in the first part of the nineteenth century, more particularly those of German and French extraction.[1]

It has been suggested that until the middle of the nineteenth century such typological thought was one way of making natural history scientifically respectable.[2] Nevertheless, there was no single definition of the word "type" held by naturalists ostensibly engaged in typological thinking. And however types were conceived, rarely was their function in making classifications and establishing relationships evident. Furthermore, prominent naturalists like Robert Brown and George Bentham were strongly averse to typological thought. Typological thought thus did not provide a satisfactory solution to the uncertainties that botanists faced when attempting to detect relationships and form groups.

Type concepts in the nineteenth century can be divided into collection, classification, and morphological type concepts. These distinctions were developed to facilitate discussion of the zoological literature, but they are also useful here.[3] Collection type concepts were concerned with how the name of an organism can be referred to a particular specimen or individual species[4]. Classification type concepts were those that dealt largely with summarizing or simplifying data, whereas morphological type concepts dealt with the order of nature and its laws (although Paul Farber, whose work I am following here, noted that these two were not always sharply distinguishable).[5] I will deal largely with classification type concepts, although, like Farber, I find them neither monolithic nor necessarily restricted to classification. Then too, some naturalists used morphological type concepts in their classificatory endeavors. (Unfortunately, the word "morphology" is not without its ambiguities; it has been used in a pejorative sense, as in "so-and-so simply looked at the morphology of the plants," which implies that attention was given only to features visible on the outside of the organism—not to anatomical or chemical characters, and certainly not using experimental techniques.[6] Morphology in this sense is opposed to anatomy at the cell, tissue, and organ level.[7])

In the first thirty years of the nineteenth century the continuities and connections between different plant organs that Johann Wolfgang von Goethe (1749–1832) and Carolus Linnaeus in particular had found in the latter part of the eighteenth century were rediscovered.[8] Several other botanists, including Augustin-Pyramus de Candolle, independently conceived the idea that there was a basic similarity between cotyledons, leaves, sepals, petals, stamens, and carpels.[9] As the morphologist and systematist Auguste-François-César Prouvençal de Saint-Hilaire (1779–1853) observed:

> Thus, a little later toward 1810, Dr Pelletier of Orléans developed for me, in conversations as instructive as pleasant, the ideas which Goethe had tried to spread in Germany; M. Dunal, then as much a stranger to the German language as M. Pelletier, exposed the same ideas in his *Essai sur les Vaccinées*, printed in Montpellier in 1819, but never distributed; and finally, M. Turpin, who knew neither Goethe, M. Pelletier, nor M.

> Dunal, developed in 1820, in his excellent *Essai d'une Iconographie*, ideas similar to theirs.[10]

Later Saint-Hilaire juxtaposed Goethe and Jussieu in a revealing way as he commented on a translation of Goethe's works:

> That which Goethe had done in 1790 for the organs of the individual plant, de Jussieu had achieved one year before for the whole of the vegetable kingdom. In classifying plants according to all their resemblances, he proved repeatedly that between different groups there was nothing but imperceptible transitions; he delighted in showing the connections that joined classes, orders, and genera the most distant one from another; it was said that sometimes he showed a kind of pleasure in unmasking certain relationships that had not hitherto been suspected and in making evident that the vegetable kingdom was a vast network in which threads intersected in thousands of ways.[11]

This perception of continuity had far-reaching implications; no longer were two organs (or groups) separate, they were aspects of one and the same thing. The degree of fusion of the styles, for example, was of no importance in the classification of the Rutaceae since there were all kinds of "nuances intermédiaires" between fusion and complete separation in the different taxa of this group, and so this allowed Saint-Hilaire to suggest a position for the family between plants that habitually had separate carpels and those that habitually did not.[12]

Two dissociable ideas are to be kept in mind. An understanding of metamorphosis could lead to a better understanding of relationships between characters and thus of organisms, so clarifying either the relationships or the "nature" of a group, its basic construction. This was basically how Robert Brown, Augustin-Pyramus de Candolle, and Asa Gray used the idea. But metamorphosis could also provide an understanding of plants that was largely independent of a consideration of taxonomic groups. For Goethe and Linnaeus, notably, the morphological type concept in its strictest sense was a quest for an understanding of the laws governing change of plant form, and ultimately an understanding of the unity of organisms and life; this was the primary goal.[13] Connected with this latter approach was the development of the "law of homology," serial repetitions of fundamentally similar structures in a plant—for example, the idea that the leaf was the typical organ of a plant from which all other organs were to be derived.[14] Similarly, in the late 1820s Karl Schimper (1803–1867) and Alexander Braun (1805–1877) drew attention to the symmetries and regularities of phyllotactic sequences—that is, the relative positions of the leaves on the stem, the scales on a pine cone, and the sepals, petals, etc. on the receptacle of the flower. This line of investigation was to persist for the rest of the century, albeit without reaching much in the way of conclusions.[15]

The two approaches did not, however, have to be separated. Candolle was much interested in discovering the basic "symmetry" or "type" of plant groups, and he also hoped morphology would provide a key to the understanding of the laws governing plant form. Such laws could be used to explain both why form as realized in genera deviated from the basic symmetry of a group and what that basic symmetry was; it was not just the nature of groups, but laws of change, that Candolle was looking for.[16] Whether the naturalist's interests inclined more to classification or to the understanding of form, metamorphosis was all,[17] but there still tended to be contradictions between the two.[18] Candolle's search for symmetries and regularities helped botanists' understanding of plant morphology and so aided their search for natural relationships. It also encouraged work that led away from the immediate study of the natural order; for example, Candolle suggested that the "dérangements" of plant symmetry were themselves subject to laws.[19] Another botanist who combined the two approaches was Christian-Horace-Bénédict-Alfred Moquin-Tandon (1804–1863; he was also an authority on leeches and freshwater molluscs, as well as a noted scholar of Provençal literature). His botanical work ranged from conventional classificatory studies and attempts to establish relationships between families and genera for which typological thought seems to have been integral, to discussions of the laws that regulated change of form in plants, as in his *Élémens de tératologie végétale*.[20]

The morphological and classification type concepts indeed are not sharply distinct, as we also find in the work of Ludwig Reichenbach (1793–1879), of Dresden, a prolific botanist and ornithologist. Reichenbach, in the epigraph of his appropriately titled *Conspectus regni vegetabilis per gradus naturalis evoluti* apostrophized "Genera are not in nature, it is true! But the method is in nature, and the method is science."[21] But the method was not Jussieu's method. For Reichenbach it was the doctrine of metamorphosis that brought order to all the heaps of plants known—Jussieu's work did form the basis of the modern system by which these plants were disposed into heaps, but these heaps remained just that, without any explanation, without the helping hand of metamorphosis. With this doctrine, plants could be assigned to their correct position in terms of a hierarchy of advancement; the limits (if they existed) by which series could be circumscribed could be established; and all individual shapes, no matter how diverse, could be reduced to their particular type.[22] Nature, it appeared, could be explained and hence understood. As Reichenbach noted, if "botanique française" was epitomized by Jussieu, the standard-bearer of "Deutsche Botanik" was Goethe.[23]

The Nature of Types

What was a type, and how did one discover it? Moquin-Tandon suggested that there was more than one kind of type and these were formed in different ways and represented different things.

> But, if one studies plant families separately, one will find some in which regularity [of the flower] is most common, and others in which it is, on the contrary, irregularity. . . .
>
> There is thus in each family a kind of particular type to which one may bring all the individuals or all the organs of these individuals. . . .
>
> Despite these "types particuliers" or family types, botanists, whenever they have talked of the *normal plan*, the *essential order*, and the *basic* ["primitif"] *type*, have always used in their comparison the regular form [a radially symmetrical flower] whatever its frequency within the family. The result is that in certain groups, as in the Cruciferae, the basic type is confused with the family type, while in others, such as the Labiatae, the two are different. In the first, a monstrous corolla is always distant from the family type and the general type. In the second, a monstrous corolla can also, in certain cases, be distant from the two types; but, in other cases, it deviates from the family type and returns to the basic type. . . .
>
> There are families made up entirely of genera in which the verticels [of the flower] are irregular, the Polygalaceae, for example. In such families nature gives no whole ["ensemble"] in the state of symmetry; the basic type in such a case is a kind of ideal being, a metaphysical fiction.

> Finally, in addition to the basic or symmetrical type, and the type of a family with regular or irregular flowers, there is a third type which has been called the *specific* type; this is the grouping of features, the assemblage of characters common to the majority of individuals that make up a species.[24]

In the Cruciferae, the cabbage family, the flowers are radially symmetrical; the family type will thus be radially symmetrical and the basic type will be identical to it. A "monstrous corolla," a teratum, will tend to deviate from both. The flowers of the Labiatae, the mint family, are usually strongly bilaterally symmetrical, and so the family type will have bilaterally symmetrical flowers. Terata may have more or less radially symmetrical flowers, so differing from the family type and "returning" to the radially symmetrical basic type. All the floral whorls of the flowers of the Polygalaceae, the milkwort family, are very strongly bilaterally symmetrical, and the ovary is asymmetrical; Moquin-Tandon apparently did not know of any radially symmetrical flowers, even in terata. Finally, he suggests that the specific type is not necessarily floral in nature, but simply the invariant form—whether of flower, fruit, or leaf—of the species.

But not everybody followed Moquin-Tandon in making these distinctions—or perhaps they were simply not aware of them. Thus Marie Jean Pierre Flourens, in his obituary of Candolle, wondered why the "primitive symmetry" of a group should always be so uncommon.[25] This question could be asked only of a the basic type of the family, not of the family type, and Candolle's symmetry was the latter kind of type. Other types were mentioned in the literature, and a complicated and largely unexplained synonymy developed. Henri Baillon noted the value of degenerate types in establishing connections between genera that were very different,[26] Christian Gottfried Daniel Nees von Esenbeck talked about types which he called "Elementärbildung," "Prototypus," and "Grundgebilde," and Jacob Georg Agardh mentioned archetypes.[27] Yet it is doubtful that naturalists were really much happier when synonymies were proposed. Did it truly matter if Elias Fries's "centrum," a genus that expressed particularly well the character of the group to which it belonged (a series of taxa related in a circular fashion), was really the same as Macleay's "type," a genus on the circumference of a circle of affinity (five taxa forming a circle) that was farthest from neighboring groups (four more circles, each made up of five taxa)?[28]

Descriptions were not the only way in which botanists conveyed the abstractions they used to simplify the variation in nature. Floral diagrams, in which the parts of the flower and their relationships to one another were indicated in diagrammatic form, were developed in the 1820s. Pierre Turpin provided in 1819 the first such examples of which I am aware in his diagrammatic representation of horizontal plans of flowers of plants like wheat and gladiolus.[29] But these diagrams were not abstractions of groups that were integral to Turpin's understanding of nature;[30] they were diagrams of flowers of individual species and they were used to clarify the positional relationships between parts. Such diagrams continued to be used, as in the meticulous drawings of calyx and corolla aestivation in Adrien de Jussieu's work on the Rutaceae.[31] (See figure 7 in the previous chapter for comparable diagrams for the Malpighiaceae).

Floral diagrams soon came into common use. Some botanists represented the floral structure of whole genera or groups of genera, as in some rather sketchy diagrams accompanying Bentham's work on the Eriogoneae.[32] Such diagrams represented types, abstractions of groups. A particularly fine example is Michel Félix Dunal's "normal plan" of the flower that represented the basic structure of the dicotyledonous flower—quite unlike that of any plant known either then or now.[33] Floral diagrams did not necessarily give a static view of nature, thus Carl Sigismund Kunth (1788–1850) used them in 1833 to help show how the distinct flower of the Cruciferae could be related to a more conventional flower type by the loss of stamens and carpels.[34]

Another way of summarily describing the type of a group was by the use of floral formulae. These, too, developed in the 1820s, although I know little about the details of their development. In 1820 Franz Peter Cassel (1784–1821) described a formula that he used to show the relative development of different parts of the flower, and that of the different members of each whorl; he compared formulae within and between families.[35] Such formulae were conceived as algebraic expressions[36] as a way of conveying a great deal of information about the flower within a very small space, using signs rather than words. In either case, knowledge would be less restricted by the bounds of language. Nicolas Charles Seringe (1776–1858) and Jean Claude Achille Guillard

(1799–1876) were particularly interested in facilitating communication between naturalists, but in 1835 they developed what amounted to a new language themselves (figure 9), needing ten pages to explain their complex formulae.[37] They must also have exhausted their typesetter. Fortunately, formulae as they developed in the late 1840s were much simpler. August Heinrich Rudolph Grisebach described a "Blüthenplan," or type, for individual families, such as,[38]

Papaveraceae: 2, 2+2, ∞, ∞̂
Rosaceae: 5̂, 5, ∞, ∞
Synanthereae (Compositae, but not in Grisebach's sense):
-̂, 5̂, 5, 1̂-

Each number, or pair of numbers, represents a floral whorl (sepals, petals, stamens, carpels). The symbols above the numbers signify fusion between different members of the same whorl, while whorls joined by the bars on the horizontal lines are fused to one another. Floral formulae could convey an indication of variation within a group by, for instance, including the range of numbers of sepals that occurred in a family.

Detecting Types

That two structures could be related to each other did not in itself indicate what the combined structure really was; for that, one needed some idea of the direction of change. The direction was generally taken to be from simple, symmetrical flowers with all parts separate, to complex, asymmetrical flowers with abortions of parts as well as fusions and adhesions between parts.[39] As Moquin-Tandon suggested, the botanical type was usually represented by a radially symmetrical flower. This explains how he could decide that the flower of *Linaria* had only one normal petal, which was spurred; there were four abnormal, unspurred petals. He claimed that the existence of the peloric form of the flower, which was symmetrical, with all the petals having spurs, allowed him to draw this conclusion.[40] Such types were in some way "perfect." Carl Friedrich Philipp von Martius observed that the type of the Amaranthaceae harmonized with the "Gesammtorganisation" of the family and was represented by a hermaphrodit-

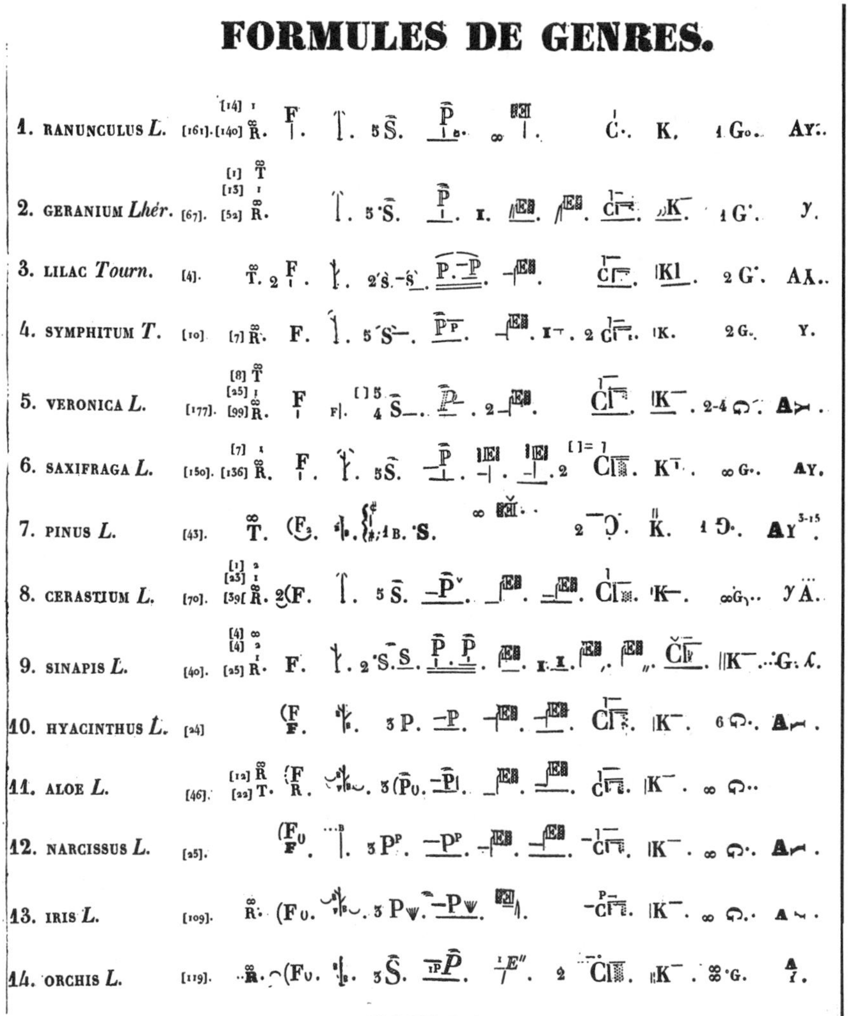

FORMULES DE GENRES.

1. RANUNCULUS *L.* [161].
2. GERANIUM *Lhér.* [67].
3. LILAC *Tourn.* [4].
4. SYMPHITUM *T.* [10]
5. VERONICA *L.* [177].
6. SAXIFRAGA *L.* [150].
7. PINUS *L.* [43].
8. CERASTIUM *L.* [70].
9. SINAPIS *L.* [40].
10. HYACINTHUS *L.* [24]
11. ALOE *L.* [46].
12. NARCISSUS *L.* [25].
13. IRIS *L.* [109].
14. ORCHIS *L.* [119].

FIGURE 9

An esperanto for plants. Source: Seringe and Guillard 1836: 18.

ic flower. Imperfect flowers were simply "stunted" hermaphroditic flowers.[41]

But another element in Martius's decision that hermaphroditic flowers were typical for the Amaranthaceae was his observation that they were the commonest kind of flower in the family. Similarly, some of Moquin-Tandon's types represented the commonest floral form within a group. Because botanists' ideas of what was common were influenced by how many plants they knew that had a particular form, the greater knowledge of northern temperate plants tended to make forms from that area typical.

Many types represented either the common or the rare condition in a group, but they were the normal state of some species or other. Other types were based on terata, abnormalities or monstrosities. Terata played a very important role in the detection of types as they provided new forms for the botanist to observe and to arrange in the continuity of form that denoted relationships. José Francisco Corrêa da Serra (1751–1823; diplomat, cleric, botanist, and notable conversationalist) wrote to Isidore Geoffroy Saint-Hilaire: "I am diverting and instructing myself with your monsters; they are friendly and frank chatterboxes which wisely recount the wonders of organization saying always much to the point and what is and what cannot be."[42] Candolle paid considerable attention to terata, to the extent that Moquin-Tandon suggested that Candolle's ideas might lead some people to think that it was only in monsters that "types organiques" were to be found.[43]

Moquin-Tandon gave types at different levels of the taxonomic hierarchy different names, and he implied that such types might be based on different kinds of characters. The specific type might represent the whole organism, while the family type was based on flowers alone. Similarly, Carl Heinrich Schultz-Schultzenstein[44] saw types at the level of class, family, and genus as being distinguished by different kinds of features—the "Classentypus," for example, having a distinctive internal organization that was evident not so much through external form, but through its physiological effects. Little was known about such physiological effects, but developmental studies were a favored tool of many of the authors referred to in this chapter. Thus Jean-Baptiste Payer[45] used developmental observations to relate the androecium of *Luffa*, which had five stamens, to that of *Cucurbita*, which had three. In a "kurz und dogmatisch" (sic) summary of studies on the Leguminosae, Mathias Schleiden and Theodor Vogel observed that the flowers were perfectly regular when they originated, fused petals were originally free, and "all parts of the flower are in their first appearance green leaves."[46] And whether or not they used developmental studies to formulate ideas of types, botanists such as Grisebach and Martius commonly included a section on "evolutio" or development in their descriptions of higher taxonomic groups.

Not all botanists were restricted by what they actually saw when formulating types. Some types might not even exist in nature. Glands might perhaps be present in plants where they could not be seen; the stamen number of a type might not actually be a number that could be counted in any living plant.[47] Such types are examples of regulatory types: transcendental ideas in the mind of the naturalist, serving only to guide one's thoughts and not having any real existence. The other types, however, are variants of a constitutive type, and they are all actually existing forms.[48] Any analysis of how types are determined thus becomes very complex.

Uses of Types

Given that types represented abstractions or simplifications, what was their role in systematic work? The answer is not easy to determine. However, a few studies show the variety of ways in which types were used. Types might simplify descriptions or to help establish relationships, as in the work of Louis Édouard Bureau (explained next). Or they might even help in identification, as in Anton Friedrich Spring's study.

One of the functions of the types (symmetries) of Augustin-Pyramus de Candolle was that of simplifying variation;[49] this helped in comparisons between families and thus ostensibly improved one's prospects for finding the true relationships in nature. Types were quite commonly used in this way, as in Bureau's work on the Bignoniaceae.

> To fix [one's] ideas and to understand these groups better, we take in each a genus which serves us as the type. This genus will be, as far as possible, that which will show the distinctive characters of a group in the simplest and most evident fashion. It will serve us as a standard ["terme"] of comparison for all genera with an analogous organization, and we will not have to do more than to indicate by which traits these latter deviate from the type.[50]

Although it is not clear how Bureau decided just what these distinctive characters were—it is likely that they were the common features of the family—having made his decision, the type then made the task of description and comparison simpler. This kind of type has a function similar to that of the characters Jussieu generalized in his families, as both made it possible to avoid repetition in descriptions. Bureau's type would be incorporated in the detailed description of a genus (for example) which would then be compared with other, briefer descriptions, whereas Jussieu's generalized characters would appear in a family description that would be compared with other family descriptions of similar length.

The Bignoniaceae are a moderately distinctive group, but some families, notably those that were families "par enchaînement," were much more heterogeneous, even polytypic. Bureau had also worked on a family of this kind, the Loganiaceae, and types here were integral to describing relationships. The related families of the Rubiaceae, Apocynaceae, and Gentianaceae were linked severally to the different types found in the Loganiaceae,[51] and Bureau removed a considerable part of the Loganiaceae to these families.[52] The existence of the polytypic Loganiaceae was essential if relationships between the more distinctive and isolated "familles en groupe" were to be established.[53]

Types could help in identification. Anton Friedrich Spring[54] (1814–1872) described "subgeneric" types at more than one hierarchical level. These types were well-known species, and Spring gave a key to these "typical" species only. The naturalist could then compare his specimen with the descriptions of the relatively few other species in that species group and so get a name for it without too much trouble. Even Grisebach felt that his "Blüthenplane" were as much an easy way of distinguishing between families as arranging these families in a system, because there were abnormal genera that deviated from their Blüthenplane.[55]

Although types were often used in establishing relationships, this by itself may not tell us very much. What *kind* of type was used? In some cases these types were abstractions. When in 1868 Adolphe Brongniart[56] summarized progress in "botanique phytographique," he observed that recent work had shown that the Zingiberaceae had five floral whorls (two petaline [or sepals plus petals], two staminal, and one carpellary), so were ordinary monocotyledonous plants, but the Marantaceae and Cannaceae had only one staminal whorl, and so with only four whorls were quite different. For Henri Baillon, who had carried out much of the work[57] on these groups, the issue was not just difference, but relationships. Because he had visualized the type of the Cannaceae and Marantaceae as having three stamens opposite the inner perianth whorl (note that no extant member of either family has three stamens), he thought that their relationships were not with the

Musaceae and Zingiberaceae, but with *Lachnanthes* and *Haemodorum*, both members of the Haemodoraceae, a family which was not usually associated with the group in which these other families were placed. An abstract type was being compared with actual genera. Bureau's types were not abstractions in this sense, but Candolle's basic symmetry of a family would tend to be an abstraction.

Similar examples are found in the work of Saint-Hilaire and Moquin-Tandon. In their joint study of the flowers of the Polygalaceae and Leguminosae, they made their comparisons largely between types alone, not between a type and extant genera. The flowers of these two families are superficially similar, both being papilionoid—that is, zygomorphic, with standard, wings, and a petaloid keel enclosing the stamens. Saint-Hilaire and Moquin-Tandon thought that when the types of the two were compared, fundamental differences became apparent; these two families were thus not immediately related.[58] They also studied the Capparidaceae and their relatives. They concluded that both the Capparidaceae and Cruciferae really had four stamens, in both families there were increases in number to six, and in the Capparidaceae to many stamens. They also thought that *Fumaria* could be related to this four-stamened type, although the Fumariaceae would not be as close because their four stamens were opposite to the petals, not alternate with them. *Hypecoum* should also be included in the group, as Antoine-Laurent de Jussieu had earlier suggested; it, too, had four stamens and so was not really a member of the many-stamened Papaveraceae in which it was sometimes included.[59] In these examples the types may correspond to floral morphologies of extant genera.

A very different approach was adopted in 1851 by Jules Émile Planchon (1823–1888) in an interesting paper in which he described a new family, the Melianthaceae. He described his idea of the type of this family, and reported that the law of symmetry had led him to find a type with a symmetrical flower.

> In both [*Natalia* and *Melianthus*], the quaternary number of the stamens is in strong contrast with the quinary proportion of the petals and calycine segments; the disc is in both incomplete and eccentric, opposite to the posterior stamens, and placed outside of their filaments, the vacant place of the fifth undeveloped stamen being opposite to the posterior petal; so that, when we come to *Bersama*, where the quinary pro-

> portion and regular alternation of all the verticils of the floral parts is a constant and usual character, we find there exemplified, by a living demonstration, what the laws of symmetry led us to conceive of the normal state of the flower of *Natalia* and *Melianthus*.[60]

The Melianthaceae were related to a group of families, including the Sapindaceae, Geraniaceae and Tropaeolaceae, in which there was a nectar-secreting disc between the stamens and the petals. But what were the relationships of the Melianthaceae within this group? Planchon compared the Melianthaceae with the Sapindaceae. In general appearance, a previously undescribed species of *Natalia* was very like the Paullinieae, a tribe of the Sapindaceae. There were poisonous plants in the Sapindaceae, and the whole plant of *Melianthus* had a strong "narcotic and virous smell," and so might by analogy be supposed to be poisonous also. There were detailed floral similarities between *Diplerisma* (another genus of the Melianthaceae), *Natalia*, and some of the Sapindaceae. There were also striking similarities between the fruits of *Diplerisma* and those of *Cardiospermum*, of the Sapindaceae, and those of the Bersameae with *Paullinia*.[61] Note that the type genus itself, *Bersama*, is conspicuous by its absence in this recitation; it seemed to have no role to play when it came to the business of establishing the relationships of the Melianthaceae. It was not discussed at all when Planchon detailed floral similarities between the Melianthaceae and Sapindaceae, yet *Bersama* alone had been mentioned when he discussed the type of the former.

Planchon's type was both a simplification of the variation in the family and representative of a living genus.[62] It is, however, clear that he used this type more as a way of relating and in some way bringing to order the variation shown within the family rather than in clarifying its relationships with other families. In the introduction to his paper Planchon had shown that continuity brought coherence to nature in the harmonies produced by the relationships of a family's constituent members, yet it presented problems by rendering what he called Linnaean delineation problematic.[63] Description became more difficult as relationships became more obvious. The rest of the paper is Planchon's solution to the problem—already evident to Linnaeus[64]—of describing families which had no defining (exclusive) characters. When it was a matter of Linnaean delineation, he seems to have been thinking of nodal points and types—a single genus could be unambiguously delineated, if not defined. But when it came to establishing relationships, Planchon treated the family as the sum of its constituent members, and he related individual genera directly, an attitude perfectly in agreement with Linnaeus's analogy between nature and a map (which Planchon had also mentioned in his introduction). Here were the harmonies of relationships that he so much wanted to discover. The problem of continuity was solved: there might not be gaps, but there could still be types. But the types, being geographically central to a state or country in a Linnaean landscape of nature, would be of no use in clarifying where one state met another; here the outlying members of the family came into their own.

Planchon's study of the Melianthaceae spells out in no uncertain fashion ideas that are evident elsewhere in his publications.[65] Interestingly, the solution to the problem he saw posed by continuity is similar to that of William Whewell (1794–1866), philosopher and Master of Trinity College, Cambridge. Whewell had noted that there might be intermingling between groups in natural history; nevertheless what connected members of a group was not a definition but a type, a particular species in the family, and these were temporary or provisional only.[66] He observed, "[A class] is determined, not by a boundary line without, but by a central point within; not by what it strictly excludes, but by what it eminently includes; by an example, not by a precept; in short, instead of a Definition, we have a *Type* for our director.[67]" The type was an organism, not an abstraction, that had all the characters and properties of the group in a marked and prominent degree; it was in the center of a crowd, not a straggler. Yet it was the stragglers that would tend to get caught up with other groups and establish connections between them.[68] At the same time—and here Whewell used an analogy between a classification and a patchily forested landscape—trees scattered on a plain did not destroy the reality of the distinct forests found on separate hills.[69]

Overall, Planchon's type seems to be a good example of a classification type concept[70] (a way of organizing data, not of establishing relationships), and Bureau's type genera in families like the Bignoniaceae could also have a similar function. But in some situations even a type like *Bersama* could be used to link groups. The little-known botanist Augustin Augier had suggested earlier[71] that variation within a family could effectively be represented by a V. The arms of the V established connections between that family and other families immediately above it (by the bases of *their* Vs), and the base made connections with the tip of an arm of a family below. Planchon's paper on the Melianthaceae encompasses a very small part of the plant kingdom, and it is possible that he, like Augier, would have been forced to convert his horizontal, planar image of variation to one with a vertical element as he extended his studies. The central genus, *Bersama*, the type of the family, might then become integral to establishing relationships with a family below it in the system.

Discussion

In general, whatever a type represented, and for whatever purpose it was conceived, it is likely to have taken on an independent life of its own. Hugh Strickland[72] in 1845 rejected the idea that types really existed in nature; he thought that they simply provided a fixed point in a group, thereby indicating groups with greater precision than definition alone could do. To him, types were "inductive," made without theoretical assumptions intruding.[73] This might be true of a collection (nomenclatural) type or of a very strictly conceived classification type, but some of the types that Moquin-Tandon described would have been unacceptable to Strickland. Many classification types in botany are also, or primarily, vehicles for understanding relationships, and ultimately for expressing the order to be found in nature.[74]

The mention of types in the context of detecting relationships became widespread in the first half of the nineteenth century. Indeed, for the most prominent botanist in North America, Asa Gray (1810–1888); who although little occupied with morphological studies himself was a keenly interested observer,[75] it was the change in our knowledge of the structure of plants, an understanding of "the laws by which the external forms are modified," that had caused the revolution in botany.[76] With respect to metamorphoses, Gray observed in 1842:

> The adoption of the theory [that of the flower being a transformed branch], accordingly, has given a new aspect to botany, and rendered it the most philosophical and inductive of the natural sciences. . . . The doctrine of vegetable metamorphosis, therefore, as applied to flowers, and to the discovery of their true structure and symmetry . . . like the

> touch of the spear of Ithuriel, causes the most anomalous structures and disguised forms of vegetable organization to reveal their typical state, and primitive character.[77]

Gray did not see these transitions as leading to continuity; instead of types degrading, they changed. And from gradual change came a structure "irreconcilable to the former type."[78] The implication is that there were groups in nature, and that the types of these groups were themselves distinct, although they could be related. Certainly, the commonly expressed belief that aberrant taxa could be related to a type[79] would encourage the idea that groups were discrete, even if the types themselves were not described and there was apparently some continuity between groups.

Nevertheless, as we saw with Planchon, "continuity" and "type" may simply describe different aspects of a similar, more or less continuous nature. Whewell's type may have been inductive, but it nevertheless yielded a very distinctive picture of nature, a two-dimensional, near-continuum of imperfectly delimited groups. The ideas of an approach to or more particularly degeneration from a type also conjure up a vision of nature similar to that held by proponents of modified continuity mentioned in the last chapter and that of Lamarck himself. At least some kinds of typological thought are not opposed to continuity.

It is hardly surprising that for some systematists the style of thought that led to the abstraction of types involved altogether too much in the way of hypothesis or speculation, and distinctions such as those Moquin-Tandon was attempting to draw were meaningless or contradictory. In 1856 Joseph Dalton Hooker (1817–1911), later to be director of Kew Gardens, observed that the reasoning that led to types being proposed varied, and that typological ideas in general were vague and arbitrary:

> It is difficult to indicate any particular genus of *Balanophoreae* which can be considered typical of the Order, though *Cynomorium* may be taken as such for the *Monostyli*, and *Helosis* for the *Distyli*. *Mystropetalon*, though in many respects the most perfect genus of the Order, cannot in any degree be considered typical of it; for it departs far more widely from the prevalent structure of its allies than any other genus does. Our ideas of what is or is not typical, are, however, vague and arbitrary; the ideal type being either the prevalent form of the group, or that which

> unites most of the peculiarities which distinguish it, or that which possesses the fullest complement of organs united in one individual, or that in which these are most complex, as well as specially adapted to the functions they perform.[80]

Moreover, as Hooker had noted a few years later, what he called the "retrogression" of types coupled with some extinction might destroy evidence of relationships,[81] implying that plant form was so plastic that types could not be recognized, or would not be useful.[82] Hooker's colleague George Bentham observed in 1871:

> We can no more set up a typical species than a typical individual. If we had before us an exact individual representative of the common parent from which all the individuals of a species or all the species of a genus have descended,—or, if you prefer it, an exact copy of the model or type after which the whole species or genus had been created—we should have no possible means of recognizing it. I once heard a lecture of a German philosophical naturalist of considerable reputation in his day, in which he thought he proved the common Clover [*Trifolium*] was the type of Papilionaceae. His facts were correct enough, but his arguments might have been turned in favor of any other individual species that might have been selected. Suppose two individuals of a species, two species of a genus, two genera of a family, in one of which certain organs are more developed, more differentiated, or more consolidated than the other; if we agree upon the question of which is the most perfect, a point upon which naturalists seldom do agree, how are we to determine which represents the common parent or model? whether the perfect one is an improvement upon or an improved copy, or the imperfect one a degeneracy from or a bad imitation of the other?[83]

An emphasis on looking at the form of the organs when they were simple and had just originated is quite in line with Candolle's attempts to discover "molécules intégrantes" in plants, the basic units from which their organs were constructed (see chapter 4), and many naturalists who mentioned types used the developmental-anatomical approach in their observations on plants. However, this would automatically make their work and ideas suspect to the substantial community of botanists more particularly concerned with classification and flora writing. Bentham himself may well have had in mind the study mentioned earlier of

Schleiden and Vogel on floral development and structure in the Leguminosae. Bentham disapproved of such work because not only did it focus on development and provide detail at the cellular level, but also the sampling of genera that Schleiden and Vogel had examined was very skimpy, to say the least; their work was largely based on a study of *Lupinus rivularis*.[84]

A related problem with work that invoked types was that it used distinctly questionable mechanisms to explain why one form was not really different from, or definitely related to, another. Thus the mechanism of "dédoublement" used by Moquin-Tandon and Saint-Hilaire to explain the increase in number of stamens in the Cruciferae, Capparidaceae, and their relatives,[85] was based on a doubtful interpretation of the basic structure of a dicotyledonous flower. And Saint-Hilaire later described what he thought was the "fleur-type"[86] of the dicotyledons. It had only a single whorl of stamens alternating with the petals, while interior to and alternating with the whorl of stamens was a whorl of nectaries. Few botanists other than his immediate followers would have accepted this interpretation.

Ideas of the symmetry or classification type of a group and morphological types were often produced in a similar way; they were based on a particular understanding of the processes governing change of form in plants, and both kinds of types led to an understanding of the variety of form in a group. Morphological types in general were also susceptible to a variety of interpretations,[87] although as there was little knowledge of variation in function within flowering plants, botanists rarely discussed the functional connotations so often mentioned by zoologists. All told, it is hardly surprising that many botanists in the middle of the nineteenth century had trouble understanding what types were or why they should have any use for them in their laborious attempts to perfect Jussieu's arrangement of plants.

CHAPTER SEVEN

Continuity and Classification

The success of Jussieu's *Genera plantarum* justified his way of looking at plants: he had described the natural method, both as method for determining the relationships of plants and as a series of classes, families, and genera. Jussieu's work entails a close relationship between his view of nature, his natural method, and the actual distribution of characters in his linear sequence of families. There is a catena of variation, with characters habitually lapping over from one group to the adjacent part of the next (see figures 3 and 4 in chapter 3), a pattern predicted by Lamarck (see figure 2 in chapter 2) and also found in the distribution of characters in plants and the animal groups Linnaeus placed next to them.[1] Yet in the early nineteenth century a change in the conceptualization of nature seemed to be underway. For naturalists such as Augustin-Pyramus de Candolle the shape of nature was quite different, and there were new questions to answer. Candolle was progressively circumscribing ever larger groups that he claimed were distinct,[2] and he was also interested in understanding how plant form arose. But as portrayed in chapter 5, these differences, although not inconsequential, seemed to have little effect; many taxonomists continued to believe in continuity, or to discuss relationships as if they did. The question must now be addressed, Did *practice* associated with continuity persist in a nineteenth-century setting?

Peter Yeo[3] has recently argued that persisting into the nineteenth century was the worldview of "conceptual plenitude," that is, belief in the actual existence on this earth of all possible forms of life. Some of the idealistic systems of nature to be discussed in the next chapter fit this notion. Yeo also distinguished a "spatial plenitude," the claim that matter exists to support life; thus, wherever there is matter there should be life. He suggested that although the idea of a single and continuous chain of being had been rejected, lively discussion persisted into the middle of the nineteenth century in England, at least, on the variety and fullness of nature. Here I suggest that continuity, another core element of the group of ideas associated with the Great Chain of Being, persisted into the middle of the nineteenth century and beyond. "Continuity-in-practice," as I shall call it, is the persistence of classificatory theory and practice congruent with a belief in continuity—despite the growing recognition that the living world had groups separated by gaps, a fact apparently incompatible with this theory and practice.

The Language and Practice of Systematics in the Nineteenth Century

When we compare Jussieu's accomplishments with the ways in which later generations of botanists through the first half of the nineteenth century described their own practices of constructing a natural arrangement of plants, it seems that there is an almost exact equivalence. Both the words used to describe taxonomic practice and also the practice itself remained largely unchanged throughout the nineteenth and on into the twentieth century. But if worldviews were apparently shifting in the nineteenth century, certainly after 1859, were systematic theory and practice really so unchanging? It is to this complex issue we must now turn.[4]

Goulven Laurent finished a study of the work of the later Lamarck by suggesting that "in passing from continuity to discontinuity, he passed from 'Philosophy' to science."[5] The issue is not so simple, however; in botany in particular there was no sharp break in belief between continuity and discontinuity. An assumption of continuity, moreover, can hardly be dismissed as "philosophy." To demonstrate the first point I shall analyze the idea of continuity as it was apparent in Jussieu's work, breaking it down into a set of methodological prescriptions for grouping organisms. It will then be possible to see which of these elements are dependent on or simply compatible with continuity, and which persisted in the nineteenth century.[6] And the issue of philosophy versus science seems not so simple. With the kinds of observation botanists at this period were using, continuity would seem to them a matter of fact, not philosophy.

To facilitate the discussion, the main elements of Jussieu's method can usefully be repeated here. (1) The natural series as a whole was built up by synthesis, the aggregation of species into genera, genera into families, and so on; intermediate taxa were essential in this process, as they allowed the system to be built up and at the same time validated it. (2) Information provided by a variety of characters was used (this has also been called synthesis); data provided by anatomical studies nevertheless played little part in this endeavor. (3) Because of the continuity of nature, the limits of groups were arbitrary, so what can be considered information storage and retrieval thus played a major role in determining group circumscription; genera and families should be neither too small nor too big. (4) The living world was visualized using analogies that suggested continuity; in discussions of relationships a rhetoric was used that is usually associated with one particular kind of continuity, the *scala naturae*. (5) There was also a distinctive language used to discuss more quotidian systematic theory and practice. (6) Finally, there was a relationship between the natural method and predictivity.

Two quotations from the middle of the nineteenth century (the first by Hooker, the second by Bureau) will show how these elements had become thoroughly embedded in the systematists' vocabulary. J. D. Hooker grappled with the problem of the relationship and relative position of the highly modified, echlorophyllous, parasitic plants of the Balanophoraceae:

> As an order, *Balanophoreae* may in one sense be considered a strictly limited one, not passing directly into any other, except perhaps through *Gunnera* into *Halorageae*; and forming a sufficiently natural assemblage of species, though, owing to causes [polymorphism and extreme simplicity of its parts] I have repeatedly dwelt upon, not easy of exact definition. Putting aside any consideration of its relationship with other Orders, and regarding it *per se*, it is not easy to say whether it should abstractly be considered as ranking high, or the contrary. Assuming that the conventional definition of perfection in use amongst zoologists is

> applicable to the vegetable kingdom, and which argues that a high degree of specification of organs and morphological differentiation of them for the performance of the highest functions, indicates a high rank, *Balanophoreae* may in some respects be considered to hold a very high one.[7]

Louis Édouard Bureau (1830–1918) faced the question of how to draw limits around families related to the Sesamaceae:

> One cannot in consequence place the Sesamaceae in the family of the Bignoniaceae without profoundly modifying the characters of that family. These two groups are moreover separated, if I am not mistaken, by characters more numerous and more important than those which separate the Solanaceae from the Scrophulariaceae or the Pedaliaceae from the Cyrtandraceae. . . . Consequently I think that I should be of the opinion of authors who see in the Sesamaceae a natural family. This family is closer to the Bignoniaceae than to all others by its quite deeply bilobed placentae, [which] make a transition to the double placentae of the Bignoniaceae, as well as horizontal ovules attached in two series in each loculus; but it more approaches the Scrophulariaceae by its general appearance; by its single placental column occupying the geometrical axis of the ovary and by the fruit and [sic] separating into parts free from the septae at the time of dehiscence; by the tendency of the seeds to become ascending, and by the order in which the ovules appear, from top to bottom, as in the majority of the Scrophulariaceae. Finally, it takes from the organization of the Verbenaceae and Labiatae a remarkable feature: that is the prolongation to the inside of the median nerve of the carpels which makes a false septum and divides each of the two original ["primitives"] loculi of the ovary into two false loculi.[8]

I describe next the position of botanists with regard to each of these elements.

Synthesis of Taxa

Systematists following Jussieu continued to elaborate the natural series by the process of synthesis for well over a century. In general, in the perpetual tension between synthesis and analysis, synthesis was seen to be the correct way to proceed when recognizing natural relationships.[9] Systematic analysis, often linked to ideas of essentialism and so considered to be both incorrect and out of date, was considered artificial, and was useful only when writing identification manuals (and in characterizing opposing schools when indulging in polemics).[10]

The role of intermediates remained crucial in the practice of synthesis. By early in the nineteenth century it had become generally recognized that families were formed in two ways. Mirbel had initially drawn the distinction between "familles en groupe," where the limits of the family were obvious and the genera were very similar to each other, and "familles par enchaînement," in which genera were linked only by intermediates, the resultant family having no obvious coherence and the two ends being quite distinct. The latter families were sometimes considered to be polytypic.[11] Jean Baptiste Pujoulx,[12] in a guide to the gardens, menagerie, and galleries of the Muséum d'Histoire naturelle, claimed that natural families would have many characters in common and would be clearly evident even to the "amateur"; they would be recognizable because of their "physiognomie de parenté." However, he questioned whether families formed by the chaining process were natural; such families might be recognizable, but only because a single character had been used as an artificial expedient to bring together their members. Martin Vahl[13] raised similar objections. Single character grouping, whether divisive (analysis) or agglomerative (enchaînement), would lead the naturalist astray. Mirbel[14] himself later clarified what "enchaînement" really involved when he distinguished between families formed by "enchaînement" in the strict sense—natural families in which several characters were used to group genera, even if the genera had no character wholly in common—and "familles systématiques," which were quite artificial, being based on only a single character. It was the latter kind of family with which Pujoulx and Vahl were dealing, but this kind of family was not usually formed by enchaînement.

There was a feeling that "familles par enchaînement" were less natural than "familles en groupe," but "enchaînement" in the strict sense was absolutely indispensable when it came to recognizing relationships. Families were formed by successively linking genera: Mirbel observed that genera or families formed by the true process of "enchaînement" could be understood only if all their members were known, because taxa at the two ends of such groups might have little similarity with each other.[15] Furthermore, it was the ends of these linking families, themselves formed by enchaînement, that in turn linked distinctive and otherwise isolated "familles en groupe" by enchaînement, so making them part of a larger grouping. As Louis Édouard Bureau remarked:

> Families that are [like the Loganiaceae] polytypic or "par enchaînement" are far from making a blemish in the natural classification; quite to the contrary, because without them such a classification would be only a clever illusion, but impossible to realize. Linkages between homogeneous groups are necessary; this role is reserved to families *par enchaînement*, and one could also say families of *enchaînement* [i.e., families that link], because the links of which they are made not only hold on to each other but form a whole which joins by its extremities adjacent families; it is indeed a chain, but an inflexible and sometimes branched chain that joins the major types, while keeping the one separate from the other.
>
> The "familles par groupe" are natural because of their organization, the others [familles par enchâinement] are natural because of their function. They justify the Linnaean axiom: *Natura non facit saltus*, in which it is impossible not to recognize the idea of a natural classification, and which summarizes so well the necessary conditions.[16]

Jussiaean continuity is more tenuous and "major types" more evident in the distinctive organization of "familles par groupe," but continuity still holds the whole together, and that was the function of "familles par enchaînement". The phrase *Natura non facit saltus* continued to epitomize nature. As we see in Bureau's work, intermediates provided evidence of relationships. For Pierre Turpin, there was no need to make abstractions at all, "because nature never fails to herself show all the intergradations ['nuances'] that can illuminate us."[17] Along the same lines, George Bentham denied that groups could be separated by sharp characters as Alphonse de Candolle had suggested; groups were joined by intermediates.[18] And Adrien de Jussieu observed, "Moreover the limits of these sections are not well determined, and certain genera establish the passage from one to the other. The thing that seems to me to be important, is that the genera are found [in the series] always placed near those with which they have the strongest and most numerous affinities; it is so that the series can be the most natural possible."[19] Groups that were distinct when their centers were compared could nevertheless be linked by their peripheral genera.

Even Augustin-Pyramus de Candolle apparently followed a similar procedure in his work on the Melastomataceae. He laid out all his material of the family on a large table, successively putting down plants most like those he had just placed. The result was a large circle in which the last plant added completed a circle and was similar to the first.[20] He recognized genera within this circle of relationships first by differences in their general appearance, and then he found precise characters that supported these differences. As shown in the previous chapter, Jules Émile Planchon discussed the type of the Melianthaceae and the relationships shown by the family;[21] he thus provided another striking example of the importance of intermediate genera in establishing relationships. Along the same lines, Alphonse Milne-Edwards (1835–1900; he became director of the Muséum in Paris) observed in 1864 that to recognize groupings within a natural group was almost impossible; whatever the characters used, the groupings were more or less artificial. Having excluded the Nolanaceae from the Solanaceae to give the latter a satisfactory circumscription, he suggested that within the Solanaceae (and within natural groups in general) there were a number of typical forms, and linked to them were "derived" forms, which might be linked to other typical forms. Indeed, he thought the study of these typical forms and their interconnections with neighboring typical forms was the best way to discern "the analogies and differences" in any group of animals or plants.[22] Even if groups were distinct, internal continuity was a reason to maintain them undivided.[23]

That groups were often linked by their peripheral members helps explain why monographing remained so important in the whole botanical endeavor. Monographs were not simply an essential part of the inventory of nature, an important enough function in itself, but they were the necessary starting point for the establishment of sound relationships, since it was possible that any member of a group might form links with a member of another group. Monographic work was particularly important in the discovery of relationships for those who eschewed typological thought, because for them the characters of a group were those of its included members. Even Alphonse de Candolle, who believed in discrete groups, observed that to describe a group as other than the totality of its included members would be like committing the fallacy of taking the part for the whole; even if a description of a genus (for example) were based on a type, one had to study all its species.[24] Similarly, George Walker Arnott noted that the character (description) of an order (family) would be longer than that of a genus; it would include more variation because it included more species.[25] Botanists continued to cite with approval the Linnaean aphorism, "The character does not make the genus, but the genus the character,"[26] which is quite in line with this position; note, however, that in this context the character is certainly not an essence.

When we examine, later, the analogies used to describe the natural system and look at the diagrams of natural relationships, we will find that they provided numerous indications of relationships, yet few of obvious groupings. The empty morphospace (the word is twentieth-century, but the meaning is appropriate) between groups was most satisfactorily filled with intermediate organisms which alone enabled relationships to be suggested unambiguously. For eighteenth-century naturalists who believed in continuity these intermediates were organisms from the unexplored parts of the world; for nineteenth-century systematists believing in more or less discrete groups these intermediates might be fossils (although the plant fossil record was less than satisfactory) as well as extant organisms.[27] In the absence of intermediates—living or fossil, organisms or remains of organisms, but not abstractions—George Bentham thought that genealogical relationships in general could not be established, as the only other line of evidence bearing on this question, the direct observation of successive generations that were genealogically connected, was necessarily severely restricted.[28] Even (or perhaps especially) in systems like the quinarian system, intermediates, whether osculant groups or not, were integral to the whole.[29] William Macleay himself was at pains to make the distinction between a hiatus, a small and unimportant gap, and a Linnaean saltus, an unpardonably big gap.[30]

Synthesis of Data

In the twentieth century in particular, systematists have emphasized that their discipline involves a synthesis of all knowledge,[31] with the variation of as many relevant characters as possible being incorporated into the natural system. This meaning of synthesis is generally in line with Jussieu's admonition not to use single characters in classification. Nevertheless, although the character base has broadened since Jussieu's time (including data from development and cell-level anatomy, and more recently cytology and chemistry), the way in which these data were evaluated has until recently remained largely unchanged.

THE USE OF MANY CHARACTERS. When linking groups, some estimate of similarity must have been obtained from evaluating appropriately weighted and relevant characters. Vague guidelines indeed, but nothing more precise can be suggested; this is hardly surprising given the earlier discussion on character weighting in chapter 5. There are few clear indications in nineteenth-century literature as to how these data were to be evaluated, and Augustin-Pyramus de Candolle was the exception when he suggested that true relationships existed if a problematic genus showed relationships with a family in general, rather than just a single part of it.[32] Another approach to this problem was by forming types and then trying to relate types of one group to the type of another, or the type of a group to an extant genus. But most naturalists found typological thought unsatisfactory, and the diversity of practice among its proponents would not make it generally attractive. When Carl Heinrich Schultz-Schultzenstein discussed higher taxa, he emphasized their types, yet even types did not do away with continuity: "It is likewise an essential feature of the truth ['Wahrheit'] of natural systems according to the inner organization, that intermediates and transitions occur, which are not simple exceptions to the rule, but exactly the rule and the law itself."[33] Ultimately continuity of form, whether provided by normal morphology or that shown by terata, guided the thinking that led to the abstraction of types. Continuity in one way or another nearly always provided primary evidence of relationships. That groups might have characters was almost a secondary consideration. As Sir Edward Ffrench Bromhead remarked, "We should throw the families into natural groups, and *afterwards* endeavor to discover some differential characters for these groups, and for series of such groups."[34] He felt he was following the true principles of inductive philosophy, with analysis preceding synthesis, and definition following knowledge.[35]

As will be shown in the next chapter, the distinction that began to be drawn between analogies and affinities focused botanists' attention on the fact that some characters were of less importance in establishing relationships than were others. But it was difficult to make such distinctions, and analogies were anyhow often considered to be of great interest in their own right. Later, in 1879, Oscar Drude observed in a revealing fashion that morphological characters determined natural relationships. However, given a particular suggestion that two groups were related, then the significance of the characters could be evaluated—that is, characters suggesting relationship could be found.[36] So if the group being studied was variable, then almost any relationship could be established.[37]

THE ROLE OF ANATOMY. Individual workers had their own hierarchies of character, preferring one set of characters over another, and vegetative characters in general tended to be treated with suspicion. In addition, in the middle of the nineteenth century, botany was seen as being divorced from the study of internal structure.[38] Jussieu had not been at all positive about the value of anatomical observations in detecting new relationships, and there was no expectation among the majority of systematists that anatomical observations would yield any knowledge of relationships that could not be discerned by a study of external morphology and the dissection of flowers and fruits. Although Hooker saw little general use for anatomical studies in systematics, he did suggest that anatomy had its place in morphology. Detailed study of structure and anatomy allowed the botanist to define parts better, and thereby reduce variation in the form of organs to a common type.[39] In general, anatomy (and development) became very important in morphological studies of the flower, seed, and fruit, and hence in establishing relationships, more particularly between groups at higher taxonomic levels.[40] Substantial advances in clarifying the taxonomic hierarchy at this level were made in the nineteenth century. Indeed, developmental observations in publications from the middle of the nineteenth century are often depicted in exquisite detail, and are only now being replaced by photographs made using the scanning electron microscope. But the number of taxa examined was inevitably low, and such studies did not provide the botanist with a general tool for grouping plants.

The Importance of Group Size

Some influential botanists wished to dismember large groups, because they thought that such groups were cumbersome to use and difficult to comprehend. Others bewailed the nomenclatural changes that such dismemberment caused; splitting caused no real increase of knowledge, and did one really need forty-one genera where previously *Begonia* (one particularly notorious example) alone sufficed?[41] In such arguments the convenience and utility of the system were nominally the issues involved.

Bentham preferred to keep families and genera large but natural, avoiding monotypic taxa wherever possible and using informal subdivisions to reach groups of what he considered to be optimal size: 3 to 6 or sometimes up to 12 taxa in the immediately superordinate taxon.[42] To achieve that goal, Bentham and Hooker (in their *Genera plantarum*) made a very complex system of largely informal rankings interpolated between the genus and the family, and the result was indeed that very few groupings above the level of genus included more than twelve subgroups.[43] Hooker argued that there should be about two hundred families, and he claimed that Asa Gray and Bentham agreed.[44] Bentham even thought that this number would not change much in the future: genera that were poorly understood and that appeared to be isolated (the two were connected) would be referred to existing families when more was known about them.[45] Alphonse de Candolle thought that most plant genera (and presumably *a fortiori* plant families) had already been discovered, but this was not true of species. He presented tables showing that the numbers of new genera described in the *Prodromus* were gradually decreasing with time, although there was no comparable decrease in the numbers of new species.[46] Rather like Jussieu, Candolle thought that some of the undiscovered species would fill the gaps separating genera. However, unlike Jussieu, he thought that reductions of previously described genera rather than descriptions of new genera would be more likely in the future.[47] This was because Candolle was bound to reduce two genera if the gap between them disappeared. Genera would tend to be large.

Jussieu, on the other hand, could decide to maintain such genera, or even to split them if too many more new species were discovered. Even genera that were perfectly clearly circumscribed, but large, could be split. Some naturalists in the nineteenth century took this position, formally subdividing a group that was too big.[48] In general, it is likely that many botanists made an informal assumption that the size of a group and its distinctness were inversely proportional.[49] However, one constraint that might deflect any such predilections was systematic tradition or the "purposes of language";[50] this favored the maintenance of well-known groups at the level at which they had traditionally been recognized and with a generally similar circumscription, whatever their size.[51]

Jussieu made it clear that the size of groups could be manipulated almost at will if nature were continuous. But the size of groups can also readily be manipulated if there are discontinuities.[52] This will hold only so long as no significance is attached to taxonomic rank; it is in principal a trivial matter to subdivide a group along existing lines of distribution, or to merge distinct groups that are nevertheless all clearly most similar one to another. Some botanists in the nineteenth century did not—in theory, at least—think that there was anything special about the different levels of the hierarchy.[53] However, Adrian Hardy Haworth discussed the issue of lumping and splitting at some detail in 1819, and he suggested that "it is not a botanist that actually *makes* genera." The botanist, rather, simply discriminates those "which the mighty hands of his own Maker have already made."[54] Haworth's dictum is not very helpful, even (or perhaps especially) when he and others used it to claim that differences in habit were as important, if not more important, than differences in features of the fructification when delimiting genera.[55] Because of (1) the difficulties in quantifying how much one group differed from another, (2) the emphasis placed on establishing relationships directly between one group and another, with the consequent blurring of the boundaries between these groups, and (3) the sheer diversity of these relationships, taxonomic practice in the first half of the nineteenth century remained very similar to that of Jussieu, with groups being arbitrary as to exclusion. Clearly there was great potential for confusion, with the limits of groups being only imperfectly related to gaps in the variation pattern.[56]

For example, John Miers discussed the relationships of the Nolanaceae, which he placed between the Boraginaceae and Convolvulaceae. Relationships were almost linear, and there were intermediates; the actual boundaries of the families depended on the weightings assigned to particular characters.[57] He thought that groups should be subdivided if they could be distinguished, but he also found that it was difficult to draw a line between groups.[58] Similarly, when one compares the diagram showing the relationships of genera of the Rutaceae (Aurantiaceae) provided by Adrien de Jussieu (figure 10) with Jussieu's own classification of these genera,[59] the major groupings that he recognized seem to have very uncertain limits.

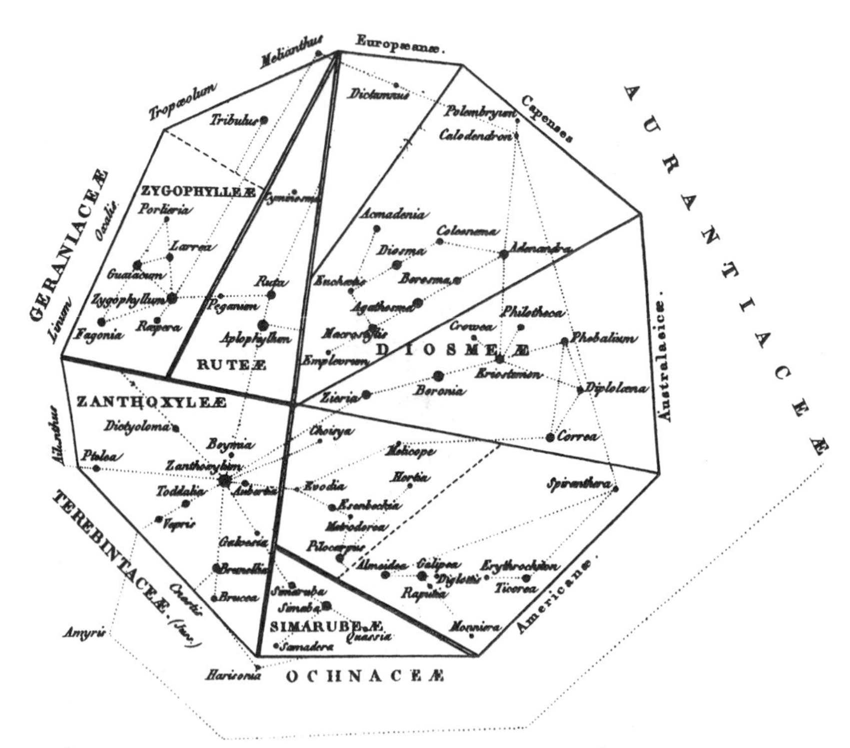

FIGURE 10

Subdivision of a reticulum in the Rutaceae (Aurantiaceae). Source: Adrien de Jussieu 1825: pl. 29.

The Depiction of Relationships

Discussions of the linear *scala naturae* emphasized the relative highness and lowness and the relative perfection and imperfection of organisms; all organisms could be arranged on a chainlike, hierarchical scale. However, the idea that relationships were simply linear broke down as more numerous characters, especially those of gross anatomy, were used in classification, and the language of highness and lowness would seem to be inappropriate for this new world in which relationships were at the very least reticulating.[60] It was a matter of frequent comment from Jussieu's time onward that the linear arrangement of families in books was artificial; however, given the nature of the printed page, this was the only way in which they could be arranged.[61]

Even if the *scala naturae* itself was no longer an acceptable way of describing nature, naturalists used other analogies, and also drew diagrams of relationships to convey their understanding of the natural world to their colleagues.[62] I shall show that botanists visualized nature in a generally similar fashion throughout the later eighteenth and nineteenth centuries (and even on into the twentieth century).[63] As Adolphe Brongniart observed in an 1868 report on progress in botany, many botanists were still trying to produce a continuous linear series so as to indicate the gradual transitions between natural families.[64] In general, relationships were depicted as being more or less reticulating, and the reticulations became even more complex as the century progressed. However, the reader might reasonably infer that relationships were more or less continuous, as with the *scala*, and certainly directly between groups. But unfortunately we today cannot assume that because two people used the same analogy they agreed in their visualization of nature, or, conversely, because they used different analogies they differed in how they saw nature. The interpretation of diagrams and understanding of metaphors and analogies is no easy matter.[65]

Carolus Linnaeus, Antoine-Laurent de Jussieu, and Augustin-Pyramus de Candolle all used the same analogy—that between the natural system and a map—when they discussed the pattern of relationships between the natural groups each recognized. For Linnaeus and initially for Jussieu this analogy meant that natural groups were adjacent; there was continuity and groups were not discrete (figure 5 in chapter 4). However, for Candolle, this analogy conveyed the idea of distance between towns;[66] there were groups (towns) varying in how far apart (distinct) they were. Around 1816 Candolle prepared a planting scheme for the botanical garden in Geneva (figure 6 in chapter 4) in which he actualized the metaphorical landscape of nature in the placement of and distances between the beds.[67] The same basic analogy, although modified again by the superposition of lines of latitude and longitude, was used by Étienne Pierre Ventenat[68] simply to highlight how a naturalist might be able to relate an unknown plant to known groups. Elias Magnus Fries[69] used a similar scheme to help clarify the distinction between two kinds of relationships, analogy and affinity. Finally, Hugh Algernon Weddell added relief to the Linnaean landscape so that it would be possible to talk about the relative advancement of groups.[70]

Jussieu also compared the natural system with bundles of sticks to convey what he conceived to be the proper idea of relationships between groups.[71] This analogy is perfectly compatible with the "bubble" diagram of relationships between plants that is believed to represent the older Linnaeus's ideas of relationships.[72] The diagram suggests a branching continuity, although imperfectly separated groups are apparent, and it looks like a cross section through a bundle of twigs.

But whether or not a naturalist believed in groups—and most naturalists were not clear on this—discussions of relationships between organisms highlighted their complexity. Candolle[73] depicted relationships in the Melastomataceae as if there were major elements of continuity between genera and reticulations between groups of genera (figure 11), although it is clear that he considered his taxa to be more or less discrete, basing his thirty-eight new genera on differences in their "organisation." Similar diagrams in which the main graphical element is circularity, with the continuity of relationships the circle implies, are described by other naturalists in the period between 1816 and 1860. As Alfred Russel Wallace remarked, circular systems were common in the 1850s.[74] Thus the cryptogamist Dominic François Delise (1780–1841) produced a pie diagram with genera arranged in a circle; this, he thought, allowed the linkages between them to be shown.[75] But, as in Candolle's diagram, Delise's circles touched other circles, and one can imagine the circles as elements of a reticulum.[76]

Other diagrams and analogies focused on a more unordered reticulum of relationships, although there might also be additional information in the diagram that provided more structure. Thus Michel Félix Dunal (1789–1856), who was a student of Augustin-Pyramus de Candolle at Montpellier, graphically represented the strengths of the similarities between groups within and related to the Annonaceae by the relative widths of the ribbons connecting the groups.[77] Adrien de Jussieu also incorporated information on the size of genera in his reticulum depicting relationships in the Rutaceae, although it is not easy to under-

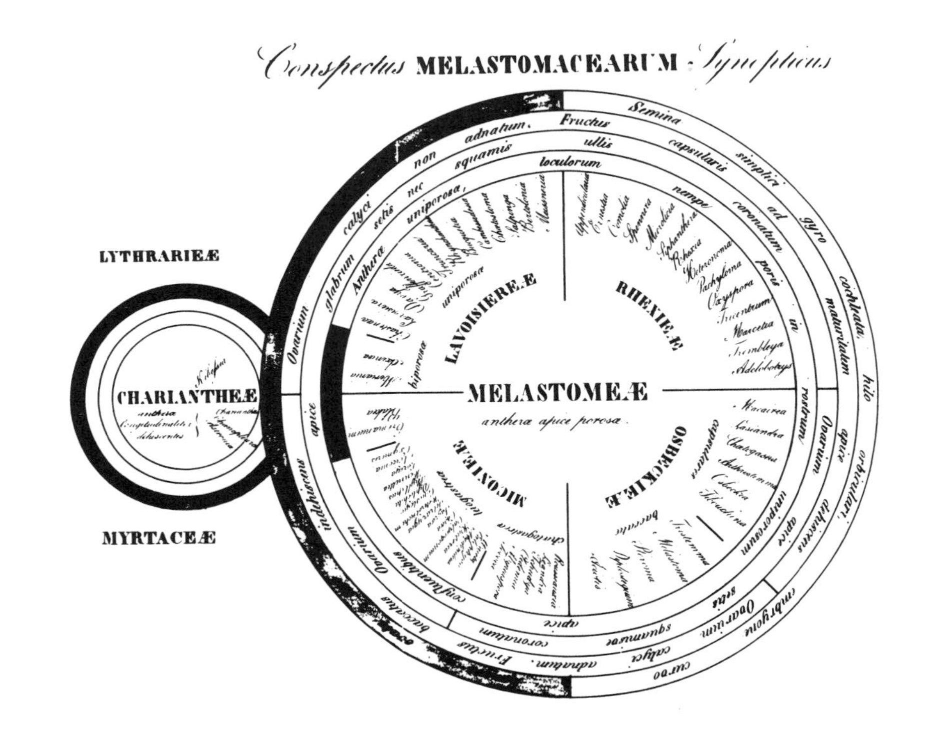

FIGURE 11

Circles of relationships in the Melastomataceae. Parts of some of the circles are painted so emphasizing the different characters involved. Source: Augustin-Pyramus de Candolle 1828: pl. 1.

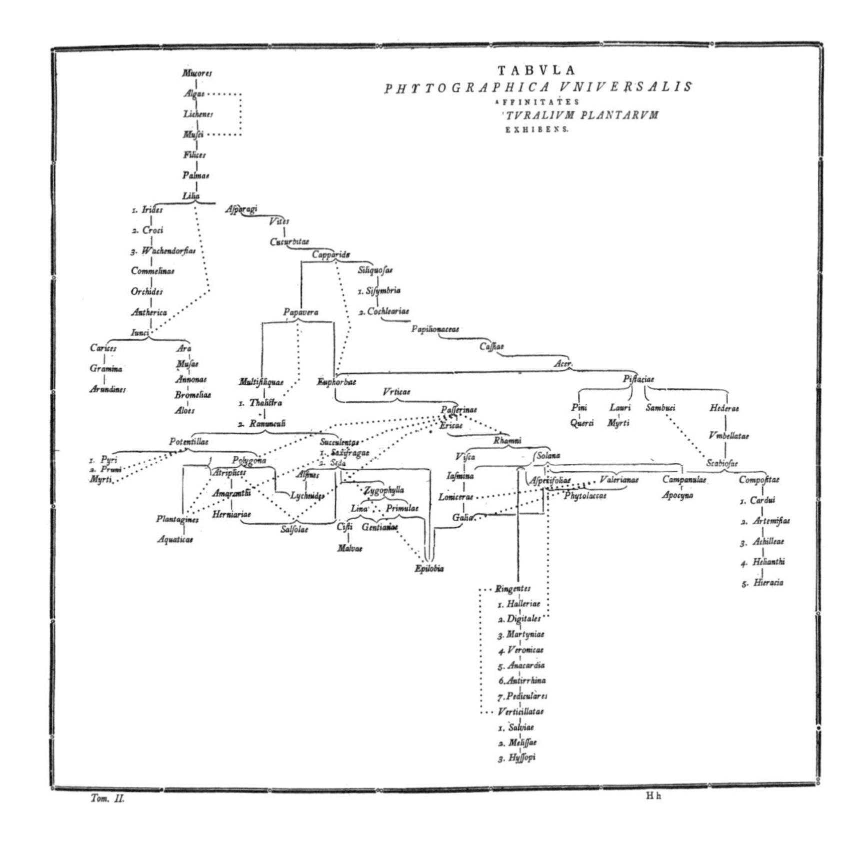

FIGURE 12

Complex reticulate relationships. Added (dotted) lines are relationships suggested in the text between pp. 459 and 462. Source: Rüling 1793: chart at end.

TABLEAU DES AFFINITÉS DES LINÉES.

Reaumuria
Hololachne.... Velezia (Caryophylleæ)
Tamarix
Frankeniaceæ ....ANISADENIA.... Sauvagesieæ.... *Ochnaceæ*
Elatineæ........RADIOLA... LINUM .... REINWARDTIA .... Turneraceæ
Crassulaceæ ***Droseraceæ, Portulaceæ***
HUGONIA.... Smeathmannia (Passifloreæ)
Erythroxyleæ.......... ROUCHERIA.... Humiriaceæ
Chlenaceæ
DURANDEA ... Ixionantheæ
Hypericineæ ... . Bonnetieæ.... Ternstræmiaceæ

FIGURE 13

An incipient grid-work in the Linaceae. Source: Planchon 1847b: 592.

stand what the relationships between the genera really are (figure 10).[78] The complexity of the relationships drawn could be quite remarkable. Johann Philipp Rüling (b. 1741), in the narrative that accompanied his diagram, made mention of more relationships between groups beyond those suggested by the rather simple illustration he provided, thereby considerably increasing its effective complexity (figure 12).[79] John Lindley attempted[80] to get around the linearity of the sequence imposed by the printed book by interpolating in the text small diagrams in which the names of families were placed above and below and before and after that of the family being described. Lindley's method would result, if all the diagrams were combined, in a complex gridwork of relationships that would encompass all plants. Jules Émile Planchon used a similar idea, and the result again approached a reticulum (figure 13).[81]

Up to the middle of the century the analogies most commonly used for the natural system allowed such multiple relationships and, at the same time, lacked any directionality. As the century wore on, relational constructs became even more complex. Many of the idealistic systems to be discussed in the following chapter suggest a nature that has relationships forming either a poorly ordered reticulum or a more regular grid, but to Henry Thomas Colebrook in 1827 this was all too simple.[82] He rejected the hierarchical quinarian system; relationships were better thought of as those between the angles of an icosahedron, or, more complex still, those between the different members of the solar system. Alphonse de Candolle suggested in 1841 that relationships could be compared with an archipelago, that is, there were islands of different sizes (groups) differing in how far apart they were.[83] But his archipelago is not very different from a galaxy, the more so because the relationships he indicated cannot be accommodated by a two-dimensional diagram. Earlier Candolle had suggested that there were limitations in the use of the map analogy; a map, being only two dimensional, was barely adequate for representing relationships,[84] and a third dimension was needed. The third dimension would simply allow more complex relationships to be accommodated; there was no fundamental change in how nature was visualized. In 1843 Adrien de Jussieu depicted reticulate relationships in the Malpighiaceae, and he compared his diagram with a map or chart of the stars, with certain genera being grouped like constellations.[85] As he noted, comets could represent taxa that did not have a fixed place in any system. In a classic study published in 1856, Hugh Algernon Weddell (1819–1877), an English botanist long resident in France, discussed families related to the

Urticaceae. Relationships between these families, he proposed, were best depicted by arranging the families on three sides of a pyramid; a number of similar pyramids could be added together and a "taxonomic relief map" produced.[86] Clearly, these complex images of nature cannot easily be interpreted in terms either of groups that show hierarchical relationships and are unambiguously circumscribed, or even of the relative advancement or perfection of nature.

At the same time the comparison of the natural order with a genealogy or a tree was surprisingly common,[87] although usually made rather casually, as by Antoine-Laurent de Jussieu himself.[88] But Jussieu never developed the genealogical motif he used to describe his synthetic method of forming the natural order. Similarly, Pierre Turpin managed to combine morphology, systematics, the loss of species making gaps that were impossible to fill, absolute continuity, the botanical landscape, nature being like the branches of a large tree, and the reality and uniqueness of each individual, into a complex discussion of nature and different ways of looking at living beings.[89] In 1844 Jean-Baptiste Payer emphasized that a natural classification was genealogical, but then he turned to a discussion of types radiating from a center and mentioned the insensible transitions between groups[90]—ideas not readily compatible with a genealogy. However, it is probable that for Payer, as for Antoine-Nicolas Duchesne eighty years before, genealogy implied reticulation, the complex crossing relationships needed to maintain a bloodline or involved in tracing descent, not a branching tree.[91]

Treelike diagrams were occasionally used to represent the characters of dichotomous analytical keys,[92] but such diagrams have nothing to do with nature. However, the general structure of logical division as exemplified in Ramean trees did suggest to John Fleming a comparison between such trees and the natural system. Fleming's idea of nature was conveyed by the successive dichotomous divisions of such trees, albeit with circular elements thrown in; species were compared to the buds or leaves of the tree.[93] Ludwig Reichenbach compared the different ranks of the hierarchy to the different parts of a tree—the trunk would be equivalent to a kingdom, the fruits and seeds to varieties;[94] in general, the ranks of genus and family positively invited comparisons with ideas of reproduction and genealogy.[95] Adrien de Jussieu discussed the tree analogy at some length.[96] For him it suggested the idea that there was a general series of plants, despite the complexity and multiplicity of rela-

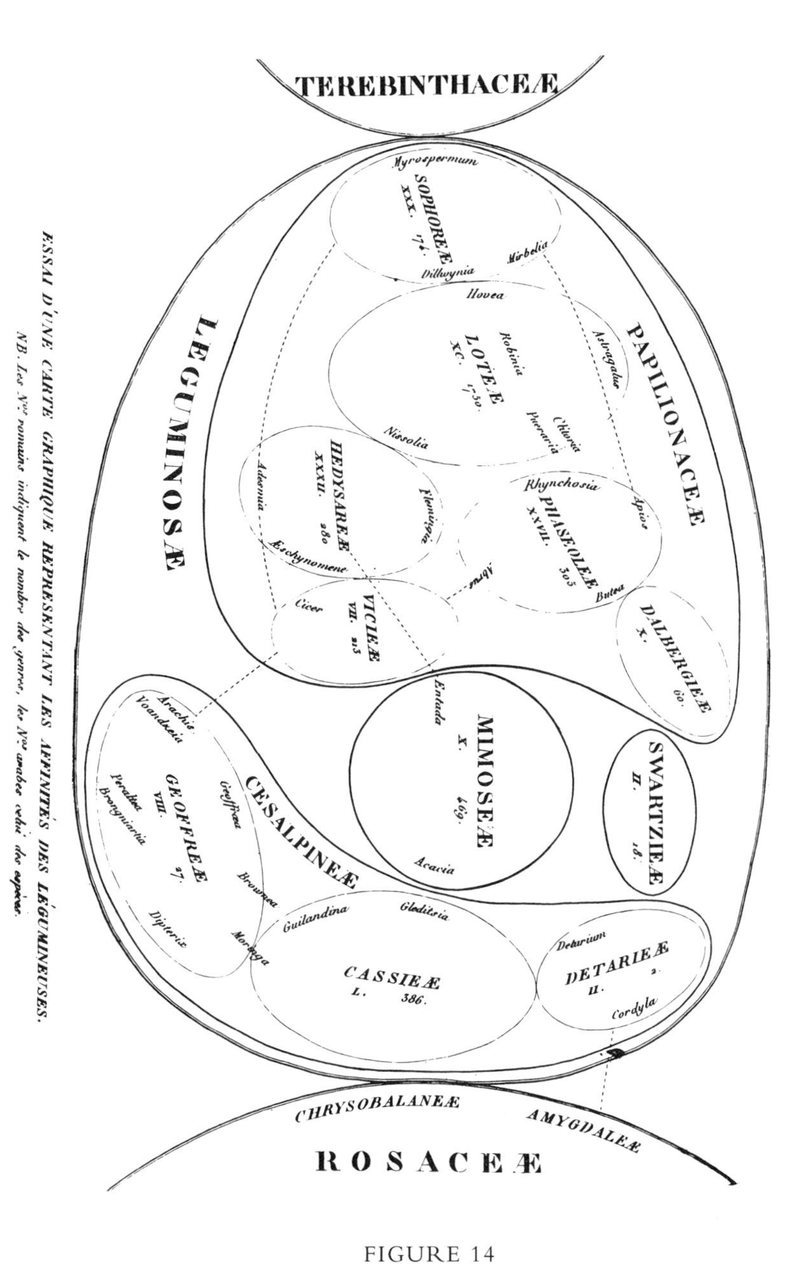

FIGURE 14

Potential dendritic relationships in the Leguminosae.
Source: Augustin-Pyramus de Candolle 1827: pl. 28.

tionships, which could be likened to the result of natural grafts forming between otherwise independent branches of the tree. He did not, however, usually think of relationships as being dendritic, and his interest in a general series of plants led him to think of parts of the trunk of the tree with their associated branches as representing groups of equal rank.[97]

Although most of the diagrams of relationships that Augustin-Pyramus de Candolle drew suggest reticulations, his diagram showing relationships within the Leguminosae differs from the others. Groups are circles within circles, albeit touching and with one or two cross-linkages (figure 14).[98] It is similar to a Venn diagram, and if extended to both higher and lower taxonomic levels, it could also be represented as a tree with species occupying only the ultimate branches. Unlike the other examples discussed in this section, Candolle's diagram can be converted directly into a classification that lacks serial elements.[99] (A number of the idealistic systems to be mentioned in the next chapter can also be converted to such classifications.)

Thus, although the comparison of a classification with a genealogy, or even the depiction of a classification as a tree, was quite common, this did not imply any common understanding of nature—still less an appreciation of nature as a tree with separate branches,[100] with the level of branching representing levels of classification. Genealogies and trees may represent different ways of looking at nature, and two authors discussing relationships as treelike might also have incompatible understandings of the natural system. Furthermore, it should not be forgotten that classifications in preliterate societies typically imply that many of the groups of organisms they recognize are somehow connected by genealogical relationships, a tendency that can only have been accentuated when the ranks of genus and family were adopted by European systematists because of the connotations of generation and relationship that came with the use of those terms.[101] In such circumstances, the nature and extent of any understanding that there are real genealogical connections between organisms becomes difficult to ascertain.

I now close this section with a summary by Constantin Rafinesque written in 1836 of how affinities were represented diagrammatically in his day: "A botanical map on a graphical plan would best represent these [affinities] by contiguity; a second mode would be by a kind of network; a third by a kind of genealogical tree; the fourth by a triple concentric circle, divided in rays, the inner circle being the cellular plants, the outer or largest the Exogenous plants or trees; the fifth by a triple series or three columns, one for each great series, with lines across to connect affinities."[102] Most of the images he mentioned accommodate multiple relationships, and while some are basically dendritic (perhaps the genealogical tree), others are reticulating (the network). Twenty five years later, when Darwin's *Origin* appeared, the same set of images was in common use. Although the analogies of botanists are more overwhelmingly arborescent after 1859, and the diagrams produced over the ensuing forty years become more tree-like, the change was more apparent than real.

On Highness and Lowness

Despite the apparent demise of continuity, or at least of a simply linear *scala*, the language originally used by Lamarck and Jussieu remained in common use; indeed, it pervaded discussions on relationships. Gustave Planchon (1833–1900) provided what is effectively a summary of pre-evolutionary thought in his *Les principes de la méthod naturelle* of 1860. Having rejected a simple linear series, or even a series of groups, as a suitable representation of nature—relationships were more complex—Planchon nevertheless proceeded to devote an entire chapter to a discussion of the hierarchical position ("dignité") of beings in the context of a serial arrangement of nature.[103] In 1837 Alphonse de Candolle had followed exactly the same sequence of topics in his textbook, *Introduction a l'étude de la botanique*,[104] while chapter 2 of Filippo Parlatore's *Lezioni di botanica comparata* (1843) is entitled "way of appreciating the grade of differentiation of a plant in the plant series." Emmanuel Le Maout observed that the progressive perfection of plants was evidence of a law, and he thought that the fossil evidence supported this because it, too, showed evidence that plants were gradually becoming more complex.[105] It cannot be overstated how prevalent this style of discussion was, whether or not the authors believed in relationships that were reticulating or circular and/or the existence of types.[106]

When in 1856 Hugh Weddell described his pyramid of relationships involving the Urticaceae and neighboring families, he observed how the Malvaceae, the family at the top of the pyramid, represented "the most perfect type," while families at the base were "degradations." These pyramids formed the elements of his taxonomic relief map, so ideas of highness and lowness were incorporated into the analogy of nature as landscape.[107] Asa Gray, commenting on Weddell's "happy illustration," observed that even if the vegetable kingdom did not "culminate," as did the animal kingdom, dicotyledonous orders could be arranged in many "short series, in groups converging on the most fully developed or representative order of each type."[108] Indeed, because the relationships of plants were so frequently discussed as being locally linear,[109] the use of "scala"-type language would seem perfectly appropriate, as it would when nature was visualized as a tree.[110] And even when relationships were not described explicitly as being linear, one cannot help but be impressed by how frequently the relationships of a group are with only two adjacent groups.[111] For example, Augustin-Pyramus de Candolle, while allowing that the Leguminosae-Loteae might be subdivided one day, doubted that the order of genera in it would be changed.[112] And Alfred Russel Wallace, having just described how to represent relationships by a branching diagram, noted that there was nevertheless a "main line" of affinities.[113]

In general, families close together in the printed (linear) sequence have tended to be interpreted as being related, no matter in which higher taxon they are placed—and this is despite the warnings so frequently made that the printed sequence was an inadequate representation of nature. Thus J. D. Hooker suggested to Alphonse de Candolle in 1884 that the latter should draw up a standard sequence of families. Candolle was quick to disavow the possibility that such a sequence could reflect nature, and Hooker then pleaded a slip of the pen—he knew that, of course.[114] Yet much later we find George Lawrence noting that the Bentham and Hooker system was patterned directly on that of the Candolles.[115] This is true (in part) only if the *sequence* of genera in the two systems is compared, but not if the sequence is disregarded and the pattern of relationships implied by the classification is emphasized.[116] Variants of the assumption that the linear sequence of taxa is meaningful *qua* sequence persisted: names close together in a key are those of related organisms.[117]

Indeed, a language implying highness and lowness, and relative advancement in a series, will tend to persist because there is a "pedagogical scala"—it is easier to fix facts in a student's memory if they are linked in some kind of linear sequence. Such a sequence then functions as a kind of memory system, with organisms being readily assigned their proper place in a linear series.[118] From this point of view, John Lindley found it appropriate to arrange plants in a linear series from perfectly organized to less so.[119] Mirbel also arranged his families in a linear continuum for his lectures, and the elderly Lamarck presented nature as being simply linear for didactic reasons.[120] There are problems here. Groups—whether or not they exist in nature—*are* needed for teaching, and they will inevitably be presented in a linear fashion, often reflecting one or another "trend"—whether or not trends exist in nature. Under such circumstances, what a student might learn about the arrrangement of nature and the general structure of relationships becomes difficult to gauge.

The conclusion is clear. Although the image of the net might seem to lend itself less to the language of the control hierarchy used in discussion of the *scala naturae* (when compared with that of the tree), there was in fact little difference in fundamental outlooks among those using one or the other preferred analogy to explain or describe relationships.[121]

The General Language of Classification

The flexibility or ambiguity of the language that Jussieu used when describing how he recognized natural relationships, evaluated characters, and described species, genera, and families allowed the same terms and phrases to be used both by naturalists who believed that there were continuous relationships in nature and by those who thought that there were discrete groups. The Linnaean hierarchy, for example, can be interpreted as a rigid class hierarchy (as did Linnaeus himself and Louis Agassiz). But it can also be seen as a much more flexible box-in-box system, as Bentham saw it—at least in theory. Other interpretations would emphasize the part-whole relationship of such a system, or, conversely, would claim that it was a mere convenience that yielded no interpretation in natural terms (the latter was basically Jussieu's position). Yet the same taxonomic ranks, albeit somewhat increased over those used by Linnaeus, were used by all. Nomenclature came to dictate the ranks that could be used, their relative positions in the hierarchy, and the terminations of the Latin- or Greek-based names of the taxa that occupied them; organisms now cannot be considered to be properly described unless their description obeys these and other nomenclatural rules.

Jussieu described groups at the level of genus and family ("ordo"), just as did the Candolles, Bentham, Adolf Engler (1844–1933), John Hutchinson (1884–1972), and others later, and all would claim that their groups were natural.

In 1874 George Bentham summarized the state of systematic botany, noting that he had "received many useful hints on the method of botanical study from the great founder himself of the Natural System, Antoine-Laurent de Jussieu."[122] Bentham also observed that his own apprenticeship was passed in the study of the works of Candolle, in particular that of Lamarck and Candolle's *Flore française* and Candolle's *Théorie élémentaire*. But when it came to approaches to grouping and ranking, the botanists Bentham named as being his mentors in the discipline held widely diverging views.

In the accompanying table I introduce some distinctions to facilitate discussion about such emotive terms as "nominalism" and "realism," and so clarify the differences between apparently similar concepts.[123] Bentham himself at this stage of his career was what I will call a hierarchical nominalist. He thought that groups from variety upwards were more or less discretely bounded; grouping was not arbitrary as to exclusion or inclusion, but ranking was arbitrary. Lamarck and Jussieu thought that the rank of all taxa at the rank of genus and above was arbitrary; the limits of their taxa were arbitrary as to exclusion although not inclusion, but they did not consider the linear sequence of taxa to be arbitrary. Lamarck and Jussieu can thus be considered limital nominalists. Candolle certainly believed that taxa were nonarbitrary as to exclusion or inclusion, but his attitude toward ranking was ambivalent, and he may have believed that taxonomic ranks as such represented some reality in nature (hierarchical realism), or he may have thought that ranking was arbitrary (hierarchical nominalism).[124]

Because hierarchical nominalists, hierarchical realists, and essentialists alike could claim that their taxa were "real" and "natural," it is not surprising that the apparently innocuous word "nature" has proved peculiarly troublesome over the years.[125] Not only does the word have several meanings at any one time, but the spectrum of its connotations has changed over time. Nevertheless, the success of Jussieu's *Genera plantarum* ensured that the goal of systematists after him would be the perfection of the natural method or system; no general system could be proposed after 1789 that was not "natural." Although Alphonse de Can-

ATTITUDES TO GROUPING AND RANKING
(GENUS LEVEL AND ABOVE)

NATURE OF GROUP	TERM	EXAMPLES
Group not integral part of nature	Nominalist (unqualified)	Linnaeus (sexual system) Lamarck (groups in keys)
Group not discrete, but integral part of nature	Limital nominalist	Lamarck, Antoine-Laurent de Jussieu
Groups discrete, but not rankable	Hierarchical nominalist	Bentham, Darwin
Groups discrete, ranks of hierarchy real	Realist	Linnaeus (genera at least), Louis Agassiz

dolle realized that "nature" was an ambiguous word, he did not see any particular problems with its use in systematics. Rather, he contrasted a nature in the sense of known phenomena experienced by many, that is, the nature of a thing (which did not seem problematic to him), with a nature in its poetic and literary connotations. However, the ambiguities surrounding the word and its derivatives were such that they were better not used.[126] This goal was easier stated than achieved, and even although it has become clear that "the nature of a thing" (or a group) is no fixed quantity in nineteenth- and twentieth-century thought, to call taxa "natural" has remained a mark of approbation.[127] Unfortunately, its use has tended to prevent discussion because it has been treated as if it were a primitive term. As the paleozoologist F. A. Bather (1863–1934) perceptively observed in 1927 when describing the use of the word "nature" in pre-Darwinian times, "not a single naturalist had a clear idea of what he meant by 'natural.' All he knew was that the other fellow's classification was unnatural."[128] So if a classification is described as being "natural," it tells us nothing about how the naturalist saw nature or delimited groups.

"Organisation" is another critical word with many shades of meaning for both Jussieu's contemporaries and his immediate successors; it is now only rarely used in systematic discussion. When I first saw this word in an early nineteenth-century French text, I thought it meant simply "structure." Yet the different meanings current at that time signified a whole variety of relationships between characters, organisms, and groups—relationships that were clarified only when development, genetics, evolution, and physiology became better understood later in the nineteenth and twentieth centuries. Terms like "analogie" and "affinité" have been much more persistent in the systematic literature, although they have a somewhat more restricted spectrum of meanings. For Jussieu, "affinité" might refer to relationships between different organs on one plant; between what might be the same organ on different plants; and between different plants or groups of plants. The next chapter will show that during the first part of the nineteenth century these paired terms came to signify two distinct kinds of variation. Analogy misled when it came to understanding relationships, while affinity did not (although affinity still also meant a more general relationship of groups). Some of the usages of "affinité" in Jussieu's early papers are no longer current, for example, the first usage mentioned above; a usage still extant, that of relationship of groups, is too imprecise for most systematists. Even the term *symmetrie*, used so much by Augustin-Pyramus de Candolle, could mean "symmetrical" (of a flower), denote a type, or refer to parallel variation between two groups.

Predictivity

That groups should be "predictive" was an important goal of nineteenth- and twentieth-century natural systems, and predictivity has remained a desideratum for classifications.[129] Various aspects of organisms may be predicted—the existence of species, genera, and families as yet uncollected, the characters of poorly known plants, the co-occurrence of features in plants, and so on, and the medicinal properties of plants related to those for which these properties are known. Frans Stafleu[130] called predictivity a modern idea, and signaled it out for particular emphasis as being an idea that originated with Jussieu himself. This then is a tie binding Jussieu to subsequent systems and distancing him from his predecessors.

However, predictivity, often of medicinally important features, had been considered a property of natural arrangements for at least two centuries before Jussieu's time,[131] and naturalists of the generation following Jussieu continued to hail its importance. More generally, success in prediction betokened an understanding of nature; if the reasons for the co-occurrence of structures (whether functional dependencies or something else) were understood, prediction would suggest that a fundamental understanding of the organization of living bodies had been attained.

Such understanding would in principle allow Cuvier's famous "predictions" of the entire skeleton of an organism from an examination of but one of its bones.[132] In a prediction of a similar kind, Augustin-Pyramus de Candolle described as an "induction" the idea that plants that resembled each other in their characters would resemble each other in their "propriétés"; fungi that grew on more than one plant species would have related plants as hosts.[133] There were other predictions. Alire Raffenau Delile[134] discussed how it was possible to predict the existence of several characters in an organism from knowledge of the presence of a single character. John Lindley, as we have seen, often mentioned predictivity, often in a more or less medical context.[135] Finally, Michel Adanson not only predicted that plants that would fill gaps in the general order of nature would be discovered, but he provided detailed estimates of the numbers of taxa at all hierarchical levels that remained to be found.[136]

Discussion

The preceding analysis suggests that what I have called continuity-in-practice did persist in an appreciable segment of the taxonomic community up to the middle of the nineteenth century—and beyond. The need to establish relationships directly between groups by means of linking intermediates is the most obvious feature in common between practice in the nineteenth century and that of Jussieu, and this was seen by several naturalists as being integrally associated with continuous nature. If there were gaps, this procedure would be inappropriate. When the difficulties in distinguishing between similarities that indicated relationships and those that did not are added to the impossibility of providing any fixed weighting of characters, the affinities of genera and families would seem diffuse.[137] Even if a family were discrete, the diverse relationships it showed immediately pulled it apart; analogies would tend to convert to affinities when each group was surrounded by several others, as in many of the representations of nature that were current in the nineteenth century.[138]

Naturalists remained passionately interested in establishing relationships, and organisms were needed for this. Hence J. D. Hooker's concern that, with retrogression from a type and extinction of some of the taxa, the morphological continuity that alone could establish relationships would be irreparably sundered.[139] Hooker observed that some families did seem to be rather difficult to distinguish from others, yet even the more discrete families were often linked by what he called an osculating genus. Worse, that genus was often large.[140] Under such circumstances, even discrete groups would lose their identity.

In general, one gets the feeling that in the nineteenth century the definition of a natural family was a family in which all the genera were related to each other; that they might also be related to other genera not included in the family was less an issue. The fundamental similarity between Jussieu and his successors—synthesis, or, less specifically, the need for intermediates when establishing relationships—rendered the very existence of discrete groups questionable. All these problems are reflected in the analogies used to describe the natural system and the diagrammatic representations of relations found in the literature.

Many botanists considered it an important task to find out where families fitted into some general scale of advancement or perfection. Thus despite the frequent disavowals of any connection between sequence and nature, connections between the two are ubiquitous in the literature. The tendency to think in terms of progressions has been accentuated by the tendency of naturalists to begin their sequences with plants they thought were simple, "primitive," or simply marking one extreme of how plant parts might be organized, and to end it with the other. Thus even though formal classifications of organisms yield a different kind of hierarchy from that of a linear scala, the taxonomic hierarchy could be, and often was, converted to some sort of scala. As we saw with Wallace, the steps that are supposed to have led to the highest form, be it plant or animal, buttercup or baboon, are emphasized, and these steps are interpreted as being between taxa; a simple series results.[141]

Other similarities between Jussieu's ideas and those of subsequent workers have little to do with specific ideas of the shape of nature. Discussion of three kinds of predictivity is prominent in Jussieu's work.[142] He predicted (1) that organisms causing the gaps in his system would be found; (2) that certain organs co-occurred in plants (the affinity between different organs), and (3) that medicinal properties would be found concentrated in particular groups. The first kind of prediction clearly depends on a belief that nature was continuous. Intermediate organisms *had* to exist and would ultimately be collected, hence the search for the missing links needed to make the Great Chain entire.[143]

The other two kinds of prediction are consistent with a belief in continuity, although the second, which is connected with the subordination of characters, is perhaps less so. In general, predictivity is almost as easy in systems in which groups are not discrete as in those in which they are. Even if groups were part of a continuum, they were a coherent part of the natural order. Organisms that were most closely related were placed next to each other and would necessarily have many features in common, perhaps even some that were at best rare elsewhere; it is these features that would be predicted. Even if continuity is branched, prediction remains possible, although it may not be such a simple matter, particularly if the branching is not regular.[144]

Similarly, manipulation of group size can occur whatever a naturalist's views as to the shape of nature. The great majority of the naturalists that I have mentioned conceded that higher taxa were not natural in quite the same way that they thought species were. By this they usually meant that genera did not form a discrete rank in the hierarchy, although individual genera themselves might be quite discrete as groups. *Ranking*, it was agreed, was arbitrary; *grouping* was not, or was less so (although we have to remember that this distinction, critical as it is, was only rarely made).[145] Groups, even if discrete, usually had groupings within them; these subgroupings could be raised in rank and recognized formally if it were so desired. Continuous nature also could be subdivided at will. For Bentham, Asa Gray, the Hookers, the Candolles, and others who preferred broadly circumscribed families and genera, the only way to ensure the generic limits they preferred was to establish them as a conventions. But the whole issue became muddy when it was linked with the debate of what was or was not good scientific practice.[146] Jussieu's families persisted in part because they, too, were treated as conventions.

Finally, we have seen there was no change in the general language used by naturalists, other than the introduction of new ranks into the hierarchy and of terms such as "taxonomie." The use of "family," "systematic," and even "systematist" also became widespread.[147] Although most of the important terms remained ambiguous, they were rarely defined; additional meanings accreted over the years.

The metaphors and diagrams used by botanists in the first half of the nineteenth century suggest a nature of groups with fuzzy boundaries, or a nature with nothing much in the way of groups at all. "Continuity-in-practice" accurately describes the practice of many of these botanists because the cornerstone of Jussieu's method, the detection of the natural order by adding groups one to another, with relationships being directly between groups, remained in place. Yet the various forms of typological thought discussed in chapter 6 that were quite common during this period might seem to counter continuity-in-practice. Augustin-Pyramus de Candolle believed that the best training for students was monographic work, because then they would emphasize resemblances between plants and gain an understanding of the symmetry of a group, whereas if they worked on a flora they would appropriately enough have to emphasize differences.[148] Yet although students frequently carried out a monographic study as their prentice work, ideas of symmetry or types and the increased emphasis this tended to place on the discreteness of groups, did not become commonplace within the discipline. Indeed, monographs were often used to clarify the totality of the variation in a group, not the nature of its type. Continuity-in-practice largely persisted.

With this in mind, issues like the fate of natural history during this period, the status of classification studies, and the idea of a revolution in systematics, all subjects of later chapters, can now be approached in a different light.

CHAPTER EIGHT

On Understanding Nature

How satisfactory did a naturalist find his understanding of the natural world? This is a difficult question to answer. If a naturalist was simply trying to provide a useful classification, the issue of understanding nature would be irrelevant. Although it was philosophical botany as it became defined in the first part of the nineteenth century[1] that took upon itself the task of such speculations, most naturalists, however, wanted more from their classificatory work than names.[2] Yet the repeated assertions by both believers in continuity and believers in discontinuity that higher taxa were the work of man, not nature—or at least as much the work of man as of nature—suggests that classifications based on natural systems would not provide a truly satisfactory understanding of the living world. Many naturalists in the eighteenth century, nevertheless, had a more or less coherent and intellectually satisfying understanding of the overall structure and patterning of nature.

As Frans Stafleu[3] observed, for a naturalist in the latter part of the eighteenth century, the natural system was the blueprint of creation, and continuity itself provided the unifying element linking all of God's productions.[4] For believers in both continuity and plenitude, whether or not the system as a whole was known, the limits of creation were in a sense known and understood,[5] and nothing fundamentally new was to be expected. No unknown laws were needed to explain the pattern of nature; all the naturalist had to do was to fill in the gaps in the series. In continuity, the linkages integral to nature gave coherence to the whole, and this coherence functioned rather as laws might in a system with less obviously regular relationships. This would be especially true when ideas of highness and lowness were associated with a linear continuity, as in the *scala naturae*.[6] The language of the *scala* did, in fact, persist throughout the nineteenth century.

Demonstrating that relationships formed a continuum simply confirmed a belief some Europeans had held about nature for about two millennia; organisms were assigned their position in a fundamentally closed system in terms of their striving for perfection toward, or their degeneration from, some perfect form—whether man or god. Thus an important part of Linnaeus's understanding of nature as a whole came from the continuity and direct connections he perceived in it, whether in the context of his system as a whole (at least those parts of it that reflected natural relationships) or in his more morphological studies in which he focused on the relationship of form largely outside of a systematic context.[7] Yet Charles Baehni[8] suggested in 1957 that although continuous nature might have been regarded as the secret plan of God revealed, and it might have solaced the soul, it would have been a useless guide for naturalists, as it could be comprehended only in its entirety. Here Lamarck's keys could mediate between the user and indivisible continuity. And in Jussieu's insistence that groups be given "characters," preferably those that could be easily seen, we see another attempt to combine comprehensibility and the indivisible natural order.

Jussieu would surely have agreed that the groups that he so carefully described and into which he placed the some 20,000 species of plants he knew did not exist in nature; that is, their contents were part of the natural order, but their limits were not. Estimates of the number of species differed: in the early 1780s Eberhard August Wilhelm von Zimmermann thought that there were some 175,000 species of plants, of which only about 70,000 had been described.[9] Nevertheless, Zimmermann's enterprise was the same as that of Jussieu—that of understanding the relationships between organisms in the context of continuity. Any gaps in the series were but temporary embarrassments for all who believed in this continuity,[10] since ultimately all the links in the chain would be found, and all the observations that had been made on organisms would finally find their proper place.

The previous chapter showed that by the late eighteenth century it was generally allowed that nature could not be represented as a simple linear arrangement. Johann Philipp Rüling's diagram of plant relationships, first published in 1766, is obviously reticulated and becomes even more so when other relationships he mentioned in the text are added (figure 12 in previous chapter).[11] August Johann Batsch's well-known diagram, published in 1802,[12] is as much like a demented spider's web as anything else. Yet the connections of the irregularly branched continuity (such as Jussieu in his later years in particular believed) could continue to suggest some degree of regularity and coherence in the system as a whole. This would especially be true when language intimating an ordered and ordering hierarchy was used. The map or landscape of nature, so frequently mentioned in the literature of the time, was a perfect representation of nature for naturalists such as Alexandre-Henri Gabriel de Cassini. The network he sketched for the Compositae was, he thought, definitely inferior in this respect, being a complex combination of abstractions.[13]

Grids and Parallelisms

For both Lamarck and Jussieu in particular there were other patterns that confirmed that the relationships they were describing were those of nature. It was by no means coincidental that the number of major groupings of plants was the same as that of animals; rather, because the number was the same, it was clear that the naturalist was using the right principles in establishing a truly natural system. Jussieu even thought that the characters used to delimit the major groups of both plants and animals were similar.[14] Thus there was an underlying symmetry when the complete systems of plants and animals were compared; the intrinsic pattern and regularity of God's design[15] was revealed.

The regularities within Lamarck's and Jussieu's classifications can be seen as indicating that systems based on repetitive elements of pattern might provide an intellectually satisfactory understanding of nature. Indeed, a variety of systems derived in this manner (with sets of circular relationships and/or regularly subdivided taxa) were proposed in the first years of the nineteenth century. Such systems combined elements of both continuity and plenitude. They might have allowed for discrete genera and higher taxonomic groups, but any gaps between known taxa were regular—nature (still) did not make unexpected leaps, as even if there was a hiatus between two groups, there certainly was no saltus.[16] Continuity thus existed, although it might be somewhat modified. Because all positions in the system that could exist would be filled, there was also a kind of plenitude.[17]

One of the most famous of such systems was the Quinarian system which William Sharp Macleay (1792–1865) proposed in 1819 to show relationships among "annulose animals"—insects were his specialty.[18] Taxa at any one level formed five circles of relationships, each circle having five taxa, and the circles themselves were joined by "osculating" taxa. There were various patterns of relationship among taxa in this system. Relationships were shown among taxa that occupied the same relative position in each circle of relationship, among taxa in each individual circle, and among osculating taxa and taxa in adjacent circles. The entire group of five circles formed a single unit in one of the circles at a higher level, and so on. Macleay later suggested that this system also occurred in flowering plants and fungi, citing the work of Elias Magnus Fries on fungi and that of Augustin-Pyramus de Candolle on the Cruciferae as evidence.[19] Macleay's ideas were adopted, although in a somewhat modified form, by a number of botanists, and the idea that relationships were circular was commonly mentioned between 1820 and 1860. Thus John Miers observed "the *Humiriaceae* will form one of those osculant relations, existing everywhere in nature, which can only be represented by the circular system, and never by any linear arrangement."[20]

Macleay saw in his system an elegant solution to the problem of comprehending nature:

> I cannot help rejoicing that the strength of this beautiful theory should be so completely brought home to the conviction of every mind, as it must be, by observing the manner in which different persons have respectively stumbled upon it in totally distinct departments of creation. We may all possibly be wrong in part, or even in much of our respective details; but however this may be, it is difficult not to believe that we are grasping at some great truth, which a short lapse of time will perhaps develop in all its beauty, and at length place in the possession of every observer of nature.[21]

But Alfred Russel Wallace[22] thought it a defect that Macleay's system "absolutely places limits on the variety and extent of creation." Similarly, Hugh Strickland[23] saw such regularities as being contradictory to nature; any meaning in the natural system would be found only by induction, not by prediction. And John Lindley similarly stigmatized idealistic systems in general as being formed "by the mere force of reason rather than attending to the facts upon which any system must depend,"[24] since the patterns came from causal laws developed by the philosophical naturalist, rather than from the simple description of nature which Lindley himself favored.

Such defects were not seen in this light by Macleay or proponents of similar systems; as Macleay observed, "the naturalist cannot have a more admirable test of his accuracy, or stronger rein on his fancy, than the parallelisms of analogous groups in contiguous series of affinity."[25] Parallelisms thus confirmed the continuous arrangement formed by the circles of relationships and osculant groups. Macleay suggested that the user of such systems would have his attention focused on different kinds of characters. Some characters provided the relationships that formed the circle, and those that linked the circular group to other such groups—concepts such as analogy, affinity, parallelism and design—were closely interconnected.[26]

These aspects of systems were very important, and the regularities they showed might still suggest the orderly hand of the Creator, with God remaining the guarantor of the natural order. It has also been suggested that circular systems in particular allowed continuity while simultaneously avoiding socially dangerous Lamarckian ideas of transformism and change.[27] But regularities were proof to some that the systems were natural. Robert Wight, writing in 1845 for nonspecialists in India,[28] saw the repetitive and symmetrical nature of circular systems, in particular that of Fries, as bringing order to the natural system; there was an "undeniable symmetry" and "just proportion" that all of nature's works bore to one another. Fries had found general parallels in the pattern of nature that embraced plants and animals and which provided him with a glimpse into the deep structure of nature.[29] There was another advantage in such systems: detailed predictions about both missing taxa and their characteristics could be made.[30] Indeed, there now seemed to be some reason for expecting to make successful predictions. In such systems the laws of nature were evident;[31] the diversity of nature was now reduced to order, to an almost mathematical regularity. For Wight, almost anything that simplified the apparently chaotic variation in nature was useful; he even felt that the floral diagrams in John Lindley's *School Botany* had an importance for botany as great as that of Euclid's diagrams for mathematics, as Lindley's work provided a geometry for botany and gave it the precision of an exact science.[32]

Such systems were ways of depicting interrelationships between organisms,[33] not classifications per se, and so it is not surprising that Gilbert Thomas Burnett claimed that connections between groups became evident when the right system was chosen.[34] As the nineteenth century opened, Augustin Augier almost inadvertently stumbled on a new way of depicting natural relationships, as a tree, that allowed him to see different kinds of relationships. Families on different branches of the tree, but in a similar position, showed the "relationship of analogy," while the "relationship of proximity" occurred between different families on the same branch.[35] Somewhat later, in 1825, Carl Friedrich Philip von Martius made similar distinctions: the Amaranthaceae were related to the Chenopodiaceae "more by *affinitas ex analogia* than *ex propinquitate*. . . . the Amaranthaceae reproduc[ing] ('wiedergeben') the Chenopodiaceae at a higher level, without being very obviously connected by a linear link."[36] And along the same lines, Heinrich Gottlieb Ludwig Reichenbach distinguished between analogy (relationships that were lateral or parallel to the group under consideration) and affinity (relationships in front and behind),[37] while William Allman discovered the very warp and weft of nature (figure 15). Allman presented his diagram in 1835 at a meeting of the British Association for the Advancement of Science. He had rejected a simple linear arrangement, because from his study of relationships "a simple four-lined warp is obtained; it being desired to make (if it might be done) the cross lines of a piece, illustrating, it may be, what modern Botanists call, analogy."[38] He used his arrangement in teaching, but what his students in Dublin made of it is not known.

How patterns in nature and the partitioning of characters interacted is shown in a paper by Sir Edward Thomas Ffrench Bromhead (1781–1855) that appeared in 1836.[39] Bromhead proposed a remarkable arrangement of plants consisting of two parallel series of families. One, the Chenopodeous race, started with the algae and finished with the Gramineae; in the other, the Thymelaeous race, the fungi were at

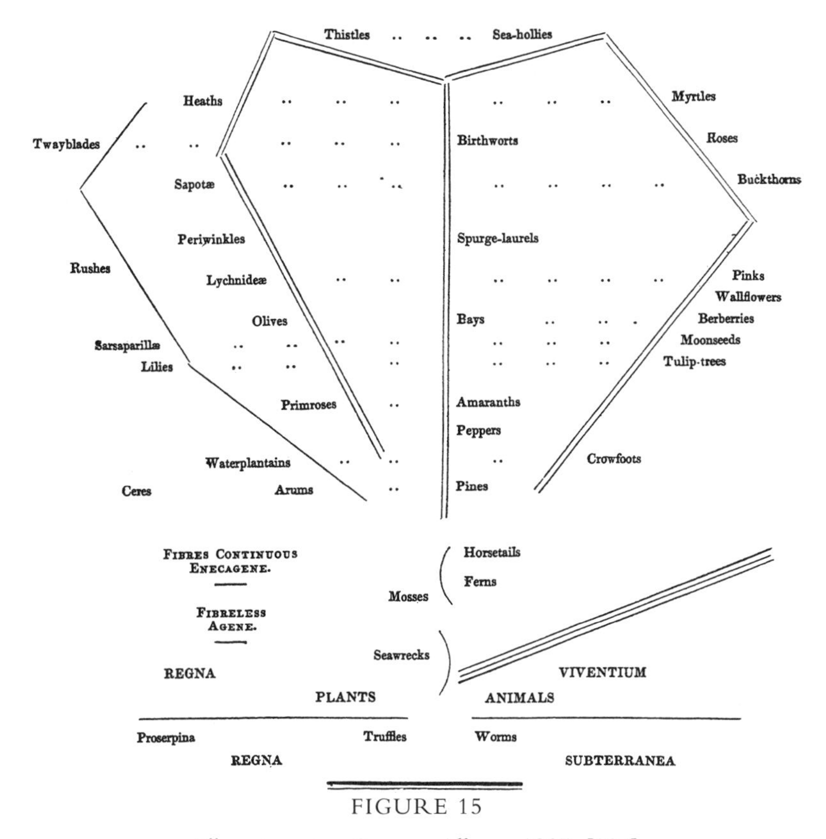

FIGURE 15

Allman's system. Source: Allman 1835: [134].

one end and the Orchidaceae at the other. Each race consisted of several groups of families, each in turn arranged across the page, with the central family of each singled out as being most important (the Chenopodeae and Thymelaeae were the central family of the central group of the two main races). Again, several different kinds of relationships became evident. Some characters were found in several families and helped in assigning a family its correct position in a group, while other characters, parallelisms, were found in only one or a few other families (figure 16). Groups in different races could be linked; thus the Orchideo-Gramineous alliance, the monocotyledons, comprised the terminal members of the two races.[40]

Bromhead was not a major figure in botanical circles of his day,[41] but

Sketch of an Arrangement of the Botanical Families in Natural Groups, Alliances, and Races.

1. Families having any material similarity of structure are in that respect said to have a *Relation.*

2. Related Families lying in the same neighbourhood are said to have an *Affinity.*

3. Families which touch or pass into each other are said to be *Adjacent,* or in *Juxtaposition.*

4. The Numerical superiority of related families, which, (with the aid of juxtaposition and progress of structure) determines the place of each family, and distinguishes the relation of Affinity from those of Parallelism and Correspondenee [see 9, 11,] is called *Joint-affinity.*

5. A *Group* is a collection of Families, having an affinity, and is named from some family contained in it:—Ex. The Orchideous Group.

6. The two great Botanical Divisions are named The Chenopodeous and Thymelaeous *Races.*

7. When the first and last families of a Group, or when the initial and terminating parts of a series of Groups or of the two Botanical Races, correspond and seem to pass into each other, they are called *Re-entering* Groups, &c., and are said to *Circulate:*—Ex. The Chenopodeous, Boragineous, Geraniaceous, Passifloreous, Nymphæaceous Groups.

8. An *Alliance* consists of a Circulating Series of Groups, and is named from the denominating family of one of the Groups:—Ex. Gramineous, Orchideous, Boragineous Alliances.—Those set out in the Table are generally to be considered as artificial divisions provisionally adopted.

9. The Groups and Alliances of the two Races, in the same numerical order from the initial Group, are said to be *Parallel.*

10. As certain Acotyledonous, Gymnospermous, Amentaceous, Apocarpellous, Apetalous, Monopetalous, Gynobasic, Albuminose, Monocotyledonous, and other peculiar *Structures,* are usually in the same parallel, and in the same stage of developement, they are likely to be treated of together, and may be named after the compound parallel Families:—Ex. The Endogenous Families would be The Orchideo-Gramineous Alliances.

11. Successions of Groups related to other successions direct or reverted, in the same or in separate Races, are said to *Correspond.*

12. Characters which give the structure of the organs in order, are said to be *Descriptive;* Characters (positive or negative) which distinguish one Group, &c. from another, are said to be *Differential.* Properties which may in certain cases be substituted for each other in a character, are said to be *Equivalents.*—Ex. Albumen and a Macropodal Embryo, &c.

FIGURE 16

Sir Edward Ffrench Bromhead on parallelisms and other matters. Source: Bromhead 1836: 251.

better-known botanists mentioned similar ideas with approval. In an arrangement of "monopetalous" families proposed almost thirty years later by Louis Édouard Bureau, families were directly related to one another, and intermediate forms established the relationships between these families. As in Bromhead's system, there were more or less linear series with cross connections between them.

> The series of monopetalous families that have a plurilocular ovary and truly axile placentation is less complete. The Bignoniaceae appear to me to form the lowest rank. They correspond quite well with the Acanthaceae in the series with parietal placentation, and I never find below them families in which the ovary contains a limited number of ovules, such as the Boraginaceae or the Verbenaceae.
>
> In sum, it is difficult to establish an entirely complete parallelism between the two series [one of families with axile placentation, the other of families with parietal placentation]; but there are however in both conditions ["termes"] which correspond in an unequivocal manner: thus the Solanaceae, a regular type with valvate-induplicate corolla aestivation, represent the Gentianaceae; the Scrophulariaceae in the strict sense, an irregular type with cochleate aestivation, correspond to the Cyrtandraceae; and the Rhinanthaceae, an irregular and parasitic type, should be placed opposite the Orobanchaceae. One could equally compare the Polemoniaceae on the one hand and the Convolvulaceae on the other; but the Polemoniaceae, which are clearly monopetalous with axile placentation, could not, because of the three loculi of their ovary, be placed in the linear and complete series which extends without interruption from the Rubiaceae to the Bignoniaceae.[42]

Other French botanists invoked parallelisms as an explanation for the distributions of characters evident in their systems. Gaspard Adolphe Chatin (1813–1901) had been criticized by Johann Xaver Robert Caspary (1818–1887) for his classification of some aquatic monocotyledonous plants. Chatin had divided these plants according to their ovule type, placing *Hydrilla* next to *Lagarosiphon*—at the same level, but in different families. He defended his decision in a paper published in 1857, and referred to Isidore Geoffroy Saint-Hilaire's idea of parallelism in justification. The latter had found that no group could be rigorously characterized or defined, and although there was no single continuous series, there were numerous parallel series, and at least some features were involved in cross-relationships in what he called "classification parallélique."[43] Caspary dismissed such ideas, observing that with such an approach, something could be made out of anything.[44] At about the same time, Henri Baillon developed Hugh Algernon Weddell's idea that relationships could be represented by arranging plants on the surface of a conical structure. Again, recurring regularities were the point of interest. Baillon suggested that plants at the same level on the cone might show parallelisms, and there were other distinctive relationships between plants that were on curves representing the bases of conic sections.[45]

The attempt was being made to distinguish between characters showing different kinds of distributions and to explain the significance of the patterns of distribution of these characters; the level in the system at which a particular kind or relationship was evident was not the focus of interest.[46] The similarities responsible for plants being placed adjacent in the series came to be called affinities; those involved in parallelisms, analogies. It is important to note that analogies were not summarily dismissed. Fries defined affinity and analogy with the help of Linnaeus's analogy of a map.[47] Affinity could be compared to lines of longitude and analogy to lines of latitude; latitude *and* longitude were needed to locate both the traveler on the surface of the globe and the naturalist on the map of nature. For Antoine-Laurent de Jussieu and a number of other naturalists, relationships that were what we might call analogical were an integral part of the system as a whole.[48] There was a pattern in the analogies, just as there was a pattern of affinities, a point made by the use of the analogy between the natural system and a geographical grid. Such parallelisms were generally found only within plants, but even as late as 1864 Barthélemy Charles Joseph Dumortier (1798–1878), then president of the Societé Royale de Botanique of Belgium, could see more general parallels between plants and animals.[49]

Hugh Strickland rejected the idea that there was a basic shape to nature, and he also dismissed analogies as being of any use in establishing relationships.[50] He thought that analogies could be distinguished from affinities if there were clear distinctions between groups.[51] Alphonse de Candolle did think that groups were distinct—in theory, at least—but he was not so clear about the distinction. He noted that there were groups in which the "éléments constitutifs" of the members resembled each other more than they resembled others, while affinities were less important resemblances, "often partial, which establish the transition of one group to another."[52] Yet it became commonplace to distinguish between two classes of characters, although "analogy" and "affinity" were frequently not defined,[53] and it might not be clear to the reader why one set of characters betokened affinities and another, analogies. Thus botanists who disliked these kinds of systems, yet who were not clear whether they too believed there were groups in nature or how groups were related, were thus hard pressed to reject in a conclusive fashion as "mere analogies" features that had been used by somebody else to suggest relationships.

Again, theory and practice seemed hard to mesh. As the young Alphonse de Candolle remarked sadly in 1830 (he was then 24), "The genera included in the monograph [of the Campanulaceae] have clear relationships among themselves; but because they are based on an ensemble of characters, this results in points of resemblance that are as diverse and numerous as the characters admitted as the basis of the classification."[54] Four years before, his father had also largely rejected the use of analogies in determining relationships, although the elder Candolle still wanted to be able to show them in his diagrammatic representations of nature. Furthermore, Augustin-Pyramus thought that related groups could be subdivided using the same analogous characters, and this comes very close to using analogous features as evidence that groups are related.[55]

Parallelisms were also important in other contexts. J. D. Hooker suggested that there were parallel series of plants and animals that allowed the ages of plants to be extrapolated from those of animals, the latter having a better fossil record.[56] There were also parallels between the fossil record, the natural system, and ontogeny—Louis Agassiz's threefold parallelism[57]—although the great ichthyologist's ideas were rarely mentioned in botanical literature. And then there are the numerical systems of German naturalists imbued with the ideas of *Naturphilosophie*: for instance, that of Reichenbach, already briefly mentioned, which is based on the number three (although he recognized eight classes).[58] Reichenbach's whole discussion is structured on the distinction between thesis, antithesis, and synthesis (with prothesis and metathesis sometimes thrown in for good measure)—for example: fungi, without chlorophyll ("Phytochlor"), could be the thesis; lichens, with enclosed chlorophyll, the antithesis; green plants, with visible chlorophyll, the synthesis. Reichenbach discussed in parallel variation under "material oder räumliches Typen" and "lebendiges oder zeitliches Typen." The result was doubtless highly satisfying to Reichenbach himself, but his system speedily sank into obscurity.

The Collapse of Order

But for some botanists such distinctions between analogy and affinity were not enough, since the natural system had no "shape," no underlying pattern. The earlier belief that yet-to-be-discovered organisms would provide linkages between groups had been lost. Here the connections between distinct plant forms that could be established by degenerations, adhesions, abortions, and the like may have dispelled some of the formlessness of nature in its entirety. Augustin-Pyramus de Candolle may have understood plant variation through such mechanisms.[59] Certainly, his taxonomic groups, like those of Jussieu, were not monothetic; they could not be defined by one or more characters that were found in all members but nowhere else in the plant kingdom. Unfortunately for Candolle, the existence of indistinct groups was not the necessary consequence of the way members of the living world were related; the cause was exactly the opposite, an embarrassment, and evidence that the botanist no longer understood the world. Although groups were distinct in theory, they usually lacked characters that provided an unambiguous mark of this distinctness; discontinuities were not satisfactorily spanned by the blurred boundaries of the groups, although such continuity was all that many naturalists would allow as evidence of relationship.

To return to a point made before, but which needs reemphasizing here: when naturalists dismissed the idea of a grand *scala naturae*, they often were silent as to what the shape of nature was to be. There are particularly good examples of this in two papers that appeared in the *Magazin Encyclopédique* in 1795. Antoine-Nicolas Duchesne was at pains to demonstrate the demise of the scala, and to this end he provided a magnificent folding "table de Pythagore" with the main groups of organisms down one side, their distinguishing characters across the top. The scala occupied the diagonal, but Duchesne showed that all the squares were filled.[60] All combinations of characters occurred, so the scala did not exist. But in the course of carrying out this exercise, Duchesne reduced nature to a great combinatorial of characters without obvious form or cause, although he suggested that his diagram might be consistent with the Buffonian idea of a group with affinities radiating from the center. In another paper, Charles-Louis L'Héritier de Brutelle also dismissed the single *scala* because each organism showed numerous relationships. However, he was much clearer about the shape of nature. It was like a globe, divided into three portions for the three kingdoms—animals, plants, and minerals—with insensible transitions between them. Then groupings of plants could be arranged on the surface of the globe. But clearly L'Héritier's ideas, too, did not preclude continuity.[61]

Adam Smith discussed a similar point in his *History of Astronomy*, which he wrote toward the end of the eighteenth century. Smith's non-philosopher dealt with coherent events or facts which the mind could easily embrace, continuity would surely be an example here. The philosopher, in contrast, had problems. Nature as a whole "seems to abound with events which appear solitary and incoherent" and which therefore "disturb the easy movement of the imagination,"[62] a good description of the way the natural world appeared early in the nineteenth century. Ultimately even Smith's philosopher wanted to establish continuity, although it would not be at the easy level of superficial phenomena. Augustin-Pyramus de Candolle was such a philosopher, yet in a fundamental sense he did not understand the "shape" of nature. He never could provide more than fragments of the natural system.[63] Despite his talk of the "symmetry" or "type" of plant groups, an understanding of the general pattern of plant relationships at even a descriptive level eluded him. He felt that botanists lagged behind zoologists in understanding the laws of relationships, although they were excellent judges of the systematic value of the characters that linked groups. As Candolle observed, "But, precisely because the task [of understanding the limits of classes] was more difficult in botany, it had been studied with more care; and if zoologists had led botanists as to the general results, they have followed them as to the delicate appreciation of laws and principles."—this was not an uncommon refrain at the time.[64] As to the application of such laws and principles, when everything depended on the "tact" of the botanist, law and principle became barely worthy of the name. Typological thought was not a generally acceptable solution to the problem of relating more or less distinct groups, even if it reduced variation within a family. There were too many ways of forming types, and all were based on principles that to a substantial segment of the systematic community were uncertain or discredited. And parallelisms, too, as evidence of some underlying order in nature, might not be of any help. As early as 1829 Adrian Hardy Haworth had questioned the significance of what he called analogies; they were far too numerous for them to be of any importance.[65]

As the nineteenth century wore on, fewer naturalists believed in continuity (although there are few clear statements as to exactly what they did believe in), so an understanding of the shape of nature as manifest in the systems being proposed was likely to have become an increasing problem. This would be exacerbated by the fact that no two systems were the same, largely because of the problems of weighting characters and distinguishing between analogy and affinity. The encyclopedia, the comforting and closed circle of knowledge, had broken up. Circular systems in general were being discarded; the shape of nature was in flux and, as we shall see in chapter 11, the way that the analogies of nature were used showed this. Broberg recently suggested that an ordered geometry had given way to a less ordered arithmetic in systematic botany, with the provision of "quick, irregular information."[66] But there was little evidence even of the latter, and if adjusting the size of groups indeed made them easier to handle, this would not amount to much if what the groups represented was unclear. Ideas of parallelism, although clearly seductive to some, were often questioned, and botanists who produced systems that were highly organized and implied an understanding of nature were in a decided minority. If relationships really were multidimensional, what could the taxonomic hierarchy mean? How could the infinite be limited by suggesting that nature was organized in any particular way?[67] The sheer complexity of relationships subverted the very possibility of classification: how could hierarchically organized groupings of genera and families be produced if the relationships between them were of such complexity that there was no way to represent them?[68] Nevertheless, these complex relationships were still this side of a totally disordered chaos.[69]

The term *organization* ("organisation" in French) that Jussieu mentioned somewhat imprecisely now takes center stage.[70] It was much in use, in particular in the period 1780–1830. As Charles Bonnet observed, organization was the "most excellent" modification of matter; "The most perfect organization is that which causes the most effects with an equal or smaller number of different parts. Among terrestrial beings, that is the *human body*."[71] For Bonnet, as with many other authors, use of the term was connected with the growing realization that there was a difference between life and nonlife, the former being distinguished in part by its complexity, by the very fact that it was organized.[72]

The meaning of the term *organization* was not always clear, however: did organization refer to some measure of overall complexity, or to a more fundamental property of living beings, or just to anatomy? Jussieu used the word eleven times in the Introduction to his *Genera plantarum*, and he also used it in his other early writings. It meant variously the commonality or perhaps even essence of a family, species, or perhaps even smaller group,[73] or perhaps some measure of general complexity,[74] or even all the characters of the plant;[75] even development might be one of its meanings.[76] But we should remember that Jussieu had asserted that characters were connected with the inside of the plant, its anatomy, organization, or both, so it is difficult to understand all the nuances of meaning that this complex of terms might have conveyed to his contemporaries.[77]

Contemporaries of Jussieu were similarly vague. Thus Adanson[78] seems to have considered "organisation" and "structure interne" to be almost synonymous: "the epidermis is a very delicate membrane, always transparent, colorless, elastic, without any obvious organization, except that one sometimes finds the barely evident pores which function in transpiration and the imbibition of juices." Lamarck[79] used the term for a combination of morphology, development, and physiology. Under the entry "organisation," Jean-Louis-Marie Poiret, Lamarck's collaborator in the *Encyclopédie méthodique*, elaborated on this usage, but Poiret did not mention the word in the body of the entry,[80] and he also used it to refer to anatomy, or perhaps development.[81] For Jaume Saint-Hilaire, the goal of botany was an understanding of the organization of plants. The botanist did not just study the names and virtues of plants; rather, he had a far more noble role to play "in seeking in the organization of plants the laws that govern them, and the degrees of analogy which nature has placed between them."[82] Étienne Pierre Ventenat, like Poiret, managed to discuss the organization of plants in detail without mentioning the word other than in the chapter heading,[83] yet it is clear that organization in Ventenat's sense largely meant the anatomy and physiology of plants. Since it was a physiological phenomenon, there was a parallel between it and the "propriétés" of plants, that is, their medicinal characteristics.[84] Somewhat later Achille Richard noted that in the physiognomy of a plant might be found a particular character that was in agreement with its "organisation intérieure"[85]—a Jussiaean idea. In the *Dictionnaire Classique d'Histoire Naturelle* he had used the word both as a synonym of anatomy and to refer to all characters of a plant.[86]

Whatever meaning ascribed to the word "organisation," its importance is indicated by the frequency with which it was used in titles. Indeed, it had become a catchword that legitimized a naturalist's work.[87] It was a key word, albeit protean in meaning, for workers in the youthful area of biology—both Lamarck and Treviranus linked biology and organization[88]—and it usually had anatomical and physiological connotations, as in Dieterich Georg Kieser's prize-winning essay for the Teylerian Society in Haarlem in 1812.[89] The various shades of meaning it encompassed prefigure many of the questions—evolutionary, genetic, developmental, mechanistic, functional, anatomical, and physiological—that would be addressed in biology later in the nineteenth century.[90] Those who made classifications might use the word, but they rarely spoke to such questions. We shall see in chapter 11 that systematists' concerns did not shift. In the 1880s and 1890s, the trees that represented relationships still show concerns with the level of complexity and parallelisms.[91] Biologists were no longer much interested in such notions.

CHAPTER NINE

Natural History and the Status of Systematics in the Nineteenth Century

Previous chapters examined what naturalists in the late eighteenth and the nineteenth centuries were doing and how they understood nature. It is now time to address the issue of how the whole classificatory endeavor fared during this time, and to assess the status of systematics within what we would now call biology. How did classification fare in comparison with other disciplines that used plants as their subject material?

To answer this question, it is helpful to know something about how studies of organisms were becoming compartmentalized. Natural history in the early eighteenth century encompassed a large part of the study of nature—astronomy, geology, and all of botany and zoology—and classificatory studies were central to it. Such studies were, however, almost excluded from the nascent field of biology in the middle of the nineteenth century; in any general ordering of botanical endeavors then, they would be ranked at the bottom.

Shifting Priorities.

Much has been written on natural history in the later part of the nineteenth century—its status then, and how it developed. Yet we are still some way from an understanding of these problems, and it is easy to see why. George Bentham considered that botany, natural history, and systematics (the making of classifications) were largely synonymous, and that the making of classifications excluded anatomy and also typological and speculative thought.[1] On the other hand, comparative anatomy has been considered to be part of natural history, and speculative thought was included as one of the methods by which natural history attacked its goals of that era. Among these goals was an understanding of evolutionary and phylogenetic problems.[2] A number of Bentham's younger continental colleagues would have agreed with this circumscription and these goals. The question then follows, How do we decide if the study of natural history should be considered to encompass or exclude excursions into idealistic thought? More generally, how can one follow the genealogy of a concept like natural history? Although there is no reason to suppose that natural history should have been regarded in the same way in 1780 as in 1880,[3] how can a historian today decide upon its circumscription—even at one particular time?

There are three main problems: (1) There is no written history of classificatory endeavors for this period, yet it is generally agreed that whatever else is included in natural history, such endeavors form a major part. (2) Botanists and zoologists approached the making of classifications in substantially different ways; one also finds differences in comparing subdisciplines within botany or zoology and among workers in different countries. (3) Natural history may be defined by the subject of its study, the aims of its practitioners, the methods they use, or some combination of the three.

It will be helpful to clarify what naturalists, in particular those interested in plants rather than animals, were doing at the beginning of the nineteenth century. It is also important to look at how natural history toward the end of the nineteenth century was regarded from a zoological point of view and to assess how natural history was seen in the previous century. Paul Farber, in a 1982 study of the practice of ornithology in the second half of the nineteenth century,[4] called attention to four main traditions in natural history: (1) nomenclature and classification, activities most often used to define the discipline as a whole; (2) a somewhat opposing tradition that attempted to construct a complete "natural history" of each species, or a complete description of each species as it existed in its interconnections with the rest of nature; (3)

comparative anatomy; and (4) physiology. On the other hand, Janet Browne[5] sees natural history in the seventeenth (and implicitly in the eighteenth) century as being primarily concerned with identification, nomenclature, and classification, using a descriptive, largely nonanalytical mode of thought.

Hans Sloane captured the spirit of the subject when he observed in 1707 "knowledge of *Natural-History*, being Observations of Matters of Fact, is more certain than most others, and . . . less subject to Mistakes than *Reasonings*, *Hypotheses*, and *Deductions* are."[6] "History" here means description, and it has little to do with time. The purely descriptive sense of the word is confirmed in etymological accounts such as that given in the Oxford English Dictionary, which includes as one definition, "A systematic account (without reference to time) of a set of natural history phenomena, as those connected with a country, some division of nature or group of natural objects, a species of animals or plants. Now *rare*, exc[ept] in Natural History."[7] Natural history was contrasted with civil history, a study of man in which time and change might well be involved.[8] Linnaeus thought that the classification of all natural phenomena was the goal of natural history[9]—which he defined as the naming of plants, animals, and minerals according to the number, shape, position, and proportion of their parts.[10] He then observed, "The first step in wisdom is to know the things themselves; this notion consists in having the true idea of the object; objects are distinguished and known by their methodical classification ['divisione'] and appropriate naming; therefore Classification and Naming will be the foundation of our Science."[11]

How was one to go about distinguishing and knowing such "Matters of Fact"? What we think of as Linnaean natural history[12] had by the middle of the eighteenth century severely restricted what was considered to be relevant to classificatory endeavors, and had largely limited itself to a study of the exterior features of the organism. However, as implied in Linnaeus's enumeration of what a naturalist observed, not all features were considered equal; size, color, and smell were largely ignored.[13] Long gone were the days when attempts were made to include in an organism's natural history everything that had ever been written about it; then the boundary between fact and fiction had been almost completely blurred.[14] Linnaean-style natural history, the "Systemkunde" or classificatory studies of Walter Baron, can be contrasted with "Wissenswerte," detailed studies of individual bodies.[15] "Systemkunde" tended to engross natural history as naturalists struggled to put some order into the masses of materials pouring into European cabinets in the seventeenth and eighteenth centuries. Of course, natural history in its early manifestations included the whole of what we today call natural science, and the term continued to be used in this sense into the early nineteenth century.[16] But by the early nineteenth century, naturalists usually worked either on plants or animals, rarely both, and through the century individual naturalists' areas of expertise gradually became more restricted in scope.

There are other common connotations associated with natural history: for example, that the study of natural history should be edifying, that it should deal with the uses to which man put the objects of the natural world.[17] Dorinda Outram[18] notes that "the ideology of natural history . . . glorified the solitary field observer and direct contact with nature." Description, classification, identification, nomenclature, a non-analytical, "Baconian" (i.e., purely descriptive[19]) mode of thought, contact with nature but not with other naturalists—these are some of the elements to be followed in discussions of natural history in the nineteenth century. Natural history as thus delimited was used to understand aspects of nature's variation[20] and to develop ideas of the natural order; from numerous separate observations, the patterns in the natural world would become evident. As James Larson put it in his discussion of Linnaean classification, "a fundamental faith in the objective existence of a natural hierarchy leads directly to the identification of natural history with inventory."[21]

Descriptive natural history—including observations on the appearances of natural objects separate from one another and when viewed in a state of inactivity, and their subsequent classification—might be contrasted with natural philosophy, which was concerned more with causes, laws, and principles, and the action of bodies on each other.[22] But still, the demarcation was not absolutely clear. In the article "Histoire naturelle" in the *Encyclopédie*, Louis-Jean-Marie Daubenton suggested that botany was one of the main parts of natural history, but he noted that attention to nomenclature had hitherto captured too great a part of the activities of botanists; he wanted to redirect these energies. Botany, he thought, should include a study of the "propriétés" of plants, of the cultivation of useful plants and the destruction of harmful ones, of their analogies, and of all the parts which concerned the vegetable economy (the last was most important).[23] But in the same article he observed, "The naturalist scrutinizes all the products of nature in his own heart; he lifts with care the veil that covers them; he looks at them carefully without daring to handle them roughly; if he has to touch them, it is always with the fear of deforming them; if he has to penetrate to the interior of a body, he divides it only with regret, he disrupts the harmony ['union'] only the better to understand the links, and to have a complete idea of the inner structure as well as the outer form."[24]

This latter characterization can perhaps be dismissed as hyperbole, yet it gives a very different understanding of the naturalist's approach, and by implication that of the proper limits of natural history. The entry "Botany" in Rees's *Cyclopaedia* stressed that botany, synonymous with the natural history of plants, included physiology, and was "not confined to the description and classification of plants, as ignorance has often been pleased to represent it."[25] Physiology had figured prominently in the entry "Plants" in Chambers's encyclopedia, which was copied in the *Encyclopédie*.[26] J. E. Smith included physiology in the purview of natural history in the address he gave at the first meeting of the Linnean Society, although he treated it only cursorily; interestingly, he observed that Duhamel du Monceau and Stephen Hales, both noted for their physiological studies, had "rendered great services to *philosophical* botany."[27] Of course, physiology might be static and descriptive, but it generally entailed studies of the inner workings of plants; it will soon be clear that "philosophical botany" and classificatory studies were largely incompatible.

On the whole, natural history as defined by many authors at the end of the eighteenth century was largely classification and as much a way of looking at nature's productions as anything else. But profound changes had seemed to be in the offing earlier in the century: for example, the change from "Naturbeschriebung," the description of a static and timeless nature, to the discipline of "Naturgeschichte," in which the goal was to detect the temporal succession and relationships of organisms.[28] Lyon and Sloan in 1981 characterized the shift in this way:

> Natural history was no longer to be an inquiry dedicated to the collection of facts, but a science concerned most broadly with the "history of

> nature." The category of "nature" itself, which for seventeenth century science had functioned as an inert, divinely ordered system of bodies in mathematically describable motions, had become a *vital*, almost teleological entity, historically changing, and endowed with self-actuating and self-realizing powers which were presumably sufficient to explain the origin of organic beings and even the apparent miraculous order that had led seventeenth century naturalists into paeans over intelligent design.[29]

Buffon, the main proponent of the new view, integrated a Leibnizian approach into his history of nature. The Marquise du Châtelet (1706–1749), another sympathizer, noted, "There is therefore no actuality, nothing but the order of successive beings."[30] An understanding of nature could be gained only by studying this order of succession. Buffon's vision of species in particular, and even nature in general, as "a constant succession of similar individuals that repeat themselves,"[31] made his natural history thoroughly historical. Despite this, he had little to offer as to how this historical understanding might be achieved. His natural history in practice largely involved the generalization of facts, their linkage by analogies, but no mathematics, and not even anatomy, that is, dissections.[32] He did suggest that fossils might provide milestones along the pathway of time,[33] but not much was made of this suggestion.[34] Hence the study of a temporally based history of nature did not find many adherents at the end of the eighteenth century; ultimately, there was no obvious way of detecting this history.[35]

Prominent naturalists like Alexander von Humboldt (1764–1859) might disparage the atemporal natural history as then practiced as nothing more than "nature study" and call for the development of a "real" natural history,[36] but to no avail; the naturalists responsible for the description of the order of nature had different concerns. William Whewell[37] used the natural history method, the science of classification as it was manifest in botany, as an exemplar of all classificatory disciplines. In botany, classification was guided by pure likeness, and was not even affected by ideas about the functions of parts; it was certainly not a historical discipline. Yet Whewell noted elsewhere that the "idea of natural affinity" was based on the subordination of characters in which functional considerations *were* involved.[38] Whewell also separated discussion of the philosophy of classificatory sciences from the philosophy of biology. Biology (he considered physiology to be a synonym) was the science of life, although he observed that the study of anatomy in zoology tended to put it with physiology.[39] Hence the apparent contradiction; classification of plants and of animals were rather different enterprises, and it was the latter that relied on subordination of characters.

And this is a pointer to where the main tension in the early nineteenth century lies. Gottfried Reinhold Treviranus and Lamarck both rather excluded simply descriptive natural history and inventory in their conceptualization of the new biology.[40] Treviranus's biology was to be a study of "the different forms and phenomena of life, the conditions and laws of their existence as well as the causes that determined them."[41] Along similar lines, he later[42] thought that the doctrine of organization, integral to the new biology, was founded on comparative anatomy, or the systematic distribution of living beings, and on organic chemistry; biology, he thought, *was* dependent on the natural system. Lamarck was less obviously inclined to allow classification a place in biology. He considered plant physiology to be the basis of all other aspects of botany, and in a definition of biology in 1803 he conceived of its proper subject matter as being the origin and the development of the organization of living bodies.[43] At the same time, Augustin-Pyramus de Candolle was circumscribing the extent of botany proper in the introduction to his revision of the *Flore française*. The study of plants involved three rather different kinds of knowledge: plant physics dealt with the living organism, and included both physiology and anatomy, although he thought that neither term was appropriate; botany "strictly speaking" considered plants as distinct beings, and was concerned with their grouping; while applied botany was that part of the study of plants that dealt with their utility to man.[44]

Many of the questions posed by the new biology were simply not of interest to most botanists who constructed classifications, although there was a partial split between the more idealistically inclined and the (for want of a better word) empiricists. Even in Britain the "philosophical naturalist" was interested in laws, not mere descriptions of groups of organisms. These laws were those of idealistic botany, or those regulating the spatio-temporal distribution of organisms.[45] In 1833 John Lindley reported[46] on "the principal questions at present debated in the philosophy of botany" to the British Association for the Advancement of Science. These questions included organography, pathology, and physiology, but not terminology, phytography, or taxonomy. Although "philosophical naturalists" by no means made up the whole community, they were the harbingers of the new order, and insofar as the word "biology" was used in the first half of the nineteenth century, it was usually linked to physiological studies.[47]

Turning to look at what practitioners of what might be called ahistorical natural history, classification, or botany in the strict sense considered to be relevant, one finds that this pursuit did not encompass all the botany that had been excluded from biology. Although it might claim to be natural history, it excluded a significant part of old-style natural history. Henri Daudin[48] suggested that botany differed from much of zoology (but not entomology) in that decisions about the limits of species could be tested directly against the organism in the environment; plants could be compared directly in the field. Such testing was done most obviously in the herborizations that were so popular in the latter part of the eighteenth century (hence the meaning of the verb "to botanize," with its connotation of activity in the field[49]), but this link between naming the plant and field observation did not persist. Buffon's descriptions included that of the whole organism and of its habits, but field studies in general gave rise to what we sometimes consider today to be natural history—the rather unsystematic and unanalytic discipline of field observation.[50] Field studies were certainly no integral part of the classificatory endeavor, whether in botany or zoology. Field experience was not seen to be important or even relevant when so much had still to be described for the first time and assigned its place in general systems of classification. Augustin-Pyramus de Candolle turned down a chance of going on a voyage around the world commanded by Baudin, and other eminent naturalists made similar decisions.[51] As Alphonse de Candolle worked through the difficult genus *Quercus*, he saw collections as providing the naturalist with a source of information that was incomparably superior to any that could be gained by studying the plant in the field. The biases of different collectors caused by their "particular habits or theoretical ideas" would be canceled out, and although there were some kinds of research which it was difficult to carry out on dried material, the overall balance overwhelmingly favored herbarium over field studies: "I do not hesitate to say that the comparison of related forms which can grow over vast territories is better done in rich herbaria than in herborizing."[52]

It has been suggested that there was a split between the field naturalists with their broader approach and closer contact with the realities of nature and the literally and metaphorically more closely focused work of the closet naturalists.[53] There was indeed such a split, but during the first seventy years of the nineteenth century it was closet naturalists who occupied most of the paid positions at universities, herbaria, and academies, and it was their approach that prevailed. However much field naturalists might extol their own work, they were unable to provide what was considered properly comparative information,[54] and they were largely considered to be amateurs and therefore of inferior scientific status.[55] As Elizabeth Keeney notes of North American botany then, amateurs as collectors "were the legs, hands, and eyes of individuals and institutions,"[56] but the naming of new plants was the job of the naturalist in his closet. Accounts of the activities of other Victorian naturalists bear this out; even when they utilized the then-popular microscope, their approach was indiscriminate and their observations largely restricted to the outsides of organisms.[57] Field studies (those in Britain in the tradition of Gilbert White) produced facts that before 1859 did not seem to be useful for anything in particular, while the new species that the closet naturalist described could at least be integrated into the Linnaean or Jussiaean systems.[58] Natural history was sometimes equated with classification, but the latter was not directly linked to field activities.[59]

The hegemony within botanical systematics exerted by those who worked in large collections was compounded by what can only be called an imperial attitude toward those in the colonies who wanted to describe species based on their own observations. This was very clearly shown by J. D. Hooker, who declared that the colonial, with his limited access to herbarium specimens, would not be able to understand the true pattern of variation; such could be seen only by those who worked at large herbaria.[60] But Hooker's attitude was not simply imperial, it was also connected with the ongoing struggle between those like himself who recognized broad species concepts and those who preferred narrowly delimited species and who often preferred to study phenomena such as hybridization. It was these latter botanists who were more inclined to make observations in the field; they were perhaps particularly numerous on parts of mainland Europe.[61] It is in the context of such concerns that Alphonse de Candolle claimed that his study of oaks allowed him to deal with the problem of the widely ranging and variable species in which he (like Hooker) believed; his work was based on copious herbarium material, while a field botanist could study variation over only part of the range of the species.

It was thus the museum worker alone who could make the broadly comparative observations on which sound generalizations could later be based. Both botanists and zoologists restricted their focus on the organism to the parts that were preserved in the cabinet, but they compensated (or in their eyes more than compensated) by being able to study so many different kinds of organisms[62]—and there were zoological and botanical gardens in which the spoils of colonial exploration grew. General comparison provided the realities and truths of the museum workers who dominated the discipline of systematics. The naturalist interested in classification retired to his collections and the description of new species and genera, and it was these collections that provided an important part of his claim to be a serious student of nature. Augustin-Pyramus de Candolle bought the herbarium of Charles-Louis L'Héritier de Brutelle, and he revealingly observed that it was this acquisition that made him a botanist (in the narrow sense defined above) in the eyes of his contemporaries: "It began to rank me among the botanists; perhaps it also contributed to distance me from plant physiology, and throw me into the description and classification of plants."[63] Only extensive collections could contain the thousands of individuals needed to make the general comparisons involved in establishing the natural order. And of course the sheer magnitude of the closet naturalists' task should not be forgotten. The number of species known increased rapidly throughout the period, the number of specimens increasing still more, and this great increase helped change the institutional base of the discipline. Thus the number of species of plants known increased from just about 5900 in the first edition of the *Species plantarum* to an estimated 97,205 in Bentham and Hooker's *Genera plantarum* a century and a quarter later.[64] An individual would have to be extremely rich to maintain the huge herbarium required for classificatory studies.

To sum up: as the nineteenth century began, botany proper and classification were near synonyms; they excluded physiology and certain kinds of anatomy and were distinct from philosophical botany.[65] And classification was seen by many naturalists themselves as an endeavor that owed little to theoretical considerations, as I shall further discuss in the next chapter. Disciplines like physiology and even anatomy were not effectively integrated with the classification of plants in the early nineteenth century, and they were often separated in introductory textbooks.[66] In his account of natural history in the *Dictionnaire Classique d'Histoire Naturelle*, Jean Baptiste Bory de Saint-Vincent detailed its subdivision into mineralogy (further subdivided into geology and crystallography) and botany and zoology, both in turn subdivided into physiology and anatomy.[67] Although anatomy was mentioned elsewhere in the encyclopedia, with zoological anatomy being particularly well discussed, it is surely not coincidental that physiology was wholly absent. The situation remained largely unchanged throughout the century. George Bentham, in his presidential address to the Linnean Society of London in 1865, suggested that botany and classification, largely synonymous, were the remains of natural history after it had divided into geology, mineralogy, and the various subdisciplines of the life sciences in the late eighteenth and early nineteenth centuries.[68] Field studies found no place in classification, being part of what in England was called natural history, but if natural history did mean the same thing as classification, field studies still were excluded from natural history so circumscribed.

The Status of Systematics in the Nineteenth Century

Those who, in the nineteenth century, described plants and constructed classifications can thus be thought of as carrying on an appreciable element of the tradition of eighteenth-century natural history, which persisted largely unchanged in the nineteenth century in terms of both its goals and methodology. In botany the outside of the organism remained the focus of attention, and botanists in 1780 and 1880 were making largely similar observations in their normal work (details of the ovules and embryos were, however, commonly added by the early 1800s—see figure 7 in chapter 5). By the time Jussieu resumed publication in 1801, Cuvier, Lamarck, Geoffroy Saint-Hilaire, and others were beginning an exciting transformation of zoological classification. But in botany, it was studies outside of naming and classification that soon came to generate the most interest. Classification in 1789 was certainly not a backwater of science; indeed it was close to flourishing. But the years after were for the most part not good ones for the discipline.[69]

Linnaeus had made the whole discipline of classification (and the classification of plants in particular) of central interest to chemists and crystallographers, as well as to those who studied organisms, but this attitude did not last. Lindley observed in 1835 that the followers of Linnaeus had insisted that the whole object of botany was naming and describing species and that Linnaeus had become "a positive incubus upon science."[70] Linnaean systematics attracted the amateur,[71] and in any event, the Linnean Society of London was seen as a bastion of orthodoxy—worse, orthodoxy in botany—and had become associated in people's minds with scientifically peripheral interests such as horticulture. It was full of "Gooseberry Societies . . . and prize-auricula-men"; botany itself was a trivial pursuit, demanding "little exertion of the higher powers of the intellect."[72] In 1864 an anonymous reviewer of Daniel Oliver's *Lessons in Elementary Botany* observed that there had been only six botanical papers in the Philosophical Transactions of the Royal Society that century, and he noted "the indifference of the scholarly classes, especially the clergy (who still so largely sway the education of the people), to the natural sciences."[73] Perhaps not surprisingly, within the Linnean Society itself, most of the activists were zoologists, and they formed the entomological and zoological clubs which finally became separate societies or developed quite independently of the Linnean.[74] Clearly, all had not been well in botany for some time.[75]

Across the Channel, the situation was similar. George Bentham's letters and addresses are full of comments about the depressed state of systematic botany; it was other areas of botany that were developing. Bentham traveled extensively on the Continent and had numerous contacts among the botanists there, so his comments are not to be ignored.[76] Camille Limoges documents clearly the same pattern for the Muséum national d'Histoire naturelle in Paris, although he suggests that the decline did not begin until around 1840, with descriptive sciences then giving way to experimental. The end result was similar in that the collections there had become almost marginalized by the end of the century.[77]

But the issue was not simply the reputation of botany among the sciences, but the nature of the support it did have. In Paris, although Cuvier's zoological lectures were not nearly as well attended as those given by René Louiche Desfontaines on botany and by Gerrit van Spaendonck on natural history drawing, many women went to the lectures on these latter subjects.[78] In the German-speaking world, too, there was also the same association of women with botany in the late eighteenth and early nineteenth century.[79] Indeed, much of the popularity of botany in this period was among women, to the extent that it could be dubbed "the female science."[80] In America William James Beal observed that as late as 1850 botany was considered to be a "pleasant and proper pastime for young school girls."[81] He thought that the times were changing then, but they were not. Even in 1887 J. F. A. Adams[82] was at pains to point out that botany *was* suitable for young men. It is surely significant that in England botany was the science chosen to show that education did not drive women insane; the results must have seemed very satisfactory, since the girls on whom this experiment was tried were "unusually intelligent, orderly and neat in their appearance and in special request as nursery maids."[83] Phrenological studies might have predicted this. The botanist and phrenologist Hewett Cottrell Watson (1804–1881) noted in 1833 that botanical heads were usually smaller in areas where the higher intellectual activities were carried out. He pursued his argument to what he saw as its conclusion: "In accordance with such general inferiority in size, we see females, and other young persons, make more progress in botany than in the other sciences first named [geology, moral philosophy, political economy]."[84]

John Lindley was nevertheless optimistic. As he wrote in 1835 in the introduction to his *Synopsis of the British Flora*:

> In the year 1820, in Great Britain, many parts of Germany, Denmark, Spain, Portugal, and Italy, Botany [classificatory natural history] was bedridden and paralytic; and so it still remains, except in Great Britain and Germany, and in the few places in those other countries in which its spirit has been renovated by the adoption of modern views and opinions. Could any one, let me ask, have got together in London, in the year 1820, a public class of Botany consisting of fifty students? Could they even in 1825? And yet the season that has just passed witnessed classes of two hundred students, attracted by the mere interest of the subject, and not brought together by academical compulsion.[85]

And at about the same time an anonymous reviewer used the flourishing state of botany in Britain to show that there, at least, the Declinists' worries about the state of science in Britain were misplaced.[86] But such optimism was unwarranted, because simply counting heads or noting the number of floristic studies being produced might not count for much among those whose opinions were deemed to matter.[87]

By the middle of the nineteenth century, taxonomists themselves did not find all was well within their discipline. In 1864 one anonymous reviewer, perhaps with his tongue in his cheek, suggested that "Human bliss is popularly supposed to be represented by a professional scientific life, by a Professor of Botany, or the Director of a botanical garden, or museum."[88] However, that reviewer found botanical studies in general languishing in Britain. The introduction of the natural system had solved nothing; indeed, its introduction was not even an unmixed blessing. Somewhat earlier in 1835 the rather eccentric Louis Henri Lefébure had complained that the Jussiaean system was always called "the" natural method, despite its perpetual changes, and plant descriptions were full of expressions such as "either . . . or," "sometimes," "rarely," "almost always," "sometimes . . . at other times."[89] Although he was certainly not a member of the botanical establishment, Lefébure was largely accurate in his criticisms: there was nothing stable. Worse, there was little tangible in the natural system. Between 1647 and 1775 some twenty-five new systems were proposed.[90] It was in this context that Jussieu himself had warned about the "unstable principles and arbitrary laws" of the systems current in 1789,[91] and he thought that his method would provide this much-needed stability. But after 1789 there was no slackening of the pace of production of new systems, some wildly different from any predecessors: Twenty-four new systems were proposed between 1773 and 1858.[92] In 1869 Dominique Clos bewailed the state of systematic botany, but even though all seemed to him to be in chaos, members of the Société Botanique de France preferred to discuss nomenclatural issues.[93]

Certainly, it was difficult to develop an understanding of nature through a study of "the" natural system. And as other parts of botany in the broad sense developed, J. D. Hooker and Thomas Thomson noted ruefully in 1855 how the better botanists were being attracted away from systematics. "Another result of the depreciated state of systematic botany is, that intelligent students, being repelled by the puerilities which they everywhere encounter, and which impede their progress, turn their attention to physiology before they have acquired even the rudiments of classification, or an elementary practical acquaintance with the characters of the natural orders of plants."[94]

A further factor to be noted in assessing the status of systematic botany is the distinctions drawn between the professionals and amateurs. In taxonomy and in field sciences in general (i.e., both closet and field natural history), professionals and amateurs were not far apart,[95] with field natural historians in particular sometimes positively relishing their status as nonprofessionals. This would not work to the advantage of taxonomy at a time when the general tendency in science was toward increasing professionalism in the sense that sharp lines were being drawn between the serious student, usually paid, and amateurs, unpaid and usually with another job that occupied most of their time. But in systematic botany, all that was needed for somebody to contribute names was access to collections, knowledge of basic botany and whatever rules of nomenclature there were, and perhaps the ability to write botanical Latin, but little else.[96] Indeed, in a retrospective address published in 1913, the mycologist William Gilson Farlow (1844–1919) noted that taxonomic studies were particularly suited to situations that existed until about 1850, or the 1880s in some countries, where there were relatively few full-time positions.[97] Certainly, a specialized education and the knowledge of complicated experimental techniques was no part of the discipline. This is not to say that there were no professional controls at all—witness the fate of the hundreds of thousands of names published in Michel Gandoger's *Flora europae* that came out in twenty-seven volumes between 1883 and 1891.[98] Rather, it means that the boundary line between professional and amateur remained blurred.

In 1887 Farlow, echoing Augustin-Pyramus de Candolle earlier in the century, had suggested that systematics was being professionalized because of the need for the taxonomist to have access to extensive collections with which he could carry out broad, comparative studies when naming plants. The way in which plants were studied was, however, not changing. Yet, in what seems suspiciously like self-justification (he was happily ensconced in collections at Harvard University), Farlow suggested that the new histological and developmental studies were most suited for nonprofessional botanists.[99] Hardly surprisingly, most of Farlow's more prominent students—professionals all—made their names carrying out just these studies.[100] At the end of the century Sir Douglas Galton reviewed science as a whole in his Presidential Address to the British Association for the Advancement of Science, and all he could see of interest in botany was a study of its fundamental physiological principles.[101] Ideas and techniques did matter, and although Mathias Schleiden and Theodor Schwann may have been wrong in the details of their cell theory,[102] it provided a unifying principle for all organisms; there is no doubt about the impact of the ideas of Schleiden in particular and of his textbook, *Grundzüge der wissenschaftlichen Botanik*, which appeared in 1842. There Schleiden paid little attention to systematics, noting that "Botany . . . is an experimental science and indeed the science of the configuration ["Gestaltung"] of matter under the form of the plant."[103] In Germany by the middle of the century, the new "wissenschaftliche" botany was coalescing around microscopical and developmental studies, the cell theory, and the methodological introduction to Schleiden's *Grundzüge*. Meanwhile, flora writers and systematists were ridiculed as being mere gatherers of so much hay.[104] Papers published in the *Jahrbücher für wissenschaftliche Botanik* (which started in 1858 under the editorship of Nathanael Pringsheim) amount to a glorification of the microscope. Carl Wilhelm Nägeli, professor of botany in Munich, contrasted morphology (which he considered to entail microscopic studies), physiology, experiment, geology, and paleontology with the outmoded and closed Linnaean and Cuvieran approach to nature.[105]

In 1880 the first edition of Charles Bessey's *Botany for High Schools and Colleges*, based largely on Julius Sachs's *Lehrbuch der Botanik*, was published. This helped to introduce "German scientific botany," as Asa Gray termed it in his review of Bessey's book, to North American audiences; this botany was largely plant anatomy and development and the study of lower cryptogams, but it had practically nothing to say about systematics.[106] Experimental and physiological studies of plant distribution also developed at the end of the century; this biological plant geography ignored floristic plant geography and paid no attention to any idea of an inventory of nature.[107] At the end of the century the institutional base of comprehensive, worldwide collections (by then largely in the hands of governments or universities) that had made possible the then-accepted style of systematic botany now came to be seen as retarding the development of a revitalized systematics. As governmental institutions, these collections had been pressed into imperial concerns such as cataloging the natural products of empire; as parts of universities, they represented a nineteenth-century research tool that was too cumbersome to discard easily.[108]

Even in the context of a more descriptive natural history, the most exciting findings seemed to be coming from cryptogamic botany. One of the advantages of algal studies was that anatomy, physiology, and systematics were necessarily intimately linked, and the prestigious word "wissenschaftliche" could legitimately adorn the page of a systematic study.[109] In 1868, in a survey of the botanical studies carried out in France during the preceding twenty-five years, Adolphe Brongniart[110] devoted almost twice as much space to studies on cryptogams as to those on phanerogams. Wilhelm Hofmeister's monumental work, which appeared between 1847 and 1862, was predominantly on cryptogams and culminated in his establishment of the phenomenon of alternation of generations in all land plants, but it had little effect on those working on the systematics of flowering plants.[111]

Discussion

Complex, indeed, is any answer to the question of what was happening to natural history in the first part of the nineteenth century. Natural history entered the nineteenth century as a largely ahistorical, descriptive, and classificatory discipline. In botany natural history was even more focused on the outside of the organism than it was in much of zoology. The attempt by Buffon to historicize the discipline must be deemed largely a failure,[112] and his effort certainly did not affect most of the people about whom I have been writing.[113] Biology as it was conceptualized at the beginning of the century was centered on developing an understanding of life, with physiological and microscope-based studies soon becoming the approaches that attracted new students. Although the word "biology" itself did not catch on until the end of the century, the most interesting work on both plants and animals was seen as being in this general area. This shift of interest, coupled with the failure of the natural method to provide a stable and readily understood classificatory and comparative framework, along with the association of classificatory botany with women, the instruction of children, and horticulture, meant that classificatory botany was held in low esteem even as the century began—a position that was only to worsen.

Richard Burkhardt observed that "it cannot be doubted that the basic concern of the eighteenth-century naturalist differed from that of the nineteenth-century biologist."[114] I agree; but I have also suggested that one of the major concerns of eighteenth-century naturalists was classificatory, and that approaches to classification in the eighteenth and nineteenth centuries were similar. An element of natural history in the early eighteenth century, "Naturbeschriebung," involved largely ahistorical description and comparison, and this approach persisted in the nineteenth century. This ahistorical tendency was particularly true of those who worked in large collections,[115] describing and classifying the immense numbers of new specimens being collected. The sheer magnitude of the task meant that field studies could no longer provide the generally comparative data that were required. For the same reason the institutional structure of the discipline was changing.[116] By the middle of the nineteenth century, moderately well-off botanists like Augustin-Pyramus de Candolle could no longer hope to amass a collection that would serve as an adequate basis for their classificatory endeavors; collections became consolidated, many (like that of George Bentham) becoming associated with the state and subject to colonial concerns, including inventories of colonial resources.

To the extent that natural history is seen as being a basically ahistorical classificatory endeavor, then in botany the history of natural history is in considerable part a history of classification. Natural history of the seventeenth century perforce largely focused on the outside of the organism—Grew and Leeuwenhoek notwithstanding—and it was basically synthetic in the sense of chaining groups; this is close to much, but by no means all, systematic botany in the nineteenth century. To the extent that systematists were against analytic (deductive) and speculative thought, they were part of the same tradition. In general, after the outlines of the natural system were laid down, new species, genera and families were fitted into these outlines. But we surely know enough now to reject the idea that classification construction is divorced from theory; nineteenth-century classifications cannot be thought of as being "Baconian." In the nineteenth century, natural history and botany in the strict sense, initially almost synonymous, tended to drift apart: botany became equated with comparative classifications made by professionals, natural history more with unsystematic field studies by amateurs. Elizabeth Keeney recently observed[117] that amateur botanical activity in the early nineteenth century, with its natural history focus on botanizing, closely resembled taxonomic practice today. However, this is not what the leaders of professional systematic botany at the end of the nineteenth century (and since) were doing. To the extent that old-style natural history was field-based and led to the description of individual species in great detail,[118] that was no part of the work of the people considered in this chapter.

In botany in the broad sense, the shift from classification to physiology and anatomy was a shift to two areas of research that were often only reluctantly considered to be an integral part of natural history (old style).[119] Biology, into which such studies would become integrated, dealt with the properties of life and with the realization that the distinction between life and nonlife was what made organisms unique.[120] But it should be remembered that classification and comparative anatomy had largely become one endeavor in zoology.[121] More speculative thought was considered an integral part of systematic work in parts of both the botanical and especially the zoological community, and historical studies, "Naturgeschichte," appealed to a few naturalists there (although perhaps particularly to those whose primary concerns were not classification). Many systematists of both persuasions were interested in biogeographical patterns,[122] and the way in which these and other studies were often carried out can also be considered within the purview of the old-style natural history.[123]

Toward the end of the century, "biology" and "zoology" became (and have in part remained) almost synonymous, describing physiological and developmental studies that became more and more based on experiments and manipulations of nature and which invoked causal analytical ways of thinking;[124] comparable studies in botany were called "scientific botany."[125] Asa Gray observed in 1879 in the seventh edition of his *Botanical Textbook* that biology and physiology had been equivalent earlier in the century, although biology had recently come to include natural history, that is, classificatory studies, as well.[126] So although biology might eventually include botany (both systematic botany and natural history *sensu stricto*), classification studies remained low in the hierarchy of the life sciences,[127] and the very place of the large collections that had been their basis throughout the nineteenth century became debatable. To the extent that systematics was able to rejuvenate itself it was largely outside collections and conventional classificatory studies, borrowing both from experimental biology and from that part of old-style natural history that had been largely ignored by taxonomists. The initial locus of this renewal was in countries like Austria, Switzerland, and Germany, and then the United States. But in the opinion of many from academia, natural history in the late nineteenth century was that part of late eighteenth-century natural history that was not of academic interest.

So in order to talk about natural history in this era, at the very least one must qualify the use of the word. There is no monolithic natural history in the nineteenth century. As with the ideas that grounded the doctrine of plenitude, or continuity, or continuity in the context of classification, the elements of natural history have to be traced independently throughout this period as they separate, rejoin, fuse with other elements, and even become extinct.

CHAPTER TEN

Stability of Classifications and Its Causes

Pierre Flourens, in his *éloge* of Augustin-Pyramus de Candolle read at the Académie in 1842, noted that two problems had occupied naturalists in the eighteenth century: the problem of method and the problem of revolutions of the earth. Flourens thought that the first problem had been solved by Antoine-Laurent de Jussieu and Cuvier, and the second by Cuvier alone. Flourens suggested that Candolle's own contributions had been more in organography, in understanding the changes of plant form and studying details of the organization of beings[1]—the important questions of the early nineteenth century. He implied that Candolle was just a follower of Cuvier and thus of Jussieu when it came to detecting relationships. Indeed, since the publication of the *Genera plantarum* in 1789 and the adoption of the natural method or system as the main goal for naturalists, subsequent workers have been seen as simply improving on the outline Jussieu proposed. The inimitable Ferdinand von Mueller, writing in 1882, noted that the natural system of Jussieu and Candolle was "in its main features so genuinously natural"[2] that he was proposing only a few modifications. In 1954 Adrien Davy de Virville could even claim (stretching the truth) that none of Jussieu's families had been reduced to synonymy.[3] But I have suggested that there are substantial differences in how Jussieu and Candolle saw the living world—for Jussieu the living world was a continuum, for Candolle groups were discrete.

Simply comparing the methodological prescriptions of Candolle with those of Jussieu and noting differences, some of which seem to be connected with their different views of nature, it is curious that the ideas of one or the other did not become dominant. Rather, the two persisted side by side. I still find it strange that there is no discussion in the nineteenth-century literature of the extent to which ideas of continuity are embedded in and integral to Jussieu's theory and practice. And there has been little discussion since; plenitude, continuity, and the *scala naturae* have been treated as historical curiosities of no relevance to a systematist's work or the natural system as it developed in 1789 and subsequently.[4]

But George Bentham may have put his finger on a reason for the absence of discussion. In his Anniversary Address as President of the Linnean Society of London in 1866 he compared the different ways in which the natural system had replaced the Linnaean system in botany and zoology.

> In zoology it appears to me that a similar change [replacement of artificial systems by natural] has been effected, not by the nominal substitution of one specific system by another, but by a gradual recognition of the principle which I believe now governs the study of all branches of natural science, but which was laid down by Linnaeus for genera only. This is, that groups of beings of every degree, from the primary class to the lowest race, are not to be limited by one character in outward form taken *à priori* as essential, but by the comparative study of every peculiarity in outward form, internal structure, constitution, and habit of life. One consequence has been the great development given in recent years to the study of animal and vegetable anatomy and physiology, and to the biological history of the individual, the species, the genus, or the class; whilst systematics, zoological and botanical, once supposed to make up almost the whole of the science, is now, by a species of reaction, often treated with contempt.[5]

The implication is that the change in botany just happened, or perhaps that there really was no change at all, while zoology evinced a greater understanding of the principles involved. Indeed, a greater variety of characters were being used in zoology than in botany. There is considerable irony here. Bentham argued against the introduction of theory in general and typological thinking in particular into botany. He claimed, furthermore, that relationships could be satisfactorily delimited if outward form alone were examined, with anatomy having no part to play in this. A clear opposition between anatomy, physiology, and the "biological history" of organisms on the one hand and systematics on the other is evident in Bentham's Anniversary Address. Karl Figlio, nevertheless, observes, "In the life sciences [in the early nineteenth century in both Britain and France], the conceptual center around which unified views clustered was the interest in organization, with comparative anatomy as its proper science."[6] An influential group of botanists explicitly rejected botanical developments of the approach that was transforming zoology, or, more broadly, studies on organisms in general.

I now want to follow up Bentham's feeling that change "just happened" within systematics and revisit the suggestion I made in chapter 7 that the practice of many systematists in the first half of the nineteenth century owed much to an implicit assumption that nature was continuous. The continuity-in-practice of these systematists is practical stasis: systematic theory remained largely unchanged, and the limits of groups also remained largely unchanged. There seems to be something of a paradox here, because many new systems were proposed in the nineteenth century[7] but only a few—for example, those of Jussieu, the Candolles, and Bentham and Hooker—were in common use.

Here I discuss some factors associated with this stasis, although they do not entirely account for it. I will suggest that a distrust of theory, a system of instruction that is similar to an apprenticeship, and a tendency to look to past masters of the discipline for justification are interconnected factors leading to stasis. Another factor promoting stability is that the major users of names, the most obvious product of systematists, are not botanists. Finally, I will suggest that the way in which issues were presented to the reader in the early nineteenth century may have made matters bearing on the theory and practice of systematics less than obvious.[8]

Theory, Classification, and the Users of Names

Systematics in Bentham's time and since is often seen as being a preeminently descriptive and even empirical discipline;[9] it is simply something one *does*.[10] Systematists as a group, although perhaps particularly those in England and to a somewhat lesser extent in France,[11] have tended to regard anything smacking of theory with deep suspicion. But in the first part of the nineteenth century many naturalists were reacting against what they saw as the excesses of the speculative thought of the *Naturphilosophen*, and against "the spirit of system."[12] The anti- or atheoretical sentiments so often expressed must be interpreted accordingly. Thus typological thought seemed problematical for many. Actual types could not usually be seen and were "hypothetical constructions," so a meaning was being read into nature which was not immediately apparent; typological and idealistic thought were often associated.[13]

Idealistic systems in general seemed to flout the proper inductive procedure by which classifications were built up from a mass of painstakingly assembled observational data. In 1828 Augustin-Pyramus de Candolle suggested that the conflict between the "theoretical *a priori* and simple describers"[14] had largely resolved itself in favor of the latter. This was of course not strictly true, since even in France what may loosely be called the Montpellier school of systematics (the development of this school owed much to Candolle himself), with its emphasis on types and symmetry, was flourishing—not to mention the more idealistic approach widespread in Germany and Sweden in particular. For Candolle, laws arrived at *a priori* lacked the sanction of observation, although in his *Organographie végétale* he observed that facts alone were insufficient and that the botanist should construct partial laws and compare them with other partial laws, as in the physical sciences; this was the only procedure that led to general truth.[15]

He neatly exposed the problem facing systematics, although he was unaware of its magnitude: "[T]hese facts [which I have discovered], stripped of any theoretical idea, will be allowed with more confidence by those who are frightened by new theories, as if to reject them is something else than to keep an old theory, usually accepted without examination."[16] The problem is that an absence of articulated theory does not mean that there is no theory.

Theoretical studies were not an integral part of the training of systematists through the nineteenth century; indeed, theory seems to have been excluded until quite recently. Nor did theory evoke much discussion in the systematic community in botany (other than occasioning negative comments). In 1875 Julius von Sachs commented on the fact that with very few exceptions there had been little discussion of the principles of the natural system after the publication of Candolle's *Théorie élémentaire*, the first edition of which had appeared sixty years before.[17] Both the immediate success and the persistence of the Linnaean system was due in part to its appeal to a considerable number of people who were distrustful of theory;[18] an anti-theoretical attitude among naturalists appears to have been evident by the beginning of the nineteenth century. Cuvier suggested in a letter to Pfaff in April 1790 that there was an unfortunate division in botany into "raisonnée et en pratique." He wondered how the two could be separated; perhaps practical botanists did not think.[19] Augustin-Pyramus de Candolle[20] referred to a similar problem when he noted the deleterious effect that people who saw only "practical applications," or who were amateurs, had on the discipline. Indeed, systematists have almost seemed to delight in giving themselves the qualification that they are "practical" or "practicing" (this is still the case). Sir J. E. Smith in 1817 characterized botany as being a preeminently practical discipline, observing with approval that "the study of botany in England, has, for a long period, been almost entirely practical."[21] And so it largely remained, but not only in England.

Moreover, the principle of the subordination of characters—nominally the most important development in systematics of this period—was in French botany a way of seeing and doing things, a practice, rather than the basis of a law or theory.[22] Along the same lines, Alphonse de Candolle described botany (systematics) as "a science of pure observation,"[23] a description that would have been applauded by many of his English-speaking colleagues in particular. Cuvier himself made much of his own anti-speculative position and was strongly seconded by Bentham.[24] J. D. Hooker conveyed a similar anti-theoretical position when he told colonial botanists to leave the business of describing new species to botanists at Kew and other large institutions.[25] John Stevens Henslow described, in very mundane terms, his vision of botany in a paper with the forward-looking title, "On the requisites necessary for the advance of botany":

> To obtain a knowledge of a science of observation, like botany, we need make very little more exertion at first than is required for adapting a chosen set of terms to certain appearances of which the eye takes cognisance, and when this has been attained, all the rest is very much

> like reading a book after we have learned to spell, where every page affords a fresh field of intellectual enjoyment.[26]

Admittedly, he was writing for the rather general audience of the *Magazine of Zoology and Botany*, but the message was that nature was basically an open book in which the meaning was clear to anybody who had mastered its essential vocabulary. (A question immediately presents itself: under such circumstances, what story did the natural system tell? I shall return to this in the epilogue.)

M. P. Winsor lays the blame for the failure of Louis Agassiz's hopes for zoological systematics on the disinclination of systematists for theory: "Agassiz's hope to raise the status of taxonomy by infusing it with theoretical issues had failed. Instead, Darwin's prophecy, 'Systematists will be able to pursue their labours as at present,' was dismally fulfilled. The widely held faith in Baconian induction allowed the majority of museum workers to do exactly as [Walter] Faxon did and hold their course."[27] The same could have been written about botanical systematics.[28] Furthermore, to the extent that theory was articulated, all too often it did not coincide with practice, a point I have already had occasion to make in the case of both Antoine-Laurent de Jussieu and Augustin-Pyramus de Candolle.[29]

Systematics was seen as a science of observation. Theory should not intrude between the observer and nature; systematic data themselves were obtained from direct observation of plants. Numerous comments in the literature are in line with this thesis. The attitude toward nature held by many French naturalists in the late eighteenth century has been described as follows:

> The word natural in [Adanson's] "natural method" therefore has a typical eighteenth-century connotation, as will be clear from the importance attached by the century to everything that was "natural" . . . . The word natural sanctified the method: everything that was "natural" had to be accepted and had precedence over man-made systems based on arbitrary criteria. The artificial system had its criteria chosen by man, the natural method gave a picture of nature in accordance with reason, facts and experience.[30]

Note the implied passivity of the naturalist—the picture of nature was *given*. More than that, the naturalist could study plants best by following nature's own sequence, because then the true natural method would result. Method and nature were thus intimately connected, almost identical.[31] It was nature itself that informed the naturalist about the shape it took. Thus Michel Adanson suggested that it was necessary to seek in nature its system, if it were true that there was one. It was by the summed effect of the comparative descriptions he was making that he became aware that plants arranged themselves ("se ranjoient") in classes and families, which could not be called either systematic or arbitrary (as he put it).[32] Achille Richard observed that "when groups or series of genera were grouped and brought together by the greatest number of characters, one has done nothing more than imitate nature which had in some way created, so as to serve as models, types of essentially natural families."[33]

Again, the context in which such statements were made is important—it was in excoriated "système" that ideas took precedence over observation. Nevertheless, it is remarkable how frequently the process by which classifications and ideas of relationship are produced have been described as "instinctive."[34] The mention by Bentham of his "apprenticeship"[35] bears on this point, although he was somewhat unusual even in the context of the early nineteenth century in that he had very little formal education. Apprentices are taught *how* to do things; they may also learn *why* to do things in a particular way, but such is seldom in the context of a well-articulated theory. Systematists learned the principles of the subject by using the work of the acknowledged masters of the discipline—people like Robert Brown, Asa Gray, three generations of Candolles, two generations of Jussieus and Hookers, and others. These principles became almost second nature to them as they identified and sorted groups;[36] eventually they came to understand instinctively how characters should be treated. As Mirbel observed, "The art of comparison necessitates a long apprenticeship. The most obvious affinities escape students. They see in most families only assemblages of plants that are very different from one another, and do not understand what the affinities can be which bring together beings that appear so heterogeneous."[37]

Of course philosophers, logicians, and systematists *have* interacted. To give but two examples, Bentham himself had an early interest in logic,[38] and thought carefully about the principles of classification.[39] Augustin-Pyramus de Candolle mentioned philosophers like Locke, Condillac, and Pierre Prévost as being important in his education.[40] But makers of classifications, writers of floras, and describers of species and genera have by and large all emphasized action over reflection.

In his vice-presidential address in 1893 on the occasion of the first meeting of the newly constituted botany section of the American Association for the Advancement of Science, Charles Edwin Bessey sensed problems:

> The fact is, that the systematic disposition of higher plants is at present a makeshift, maintained by conservatism and a reverence for the time-honored work of the fathers. It is unscientific to let our practice drag behind the present state of our knowledge: it is far more so for us to cling to the opinions of our fathers through mere reverence, long after we know them to be untenable. It is not to the credit of our science that for the second time she has persistently held to a system through such considerations.[41]

Systems were makeshifts, maintained by convention (although Bessey did not put it that bluntly). Systematists tended to justify their work by reference to earlier classifications in part because it was so difficult to justify rejecting or accepting particular realignments or groupings unless such had been based on single characters and so were "obviously" flawed (see chapters 5 and 7). The natural families that Antoine-Laurent de Jussieu held up for exhibition as examples of unquestionable natural order were groupings that all naturalists in the past had recognized, and these families provided one of the main proofs of the validity of his method because they remained undivided.[42] The detection of the natural order was justified in a similar fashion by Linnaeus and Ray,[43] to mention but two names among Jussieu's predecessors. Auguste-François Saint-Hilaire[44] even proposed that the norm for estimating how much difference was needed between two groups if they were to be recognized as families should be set by Jussieu's families. Although Saint-Hilaire seems not to have understood what Jussieu was trying to do, his contribution to the debate further demonstrates that family limits were being set up as conventions.[45] Although long-established usage of a name in a particular way does not necessarily convert to its biological significance,[46] the "correctness" of Jussieu's families and of some other groupings he recognized was shown by their persistence, and this persistence became a vindication of Jussieu's own work and the practice of later systematists.[47]

Given the atheoretical orientation of taxonomic work and the tendency to evaluate classifications by reference to earlier work, one might ask, how do classifications change? The short answer is, only with difficulty. The difficulty is accentuated by what are often considered to be the deleterious nomenclatural consequences of changing ideas of relationship,[48] in turn connected to the need for classifications to be useful to a wide variety of people.[49] Sir J. E. Smith, accordingly, claimed in 1819 that botany (classification) had long been "practical" in England.[50] The nature of the problem becomes clear: the general users of botanical classifications have played an important role in determining the allowable changes in the system. The names given to plants were (and are) in wide currency among people without any pretensions to being scientists, let alone botanists or systematists, and who were mostly interested in names as names, not for comparing organisms. François Dagognet suggests that Jussieu linked the names of plants to their properties, so making classifications and commercial interests coincide; exotic plants could be replaced by their local relatives.[51] But even in horticulture, other than serving as imperfect guides in predicting success in grafting,[52] ability to hybridize, or the preferred cultural conditions of a new plant, the main function of names was to provide an unambiguous tag for a plant.[53] In the late nineteenth century the previously close linkage between botany and medicine was becoming less important. Imperial functionaries needing to know about the productions of the colonies and gardeners growing vegetables and flowers were all mainly interested in knowing what a particular plant "was"; the natural system seemed to have little to offer them.[54]

It is not unreasonable for the burden of proof to lie with those who wish to change names, and at one level, Alphonse de Candolle's advice in this context,[55] "*in dubio abstine*," is perfectly reasonable. However, some botanists discussed what they considered to be the real relationships of plants, yet did not change the names of the plants to agree with these relationships;[56] there was the nagging fear that the users of names would be upset. Brent Elliott has recently observed that in the early nineteenth century the use of Latin names by nonexperts acted as a brake to changes in names that were seen by experts to be needed if classification were to reflect relationships.[57] Despite this acting as a spur to the introduction of English names, names used by expert and nonexpert were not dissociated, so botanists could not change Latin names with impunity. At the level of family, additional constraints came from factors such as the systems adopted in major herbaria. It is difficult to shift around a big herbarium; indeed, for a herbarium to function efficiently as a repository, one simply wants to know where to find things, and the particular system followed is immaterial.[58] Colonial herbaria adopted the same family limits and followed the same system as did the home institutions, and so did floras and other taxonomic work coming from those institutions. In this department, the great herbarium at Kew, England, served as a model for institutions elsewhere.[59] Nevertheless, it is by the sequence and circumscription of families adopted in floras and herbaria that many botanists gained their understanding of the order of nature.[60]

The main function of floras themselves was to provide a practical and ready access to the correct names of plants. The writing of floras, more encyclopedias than anything else and usually utilizing directly only a subset of plant characters, has long engrossed the activities of the systematic community.[61] Of course, even an expert will want to use the most obvious features of an organism in identification. Yet although the distinction between recognizing organisms in keys and understanding their relationships in natural classifications was often made, a feeling persisted that taxonomic groups had to be readily recognizable. To the extent that this affected the circumscription of groups, there would be problems for naturalists in their attempts to understand nature; the use of few or even single characters to delimit groups would also be a likely result. As T. A. Sprague observed in 1925, "The growth of the natural system has been continuously retarded by the desire for easily recognizable groups."[62] This outlook contrasts with that of Condorcet, who had preached a century and a half before that it was usefulness to the science of botany, rather than to botanists themselves, that was important. He had observed that cotyledons might be visible only when no other feature of the plant was, and so were useless in identifying plants; that did not matter.[63] Similarly, Buffon warned that the search for single-character natural classifications was fated to fail,[64] but his warning was seemingly to no avail. The problem was made more serious because of the close connection between amateur and professional, as discussed in the last chapter; the amateur in particular would want taxa to be readily recognizable.[65]

In 1836 Asa Gray discussed recent progress in systematics with optimism:

> About sixty years have elapsed since the death of Linnaeus, during which period the science has undergone a revolution quite as complete as that effected by the great reformer of systematic botany himself. We have now a more perfect and intimate knowledge of the structure of plants, and so far understand the laws by which their external forms are modified, that natural groups are defined as accurately as artificial assemblages; and thus the chief obstacles that impeded Linnaeus are now surmounted.[66]

The morphological "laws" to which Gray was referring may have been satisfying to him, but families and even many genera could not be diagnosed simply. Despite his hopes, it was very difficult to relate the variable external form of a family to some inner invariance.

Muted Principles

An entirely different set of factors may have contributed to the stability of classifications. These concern ambiguities in the work of Charles-François Brisseau de Mirbel and Augustin-Pyramus de Candolle in particular that are apparently connected in part with the personal and professional relationships of these men with each other and with Jussieu, and in part with how Candolle was served by his translators. Add to these the inconsistencies in their work noted here in chapter 4, and the principles informing their studies can be hard to grasp, even for a willing reader; the issue of continuity *versus* discontinuity, and its consequences, becomes obscured.

Jussieu was widely acknowledged to have outlined the natural system, and the interpretation of the history of the discipline reflects this. Jussieu was also a senior and well-connected member of the Institut, and any criticism of his work in his lifetime (in France, in particular) would tend to be oblique and veiled. As recounted in chapters 4 and 5, when Mirbel discussed the use of anatomy in the development of the natural system, he made Linnaeus bear the brunt of his criticisms; Theophrastus became the *fons et origo* of the anatomical method. In addition, the conveniently dead Ventenat was assigned most of the blame for the rigid hierarchy of characters he and some other followers of Jussieu had proposed, although Jussieu's own approach to weighting did not allow much flexibility.

Candolle, in the course of his revision of the major groupings of the

Compositae, proposed an approach to grouping substantially different from that advocated by Jussieu, but this is not evident in Candolle's papers on that family. Yet there he claimed that what he was doing was the usual practice of naturalists.[67] Even the title of the *Théorie élémentaire* was understated;[68] its publication was delayed for some years as Candolle attempted to become a member of the Institut, and we do not know if there, too, criticisms were deliberately muted and positions understated. Candolle was not well-served by the translators of the *Théorie élémentaire*, and even the posthumous third edition, edited by his son, came in for criticism over its editing.[69] English-speaking readers depended ultimately on Kurt Polycarp Joachim Sprengel's translation-cum-adaptation of the second edition. This translation, however, was full of misunderstandings of Candolle's ideas, and the latter disassociated himself from the translation.[70] Importantly, Sprengel was less clear than Candolle over the important issue that groups in nature were discrete, although it should be remembered that it was the systematic side of Candolle's work that least interested his contemporaries. Other works purporting to present Jussieu's "system" also largely ignored the issue of what his groups represented in nature.[71] Later criticisms did little to clarify matters. For example, when Joseph Duval-Jouve[72] attacked Candolle for what he took to be his position on the use (rather, the nonuse) of anatomy in systematics, he selectively quoted Jussieu to show that the latter supported its use.

Finally, we return to the basic ambiguities of classifications. Candolle and Jussieu each produced a similar formal classification. Although Candolle was a hierarchical nominalist, or perhaps a realist, there is a hint of continuity in his discussion of natural groups. Moreover, he did not develop a fundamentally different way of forming groups: the new order was promised, but not delivered; support of anatomy was not whole-hearted, and his own methodological prescriptions were sometimes ignored. Hardly surprisingly, there are internal inconsistencies in his work. For example, early on in the *Théorie élémentaire* Candolle suggested that the order of nature and that of books might be the same; yet later at the head of his listing of plant families he stated most emphatically that the two could not be equated. (In 1931 Walter Baron read only the former entry and concluded that Candolle believed in a linear arrangement of nature.[73])

On the other hand, Jussieu was surely a limital nominalist who did indeed believe in a continuous nature, but the cursory reader of his *Genera plantarum* might suppose it to represent a classification of discrete groups, not an arrangement. Jussieu had asserted that there was a connection between the interior and the exterior of the plant, even if he thought that the former did not need to be investigated.

Discussion

As Charles Bonnet concluded three centuries ago, for a believer in continuity, observation *had* to stop at the outside of the organism; even dissections were to be avoided.[74] Bonnet's approach to observation was similar to that of Jussieu and other early practitioners of the natural method.[75] Theory, a lens, or (worse) a microscope could not intrude between the systematist and the organisms being examined. The microscope almost necessarily rendered questionable what could be seen with its aid because not only was the integrity of the organism destroyed but observers who used microscopes were suspect of paying too much attention to theory. Synthesis of groups was prescribed, yet at the same time it was difficult to reject characters as indicating some sort of relationship; apparent continuity of relationship was the almost automatic result.

Since Jussieu's time much of systematics at the generic level and above has entailed the implementation of the research program he laid down. As largely the same sorts of observations were made, the data were analyzed in the same way, and the same conventions were used for circumscribing groups, it is not surprising that the structure of the system remained relatively little changed.[76] Moreover, most of the gaps that marred Jussieu's continuous series were never filled. There *are* gaps in the variety of forms that make up the living world. Jussieu, with his interpretation of plant morphology that is different from ours, and his lack of knowledge of anatomy and development, might not have seen the same gaps as we see today (or as we might see using twentieth-century ways of seeing, yet looking only at the plants with which he was familiar), and we would certainly interpret these gaps differently. Nevertheless, when drawing up classifications Jussieu was constrained by those gaps that existed; he would not allow major gaps to break up his groupings.

Thus the limits of at least some genera and families were the same for many naturalists, whatever their views of the world. A good example is provided by the Nyctaginaceae. As Jussieu understood that family in 1803, it was quite discrete and without immediate relatives directly linking it to other groups. But he confidently expected that these links would be found. They never were.

Thus insofar as the thesis advanced here—that systematics has remained Jussiaean in spirit—is correct, stability of practice and classifications may not bear on the issue of how effectively those classifications reflect patterns in nature, or what patterns they reflect. There are several linked elements that tend to cause stability: (1) a distrust of theory, with resultant appeal to patterns "evident in nature"; (2) the apprenticeship system by which systematics was learned; and (3) the constraints exerted on professional systematists by amateurs and others who would not even call themselves botanists. For such people, the less that the names of plants changed, the more useful they were; this led to a reluctance to change names and an emphasis on the need of all taxa to be readily recognizable. Such linkages have provided a formidable if much underestimated restraint to the development of ideas of relationship.

The final element discussed here that may, in part, be held accountable for the stability in systematics that lasted more than two centuries is that alternatives to Jussieu's approach that were advanced in the first part of the nineteenth century were not clearly presented. If it was through Jussieu's results, his classification, that he was "understood" by subsequent generations of workers, there was a powerful reason for accepting his approach: it appeared to work since at some levels the system was stable. Furthermore, because the terms Jussieu used continued in use (see chapter 7), it was easy to believe that systematists were all striving toward the perfection of the same natural system.

CHAPTER ELEVEN

Revolution and Change?

Jussieu's *Genera plantarum* was associated with a fundamental reorientation of systematics after its appearance in 1789. There was no further major change in the ensuing seventy years. A. G. Morton drew an analogy between change in science and change in society to emphasize what he took to be the revolutionary character of Jussieu's work.[1] Jussieu himself used this analogy—but to deny that he had instigated any fundamentally new approach: "In the same way in the history of science as of people, [there are] no unforeseen and unexpected changes, although they appear so, but all was ready before, they are brought to a conclusion gradually and by degrees, debito tantum tempore absolvuntur."[2] This remark terminates the text of the appendix to the posthumously published Introduction to the *Genera plantarum*, in which Jussieu sketched the development of his own ideas. It is almost literally his last word on the subject.

It is surprising that Jussieu should suggest that society changed gradually—after all, his life spanned the beginning of the Revolution of 1786 and the succession of governments of varying political hues in the ensuing years.[3] But his reading of the analogy between science and society is in many ways a fair estimate of the relationship between his work and that of his predecessors. (In this assessment I differ with most other commentators, who have preferred to see Jussieu, or the Jussieus, uncle and nephew, as revolutionaries.[4]) Here I discuss the general development of systematics over the period from 1735, when Linnaeus started publishing, to 1900, when Mendel's laws were rediscovered. This will entail a brief survey of botanical systematics in the period immediately after 1859. I can then interpret the significance of my suggestion that systematics in the preceding period was extensively informed by continuity-in-practice, and of my analysis of the reasons why classifications did not change then—reasons that simply reinforced this continuity-in-practice. I will then be in a position to reassess the development of systematics during the whole period 1735–1900 from the point of view that major conceptual shifts were taking place in the discipline.

Systematics Between Linnaeus and the Twentieth Century

Jussieu's work includes descriptions of natural genera and families, and also some natural higher groupings;[5] all known genera were treated, although a few of unknown relationships were not placed in natural families. However, Jussieu's ideas of arrangement, classification, and relationships were similar to those of many of his predecessors, particularly those of Lamarck, and they were based firmly on the principle of continuity—as he emphasized to the last, "natura non facit saltus."[6] Jussieu introduced no new ways of looking at plants, although the characters of the fruit and seed that he and Joseph Gaertner paid special attention to brought a suite of new observations to the estimation of relationships. In his apparent emphasis of external form as the key to understanding relationships, his approach is like that of Linnaeus and Bonnet;[7] his view remained largely fixed on the outside of the organism, and on those internal differences that could be readily seen on dissection of the flower or seed. Finally, he did not propose any new way of evaluating data, although he judged that one of the approaches used by Linnaeus, that of analysis, was impermissible.

The way in which Jussieu looked at plants, and in particular the theory he brought to the construction of the natural series, affected the groups he recognized.[8] Note that I am by no means suggesting that the shape a classification takes is determined entirely, or even largely, by theory; chapter 10 showed that theory constitutes but one of the elements that constrain classifications. There I contended that the gaps Jussieu perceived in the living world tended to bound his groups. I also suggested that natural systems through the first half of the nineteenth century tended to reflect Jussieu's view of the world because those who made the systems largely followed his principles or, more accurately, his practice. Hence come some of the similarities in both the circumscription and the relationships of the families that were recognized in 1789 and 1889 (and even in 1989). Nevertheless, because simple differences and similarities were not the only factors that constrained the groupings proposed by Jussieu, some of these groupings were subsequently ignored; thus his Corymbiferae, Cinarocephalae, and Cichoraceae have sunk into obscurity. Through the nineteenth century, plant form gradually came to be interpreted less as a complete continuum,[9] although Jussieu's practice of forming groups by synthesis persisted, and taxa at all levels of the hierarchy remained directly linked. But in a shift from this practice, the recognition of groups came to be justified by how much they differed, even if the amount of difference needed for recognition tended to be inversely proportional to the size of the group.

With the introduction of evolution into the equation, little changed. I noted in chapter 7, for example, that trees were occasionally used by botanists in even pre-Darwinian literature to depict relationships, and comparisons between classifications and genealogies were made, albeit casually. After 1859 Jussieu's natural system, modified in detail but not in principle by his followers, was simply interpreted in evolutionary terms; after all, it was the same natural system that was involved.[10] John C. Greene has suggested that systematists gradually relocated and transformed their discipline within the broader framework of evolutionary biology.[11] But is this true of those systematists working at the higher levels of the classification? Many terms were simply given new meanings—a change encouraged by Darwin himself;[12] archetypes, for example, became ancestors. At lower levels of the hierarchy there were substantial changes in botany, especially toward the end of the century; these changes had begun earlier and were more pronounced in zoology. Seminal works like Richard Wettstein von Westersheim's *Grundzüge der geographisch-morphologischen Methode der Pflanzensystematik* appeared then, and the focus switched to field studies, experimental methods, and variation at and below the species level.[13] The "modern synthesis" of the 1940s also emphasized species; it, too, was more zoological than botanical in its inspiration.[14]

At higher levels, tree-type diagrams became more frequent. The first detailed botanical evolutionary trees of which I know—although they barely warrant the name—were those by Giacomo Guiseppe Federico Delpino (1830–1905), director of the botanical institute at Naples. His diagrams of 1869[15] (figure 17) are much like reticula. He represented genealogy by series of taxa directly connected to one another,[16] and he suggested revealingly that tree diagrams were advantageous because they could show relationships between one genus and at least three others, while in a key each genus was adjacent to only two neighbors.[17] Many of such early phylogenetic arrangements show elements both of reticulation—perhaps the multiple relationships Delpino wanted to show—and of highness and lowness. For example, Franz Buchenau (1831–1906) produced a tree that was a reticulum and yet which also included taxa joined directly: his "Urtypus" was essentially *Juncus poiophyllus*—that is, it was an organism, not an abstraction, and it was joined directly to other groups—and the species he placed in subgenus Septati were of independent evolutionary origins.[18] Russian botanists produced similar diagrams. Nicolai Ivanovich Kusnezov (1864–1932) showed relationships within *Gentiana* in a diagram in which members of the one group might have independent origins and in which the relative age of groups and the strength of the relationships between them were indicated (figure 18). Kusnezov specifically rejected the idea of depicting relationships as trees in favor of the reticulum he actually drew.[19] Olga Tchouproff worked on the Acanthaceae for her doctoral studies in the early 1890s. She denied that her diagram of reticulating relationships in the family, which showed groups directly joined, indicated real relationships; she thought groups might have multiple tendencies.[20] Even treelike diagrams might have nothing directly to do with descent. In 1872 Alexander Anrejewitch von Bunge (1803—1890) drew an elegant tree of relationships in *Acantholimon*; groups were joined serially to other groups and there were a few reticulations (figure 19). Bunge denied that his tree had anything to do with descent; it was simply a "Tagesordnung," so those using it should beware of indulging in fantasy and error if they interpreted it in any other way.[21]

Diagrams appearing in publications of the Linnean Society of London show the same complex of ideas. George Bentham showed relationships within the Compositae in the only diagram of this kind he drew in his career. The diagram is circular, there are reticulations and

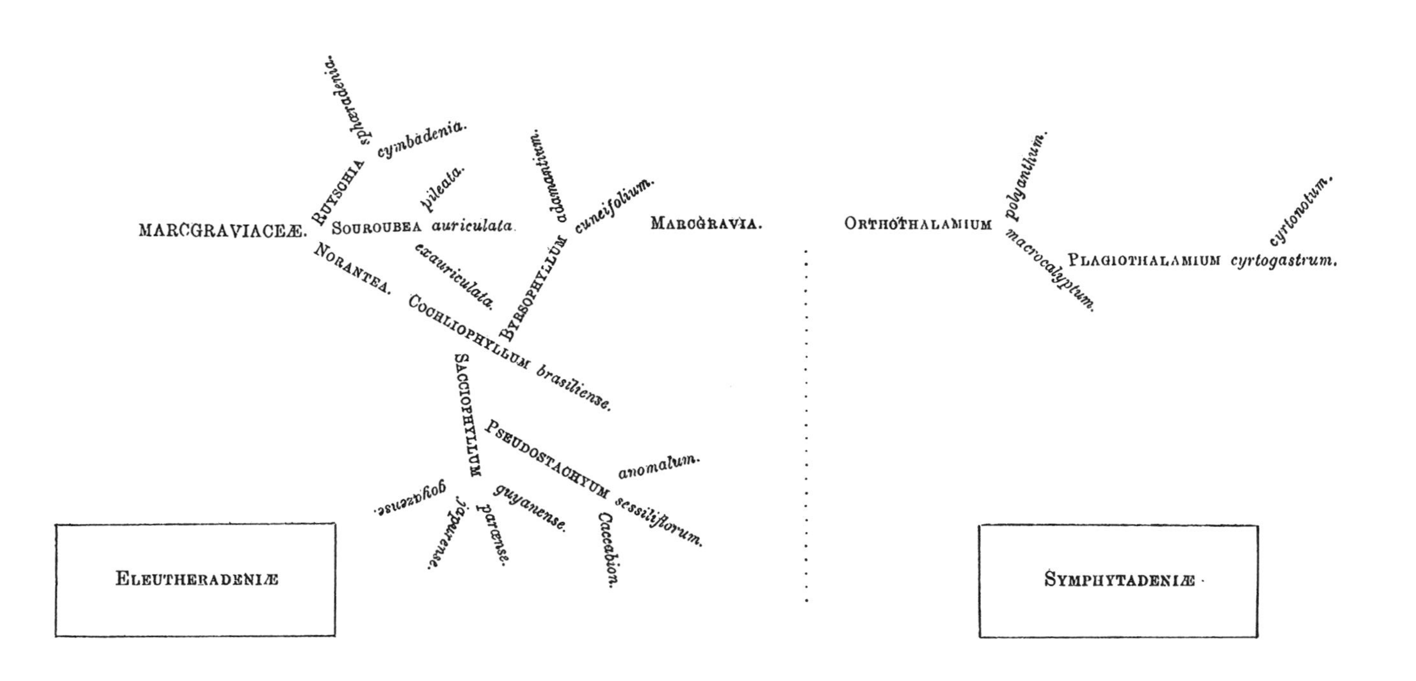

FIGURE 17

Delpino's idea of genealogical relationships. Delpino was working on the Marcgraviaceae. Source: Delpino 1869: 268.

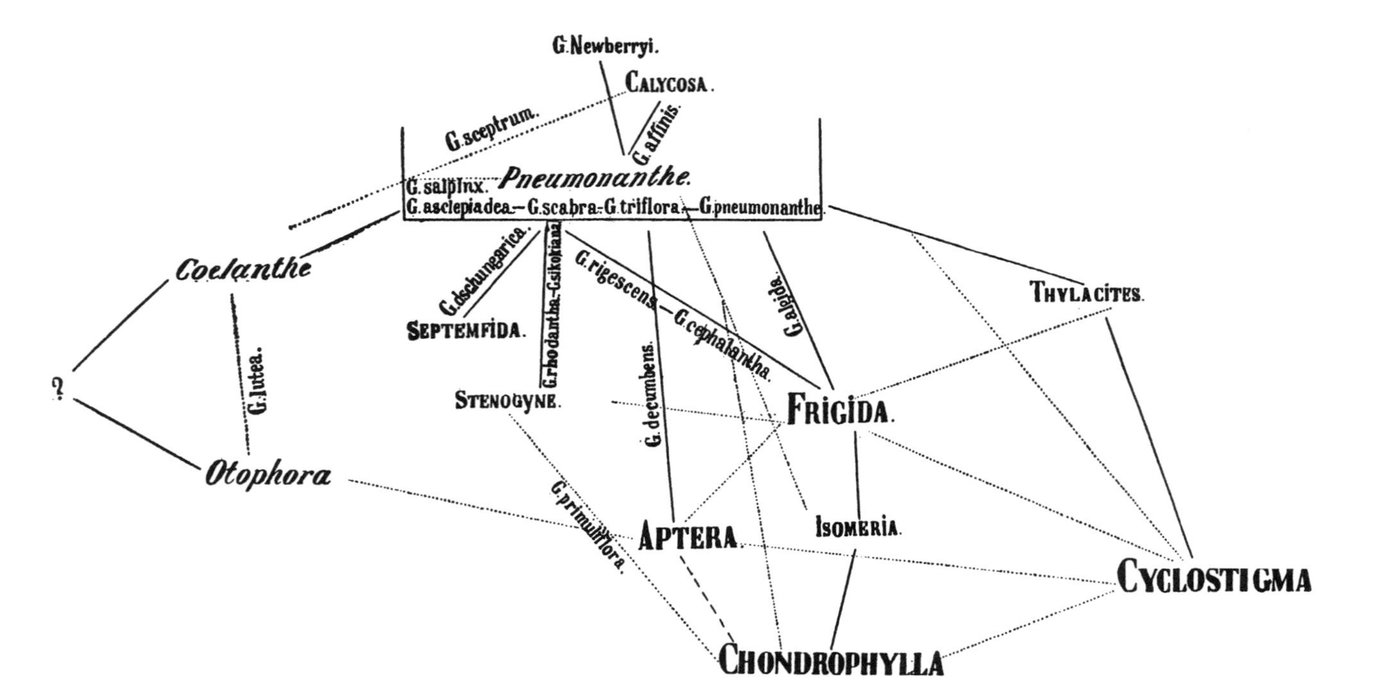

Verwandtschaftsschema der Sectionen der Untergattung Eugentiana[1]).

FIGURE 18

Kusnesow's idea of relationships in Gentiana. Source: Kusnezow 1894: 84.

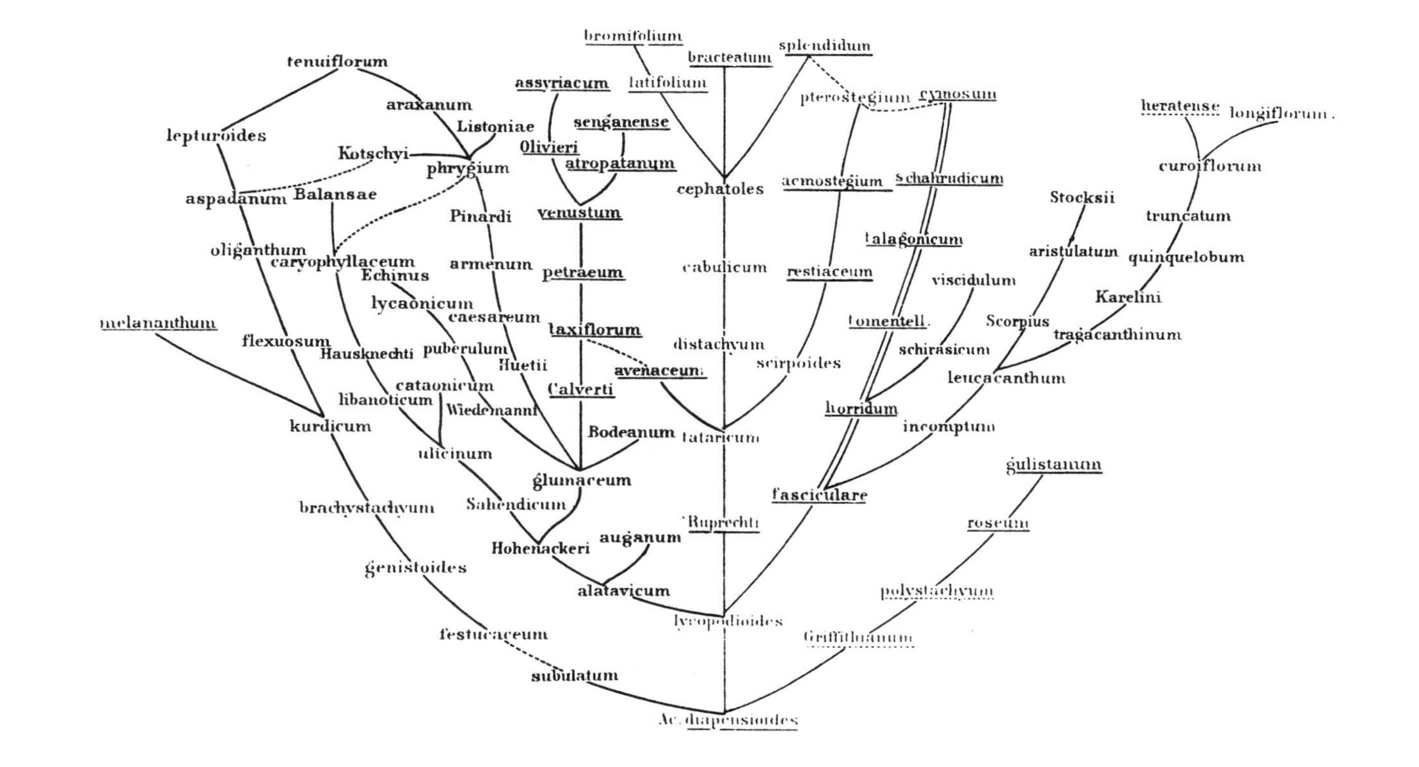

Stammtafel der Arten der Gattung Acantholimon Boiss.

FIGURE 19

A tree in the Plumbaginaceae. The lines, underlining of the species names, and the species names themselves are variously colored, so indicating where the plants grow. Source: Bunge 1872: [pl. 2].

direct linkages between taxa, yet at the same time there is axis, since old is at the bottom of the diagram and young at the top (figure 20). All in all, Bentham's diagram shows little conceptual advance over a similar diagram prepared by Alexandre-Henri Gabriel de Cassini fifty years before.[22] A. W. Bennett's diagram, which encompasses all plants, was more treelike; it does not show reticulations, but it includes direct linkages between taxa, ideas of highness and lowness, and taxa as levels of advancement along the tree (figure 21).[23] Indeed, the linkage of evolution with ideas of progress, of increasing advancement were generally popular.[24]

There are a few evolutionary trees lacking both reticulations and direct linkages between extant taxa, and most were published by German-speaking botanists.[25] Before 1859 such botanists would have presented their work in terms of types and deviations from types; conversion of types into extinct ancestors would naturally lead to the production of trees.[26] In 1872 Eduard Adolf Strasburger (1844–1912), professor at Jena and later at Bonn, provided a treelike diagram of relationships within the Gymnospermae—although genera occupied the ends of the branches, higher taxa represented levels of advancement, and three separate lineages coming from within the Gnetaceae gave rise to the angiosperms.[27] Two years later Heinrich Gustav Adolf Engler (1844–1930; he became perhaps the most powerful figure in the German systematic community at the end of the century) discussed the importance of evolution for systematics. He thought that "Descendenztheorie" would invigorate systematics, although it had been dismissed by the most prominent workers in the preceding decade and was feared by most descriptive botanists and zoologists.[28] In his work on the Ochnaceae, published in the same year, Engler drew a tree viewed from the side; genera, with their relative sizes indicated, were placed at the end of branches, but at different levels, indicating their relative advancement (figure 22).[29] Later he depicted relationships in the Anacardiaceae as a tree seen from above, with concentric circles showing how far genera were from the center of the diagram, again indicating how advanced they were.[30]

There were various advantages in depicting relationships as a tree. As the zoologist Edwin Ray Lankester observed, "In the [genealogical] tree, as in a family pedigree, no arbitrary arrangement is admissible."[31] The tree might not be arbitrary, but classifications based on it were less

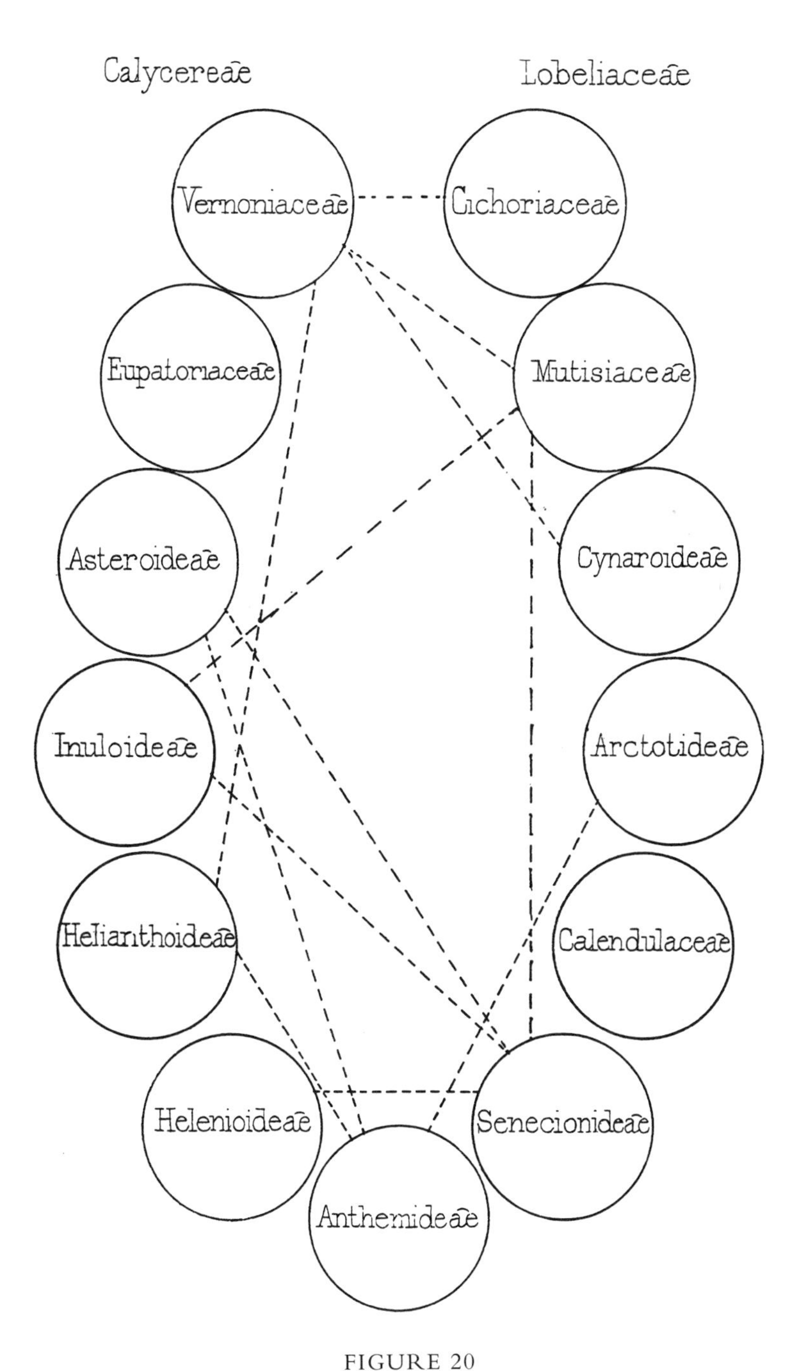

FIGURE 20

Reticulate relationships in the Compositae. Source: Bentham 1873b: pl. 11.

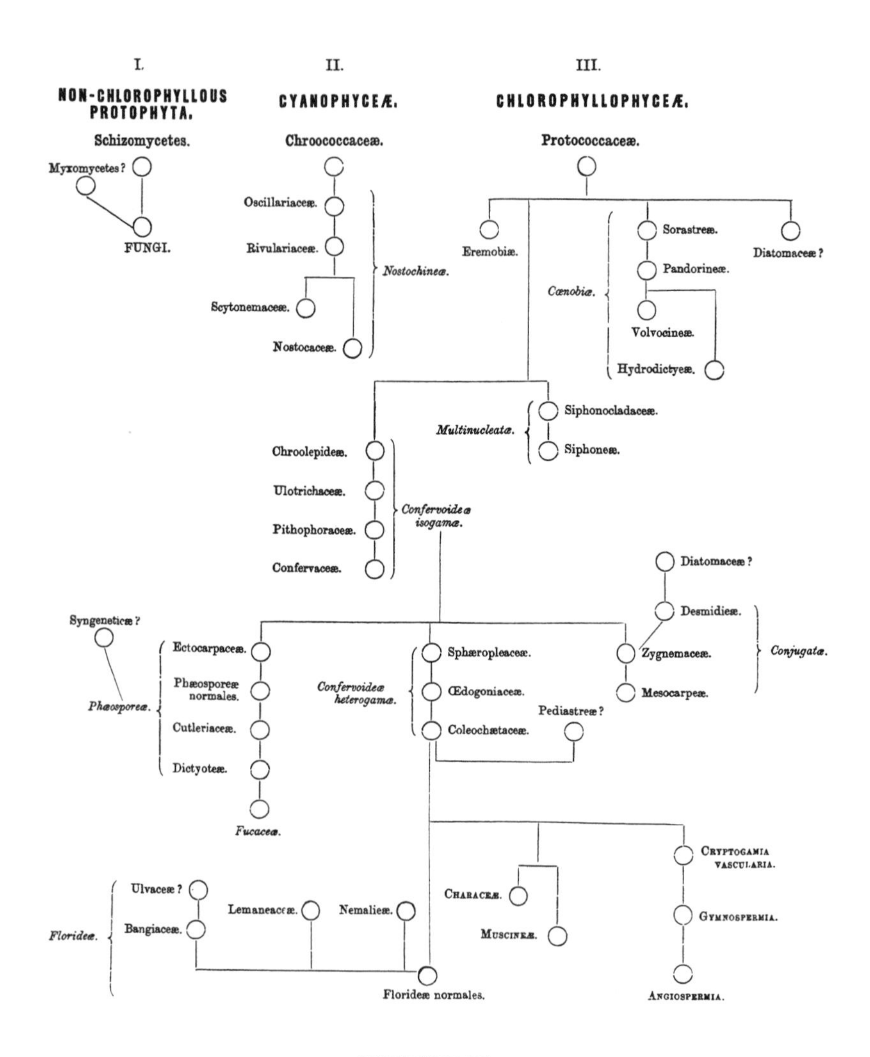

FIGURE 21

Bennett and relationships in plants. Source: Bennett 1888: fig. facing p. 60.

constrained. Indeed, as was common for the time, Lankester was interested in taxa as representing levels of organization, not lineages.[32]

An exchange between Édouard Marie Heckel and Richard Wettstein in 1894–1895 addressed another issue that I have already alluded to. Heckel had depicted relationships in the Globulariaceae as

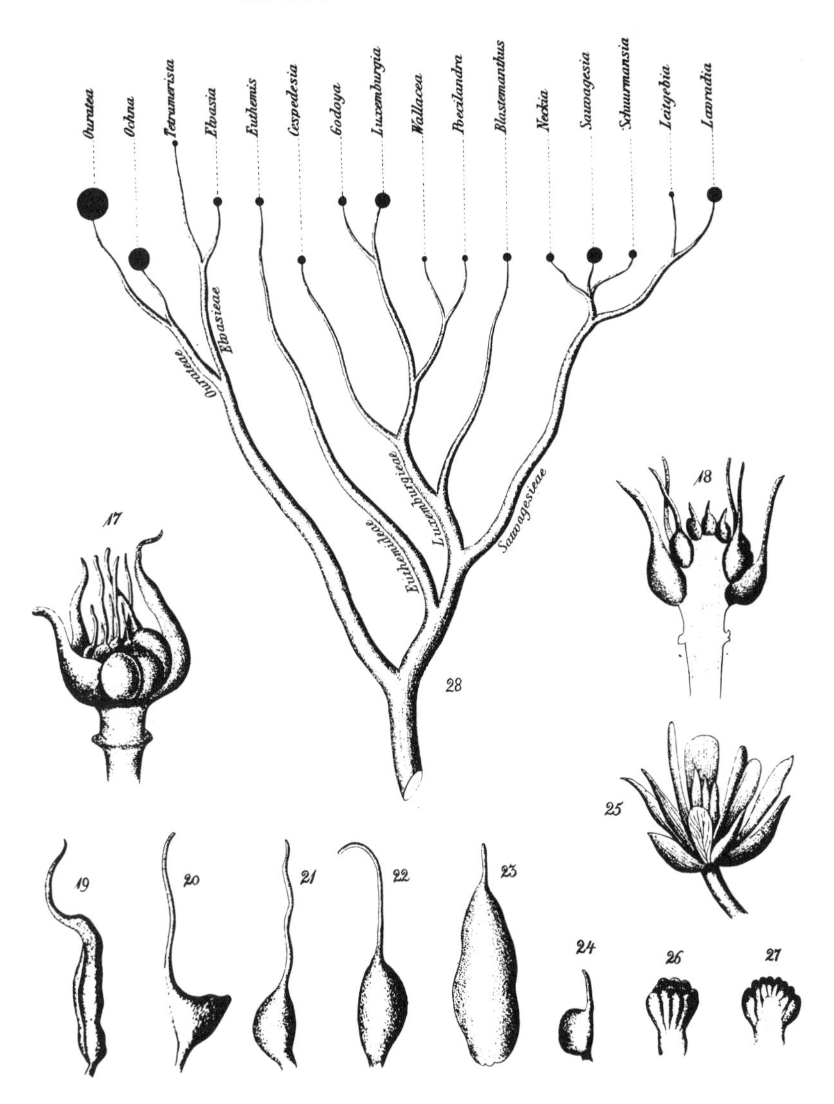

FIGURE 22

Dendritic relationships in the Ochnaceae. Source: Engler 1874: pl. 13.

if all extant species were joined directly in series. Wettstein specifically objected to this aspect (and many others) of Heckel's work; small groups of species were related not to each other but to extinct "Stammarten."[33] Similarly, taxa in Engler's diagrams are at the ends of the branches and there are no linkages between branches once they separate, so taxa are perfectly discrete and do not directly join each other. However, there were few imitators of either Engler's or Wettstein's approach,[34] and Engler himself seems not to have published evolutionary trees later in his career.[35] Even Wettstein did not show relationships between the different "Stammarten" that he recognized; he could follow history for only a short distance into the past.[36] It should also be noted that although trees depicted genealogical relationships, characters that might serve as evidence for the genealogy were rarely shown. Such trees were often based on ideas of character progressions or tendencies.

All the authors I have been discussing were working on groups of extant plants; perhaps paleontologists found tree construction a simpler task. Heinrich Solms-Laubach provided several examples of what he described as genetic relationship or filiation established from the geological record in his book *Paläophytologie* written in 1887.[37] Yet at about the same time, Heinrich Goeppert (at 81 looking back at a long life in paleontology) observed that the prototypes, or organisms linking major groups, had still not been discovered.[38] So even some paleontologists were in trouble. Although they were accustomed to reading paleontological history directly from the fossil record,[39] fossils being ancestors,[40] not much progress was being made. And toward the end of the century, the paleontologists' search for regularities and laws of evolution, ideas of highness and lowness, parallelisms, convergences, and levels of advancement led to the evolutionary trees they produced becoming much more tentative.[41]

Neontologists were indeed in worse shape. In 1880 the elderly Alphonse de Candolle, then 74 and the doyen of French-speaking systematists, summarized the use of diagrams depicting relationships. Most of these diagrams were reticulating, and although Candolle had accepted evolution early in the 1860s, he inclined to the view that relationships were simply too complex to be accommodated by any diagrammatic representation.[42] In 1873 J. D. Hooker, one of Darwin's staunchest supporters and his confidant since the late 1840s, found it "astonishing how few [families] are absolutely limited."[43] He described his general ideas of relationship in a remarkable passage:

> The Cohorts [groups of families] may thus be fancifully likened to the parti-coloured beads of a necklace, joined by a clasp, the beads touching at similarly coloured points of their surfaces. The position of each bead in the necklace is determined by the predominance of colours common to itself and those nearest to it; whilst the number and proportion of other colours which each bead presents, indicates its claims to be placed elsewhere in the necklace; in other words, such colours represent the cross affinities which the Cohorts display with others remote from the position they occupy.[44]

Both Candolle and Hooker come close to directly contradicting Darwin's ideas of classification as being a group-in-group arrangement of organisms. Darwin himself had observed, "This classification is evidently not arbitrary like the grouping of the stars in constellations."[45]

Even if the image of the natural system as tree or shrub were accepted, a systematist might prefer to view the evolutionary tree from above, but looking at the tree in summer, so emphasizing the foliage not the branches. The result is very different from the branching trees drawn by Engler and Hallier. Thus when George Bentham in 1873 described such an evolutionary tree, he found no need to talk about either ancestors or real relationships, both of which he considered to be entirely speculative given the current state of knowledge. Ancestors remained elusive—whether they were based on some kind of abstraction (i.e., types) or drawn from fossils, direct evidence from breeding, or ideas of late tertiary and quaternary climatic change. The trunk and main branches were obscured, and Bentham's ideas of relationships remained unchanged from those of over twenty-five years before.[46] Thus the massings of foliage that he saw in 1873 were directly comparable with the massings of trees growing in his pre-evolutionary landscape of relationships of 1856, itself conceptually very similar to a Linnaean map of nature.[47] Under such circumstances, it would be impossible to guess the shape of an evolutionary tree from the hierarchical structure of a classification.

So even if the analogy used for the natural system changed, there was often little change in the conceptualization of nature.[48] Many of the diagrams published seem to have been conceived as reticula, maps, or similar structures, with a temporal element simply imposed, or at least they were produced by the same methodology that tended to yield these figures. A classical, albeit somewhat later, example is Charles

Edwin Bessey's diagram of relationships. This looks like a prickly pear;[49] the major groupings of plants, the cactus pads, are directly joined one to another. Lucas Rodrigues C. cut down Bessey's "tree" and provided instead a horizontal, archipelago-like schema of relationships with a new set of connecting branches, a genealogy. But he cautioned that no "factual value" should be attached to it.[50] Archipelagic schemata—usually destitute of isthmuses—remained popular through the twentieth century, and they allowed taxa to have a greater variety of relationships with others than would have been possible if all were directly joined by branches of the tree that they terminated.[51] Groups were certainly not reducible to types or ancestors, and, importantly, there was still no agreed way of distinguishing between analogy and affinity (homology). Relationships were of considerable complexity, groups could not be defined without exceptions being entered, and there were transitions between groups.[52] All in all, an archipelagic schema remained a good way to indicate relationships.

When the natural system was visualized, groups were frequently joined directly; "the great vice of the comparative method."[53] This vice is the result of the direct conversion of the maplike or reticulate representations of continuous nature (produced by synthesis which predominated in the first half of the nineteenth century) into a tree form by simply superimposing a vertical axis. Intermediate taxa remained integral for detecting relationships; linking groups became evolutionary intermediates and groups joined each other directly.[54] Under such circumstances a botanist—or a zoologist—would still try to find the organisms that represented the missing links between groups and to convert the continuum or quasi-continuum of direct morphological relationships into a genealogy.[55] Charles Bessey summarized the situation as he saw it in 1893:

> It is now a full third of a century since a great light was first turned upon all biological problems by the formulation of the doctrine of evolution by the master mind of Darwin. In its light many puzzles have been solved, and many facts hitherto inexplicable have been made plain. We now know what relationship means, and we have given a fuller meaning to the natural system of classification. From the new point of view a natural classification is not merely an orderly arrangement of similar organisms. It is an expression of genetic relationship. The present similarity of two organisms is not enough to determine their relationship, or

> place in a system. Common origin must be inferred in order that relationship shall be assumed.[56]

Nothing could be clearer: the idea of evolution gave botanists a way of understanding natural classifications. But Bessey was no more able to show how to recognize similarities that indicated the "genetic relationship" underlying the natural system than was Buffon 140 years before.

No new method of evaluating data was proposed by botanists in the latter part of the nineteenth century. If botanists had turned to Darwin himself, they would not have been able to extract any better guidance.[57] Rather, they were left with vague ideas of a general equivalence between morphology and genealogy, the same equivalence that had been made between the historical element of Buffon's "histoire naturelle" and simple similarity relationships.[58] However, new suites of characters were brought to botanists' attention, and the early hopes were always that new characters would at some point enable them change the discipline.[59]

Augustin-Pyramus de Candolle cautioned that the world was not the way Antoine-Laurent de Jussieu saw it, also suggesting that new kinds of information (from comparative anatomy) be integrated into the system. But nothing much came of his suggestion. His "revolution," by which taxonomy, anatomy, and physiology were to be integrated, failed.[60] In 1871 Joseph Duval-Jouve[61] claimed that "histotaxie" (comparative cell-level anatomy) provided characters that were less plastic than others, being less subject to environmental modifications. Such anatomical studies did become more common in the 1870s. Ludwig Adolph Timotheus Radlkofer, professor of botany at Munich and a student of Matthias Schleiden, delivered an address in 1883 during festivities surrounding the birthday of Ludwig II of Bavaria. He forecast that the ensuing century would be dominated by the anatomical method ("die Methode der mikroscopisch-anatomischen und mikrochemischen Untersuchungen") in systematics; the Linnaean approach must be transformed, and this was the way to do it.[62] Although numerous and extensive studies using comparative anatomy were produced during the next thirty years or so, fewer appeared after the First World War, and these early studies remain the only ones available for many groups.[63] The great similarity in the description of the Malvaceae by Adanson in 1763 and by George Lawrence in his *Taxonomy of Vascular Plants* almost two hundred years later[64] suggests how little the observational basis of systematics had changed. Lawrence was writing a textbook and naturally would tend to simplify, but the systematist's world remained largely that of the surface of the plant.

Yet as the nineteenth century drew to a close, there were hints that a change might be in the offing. This change can be followed in the writings of Alphonse de Candolle. Alphonse had begun his botanical career very much as his father's son, wrestling with the issue of symmetry and its implications for systematics. In an introductory textbook that he wrote in 1837 when he was barely twenty-one, he finished his discussion on the development of botany with some comments on symmetry.

> Another characteristic trait of our time is the search for *laws* that regulate the form of organized beings.
>
> The symmetry of organs is recognized in principle. One is occupied in researching apparent aberrancies, by means of the *fusion* of neighboring and analogous organs, of the *abortion* or incomplete development of certain parts, of the *doubling* or additional development of organs, finally of their *variations* (metamorphoses) of forms and functions, which can influence neighboring organs. The law of symmetry has become a general principle in natural history, like attraction in physical sciences, like determinate proportions in chemistry, by which one explains anomalies by secondary laws, or by distant consequences of the same great principle.
>
> Natural groups are reduced ["ramenés"] by observation to more regular ideal types. By the comparison of these types and their variations one will one day understand the plant kingdom in all its modifications and with its so complicated affinities. . . . The informed public no longer regards this science as a study of words, but rightly as a true science, which has its theories and its facts, its hypotheses and its laws.

Alphonse treated the existence of types as a kind of promissory note that nature would be understood. But the use of types divided rather than unified systematics, and the first characteristic of "our time" (the late 1830s), which was the union in a single science of organography, physiology, and taxonomy, seemed ever in the future. Indeed, even then the life sciences were in the process of subdivision—with botany separating from zoology, and physiology from classification studies.[66]

In the early 1880s Alphonse, then one of the most highly esteemed systematists in Europe, had a glimmering that a revolution in systematics was developing, one in which temporality (phylogeny) would be incorporated and which would cause a major upheaval. In letters to George Bentham and J. D. Hooker, he grappled with the consequences of his realization that cell- and tissue-level anatomy might indeed be important in systematics, and that genealogy was also to be reckoned with. The third botanical generation of Candolles was in part responsible for this. Casimir de Candolle (1836–1918), son of Alphonse, had in 1879 published a study on the comparative anatomy of the leaf.[67] By then, it had become apparent to Alphonse that the northern beech, *Fagus*, and the southern beech, now called *Nothofagus*, which appeared to be closely related, were in fact very different anatomically. Moreover, Alphonse interpreted anatomical characters in the light of Haeckel's law of recapitulation;[68] anatomical characters appeared before the floral characters, and so might be more important than other characters in determining relationships. If the idea of evolution was added to all this, a period of change was in the offing. Alphonse expressed his problem in a letter to George Bentham:

> Perhaps there will come a time when by the help of genealogies one will have on top of that more than conjectures. I assume at present that two forms similar enough to be put in the same genus for the time being may sometimes come from two stocks, originally far from each other, by subsequent chance modifications having brought them together. For example the Fagus of the northern and southern hemispheres do not differ enough for me to do other than make two sections based on external characters, but look, see what my son shows me in the internal vessels of their petioles of two different types. From other facts which he has assembled in several genera, I was in doubt of the importance of this character, however, for Fagus, very disjunct geographically, that appears to one as an indication of different origins.[69]

Similarly, Alphonse conveyed his concerns to Hooker:

> I think that we are at the beginning of a major confusion over the series of Phanerogams, after which some great botanist will come along and make a new order. I will explain myself—because it is necessary to leave this Apocalyptic style.
>
> It is not possible to ignore indefinitely anatomical characters in the formation of natural groups. One begins by linking them with families, genera, and species, but these are mediocre or *badly done* trials, because the anatomists look only at their microscopic findings and furthermore

> have not yet done the *complete* anatomy of more than 15 or 20 Phanerogams. They proceed randomly, taking for the character of the family that which is from a genus, a generic character that which is from one or two species. At Paris, M. Vesque, M. Poisson, etc., have made anatomical characters of certain families, but still they have not looked at all organs and have not combined anatomy with external characters, as it will be necessary to do some time. My son's memoir on the vessels of leaves is a start, in this sense, but he will have to do similar ones on other organs in many genera and species. Despite the lack of confidence in real generalizations in anatomy, M. Sachs[70] and others don't hesitate to make classifications of Phanerogams. This will soon be the confusion of which I spoke. It will last perhaps 20 or 30 years. . . . It will be in a *mixture* of *philosophical considerations* on *natural history* that I have some idea of drawing up [the new series].[71]

> Anatomical characters have indeed caused mistakes, perhaps because they have not yet been enough studied to . . . estimate their relative value. Vesque appears to me to have touched on original points on this question, while using a language that is often baroque and following too hypothetical ideas. His memoir in the Annales Sc. Nat.[72] is strange. One thing appears to result from the modern ideas of evolution: that *the first organs formed have the most importance*. They are uniform, necessarily, in the major groupings of plants. It is this that one has seen, without knowing the reason, when one attached great importance to the cotyledons. In general, vegetative organs, which precede those of the flower, should be enhanced in the characters of classes, families, and genera, as they are formed first.[73]

A saviour was in sight. A version of the so-called law of recapitulation, "ontogeny recapitulates phylogeny," was to provide the solution to the problem of detecting phylogeny. As Ray Lankester observed, development provided the trunk of the evolutionary tree.[74] But even here little progress was made in botany, and much of higher-level systematics in the 1960s would have seemed very familiar to Candolle and his predecessors.[75] Plus ça change, plus c'est la même chose.

The idea that classifications are based on categories that have common properties, and the commonly held belief that most taxonomists in the eighteenth and early nineteenth century were essentialists, must now be reintroduced.[76] Systems have part–whole relationships and lack essences, as has been emphasized in recent developments in phylogenetic systematics. Insofar as there has been a change in systematics, however, it has not started from classification in this sense, but almost paradoxically has proceeded from systematics to something more like classification and then back to systematics again—although the two systems involved are very different.[77] Jussieu's method was preeminently an attempt at a systematization of nature, although the part–whole relationship was atemporal and the whole was indivisible.[78] In phylogenetic systematics, the part–whole relationship is temporal, and the whole as it exists in our own slice of time can be divided. In the intermediate, "classificatory" stage, polythetic groups were described, but there was considerable ambiguity as to what these groups represented and whether or not they could in principle be defined. It seems to me questionable to insist that the groups described by many of the people I have been discussing were believed to have essences in any ordinary (i.e., classical) sense. We also have to remember that the job of taxonomists was to name, describe, and identify organisms, so they would automatically emphasize characters whatever their view of nature and characters. Thus Pierre Turpin might believe that not only was nature continuous, but characters that distinguished local areas of this continuum themselves shaded away;[79] not only groups, but characters, too, had fuzzy edges. This did not stop him from describing plants.

There is, then, a general continuity in the way botanists approached their studies of the natural world in the period 1735–1900. What I have described as continuity-in-practice persisted after there was no longer explicit belief in continuity. We begin to see the magnitude of the problem that Frans Stafleu adumbrated when discussing Linnaeus's fragmentary natural system; it was "the tribute Linnaeus paid to the principle of plenitude [sic]."[80] Forty years earlier Henri Daudin had suggested, "In sum, and by whatever method that can reconcile or blend the exigencies of systematic arrangement and the idea of real affinities, it is clear that Linnaean natural history, far from beating back the traditional belief of a universal continuity of living beings, profoundly assimilated itself with it [the idea of continuity] and revived it without contradiction."[81] The search for the natural system occupied only a small portion of Linnaeus's time and publications, so busy was he cataloguing the stream of undescribed plants and animals being sent to him from all over the globe. The development of the sexual system and the rather analytical description of genera and species occupied most of his time and energy, and so it is clear why plenitude should seem to have "only a speculative and rhetorical importance"[82] for him. However, even with Linnaeus it is evident that the arrangement of at least parts of his general system is informed by—or is at least consistent with—ideas of continuity.[83] Antoine-Laurent de Jussieu also produced a natural method which both in theory and practice is closely integrated with his belief in continuity, and what Stafleu and Daudin had to say of Linnaeus is equally applicable to him.

M. P. Winsor[84] suggested, although not without qualifications as to the work of Thomas Henry Huxley, that in zoology the enterprise founded by Linnaeus had led to the development and refinement of a generally satisfactory natural system, largely with the help of a thoroughgoing comparative morphology (including in this context gross anatomy and development). The comparative anatomy championed by Vicq d'Azyr, Cuvier, and others, as well as observations on embryology and development that were so popular in the nineteenth century,[85] led to changes; distinct groups became evident.[86] Jussieu's *Genera plantarum* and the subordination of characters enunciated there were recognized by both botanists and zoologists as being major achievements, but they were simply way stations in zoology. However, in botany there were a few suggestions that the synthetic approach was not the most suitable; the botanist's view of the world did change somewhat, but method and continuity persisted. As was widely recognized by the end of the first decades of the nineteenth century, zoologists had moved to the forefront of systematics, and the wealth of information disclosed by comparative anatomy was partly responsible. Lamarck's complaint—"our knowledge of the organization of plants is much less advanced than that of the organization of a great number of known animals"[87]—was common, and the zoological system appeared to be more securely founded than the botanical system. Plants were seen to be less easy to work with than animals, their groups were less obvious.[88] Not only was there simply more to be seen on dissection of an insect, or the animal in a molluscan shell, than of the flower of a buttercup or even an orchid, but no agreement on a calculus of characters was in sight in botany.

In 1893 it seemed to Bessey that there had been two occasions of stasis in systematics—the first entailed by the grudging acceptance of the natural system by some botanists, and the second by the failure of evolution to have much of an effect on the discipline.[89] But it was the idea that nature was continuous that was at the heart of Jussieu's natural method, and this idea was only partly supplanted by the belief that groups were discrete when *On the Origin of Species* was published.[90] Thus it was not simply early nineteenth-century classificatory practice that had become embedded in evolutionary systematics, but a practice with considerably more ancient conceptual roots based on a worldview that presumed continuity between all living forms. Jussieu's unswerving belief in continuity might seem to make some of his methodological prescriptions of dubious relevance for systematists working in the nineteenth century, let alone those active in the twentieth century. Yet Mayr signaled out for attention in his history of systematic thought the importance of the idea of synthesis, which he claims had crept "virtually unnoticed" into systematic methodology early in the nineteenth century.[91] Yet the initial impetus for the spread of the synthetic method was given by Lamarck and Jussieu in the previous century; both were believers in continuity, and continuity was the essential precondition for synthesis to succeed. Continuity obligated synthesis; conversely, synthesis was questioned by those who believed the natural world was interrupted by gaps.

Paradigms and Epistemes

This gloss on the history of systematics bears directly on discussions of paradigms, epistemes, and change. I have argued that one element in the eventual acceptance of synthesis and other ideas originally connected with continuity was almost inadvertent and that it arose from theoretical confusion (or apathy). Our acceptance of synthesis did not come from discussions as to the most appropriate way of recognizing relationships in a world increasingly regarded as comprising discontinuous groupings. How then can one think of systematics in the context of paradigms or revolutions?[92]

At one level, systematic thought in the period 1789–1900 and beyond (focusing, here, on higher levels of classifications and on plants more than animals) seems to have been in an almost classically paradigmatic phase. All naturalists worked to perfect "the" natural system. There was little discussion over fundamental questions such as the subject matter and the methodology of the discipline.[93] Systematic practice then owed much to a belief in the continuity of nature. Of course, there was discussion over matters such as typological thought, the place of cell- and tissue-level anatomy in the discipline, and at the end of the century the value of evolutionary diagrams, but these issues did not really disturb the discipline.[94] Use of a mounted lens became widespread by botanists in the 1820s and 1830s, and this was perhaps the major technological advance for the general systematist during the whole period. The use of cell- and tissue-level anatomical data (especially after the 1870s) promised to be the major innovation in terms of characters utilized. Although the constraints of the hierarchy within which systematists describe their findings may well hide major differences in thought, I have attempted to get beyond this hierarchy and to enquire about its construction and interpretation. In so doing I still find little evidence of fundamental change.

The case of systematics was particularly interesting to Thomas Kuhn because it seemed to be a science that had no or almost no symbolic generalizations[95]—indeed, the general conceptual core of systematics during the whole period covered here is remarkably slight. This befits an activity that can claim to be one of the direct descendants of seventeenth-century natural history, that is, of an approach to the study of nature that focuses on cataloguing using an effectively nonanalytical approach. Many naturalists would have been hard pressed to identify any conceptual core to their work at all; their job was simply to produce a classification. C. C. Gillispie accurately captured the mood of the discipline when he described the procedure followed by French botanists at the end of the eighteenth century as they grouped genera into natural families: "The principle of subordination of characters was a method of analysis, therefore, or more loosely, a way of seeing and doing things, rather than the basis of a law or theory. The history of science offers no more striking illustration of actual procedures forming a tradition of research within a scientific institution."[96] But what was involved was more than an institution—more than the Jardin du Roi and its successor, the Muséum national d'Histoire naturelle. It was the whole discipline, yet its practitioners could not agree over how characters were to be evaluated and used in classifications, and on other basic matters. So what the classification that resulted from such activities represented, other than a very ill-defined "nature," was unclear even to these practitioners.

As Michel Foucault[97] observed, it was a positive requirement of botanical method of this era that nature should be deemed continuous; Jussieu certainly would have agreed. This was the conceptual core of the practical procedures that Gillispie singled out for attention. The natural method as it was initially developed was tied to continuity and to a particular way of seeing the world—whatever its later practitioners might have felt—and in botany it remained largely focused on the outside of the organism. In this respect the practice of systematics then was similar to Foucault's idea of natural history in the classical period, traversing "an area of visible, simultaneous, concomitant variables, without any internal relation of subordination or organization."[98] Within botanical systematics there was no real move to looking below the surface of the organisms in order to understand the hierarchy of functions between characters; although characters were subordinated, it was usually on an ad hoc basis, and if they were rigidly subordinated, the results were usually considered to be decidedly less than satisfactory; and there was little interest in investigating what made life possible and in how to distinguish life from nonlife.[99] There was thus no transition in epistemes,[100] no transformation of natural history into a biology that looked at the workings of life and living organisms. In particular, there was no change from "Naturbeschreibung" to "Naturgeschichte."

The very word biology, coined at the end of the eighteenth century by Lamarck among others (but not to gain common currency for over fifty years), suggested the beginning of the breakdown: a qualitative gap between life and nonlife had become evident.[101] It might seem (with the advantage of hindsight) that the botanists early in the nineteenth century who explicitly rejected continuity and who simultaneously questioned synthesis were groping toward a methodology compatible with a new interpretation of the living world that was then emerging. But nothing came of Augustin-Pyramus de Candolle's rather half-hearted suggestion that a more analytical way of processing data of systematic importance was needed; his focus on the symmetry of groups never became an integral part of systematics.

Both before and after 1859 few naturalists articulated the theoretical consequences for their discipline of the existence of gaps in the living world, and a number of distinctive ideas and terms used by Jussieu remained in common parlance. Despite the brave talk about revolution early in the nineteenth century, there was no abrupt transition when it came to the classification of nature. Yet it could still be argued that the revolution Jussieu attempted was complete—it just had little effect. Foucault suggested that keys became necessary because natural history gave way to a discipline that looked below the surface of the organism,[102] and the new groupings that became evident conflicted with the superficial appearance of an organism. Lamarck in 1778, although not obviously a proponent of anatomy then, heralded the change, and Jussieu, more interested in gross anatomy, brought the change to its conclusion in 1789.

But historical assessments ranging from "no change" to "revolution" or "epistemic rupture" are alike unsatisfactory.[103] To understand the development of systematics, one has to make sense of how the naturalist observed the organism in light of the way nature was visualized and in light of the nature of the organism. And in botanical systematics what seems to be important is the fact that flowering plants did not readily yield to the needle and blade the secrets of their relationships; it was perhaps because of this that ideas of continuity tended to persist. There are also more external factors to understand; some of these were discussed in the previous chapter. Notably, the desire of many botanists to be "useful" led them to seek stability almost whatever the appearance to the contrary. At the end of the nineteenth century, systematists seem to have turned even more toward floristic studies and inventory;[104] this would not only further alienate them from other biologists but also would make the prospect of major changes in the discipline less likely.

Persisting Ambiguity

We can now better understand the conflicts between the continuous series and the hierarchy that consisted of exactly characterized genera and species,[105] and between natural affinities and practical disposition.[106] Both conflicts were evident in naturalists' efforts to classify a nature perceived as continuous. Henri Daudin was mistaken when he suggested[107] that Cuvier and Jussieu had integrated the (analytical) Linnaean system with the natural method in their discovery that natural relationships could be satisfactorily represented by a hierarchy. Linnaeus, Lamarck, and Jussieu linked organisms so that the continuous relationships of nature would become evident. Even a *natural classification* would have seemed an oxymoron to them; there was nowhere to divide nature.[108]

Linnaeus was unable to provide descriptions, let alone definitions, of his natural groupings of genera,[109] although the genera themselves could be both characterized and identified. For Lamarck, characters and groups were only incidentally coextensive. Jussieu concentrated on higher groups, but these were part of the continuous natural order, and he paid much attention to characters, although they rarely characterized his groups. Furthermore, he tried to link inconspicuous but important characters with conspicuous but less important differentiae.[110] But (as presented in chapter 2) Lamarck had dissociated the identification of organisms from the recognition of natural relationships. The dangerous separation between series and hierarchy that Daudin saw as being healed is dangerous only insofar as one thinks nature should be readily understandable.[111]

Jussieu's ultimate goal was a natural arrangement of plants.[112] To the extent that his groupings came to be interpreted as a class hierarchy that existed in nature, it would represent a failure of his intentions. Thus, to the extent that his nineteenth-century successors persisted in interpreting what we think of as a class hierarchy as a modified series, they were following the spirit of his method. So if we accept Daudin's point of view, Jussieu can be considered a success only by taking his "classification" for what it was not. Most of his successors did not have a clear idea of the shape of nature, let alone a methodology that was articulated with it. The Candolles, rather exceptional in this respect, had trouble finding characters for families that they believed were discrete. In fact, the discovery of characters defining groups was a major goal for nineteenth-century systematists, even for those who seemed to believe in a nature that was more or less continuous.[113] But groups obstinately remained polythetic, with the consequence that their boundaries were fluid.[114] The main way to recognize relationships was by finding intermediates—whose very existence called into question the distinctness of the groups they joined.

Jules Émile Planchon nicely captured the conflict between continuous nature and definable groups in the introduction to a paper he read in 1848 at the Linnean Society of London. He felt that he had solved a long-standing problem, that of relating the idea of groups to the notion of continuity. But although he described clearly the continuous nature of relationships, the "happy and correct delineation" of groups to which he refers seemed to elude him.

> The task of naturalists, in tracing out the affinities of beings, is not unlike that of putting together the many and various parts into which a picture has been purposely cut. If of these parts a great number are wanting, the difficulty of arranging the existing ones will be increased; groups will form themselves either isolated, or connected only by narrow links; and perhaps some parts, finding no immediate neighbour, will be forced into unnatural connections. But in proportion as the missing pieces are collected, gaps will successively be filled and projecting angles find their corresponding sinuses; until at last, by the connection of all its parts, the picture will assume its perfect integrity. Such would be the progress and such the end of systematical natural science, if, according to an ingenious suggestion of Linnaeus with respect to plants, the juxtaposition of countries on a geographical map were a faithful pattern for the representation of the affinities which connect in one harmonious whole the innumerable objects of nature. Now, although such a disposition of natural tribes is but a degree of improvement over the imperfect linear series; although the outline of no group can be traced with mathematical precision; still every real advance in knowledge, every new object which is carefully compared with the mass of others, serves ultimately either to fill up intervening chasms, or give a new direction to the outline of some group; or perhaps becomes itself the central nucleus around which parts, until then floating without determinate station, will gather themselves into one homogeneous mass. That such is the usual march of natural methods I need not say in a place where the memory of Linnaeus, although justly connected with the most ingenious of systems, is no less so with the happy and correct delineation of the groups which animate the vast picture of organized creation. The subject which I submit to the enlightened judgement of the Society naturally suggested the foregoing reflections, since it offers a striking example of the use which the natural method makes of new materials to improve (as I venture to hope) the arrangement of the old.[115]

Nature was continuous, and the more the continuous natural order was known, the more groups lost their definition. Yet it was only through the outlying members of natural groupings that further natural connections, direct continuity between one group and another, could be established. So Planchon solved his problem by conjuring up a type for the new family that he was describing, a type that reduced the variation in the family to reasonable bounds. More precisely, it eliminated it; indefinable nature was defined. Relationships, however, were mediated not by the type, which was almost central in the family, but by the genera that were most different from it and were on the periphery.[116]

This conflict does not result simply because Planchon interpreted Linnaeus in an idiosyncratic fashion. Linnaeus denied that defining characters were known even for the most natural (and hence apparently definable in the strict sense) groups such as the Umbelliferae; the naturalist should emphasize transitions since natural families could not be given a character.[117] And from the very beginning of the *Genera plantarum* Jussieu's belief in continuity and his concurrent description of taxonomic groups at different hierarchical levels set up similar conflicts. The close of the Introduction leaves the reader in some confusion:

> Meanwhile, he [Bernard de Jussieu] much wanted to exhibit another distribution in the public garden, one which, while somewhat systematic, and not subverting the natural orders, would accommodate easy study by means of established and multiplied classes. Therefore, this method was carried out at the school [of botanists] of Paris; in [this method] the orders and the primary characters of the seed and sexual organs are preserved, and in addition it supplements the convenient primary divisions of Tournefort. Happy the author [Antoine-Laurent] who, on the one hand, mindful of the work of true botanists, has made clear the natural laws of affinities, as well as transmitted to posterity the accurate definition of orders and genera, and, on the other, for the sake of beginners has arranged orders and genera sometimes with the most certain of signs, sometimes supported by others that are general and obvious together . . . .[118]

Taxonomic groups with definite, let alone essential, characters on which Jussieu laid so much emphasis and continuity, the basis of his nature, would seem to have little to do with each other. Yet the use of terms like "organisation" by Jussieu in connection with these groups would tend to subvert the indivisible continuity of which these groups were artificially delimited parts.[119] For out of his continuous order—continuous at least at the level of the catena of characters that distinguished it—came discontinuity and hierarchy. These arose from the differences and necessary relationships between ranked and subordinated groups and their characters. Believers in continuity like Jean Robinet could envisage a single prototype, and the fullness of nature was the result of all possible permutations on this plan.[120] A yet finer grid of discontinuities and types might result. Foucault described the results of the systematization of nature, "[categories] will correspond (if they have been properly established) to areas that have a *distinct* existence on this *uninterrupted* surface of nature; they will be areas that are larger than individuals but just as real."[121] Although strict continuity might allow neither the existence of groupings nor even the reality of individuals, both the subordination of characters and the reification of taxa would suggest that subdivisions of the continuum were discrete and characterizable groups, perhaps even with typical members.

The ambiguities and discrepancies evident in the *Genera plantarum* became far more pronounced over the years, yet they were never explicitly resolved. Most of its readers, largely ignorant of changing worldviews yet aware that systematic practice had not changed, would be able to interpret its author as they wished. Yet Jussieu was a limital nominalist,[122] even if he emphasized the characters of the groups that he described. In contrast, Augustin-Pyramus de Candolle was a hierarchical nominalist, or perhaps even a realist, although he could not provide definitions (in the strict sense) of the groups he recognized. Both men described taxa at various levels of the hierarchy in the natural system.

In Adam Smith's discussion of the differences between lay and learned in their appreciation of nature, he mentioned the ease with which "the learned give up the evidence of their senses to preserve the coherence of the ideas of their imagination."[123] But in the case of botanical systematics the evidence from the senses and the ideas of the imagination were interconnected. Emphasis on external form would make any distinction between characters that showed analogy and those that showed affinity difficult; relative, let alone absolute, weights for characters were hard to assign. The natural system was built up by synthesis, and relationships remained diffuse and more or less continuous. Continuity would tend to be evident when external form was emphasized.[124]

Ultimately, many botanists rejected the notion that groups had some sort of quiddity; the extent of a group, not its center, would be their interest. As B. L. Robinson observed in 1906, the type method might give a botanist the latitude and longitude of a botanical archipelago, but not its extent; the job of the botanist was to sound the waters and find where the deep channels lay.[125] The reader was not told how the botanical survey vessel went about its task; indeed, without the idea of the existence of a real center to the archipelago being added to that of deep channels,[126] those channels would probably remain undiscovered.

Alphonse de Candolle, just like his father, believed in discrete, definable groups, and, also like his father, he believed that size was irrelevant when determining group rank or circumscription.[127] In his view it was axiomatic for families and classes to be distinguished from one another in as permanent and universal a manner as were genera and species. But when it came to the real world, in this case the distinction between the Loganiaceae and Apocynaceae, he was reduced to saying that the variation in these two families was fundamentally different. And Alphonse went on to say that although there were no constant characters separating the two families, it was up to botanists to verify and demonstrate the constant differences he believed should exist![128] This was hardly satisfactory.

We see systematists in something of an impasse, caught between the extremes of continuity, on the one hand, and groups everywhere, on the other. As librarians of the book of nature, they were inevitably drawn to identify, categorize, name, describe, and even define. Yet definitions—even in a loose sense—of genera, and still more those of families and other higher-level groupings, remained elusive. When it came to the development of the natural system, although systematists after 1859 recognized that evolution when combined with extinction might produce groups, most did not see how genealogical relationships could be established. They still preferred to detail direct connections between extant groups—continuity in the present—in lieu of establishing a genealogy involving extinct ancestors—continuity in the past. There seemed to be no way out of the impasse; worse, neither biologists nor gardeners were sympathetic to their plight. Insofar as either was interested in what systematists did, they had different, even apparently conflicting, needs: a genealogy for the one, names that did not change for the other.

Epilogue

Gerald Holton maintains that themata, recurring styles of thought springing from the ground of human imagination, play an important part in shaping our discourse.[1] One thema–antithema couple that pervades this book is that of continuity and discontinuity. This approach is very close to the complementary ways of seeing in systematic biology identified by Olivier Rieppel: *natura naturans* and *natura naturata*, becoming and being, and process and pattern.[2] To these can be added synthesis, equated with a vision of the "tout ensemble," and analysis, with its emphasis on differences.[3] Taxonomy is often seen as a discipline that stresses discontinuities; at least this is true of the person whom one regards somewhat pejoratively as "a mere taxonomist"—someone hidden away in his or her office, working in solitary study on often very complex hierarchical classifications, but not really integrating this work with biological thought. It is thus ironic to think that the foundations of the discipline were laid by Jussieu, a staunch believer in continuity,[4] and whose method entailed detecting relationships by establishing direct linkages between groups. Admittedly, in his explanation of his natural arrangement, Jussieu tended toward the ecological-functional style of explanation of Cuvier—not the evolutionary connections seen by Lamarck.

Detection of direct relationships remained integral to systematics into the twentieth century. Even when the natural system was depicted as a tree, whether seen from above or from the side, groups usually were directly joined, "the great vice of the comparative method."[5] The maplike or reticulate representations of nature that predominated in the first half of the nineteenth century had been converted into a tree form by superimposing a vertical axis. Relationships were still discerned in a manner appropriate to a world in which there were no gaps; intermediate taxa remained essential for this process, and systematists saw as one of their main tasks the discovery of further direct connections between groups.

Nevertheless, in the nineteenth century the idea of a nature of continuous relationships was gradually replaced by that of a nature with more or less discrete groups, although this transition was by no means complete by 1859. But it was to take botanists almost another century to begin active development of a theory of systematics that was compatible with the seemingly new world that became apparent in 1859 and the years leading up to it.[6] Only recently has botanical systematics grappled with the consequences of articulating Darwin's nature—history, genealogy, not simply gaps—and systematic methodology, consequences Miller made clear in the 1930s: "The doctrine of evolution is essentially historical doctrine, for it conceives the uniformities disclosed by theoretical analysis to be features of a temporal evolution that can be recovered by way of historical reconstruction."[7] Yet the notion of a Jussiaean revolution in botanical systematics, and of a Jussiaean foundation of "the" natural system, had become an integral part of the history of the discipline by the first part of the nineteenth century. Antoine-Laurent de Jussieu has retained his position as founding father, although the extent of the relationship between his beliefs as to how nature was organized and his method, both as theory and practice, has not been understood.

It has been all too easy to assume that the development of systematic botany since 1789—and before—has been simply the progressive refinement of *the* system. Classifications by Adanson and Linnaeus, and even those of Ray, Cesalpino, and the herbalists, have been compared for resemblances; all these classifications showed, more or less clearly, parts of nature in the groupings that they had in common.[8] Jussieu and his followers perceived a number of families that represented indissoluble and self-evident parts of the natural order, in considerable part because genera in these families had always been recognized as being related. As Alphonse de Candolle noted, "the whole world knows that these groups are natural."[9] But in the 1990s the circumscription of several of these families is being challenged by those who believe that the genealogical integrity of groups is of paramount importance.[10]

For Jussieu theory, practice, and view of the world were all related; he thought that parts of continuous nature were already evident, and he proceeded on the assumption that it was simply a matter of time before the undivided whole would become apparent. Jussieu's world had little to do with that of Augustin-Pyramus de Candolle, let alone that of Darwin, but through the persistence of continuity-in-practice the general approach of most nineteenth-century systematists remained similar to that of Jussieu. One signal that all was not well was the unending proliferation of new systems and the failure to develop any way of evaluating characters and groups other than by trying to achieve a consensus with present and past masters of the art.

But continuity-in-practice is by no means the whole story, and I have not provided a single, satisfactory explanation of the course of systematics in the late eighteenth and nineteenth centuries. There have been continuing intrinsic problems caused by the plant as subject—the relative simplicity and lability of its structure, its distinctive development, perhaps even its basic construction.[11] We also must remember that many of the genera and other higher taxa recognized by naturalists in the eighteenth and nineteenth centuries are likely to be maintained even in a system reflecting only genealogical relationships; gaps, for whatever reason they were recognized originally, may reflect separations of lineages.[12] The sociological forces that have helped mould the shapes of the classifications also need more attention,[13] and the issue is not simply the lowly status within the scientific hierarchy of those who made classifications. In this last context, we may find classifications and names persisting for no good scientific reason—particularly when there are a number of equally acceptable classifications and when a user, not a systematist, is effectively allowed to moderate between them. Names become conventions, and they continue in use with a particular circumscription because they always have been used in that way. The stability of a classification may then owe little to the accuracy with which that classification reflects nature, however nature is defined. It is not simply that such classifications are made—that they are artificial, made by humans, rather than found, existing in nature—but also that there is no direct appeal to nature when they are evaluated.[14]

Overall, it is evident that new observations are not integrated into classifications in any direct and secure fashion, and that classifications do not represent simple summaries of cumulated information and knowledge of systematists. Certainly, the British and colonial governments, who between them bank-rolled the so-called Colonial Floras (one of the major products of the English-speaking systematic community in the century after 1860[15]) had minimal interest in understanding the relationships of plants. They wanted an inventory of the natural products of the Empire.[16] Botanists on the German expeditions active mainly in Africa and the tropical southeast Asian region at the end of the nineteenth century (the expeditions were searching for new sources of guttapercha and rubber) do not seem to have invoked a knowledge of relationships as being integral to their endeavors, although the ability of plants to produce latex of one sort or another is connected with their affinities.[17] Those expeditions seem to have been more like many current programs that screen plants for their potential use as anticancer agents—very much hit-or-miss affairs.

But what *are* names? They can serve as a finder's tool for locating all that is known about a subject. Is that all, or is that something which is of itself tremendously important? Important, yes, but systematists have always been interested in relationships as well. They have seen themselves not only as librarians—knowing what something was, where it was to be found in a collection, and where to find out what was known about it—but also as integrally involved in the explication of nature's diversity. If we return to the 1780s when Jussieu was working on his *Genera plantarum*, we find the whole scientific community occupied in developing nomenclature and terminology appropriate for each discipline. Maurice Crosland summarized the aims of the reformers of chemical nomenclature in his quotation from the *Méthode de la nomenclature chimique*, in which Antoine-Laurent Lavoisier had a major hand; this book appeared in 1787, two years before the *Genera plantarum*.

> Now, if languages were true instruments which men had formed to assist the process of thought, it was important that these languages were the best possible. The perfection of language was most important to

> those beginning to study science, for once false ideas had become established, they became prejudices which were difficult to discard[. As Guyton de Morveau, Lavoisier, and their colleagues wrote in the *Méthode*]: "A well-made language . . . will not allow those who profess chemistry to deviate from the march of nature; it will be necessary, either to reject the nomenclature, or to follow irresistibly the path which it indicates." If language were considered merely as a collection of representative symbols, it would follow that an imperfect language could also transmit false ideas, even if the original idea were not false. From this point of view, the perfecting of the language consists in expressing in words the facts—no less and certainly no more.[18]

In the late eighteenth century, it was generally agreed that Linnaeus had provided a language for botany, and Crosland rightly concluded that Linnaeus's main claim to fame lay in his binomial nomenclature that had been applied consistently to all organisms.[19] Linnaean binomials formed the inscriptions in the book of nature, but the Linnaean system as its table of contents was a failure, even when detailed descriptions were provided. It was the so-called natural system that was to guide the reader around the book.[20] Yet the natural system, too, failed because the names of plant groups were not symbols expressing known or knowable quantities analogous to those used in mathematics,[21] and they had no role to play in laws like those of chemical combinations and angles of diffraction.[22] Moreover, the laws of morphology in which Augustin-Pyramus de Candolle and others were interested did not materialize, and evolution, let alone phylogeny, did little to make the shape of nature as perceived by systematic botanists comprehensible. The method Jussieu extended to the level of family and beyond was based on continuity, so the very idea of a language—in this case, names of genera and families—expressing the "facts" of nature becomes still more problematic.[23] From this point of view, the language of botany (the names of organisms in the context of a classification) communicated less than systematists might have wished (or thought).

One goal of the reform of scientific nomenclature was to connect names and things, and by providing a sound basis for discussion to aid in the discovery of regularities and laws in nature. The conceptual background Linnaeus brought to his nomenclature and classification was largely that of an essentialist, and the intent of the nomenclatural reform he achieved in botany and zoology can be seen as appropriate for disciplines like chemistry and crystallography, even if the binomial nomenclature was not. Although plant taxa are no longer considered to be part of a continuum in the here-and-now, the nature of species, genera, or other taxonomic groupings of organisms has turned out to be very different from that of a "species" of salt or crystal.[24] Whatever different species or higher taxa of plants are, they are not like samples of the one salt that can be produced in different places, perhaps even using different synthetic pathways. All plant taxa are unique,[25] and, as George Bentham observed, a variety was not really different from a family.[26] Any idea of a language of nature in the sense advanced in the later part of the eighteenth century thus vanishes.

The problem that systematists (unknowingly) faced as they attempted to locate botanical classification within the new framework of evolutionary biology was enormous.[27] Much of nineteenth-century systematics assimilated a very distinctive idea of nature because the *Genera plantarum* had been accepted into the canon of systematic literature. The discipline was thereby predisposed to rediscover the kinds of relationships discovered by practitioners of natural history in the eighteenth century. Despite systematists' pride in the empirical nature of their discipline, and despite their suspicion of theory, they had placed a veil between themselves and nature. Nature was not to be apprehended directly in the way that systematists thought possible. Furthermore, nineteenth-century systematists had a clear idea neither of the shape of nature nor of what their discipline was about, other than enumerating nature's productions and detecting "relationships." Their problem was compounded because they allowed a heterogeneous and ill-defined group of users of plant names to have a say in the names they themselves used, and thus to fix the limits of taxa in the natural system.[28] The "natural" system took on a life independent of the ideas of the naturalists cataloguing nature.

The story that is unfolding makes sense of the halting development of a discipline that has as its proper subject an understanding of relationships of organisms, but which has all too often been seen as ignorant of how to go about discovering the laws of the universe. Indeed, the laws that the Marquis de Condorcet, Antoine-Laurent de Jussieu, Augustin-Pyramus de Candolle, and others wished to find are still unfound—and still relatively little is known about plant development. Nevertheless, progress is being made in detecting the genealogy of life, and a grammar and syntax proper to a biological language in the context of genealogy are being developed.

As a systematist, I have been alternately depressed and elated as the themes of this book developed. I am depressed when I reflect on the long history of the confusion that is still evident and pervasive even now: systematics is a discipline that has not clearly separated the development of knowledge of plant relationships from the provision of identification services for humanity, and its practitioners have not been silent about their disdain for theory. Systematists have, in part, themselves to blame when they complain about the low esteem in which their discipline has been (and is) held. I am elated, however, by the prospect that the discipline does have a chance to open up: systematics can free itself of its undue reverence for tradition if systematists come to understand more about the development of that tradition in its historical context.

APPENDIX 1

Translations of Jussieu's Early Works

This appendix includes translations of four of Antoine Laurent de Jussieu's early works. All are important for understanding his philosophy, yet none is readily accessible even in the original languages (French and Latin). Furthermore, to my knowledge, none of these works has yet been translated into English. The first two pieces were translated by me, the third by Susan Kelley, and the fourth by Susan Rosa.

The first three articles translated here were published originally in French, but they appeared in periodicals that are not easily obtainable by most botanists today. The fourth, the Introduction to the *Genera plantarum*, is more generally available, but, being written in Latin, is little read. I have also included a brief commentary on the posthumously published edition of this Introduction, with translations of a few passages that are particularly germane to the purposes here.

I have burdened these translations with only minimal commentary, as the whole book (particularly chapter 3) is an extended discussion of the contents of these works by Jussieu. Page numbers from the original publication appear in brackets in the appropriate places in the translations; this will aid consultation with the original. Jussieu's footnotes have been converted to endnotes and integrated into the numbering system used for my own notes within the translations; an asterisk (in both the text and the list of notes) distinguishes Jussieu's original notes from mine. The first note by Jussieu does not appear until note 44 in the sequence.

Major punctuation marks (colons, semicolons, and ellipses) have largely been kept, but commas and word order may have been changed and serial commas have always been inserted, in accordance with American publishing standards. "Analogie," "affinité," and "essence," their variants and their Latin equivalents, have been translated literally because of the importance of the different kinds of relationships between organisms and structures to which these words refer. "Rapport" is translated as relationship and "primitive" as fundamental unless otherwise noted; no other words are similarly translated. "Méthodistes," referring to people who used single characters for delimiting groups, and "organisation" have also been untranslated (except that the spelling of the latter has been Americanized). Jussieu's italics have been kept except where these are used for references and quotations of Linnaeus's Latin; the latter have been translated. Generic names have been silently capitalized and italicized, formal Latin names added to vernacular names, and informal family names converted to their formal Latin equivalents (in the translation of the Introduction to the *Genera plantarum* using the family names as they appear in the body of the work); the single exception is the "renoncules" of Jussieu's paper of 1773 which I have translated as ranunculi, rather than Ranunculaceae, to preserve some of the flavor of the original. Spellings of personal names have been changed silently where these differ from those used elsewhere in the book. Jussieu's capitals I have kept, although these are not consistently used in the different articles, or even in the different printings of the Memoires of the Académie royale ("Botaniste" and "botaniste" is a good example). Additions, and words the translation of which presented problems, are enclosed in square brackets in the text. Other annotations, mostly brief explanations of difficult passages or full citations for the references Jussieu makes, are included in the end notes.

Translating the Introduction to the *Genera plantarum* presented special problems. One of these concerns the real significance of the botanical terms Jussieu used. For example, in the Introduction Jussieu used the term "corculum" as equivalent to the whole of the embryo, and in the second edition of the Introduction he made his equation of the two explicit; in earlier botanical authors the word refers to the seat of life in the embryo, the plumule which contains the *punctum vitae*.[1] Although "corculum" barely seems out of place in a Latin text, this would not be so in a French text, and in the latter situation Jussieu used "embryon." Jussieu uses two pairs of words, "animales" and "animanti," and "plantes" and "vegetanti," apparently interchangeably, although the two members of each pair give different impressions; these words have been translated literally. Other words may more clearly be used in different ways: "cortex" may mean the bark of a tree, or one of the two basic tissues in a particular conceptualization of the plant body (the other tissue is the medulla, which is enclosed by the cortex). Here different meanings are indicated in the endnotes. It should also be remembered that "ordinatio" has been translated as arrangement and "distributio" as distribution and that there is no Latin word that unambiguously means classification in the technical sense of the plant and animal classifications discussed here ("classificatio" does not exist in classical Latin).

A problem both in translating and in understanding these texts is that the late eighteenth century was a period of development and codification of botanical terms. At the same time, Latin lost its position as the medium for general communication of ideas. Jussieu used "ordre" and "famille" interchangeably for the rank that is now called family (although only "ordo" when he wrote in Latin), and class for any of the three ranks above the level of family. But as Roselyne Rey has noted, in the late 1700s these and other words that came to be commonly used in taxonomy had not yet acquired their current meanings; indeed, some of them, in particular "famille" or "family," might carry connotations of generation or genealogy to which some naturalists objected.[2] The Introduction in particular reads as a document that owes much to an early eighteenth-century view of the plant world. Jussieu uses many archaic words, some of which do have modern meanings, but his view of plant construction and development (for example) is very similar to that of Linnaeus and other earlier authors. I believe that Susan Rosa's translation, which involves a minimum of interpretation, may help in the much-needed task of relating Jussieu's work to that of his predecessors. To this hoped-for end, I offer the following translations.

EXAMINATION OF THE FAMILY OF THE RANUNCULI

EXAMEN DE LA FAMILLE DES RENONCULES.

Mém. Math. Phys. Acad. Roy. Sci. (Paris) 1773: 214–240.

Translated by Peter F. Stevens

[214] The plants spread over the surface of the earth have similarities and differences among themselves that are based on the position, number, and arrangement of their parts, and these similarities can, from one point of view, be compared with the affinities that the Chemists recognize in the mineral substances subject to their scrutiny. Chemical affinity is that disposition, more or less strong, that two bodies have to combine, [but] it is not the same in all; some combine closely and readily; others with a weaker attachment, and can be separated by an intermediary; others never unite, or with great difficulty. Plants show almost the same slight differences, the same gradation; they have characters by which they are brought together, they have those in which they differ: the combination varying from one to another, has helped botanists to make classes, sections, orders, families, genera, species. The principles of the science have been condensed into a Table ["tableau"],[3] and this Table could be in Botany what the Table of affinities is for Chemists.

Naturalists, who are occupied in the examination of plants, and the necessity of classifying them, have followed different routes to achieve the same end: each has made up his characters from the parts that appeared to him the easiest to observe, or the most suitable to form the basis of an arrangement: some have preferred the fruit, others the flower, this one the corolla, that one the stamens. Without stopping to evaluate their work, I will observe that all have imitated, as far as they could, the order of Nature; [215] that some of them have sought that order; but given the impossibility of finding it, they have produced systems which can be considered analytical tables ["tables raisonnées"], which bring together from a single point of view the material for its construction, until a genius more fortunate or richer in observations undertakes to work them up.

In an artificial method one easily finds the characters of classes ["classiques"][4] as they are taken from a single part [of the plant] and as they depend on the whim of the author; after having decided upon his classes, he can without further thought place there all genera which have the characteristic feature; but the result is firstly that his characters are not invariable; an aborted stamen or an extra one embarrass the followers of the sexual system; in another method, the regular corolla is often confused with that called irregular: in the second place, analogous plants are often separated because they differ in a single feature, while a character common to some plants, otherwise very different, joins them in the same class. It is for this reason that M. Tournefort is forced to join the cinquefoil [*Potentilla*] and buttercup [*Ranunculus*], and to separate the columbine [*Aquilegia*] from the latter, because of [the constraints of] his arrangement; that the system of M. Linnaeus gives us the sorrel [*Rumex*] near the colchicum [*Colchicum*] in a class different from that of the knotweed [*Polygonum*].[5] An author who is a Méthodiste buys at some price the ability to form classes easily, because his work cannot fail to be defective in some way, and the perfection of his work can only be measured by the mistakes he knew should be avoided.

The person who seeks the natural order does not have the same ability [to put plants in classes], because the characters do not depend on his choice, but he has a slighter risk of falling into the same trap [of placing unrelated plants together, and vice versa]. It is only after having examined all the genera that appear to belong to a family that he tries to make a general character from them; nothing escapes his researches; he observes with care all the parts of the fructification without neglecting the others; when he has assured himself by repeated observation of the [216] nature of all these plants, he establishes an analogy between them based on the resemblance of several parts, and the character of the family becomes the result of a speculation equivalent to those in the most abstract Sciences. A man of talent can make systems, he can vary them *ad infinitum*; but the natural order will only be the work of a consummate Botanist in whom the patience for examining the smallest details equals the genius in drawing consequences from them, for forming connections, in a word, for making Botany not a Science of memory and names, but a new Science, which has its combinations and affinities as in Chemistry, its problems as in Geometry.

Some highly esteemed Scholars have already outlined this work; I will only mention the more important. M. Linnaeus has given the *Ordines naturales*;[6] M. Bernard de Jussieu, my uncle, established in the royal garden of the Trianon a series of natural families which has only distant similarities ["rapports"] with the arrangement of the Swedish botanist.[7] M. Adanson, in his *Familles des plantes*,[8] followed a plan different from the arrangement of the Trianon, but which is more like it than that of M. Linnaeus. It would not be at all proper to praise these authors here, all three are living and members of this Academy; the description and analysis of their works would need a more complete panegyric. My object today is only to examine a series of plants which M. Linnaeus places together in his twenty-sixth order under the name *Plantae multisiliquae*, and which, at the Trianon and in the work of M. Adanson, bears the name *Famille de renoncules*.[9] This examination consists of (1) the determination of the true characters essential to this family, and which distinguish it from all others; (2) the assignment of the place it should occupy in the series of the natural orders: two equally interesting points which need many details [to understand], and which present some difficulties to surmount, some problems to resolve; on the one hand, one must examine individually all the plants of the family, recognize the structure and position of each of their parts, and establish their general relationships; on the other, considering them [217] together and from a single point of view, it is necessary, by a comparison between all families, to determine those which have most affinity with the ranunculi. I will limit myself, in this Memoir, to the first part, to the examination of the characters of the family.

These characters are: (1) an embryo with two lobes or cotyledons, concealed near the funicle of the seed in a cavity almost at the apex of a horny body, which occupies almost the whole interior; (2) a calyx of several parts which is sometimes absent; when it is present, it is always borne on the support of the pistil [= receptacle], in the same way as the corolla which is usually made up of several petals; (3) an indefinite number of stamens borne on the same support; their anthers are of two loculi separated and attached by their entire length on both sides of the apex of the filament which supports them; these loculi open longitudinally into two valves and allow the globular powder to escape; (4) the pistil is made up of several ovaries borne on a common receptacle and each surmounted by a style terminated by a simple stigma; (5) these ovaries become on ripening so many capsules, either single-seeded or many-seeded; the latter open into two valves on the inner face, and to the edges of [the valves] are attached several seeds; the single-seeded capsules never open, and can be regarded as the outer covering of the seeds that they enclose; (6) the leaves are alternate in almost all genera; their base is never accompanied by stipules, it is sometimes enlarged and forms a half sheath, or even sheaths almost completely around the stem, which [itself] is usually herbaceous.

Among these characters, some are constant, others can sometimes vary: each of them taken separately is found in one or several other families; but their combination is found only in that of the ranunculi: it is this combination which makes its essential and invariable character. M. Adanson referred to them all in his account,[10] [218] adding other details which are suppressed here as [being] less interesting. One can, like him, divide this family into two sections; the first will include *Clematis*, *Atragene*, *Thalictrum*, *Anemone*, *Adonis*, *Myosorus*, and *Ranunculus*, which have single-seeded capsules; *Paeonia*, *Caltha*, *Helleborus*, *Isopyrum*, *Trollius*, *Aquilegia*, *Delphinium*, *Aconitum*, *Garidella*, and *Nigella* will be placed in the section with many-seeded capsules. These are almost all the genera which the Authors mentioned have referred to the ranunculi, if one excepts only one or two that they have either added or removed for particular reasons which would be easy to explain. This family bears the name of one of its most common genera; that is most suitable, as one lacks simple names that briefly indicate the principal characters of a family, like the names of umbellifers and composites; those of the ranunculi, roses, pinks, and lilies convey without difficulty to the mind the idea of the families to which these plants give their name, and are much preferable to the obsolete and barbarous terms used by some Authors. One cannot sufficiently clarify the language of a science which, already difficult by itself, becomes still more so by the shackles which an obscure and idiosyncratic idiom give it.

Nomenclature ought not to be neglected, but the study of characters is a more important part of Botany; some are general, others particular; there is one that is essential, another which is not; the distinction between families is not always based on the same parts, nor on the same number of parts; they vary in their relationships. Some, such as the Labiatae, Umbelliferae, and Leguminosae, have characters so simple and at the same time so uniform in all their genera that they have not escaped [recognition by] any Botanist; and in all methods they form classes, or at least well-characterized sections. But this uniformity, which is very useful for determining families reliably, is an embarrassment when it is necessary to distinguish genera that [219] are easily confused, having only minute differences: it is perhaps an advantage that the number of these families that are so simple is not great. Nature, in varying its productions, makes the distinction of the genera in the other [families] easier; but that cannot be done without increasing in the same degree the difficulties in the research on the families themselves, because a larger number of special characters is then needed to make the general character.

There is an example of this in the family of the ranunculi which, although complicated at first sight, is nevertheless very natural; the agreement of the three illustrious Scholars [mentioned above] in the enumeration of its genera is already a strong inference ["induction"] in its favor; the precise and detailed analysis of each of its characters lends support to it. In order to make [the analysis] more sure and at the same time more interesting, I thought that it would be relevant to add some general ideas as to the way to be followed in research on the natural order, and to recall some of the principles that establish a true affinity between the different parts of the fructification.

The embryo contained in the seed is the first part of a new individual; it is the most essential part, the most widespread in plants; their principal organs cooperate in producing it, in preparing the nutriment suitable for it, and maintaining it throughout its development and perfect maturity; they were formed for this purpose alone and dry up after having accomplished it. An apparatus so admirable proclaims its importance, and seems to say to Naturalists that it is in the embryo that they should seek their best characters; the number of its parts and the mechanism of its germination provide differences that are striking enough to divide plants into three large orders, which one will call *classes*, without attaching to them the same significance as the Méthodistes [would do]. The embryo of a large number of plants is made up of a radicle, plumule, and two lobes or cotyledons enclosed in a double membrane; [220] when it begins to germinate, the membranes expand, the lobes transmit an elaborated juice to the radicle and plumule; these latter enlarge; the one is directed downward to form the root; the other, destined to become the stem, raises itself above the ground with the two lobes which are most frequently changed into leaves called *seedling leaves*; these plants are called *dicotyledons*. In those that have the name of *monocotyledons*, there is a single lobe which allows the plumule and radicle to escape: these two parts do not always come out from the same place or in the same way; the differences that they show may become a subject of careful researches. Finally, others in which germination appears to be more simplified have no obvious lobe; in these plants, the body one takes to be the seed develops by simple extension of various points of its surfaces: these beings have been regarded as imperfect, because they do not have all the parts one finds in the others; also, in their arrangement, one does not follow the same rules. Along with some Authors, we will call them *plants without lobes*, or *acotyledonous plants*.

Cesalpino, and others after him,[11] were aware of and adopted these characters in their methods; but they did not distinguish the three classes as they are at the Trianon: this fundamental division, which appeared natural and constant from general observations on germination, was not allowed by MM Linnaeus and Adanson in their natural orders. One even finds monocotyledonous and dicotyledonous plants combined in some of their families;[12] this mixture is, however, rare; in general they never confuse the different beings that make up these classes, but they do not distinguish enough the different families that ought to be brought together. Should one, with these Authors, divide the plant kingdom simply into families, or would it be better to have classes of which families would be subdivisions? This second opinion is more natural, and is even confirmed by the analogy that exists between plants and animals. [221] Germination in the one and incubation in the other present almost the same phenomena; the heart which is found in all animated beings corresponds to the plant embryo; it is, as is the latter, formed first, and by the diversity of its structure makes up classes which each includes several families. Quadrupeds, whales, and birds have a heart made up of two ventricles and an equal number of auricles; that of fishes and reptiles has only one ventricle and auricle; in worms and insects there is a single ventricle, but no auricle: these differences in the structure of the heart should produce some [differences] in the development of the fetus, and the classes thus formed by Nature are readily distinguished by their general appearance,[13] a manner of being which is particular to them; instead of being isolated like the classes of the Méthodistes, their extremes are joined in such a way that one can, up to a certain point, establish an almost insensible gradation between a polyp and a man: the classes of plants, which are no less natural, likewise have a general appearance which is restricted to them and intermediate genera which unite them; two kingdoms, which in other respects have so many properties in common, can still further resemble each other in their major divisions.

The position of the embryo in the seed provides a second essential and constant character. There are plants in which the seed contains only the embryo; in others, it occupies only a part, and the rest is filled with a body of another nature ["substance"]. It is either included in this body, or placed on one side, sometimes at the base, sometimes at the apex, near the funicle or at the opposite end: this character never changes; all the seeds of a plant, all those of a genus have the embryo placed in the same way; this uniformity is even met with quite generally in all the plants of families recognized as being very natural; the Compositae have the seed filled by the embryo; the Umbelliferae always have a compact, hard, almost horny body which encloses the embryo at its apex; that of the Gramineae [222] is placed against the base of a mealy body which it does not penetrate at all. Does not nature, in showing us this character that is constant in known families, show us that it exists similarly in the others, and that none can be natural if the position of the embryo is not uniform in all the genera?

That of the ranunculi does not deviate at all from this basic principle; its embryo is always dicotyledonous, very small, and placed in a cavity let into the upper part of an almost horny body which occupies the whole interior of the seed. It is further observed in this family, and in several others, that the embryo in germinating does not develop its lobes immediately; they remain enclosed in the body that covers them like a skull cap, and it rises with them above the ground; after some time, it falls by itself. Its function is apparently to protect the very tender growing point ["germe"] from direct contact with the air until, having become stronger, it can dispense with this help: these different observations, which I have carefully verified, are further based on the testimony of my uncle. I could also cite that of M. Adanson who, in his characters, attests to the existence of this body and never forgets the position of the growing point.[14]

One could ask, if the families in each class that have some similarities in this last character should be brought together? Some observations make this opinion likely, but contrary observations seem to destroy it: this question is a real problem which will not be resolved until after examining carefully the insides of all seeds, and establishing a comparison between families based on this single character.

The embryo provides, as one can see, invariable distinctions: the number of its lobes is the basis for the formation of three classes; its position in the seed is at least uniform in each family: these considerations separate from the ranunculi *Alisma*, *Damasonium*, and *Sagittaria* which are monocotyledons and have no horny body.

From differences in the conformation and the nature [223] of this body, *Fraxinella*, *Ruta*, and *Harmala*, which M. Linnaeus had included in his twenty-sixth order,[15] differ further. *Nigella* and *Garidella*, placed in the rockroses [Cistaceae] by M. Adanson,[16] go better in the ranunculi, because they have the horny body and the growing point at the tip; the following details will provide new reasons for separating the first and bringing together the others.

The calyx and corolla, which enclose the organs of fructification, are the two most prominent parts of the flower; also, they were necessarily observed by all Botanists and have provided them [the botanists] with characters more obvious than the preceding ones that were taken from the seed; in some plants, one of the parts does not exist; in most acotyledons both are missing at the same time.

Of the calyx one can observe its presence, position, and the number of its parts. We have just said that it is not always present; several genera, and even entire families, are without it; in some, it is replaced by spathes or other special membranes. The calyx, when it is present, must always begin at the support of the pistil [= receptacle], being a prolongation of the cortex of the stem, and serving to enclose the essential parts of the flower: it is from this point that it normally diverges to form a distinct part; sometimes, however, it is intimately united with the base of the pistil, or even, continuing its attachment even higher, covers it all or in part. The pistil thus covered over is in no way placed under the calyx, as M. Linnaeus would have it; one could still less say with M. Tournefort that it is the calyx itself:[17] it is only enclosed in that part which achieves with it an intimate union and becomes, so to speak, the skin of the fruit; the calyx is then necessarily monophyllous, and if it divides, it is only above the fruit when it ceases to adhere to it. When it is attached to the support alone, it can enclose one or several flowers; it can moreover be of a single, variously shaped part, or of several parts that fall off separately. M. Vaillant [224] observes that in complete flowers, that is, flowers with all their parts, the calyx is always monophyllous when the corolla is monopetalous:[18] one might add that the calyx is again monophyllous every time the stamens or the corolla adhere to it; these different axioms would appear to be demonstrated by observation, and give rise to the formation of a new one; that is, the calyx can have many parts only when the corolla is of several petals, and when these petals, as well as the stamens, are attached to the support of the pistil.

Such are the distinctive generalizations about the calyx; its presence, which is a constant feature of the Cruciferae, is by no means so in the ranunculi. Several genera, such as *Thalictrum*, *Clematis*, *Anemone*, and *Caltha*, have no calyx; but in its absence the anemone is furnished with a kind of covering of two and three leaves or more, which is hardly separated from the flower in one species, and much more in the others. One sees also in two species of *Clematis* a monophyllous covering shaped like a cup ["godet"] beneath each flower for which it served as a cradle; the one is the *Clematis foliis pyri incisis, nunc singularibus, nunc ternis. Inst. clematis cirrhosa. Linn.* of which M. Adanson made a separate genus, *Muralta*;[19] the other has not been mentioned by any author; I exhibited it last year in the Jardin du Roi under the name *Clematis balearica sempervirens foliis tenuius laciniatis:* it comes from the island of Minorca and differs from the preceding in this more substantial covering, smaller flowers, and much more divided leaves; perhaps it is only a variety.

Either in the clematis or the anemone, these coverings might be taken for the true calyx, considering the part of the stem between the calyces and petals to be an elongated support. M. Adanson is of the opposite opinion,[20] because in some anemones the same covering allows several peduncles, each terminated by a flower, to escape from its midst; that would actually not agree [225] with the general character of the family which allows only one flower from each calyx. Whatever the nature of these parts may be, they at least serve to establish a relationship between the anemone and the buttercup, between the plants of this family which do not have a calyx and those which do have one: in the latter it is always separate from the pistil and is attached to its support; it is usually made up of several pieces of which the number is not always relative [e.g., the same number or half] to that of the petals, nor the same in all genera; this number is also one of the reasons by which M.

Adanson proves that the covering of *Muralta* is not a calyx; the same character also suggests, following the preceding axioms, that it should be linked with the corolla and stamens.

The corolla forms the foundation for the method of the celebrated Tournefort, who considers its presence or absence, the number of its parts, and its regular or irregular shape:[21] these characters, very well conceived for a systematic order, often become useless or at least insufficient in the natural order because they are neither general nor even constant enough; the Labiatae always have a corolla, the Polygonaceae never have one; the Euphorbiaceae on the contrary have flowers with and without a corolla; even in the Cruciferae, where the presence of this part would seem to be an invariable feature, some species lack it. It can thus be concluded that this character is not always definitive, and that if plants, otherwise very analogous, differed in this single point, it would not be a reason to separate them; the absence of a corolla would not remove a genus from the ranunculi which had all the other characters [of the family]: this is the more necessary to observe [in] that one could argue against the existence of the corolla in *Thalictrum* and *Clematis* by taking for the calyx what are called petals in those genera.

As long as one does not have an accurate definition of these two parts, one will often confuse them; the same family gives us a new example in *Aconitum*, *Delphinium*, and *Aquilegia*, where Linnaeus has taken the petals, which [226] are irregular, for nectaries, calling the corolla what we call the calyx. He made the same changes in *Nigella*, *Garidella*, *Isopyrum*, *Helleborus*, and *Trollius*, which also have remarkably formed petals; but this terminology does not seem to be correct, because the supposed nectaries of the columbine [*Aquilegia*] change themselves in a double flower to a kind with ordinary petals, and because these nectaries and those of *Trollius* differ very little from the parts that M. Linnaeus himself called petals in *Myosurus* and *Atragene*.[22] M. Adanson, who does not agree with that author, used with success the parallel between the nectaries of the hellebore and the petals of the buttercup [*Ranunculus*], which resemble each other in several points: this parallel is still more striking and more decisive between these same petals and those of *Nigella* or *Garidella*, which both have on their limb a cavity covered by a scale; it provides at the same time a new proof of the affinity of these two genera with the family of the ranunculi. The two petals of the monkshood [*Aconitum*], although of a very bizarre shape, do not differ from the two upper ones of *Delphinium* except by a shorter spur and a longer limb; from the second genus to the columbine the transition is easy; the petals of *Myosurus* and the horns of the columbine and of the hellebore [*Helleborus*] are quite analogous and differ solely in the position of their limb; they correspond to the petals of *Nigella* and *Trollius*, which we have already compared with those of *Atragene* and *Ranunculus*. This gradation, of which examples are found in other families, proves that the nectaries of M. Linnaeus are true petals,[23] that there can be and actually are in the ranunculi regular and irregular flowers, and that the form of the corolla gives at most only generic characters.

The number of its parts [of the corolla] provides more distinctive characters, but always insufficient and sometimes variable; it is monopetalous in several families, polypetalous in others: among these latter, some, like the Umbelliferae, have a fixed number, others, like the ranunculi, have an indeterminate number; although the latter usually [227] have five petals, at least twelve of them, however, are to be counted in *Helleborus* and *Atragene*, nine in *Trollius*, six in *Pulsatilla*, four only in *Clematis*, *Thalictrum*, and some species of *Delphinium*, two in the monkshood; there are even [species] of *Delphinium* in which the two lower petals are suppressed and the two upper united into one; this single petal, placed on one side of the flower, ought not to be confused with a monopetalous corolla which, no matter what, always surrounds the pistil: it is incompatible, as one saw before, with the plurality of pieces of the calyx, and consequently cannot occur in the ranunculi.

The main character taken from the corolla, its attachment, was neglected by M. Tournefort and his predecessors; it is not always mentioned in the genera of M. Linnaeus [*Genera plantarum*]: the work of M. Adanson [*Familles des plantes*] is the first in which it is mentioned more constantly. The corolla can be attached to the calyx, to the pistil, or to its support: these different insertions, combined with those of the stamens and with the position of the calyx, provide general characters in the natural order that have the double advantage of never varying and being easy to grasp; they help not only in distinguishing the families, but also in establishing their relationships, in forming the chain that unites them. This is not the place to pass [the families] in review [as they occur] in the catalogue of the garden of the Trianon,[24] which is based on this plan; the Master who has outlined it could better than anybody else give us the map ["carte"] of it. It is enough for present purposes to know that the insertion of the corolla is the same in all plants in a natural order, that this uniformity must be considered as an essential condition for forming families, and that again it occurs in that of the ranunculi, which always has the corolla attached to the support of the pistil. If one allows these principles, it is consequently necessary to bring together *Sagittaria*, *Alisma*, and *Damasonium*, which have only three petals attached to a monophyllous calyx which is likewise divided into three parts, in another order; by this number and attachment, as well as by characters observed in the seed, they have more [228] relationships with *Triglochin* and with other plants near the Juncaceae, in which the supposed petals are rather part of the calyx.[25]

The stamens, which are the male parts of the flower, were regarded at the end of the last century as excretory tubes hardly essential to the economy of the plant; their true function was shown by modern authors, and the ingenious system of M. Linnaeus has spread this knowledge by presenting them from different points of view, in considering their number, attachment, proportion, fusion, whether by the filaments or the anthers, and position relative to the pistil.[26] His arrangement is well executed, but he can not avoid the defect of all systems, it deviates from the order of nature which allows only characters taken from the ensemble of parts; thus it is that the number, whether definite or indefinite, can never suffice, and even often becomes useless in forming a family. What distance is there not between the elm [*Ulmus*] and parsley [*Apium*, now *Petroselinum*], which both have five stamens? what affinity on the contrary between *Andromeda*, which has ten of them, and heather [*Erica*], which has only eight, between *Hypecoum* and the greater celandine [*Chelidonium*], of which one has four and the other an indefinite number, that is, more than twelve?[27] A single stamen is indeed to be counted in the Cannaceae, two in the Oleaceae, three in the Iridaceae, four in the Labiatae, five in the Boraginaceae, six in the Liliaceae, ten in the Leguminosae, an indefinite number in the Malvaceae, Cistaceae, and the ranunculi: but these generalizations allow some exceptions; the number can vary sometimes in the flowers of the one plant, as I have observed in *Trientalis*, sometimes in the species of a genus, as has been noticed in vervain [*Verbena*] and valerian [*Valeriana*]: it deviates sometimes in two or three plants from the character common to the family: rice [*Oryza*] provides an example of this in the

Gramineae. *Myosurus*, which has only five stamens, and *Garidella*, which has ten, are placed among the ranunculi despite this difference, and cannot be separated from them, having all the other characters of the family: moreover, these two genera can be regarded as having an indefinite number, because they [229] often have more than five or ten stamens, according to M. Linnaeus[28]; I have counted seven of them in *Myosurus*; then they are no longer alternate with the petals nor opposite the parts of the calyx: this regular arrangement is not generally noticed in the stamens except when they are determinate in number and related to that of the parts or the divisions of the calyx and of the corolla.

It is only with a determinate number that one observes the ratio ["proportion"] of the stamens [to the petals] and their fusion by the anthers: likewise, these two characters, restricted to certain genera and some families, are foreign to the subject of our discussion. We will not even discuss fusion of the filaments, which is equally found in [plants with] an indefinite number [of stamens], other than to add that it never occurs in the ranunculi where the stamens are always separate.

It would also take too long to explain in detail the position of the stamens relative to the pistils, to distinguish hermaphroditic flowers from those which are not, to show [plants with flowers of different sexes] sometimes forming separate families, sometimes occurring together in one, to prove, by many examples, that the separation of male and female flowers usually does not provide other than generic, or sometimes simply specific, characters. They are generally hermaphroditic in the ranunculi, and only two [species of] clematis and one of *Thalictrum* are known in this order which have the sexual parts separated on different stocks: when *Anemone ranunculoides* L. has more than one flower, the supernumerary is male. The abortion of the pistil or stamens is the sole cause of this singularity, and it is an insufficient [reason] for separating these species from their genera, still less from their family; if *Sagittaria*, which is dioecious, did not differ in other characters, it would similarly be included.

The attachment of the stamens is one of the general and essential characters that was unknown to the Ancients, which even many Modern [authors] have not developed enough; they are attached either to the calyx or to the pistil, but more often to the support of this same pistil or to the corolla; furthermore, because [230] these four parts have a definite area, the stamens can arise on each from different points which are always determined. The first three insertions are essentially distinct and [their co-occurrence in a family] is incompatible with the natural order; the fourth, on the contrary, is compatible ["correspond"] with the three preceding, and can be linked separately to each of them. Thus a single plant cannot have some stamens borne on the calyx and others attached to the pistil or its support; it has been shown by repeated observation that one of these characters excludes the two opposites; this proposition holds equally for a whole family; these three characters are never found together in the same [family]; they are never [found together] in those [families] which are generally thought to be natural; the stamens are always attached to the support in the Cruciferae, to the calyx in the Leguminosae, and on the pistil in the Umbelliferae. Their insertion on the corolla is not subject to the same laws; it is in fact constant in the Labiatae and Boraginaceae: but this character holds up only in a certain number of families: that of the Crassulaceae has some stamens adherent to the corolla and others borne on the calyx; one even finds in the flower of the pink [*Dianthus*] five stamens attached to the petals and five to the support of the pistil. The explanation of this singular occurrence will not be difficult if one admits on principle that the insertion of the stamens on the corolla must be deemed the same as the insertion of the stamens on the part that at the [same] time supports the corolla. This principle, which seems to be true, solves several problems; it makes the difficulties that pinks and their congeners can cause in the natural order vanish; it provides the means to reconcile the houseleek [*Sempervivum*] and *Cotyledon*, the dogwood [*Cornus*] and the elder [*Sambucus*]; it indicates a new similarity between the Rubiaceae and the Umbelliferae.

If in remembering here what was said above about the calyx, one adds that an indefinite number of stamens can occur only when [stamens are] inserted on the calyx or on the support: if one further observes that in monopetalous flowers the [231] stamens are almost always borne on the corolla, that in polypetalous flowers they are rarely adnate to it, that in all other cases the stamens and corolla are usually attached to the same part; these facts, already known, added to the preceding principles and axioms, may be the basis for a theory of the insertion of the stamens. The actual degree of affinity that exists between the stamens, calyx, and corolla, between the insertion of the one and the number of parts of the other will then be determined; the stamens will [be used] to explain why it is more difficult to find monopetalous corollas in the Malvaceae than in the Leguminosae; it will moreover become apparent that there are characters so analogous that they usually occur together, and never vary without the other: the insertion of the stamens on the corolla in certain monopetalous trefoils [*Trifolium*] is proof of it.

We have already observed that a calyx of several parts always indicates stamens attached to the support of the pistil; this character is indeed constant in the family of the ranunculi; it never varies in any species, not even in those [species of] *Delphinium* which have only one petal: this single petal is then very different from a monopetalous corolla which, moreover, would not agree either with the calyx of the ranunculi or with their indefinite number of stamens. It would perhaps not be impossible that this [monopetalous] corolla should exist in some plant of the family, but the calyx would at the same time have to become monophyllous and the stamens reduced to a definite number and borne on the corolla, these conditions make the variation more uncommon; in general, the more there are characters dependent the one on the other, the less they will be likely to vary: deviations from nature must be less frequent in proportion to the number of rules to which they are exceptions ["derogent"].

These are in brief the principal ideas which complete the account of the attachment of the stamens; the natural order can also draw several distinctive features from the anthers, from the number of their loculi or chambers, from the way in which [232] they open, from their arrangement on the filament that supports them, from the shape of the seminal powder that they enclose. The functions that the anthers discharge in the economy of the plant must merit them a particular distinction. M. Adanson, in his characters, always described them, and they are sometimes advantageously used when placing a plant in its true family. Thus in 1739, the author of memoirs on *Lemma* [= *Marsilea*] and *Pilularia*[29] placed these two genera near the ferns because like [the ferns] they had anthers with a single loculus, which opened transversely into two valves. The character of the ranunculi is always to have two distinct loculi [that are] applied for their length to the extremity of the filament that separates them; they open longitudinally into two valves and allow the powders, [the grains of] which, according to M. Adanson[30] all have a globular shape, to escape. In a memoir of M. Geoffroy the Younger printed in 1711, it was seen how much the shape of the seminal dust can vary:[31] this author also further observed that it is the same in all congeneric plants; it is to be presumed that this character should always be uniform in a family: observation alone could destroy this probability, or change its likelihood ["certitude"].

It is the pistil to which all parts of the flower have reference; placed in the center, it is, as it were, the meeting place of the calyx, corolla, and stamens; its position determines several of their characters and can equally form the basis for some of the principles which have been deduced from their insertion; we have seen [the pistil] sometimes separated from its coverings, sometimes thrust into the calyx and intimately united with it, sometimes bearing the stamens or the corolla, but most often raised above the point of their origin. These arrangements are variable and persist in the fruit after fertilization; each is connected with a particular arrangement of the corresponding parts; the presence of one announces that of the other, they cannot exist apart: thus if in a complete flower the stamens [233] are borne on the pistil, I can conclude from this that they are definite in number, that the corolla is polypetalous, that it, too, is borne on the pistil, that the calyx is intimately united with it, that it is a single part; I will even add that the number of its parts must be proportional, that the stamens are alternate with the petals and opposite the divisions of the calyx. These consequences follow naturally, but they are not all entirely true, because the principles on which they are based sometimes have exceptions: there are few rules in which these are not found; the rarer these exceptions are, the less [chance that] the consequences are incorrect. A probability which is equivalent to one hundred to one is almost a certainty.

Aside from these common characters, the pistil has those that are particular to it; they are taken from the number and relative position of the parts that make it up: these parts are the ovary, the style which occurs on top of it, and the stigma which terminates the latter. This arrangement is at the same time most regular and most usual, but it is not the same in all plants; sometimes there is no style, and the stigma directly crowns the ovary; sometimes this style, instead of being continued from the summit of the ovary, leaves from one of the sides; sometimes also the stigma is prolonged downward on the style and is almost confused with it. The number varies even more than the arrangement; often there are several ovaries in the same flower, several styles on a single ovary, and when [the ovary] is single, several stigmas at the end of a single style.

These differences are easy to grasp; the detail of the characters resulting from their combination would be superfluous. If we go over the main generalizations, we see that the ovary and the stigma are the only essential parts of the pistil; that almost all monopetalous flowers have a single ovary; that among the polypetalae themselves, several ovaries are concentrated in the smallest number [of genera]; [234] that in these, each ovary has only a single style and stigma. In seeking further to find some profit for the natural order from the observations, one might recognize that the absence of a style must be considered to be a purely accidental difference, that a single ovary characterizes the majority of families, that in some it is balanced by many, that in yet others, it is many [ovaries] which preponderate: this last character is general in some families, but never essential, because the number of ovaries can sometimes be indefinite, or reduced to three, two, or even to one, not being in such cases subject to any fixed rule. In the ranunculi, for example, which are usually distinguished by [having] many ovaries, some genera, like the anemone, have many; others, like the paeony, have fewer; the number varies equally in the two genera, but the variation is easier to recognize in the paeony because there are fewer ovaries. It is the same for all other genera; one counts in *Delphinium* sometimes three ovaries, sometimes a single one: this single ovary in no way diminishes its similarity with the ranunculi; it will serve rather to connect this family with *Actaea* and *Podophyllum*, of which there will be mention when discussing the fruit.

When the ovaries are few in number and arranged regularly around the receptacle that supports them, they sometimes are attached and blended together almost into a single ovary; this adhesion is perhaps the reason that made M. Adanson remove *Nigella* and *Garidella* from the ranunculi, because they have, according to him, only a single ovary surmounted by several styles;[32] but these styles, three, five, or ten in number, are normally distinct and with a pin one can easily separate the ovary into as many parts: each can be regarded as a distinct ovary surmounted by its own style: the separation of the capsules in a species of corn cockle [*Agrostemma*, of the Caryophyllaceae] proves this still better and demonstrates the analogy of these two genera with the ranunculi.

[235] Many plants of this family have no style, or at least it is so short

that one cannot distinguish it from the apex of the ovary. Some others, such as the pasque flower [*Pulsatilla*, or *Anemone* s.l.] and *Clematis*, have quite long styles, but furrowed on one side for their entire length; this furrow appears to be a prolongation of the stigma, and in some species allows a large number of hairs arranged like the plumes of a quill to escape when the fruit is mature: it appears that throughout this family the stigma always extends over the style.

It is [the style] that is said to transmit the seminal liquid contained in the dust of the stamens to the ovary; this ovary, thus fecundated, begins to grow and becomes a ripe fruit which bears the names *capsule*, *siliqua*, *pod* and *berry*, etc., according to the form which it takes and the material of which it is made. These differences do not provide anything constant for the natural order: one gets more benefit from the number of loculi, the way in which they open, from the arrangement of the partitions that separate them, and from the position of the placentae that bear the seeds. There are only a few families in which the number of seeds is fixed and provides an invariant character; but in that case the seeds are naked, they alone make up the fruit, and the calyx is like a capsule for them, as in the Labiatae and Gramineae, etc. One finds here no linkage between characters; nothing indicates in the flower how many seeds or loculi the fruit should have: the number of styles and stigmas, which would appear to be a natural sign, is often uncertain or even contradictory. It has been observed that only when there are several ovaries each usually becomes a single seed or a fruit with a single loculus: that is evident in the ranunculi in which the ovaries change into so many capsules, which, opening by two valves on the inner side, allow several pendulous seeds to be glimpsed. In the single-seeded section the two placentae which edge the valves are united into one; in that case, the capsule does not open, [236] and it could simply be taken for the outermost covering of the seed.

Ruta, *Harmala*, and *Fraxinella*, which M. Linnaeus has referred to this order,[33] have a single ovary surmounted by a style [which is] terminated by a simple stigma that becomes a capsule with several chambers; this last character is another reason for their exclusion. *Actaea* and *Podophyllum* are the two genera that have most affinity with the ranunculi; they indeed have a single ovary which becomes a berry, but it is [made up] of a single loculus full of seeds which are borne on a placenta lying on one of the walls of the berry; moreover, their seeds have an embryo at their tip, and their horny body has the same function in germination. It is immaterial that these genera form a distinct section in the ranunculi or make part of a neighboring family; their relationships will always be the same; they will equally serve to establish an easy intercourse between two orders: this transition, which would be considered as a defect in systems, is a [mark of] perfection in the natural order.

The affinity which exists between the parts of the fructification is evident in some general principles scattered in this Memoir; diverse degrees of this affinity have been recognized: all these characters do not have the same value, the same power for uniting or separating plants. Some are fundamental, essential by themselves and invariable, like the number of lobes of the embryo, its placement in the seed, the position of the calyx and pistil, the attachment of the corolla and stamens, they provide the principal divisions. The others are secondary, they vary sometimes, and only become essential when their existence is tied to that of one of the preceding ones; it is their combination which distinguishes the families.

Among these characters of the second order, one could place those that make up the arrangement of the flowers, and [also those taken from] the consideration of other parts of plants; the root is attached to the ground and extracts from it the appropriate juices for their nourishment; its upward prolongation [237] forms the stem which is properly the plant body; one finds no features in these two parts sufficient to characterize a family; the common distinction between herbs and trees, used by M. Tournefort,[34] cannot be admitted in the natural order which often joins prostrate plants and tall trees, *Bignonia* and sesame [*Sesamum*], the Judas tree [*Cercis*] and the bean [*Phaseolus*]. One combines in classes of animals the musk deer and the giraffe, the eagle and the merlin, be it that their sizes are very different: that must be the same for plants; but in both kingdoms, these beings that are so disproportionate in size do not follow each other immediately, they are always separated by a series of genera or species which in each order form a gradation from the smallest to the largest.

Leaves provide more general characters, and sometimes even corresponding to those of the flower: in some families their arrangement is indeterminate, but it is uniform in some like the Labiatae, Cruciferae, and Gramineae. Among the ranunculi there are only two genera (*Clematis* and *Atragene*) with opposite leaves and a rather woody stem;

in all others they are alternate and their base is usually enlarged to form a half sheath, or even an almost complete sheath, around the herbaceous stem; they are accompanied neither by a tendril, as in the Cucurbitaceae, nor by stipules, as in the Malvaceae: the presence or absence of these two parts is often of use to distinguish families without the aid of the fructification. One knows, moreover, how much the arrangement of flowers is helpful in recognizing the Umbelliferae and Labiatae; sometimes [the flowers] arise from the axil of a leaf, sometimes from another part of the stem, and sometimes they terminate it. These three conditions occur together in the ranunculi which consequently cannot derive any distinctive features from it.

We suppress all the less important details on the development of the young shoot, on the aestivation [238] and dissection of the leaves, on the shape of the hairs which sometimes cover them, on the general arrangement of the foliage; none of these parts, at least in the ranunculi, provides characters constant enough to detain us; they contribute only to make up what is called the facies ["caractère habituel"], that is to say, the general appearance of plants which is rather similar in all analogous genera: no description can give a true idea of this character which is recognizable at a glance, but everybody accustomed to looking at plants will grasp it easily.

All these means of discrimination, sometimes used by the old Méthodistes, have been neglected in modern systems. They have not been thought appropriate except when designating species, and one has in general allowed this assertion. It is true that the basic characters of any order must always be taken from the fructification; but at the same time it is necessary to look at those [characters] that make up the other parts, like the accessory characters which announce the existence of the preceding, although their minute size or position sometimes hinders their being noticed, which in an individual without flowers often suffices to determine its genus, or at least its family. Thus in animals the external arrangement of parts indicates the number of ventricles of the heart, and other distinctions of classes and genera.

There is no need to extend these reflections to assure oneself that the natural order alone can give invariable characters that are appropriate for recognizing a plant in all conditions, furthermore, it offers more substantial advantages; the virtue of a plant depends on the particular development of its parts, of a proportion determined by the principles of which it is composed; the same development, the same combination, is found in all congeneric plants which are consequently endowed with the same virtues, as daily experience teaches us. The different sages [*Salvia*] are used for [239] the same things; the polygala from around Paris has been substituted with success for that of Virginia, known by the name of *seneca*; the root of goose grass [*Galium aparine*] gives a red dye like that of madder [*Rubia tinctoria*]; further, one recognizes in general an aromatic virtue in the Labiatae, bitter in the Rosaceae, anti-scorbutic in the Cruciferae, diuretic in the ferns, narcotic in the Solanaceae and the Papaveraceae, caustic in the ranunculi. Many plants of this latter family are known for their harmful effects; but used externally by the masters of the Art, they are usefully employed in removing the proud flesh of wounds, in making poultices and drawing the humours toward the skin; their harmful property then being turned to the benefit of humanity. If, as one can presume with some reason, all the plants of a family have almost the same virtue, the same quality, should not one put aside every system so as to be attached only to the natural order? Botany is not worth taking our time unless it can become useful for man, in making for him the choice of healthy food, in increasing the reliefs that can soothe his ills, in contributing to the perfection of the arts. Studying families is one of the principal means of enriching the materia medica with a large number of useful plants.

This order has another advantage over any artificial arrangement; it alleviates the memory in the study of genera, in which the principal characters are always included in those of the family in general. When one knows that the paeony [*Paeonia*] is among the ranunculi, one knows the position of all its parts, and the genus will be determined by adding that it has a calyx of five leaves, as well as regular petals, many stamens, two ovaries or sometimes more [which are] immediately crowned by a large and flattened stigma, [and] which become oblong, velvety and many-seeded capsules.[35] It is the same with other genera, which I will not try describing here, that are found in elementary books. This example [240] suffices to prove the value of families in generic descriptions.

Botany, considered from this point of view, shows us on one side a way of simplifying study and recognizing the virtue of a plant by its characters; on the other, it discloses to the mind a vast area to discover, a source of new knowledge: a double perspective equally pleasing for the person who seeks his own satisfaction in the dedication of his work to the general good of Society.

OUTLINE OF A NEW ORDER OF PLANTS ADOPTED IN THE DEMONSTRATIONS AT THE JARDIN ROYAL

EXPOSITION D'UN NOUVEL ORDRE DES PLANTES ADOPTÉ DANS LES DÉMONSTRATIONS DU JARDIN ROYAL.

Mém. Math. Phys. Acad. Roy. Sci. (Paris) 1774: 175–197.

Translated by Peter F. Stevens

[175] Among the sciences that have natural bodies as an object of study, Botany, of which the extent is most considerable, needs for this reason more method in the arrangement of the beings subject to its examination. The number of Authors who have successively occupied themselves with this work demonstrates its necessity; it shows at the same time the inadequacy of their work and the difficulty of finding the natural method. The plants spread over the globe appear to form a continuous chain with one another, of which the two extremes are the smallest herb and the tallest tree. One will ascend by an imperceptible gradation from one to another by arranging in sequence those [plants] in which affinity is marked by a great number of relationships. This order, which is Nature's, does not interest the Natural Philosophers alone; it has a more substantial use. Reasoning, confirmed by experience, shows that plants agreeing in their characters also have the same properties; of the kind that, once the natural order is established, one can determine their properties by external features. A subject of such importance should be worthy of occupying the attention of botanists who would wish to join with the title of Savant that of useful Citizen.

The ancients had no idea of the natural order; more occupied by the virtues of plants than by their perceptible characters, [176] they neglected the study of the latter, and were satisfied to separate herbs from trees. Their better-educated successors felt the need for a more exact approach. They examined the plants with care, and soon established that their principal characters should be taken from parts of the fructification, alone suitable for giving general results. But in wishing to regulate the science, in deciding that a single one of these parts must serve as the basis of a method, they lost their way. This principle gave rise to several methods, some [based] on the fruit, others on the calyx or the corolla, which were successively adopted and then sank into oblivion, when M. Tournefort published his at the end of the last century.[36] This excellent author, who was looking only to facilitate the study of the science, chose the corolla because it is very prominent and easy to observe. His arrangement would be more natural if he had been interested only in essential characters, and if he had devoted more precision to the delimitation of genera and species. Botany, which had taken a new shape under his hands, was still susceptible to a yet greater degree of perfection.

M. Tournefort, along with the majority of his contemporaries, did not know the function of the stamens; the discovery of their fertilizing capability produced a revolution in the science,[37] and settled the discussion over the nature of certain parts. It was acknowledged that only the sexual organs, the pistil and the stamens, essentially made up the flower, that the corolla bore that name incorrectly,[38] and was nothing but a colored envelope. The stamens, better understood, were observed more carefully, they were made part of the generic characters, and became in 1737 the basis of an ingenious system conceived by M. Linnaeus, the celebrated Swedish Naturalist.[39]

This system, like the preceding one, had its beauties and imperfections, its clear parts and its weak side. The author wanting, according to the received principle, to draw all the divisions of his classes from a single part, and so as to increase their number, was obliged [177] to use indiscriminately characters that were essential and those that were not. If progress in botany is to be measured by the number of steps it makes toward the order of Nature, the system of stamens, more difficult and less natural than the method of M. Tournefort, seemed to the same extent to be less advantageous; its characters are sometimes less obvious, its course always constrained, its classes full of dissimilarities; but it is preferable in that it has more precision, more perfect generic and specific descriptions, a simpler nomenclature [that is] less fatiguing to the memory, in that it gathers together under stated relationships all known genera, whether old or new. This system, as in earlier methods, can be regarded as an analytical table ["table raisonnée"] where plants are arranged following conventional features that allow Botanists to understand one another. It has been adopted by many of those who would prefer more to follow a ready-made order, rather than to think of a new one or to reform older ones.

M. Tournefort, however, has kept some adherents; the method that he himself established in the school at the Jardin Royal is still there, despite changes that have happened unexpectedly in the science; one would have been able to make it more practicable by correcting defective genera, adding those that are absent, and simplifying the nomenclature. Science would have gained by that, the demonstrations would have been clearer, study easier: [but] that change would not have taken place without increasing the extent of the Demonstration [which was] too confined for the number of species it needed to contain, and without carrying out a complete transplantation, always costly and often risky for the plants moved. M. le Duc de la Vrillière [Louis Phelypeaux, Comte de Saint-Florentin], who loves the Sciences and patronizes Savants, instructed by M. de Buffon on the state of that garden, was willing last year to submit a project to the King [Louis XVI] which would lead to the embellishment of the place and of its public usefulness. His Majesty, at all times favorable to Botanists, gave them on that occasion a [178] new sign of his favor in permitting the execution of the project that was presented to him; the area was doubled and marked out with new flower beds. Nothing remains but to decide on the order that should be established in the demonstrations, on the method that should be preferentially adopted. That of M. Tournefort needs changing: the principal merit of that of M. Linnaeus is in the genera and nomenclature. It was necessary either to correct the first using the second, or to make a new one that combines the advantages of both, but having also its own [advantages]; the first project seemed insufficient, the second better met the goal of public utility.[40]

This most perfect method is that which is closest to the natural order, or, rather, it will be the natural order itself, of which all artificial methods are nothing but very imperfect imitations. The route to be followed in finding it is different from that we have followed up to now; it is true that the advantages that must result from success are balanced by the difficulty of the enterprise: the natural order is like the philosopher's stone of the Chemists.[41] The impossibility of uniting, or even of being acquainted with, all the plants which must make up the general chain will always be an insurmountable obstacle, and will leave gaps that are difficult to fill; but if Nature has scattered the material intended for the construction of this order, she allows us to catch at least a glimpse of the principles on which it is based. Among the characters that plants provide, there are some [that are] essential, general and invariable, which must serve as the basis of the order that we seek. They are not arbitrary, but based on observation, and are not to be obtained except by proceeding from the particular to the general.

After having determined successively species and genera, those [genera] that resemble each other in many parts are brought together: these incomplete groupings ["approximations"], better known as *Families*, are easy to obtain, and Nature herself seems to have favored us in this work in providing some very obvious ["simples"] families acknowledged by all Botanists.[42] There are [179] seven main [families]; the Gramineae, Liliaceae, Compositae, Umbelliferae, Labiatae, Cruciferae, and Leguminosae: these families, precious for us because they are the basis of our studies, have a second advantage. We can, with their help, understand and appreciate the true principles of the natural order: everything that tends to separate two plants that are included in [one of] these families will be rejected; one will keep only those that draw all their authority from them.

This is what has to be admitted; *Any character that varies in a particular case cannot have general value.* To help this first principle, all characters that are neither general nor constant enough are discarded; roots, stems, and leaves which may vary within one family are set aside: the Liliaceae have tuberous and fibrous roots; there are to be found [some species] in the Leguminosae with herbaceous stems and others with woody, which destroys the famous distinction between herbs and trees; finally, the Compositae have [both] alternate and opposite, and simple and pinnate leaves; the absence of any of these parts in certain plants becomes a new reason for their exclusion.

Thus it is only parts of the fructification, and above all the most essential, which can provide the fundamental characters of the natural order; thus the calyx and corolla may still be laid aside because these two coverings of the flower can be absent either individually or together from a plant without its consequently being the less perfect or the less able to reproduce itself. This latter characteristic, which implies all the others, constitutes the true perfection of a plant; it is inherent in species formed by Nature, and resides in the sexual organs. If sometimes damage or abortion has suppressed these organs in a particular individual, it is a perverted being which has lost the reason for its existence, a monstrosity which is no exception to the general order.

The pistil and the stamens are thus the essential and principal parts; they cooperate in forming a new individual, [180] which begins a new life in germination and which is not accounted perfect until it has produced another individual like itself. The reproduction of beings is the object of the works of Nature, the highest point it can reach; it is also the limit ["terme"] which must confine our researches. The apparatus of fertilization, the importance of the organs that carry it out directly, and even of those that are only secondarily involved in this function, the uselessness and the loss of the organs once they have produced their effect, everything proclaims that the seed, formed at so great a price, is the essential part par excellence, the basis for the fundamental divisions of the natural order.

We will not talk at all here about the general differences noted in the germination of the seed, of the development of the embryo that it contains, differences that are constant and invariable and which separate plants into three large orders[43] as easy to distinguish by their general appearance as by their particular characters. In my first memoir on the ranunculi (Jussieu, 1773, p. 414 [1777:221]) is seen the evidence that supported the distinction between acotyledonous, monocotyledonous, and dicotyledonous plants: these evidences, based on general observation, upon agreement with known families, on the analogy of the two organized kingdoms, and on the conformity of the heart of the animal with the embryo of the plant, would appear to leave no doubt as to the strength of this fundamental division.

The acotyledons, which are the simplest in their structure, form the first and smallest part of the chain; they are followed by the more numerous and better organized monocotyledons; the dicotyledons, which follow, have a more complete organization and are much more numerous: the extent of these classes, and especially of the two latter, necessitates secondary divisions in each, the characters being both essential and general. Before looking for [secondary characters] in other parts of the plant, it is necessary to see if the embryo, which provided the first [181] distinctions, could also give those of the second. Germination, independently of the fundamental differences already mentioned, gives special features in each class, especially in the monocotyledons. The other main characters [of the embryo] are its position in the seed,[44*] and the position of the latter in the capsule. We have elsewhere observed (Jussieu, 1773, pp. 415–416 [1777:222–223]) that the first was sufficiently uniform in the known families; the second also appeared constant, but there are not enough observations to generalize this double assertion and to place these characters in the second rank. Having obtained only individual results thus far, we have to leave the question undecided, to wait until new studies and a careful comparison of all seeds have thrown more light on this important matter.

The choice of the parts which, in default of the seed, must serve as the basis for the subdivisions is not unclear; it is decided in favor of the sexual organs by the close relationship that they have with the seed, a relationship based on their influence on its formation, on the necessity of their previous existence [for its production]. As in the economy of the plant, they have importance only when they join forces, in the same way that their conjunction in the natural order is needed to form reliable characters: one knows how much the number and proportion of stamens can vary in a single family, similarly with the shape, texture, and number of loculi of the pistil: these two parts individually provide many characters, but the one that results from their respective consideration is the only one constant in known families, consequently the only [182] one that is permissible; this unique character is the position of the stamens relative to the pistil, or alternatively, the insertion of the stamens; it has not been grasped by most botanists. M. Linnaeus, who in his system considered stamens from every point of view, seemed to make little of the way in which they are attached, and used it to characterize only three of his classes;[45] M. Adanson used it more consistently in the characters of his families.

Stamens may be borne on the pistil, or attached to its support; they may also arise from the calyx or corolla.[46*] By examples taken from known families we proved, in the Memoir already cited, that of the four insertions, the first three are fundamentally distinct and [their co-occurrence in a family is] incompatible in the natural order; the fourth, on the contrary, follows other laws, it is compatible with the three preceding and can be linked separately with each of them.[47] A mixture of insertions on the pistil, on the support, and on the calyx has never been seen in a family, still less in an individual; each plant, each family, has its own that is appropriate and constant. It is not the same with the fourth insertion, which mixes indifferently with any of the preceding in the same family, or, even more remarkably, in the same flower, as in the pink. This is explained by the following proposition, taken from the same Memoir: *the insertion of the stamens on the corolla must be deemed the same as the insertion of the stamens on the part which at the [same] time supports the corolla.*[48] This distinctive proposition comes from the fact that the corolla bearing the stamens is directed to the point they would have occupied had they not been adnate to it. In this case, it can be regarded simply [183] as an intermediary support, compatible with each of the three main insertions; its existence then becomes necessary, and its own insertion substituted for that of the stamens acts as an essential character. It results from this conformity that a corolla bearing stamens must have three insertions as dissimilar from one another as are the three corresponding insertions of the stamens [alone]; this is confirmed by observation. On this supposition, a single family will never exhibit corollas borne on the pistil while others are attached to the support; they all have the same insertion in analogous plants.

These propositions should perhaps merit a more extended discussion; but without going into details that anybody could easily provide, we will observe that this simple outline suffices to indicate the way that will lead to the discovery of the natural order. There are in plants, as in animals, fundamental classes which include secondary classes; both are based on general and invariable characters which cannot be taken other than from organs that are the most essential for life, for the reproduction of the species; all beings that differ in the structure, position, and function of these principal organs must be separated; from there the first divisions of the animal kingdom [are taken] following the inspection of the heart, of the number of its ventricles and auricles. The organs in the animal economy that take the second rank after it will provide the second divisions, and so on; this principle, from which one will never deviate without making a mistake, is the basis of all study of organized beings; consequently, one cannot content oneself with the examination of external parts, of those parts which at most provide characters of the third or fourth rank; in both kingdoms the methods based on these characters always deviate from Nature. These truths did not escape my uncle, and the arrangement of his families in the garden of the Petit Trianon[49] proves that he understood them well; [184] his order is more natural than those published up to now because it is simple in its general divisions and conserves the families in their entirety. One finds there the three fundamental classes characterized by the embryo; the acotyledons are arranged following the more or less marked appearance of the parts of the fructification; in the monocotyledons the Author is guided by the insertion of the stamens, and passes successively in review [plants with] the stamens borne on the pistil, those which adhere to the calyx, and those which are attached to the support. The dicotyledons are similarly divided, observing that when the corolla bears the stamens, it is its insertion that becomes the decisive character to refer the plants to one of the three insertions of the stamens.[50] This is almost the plan my uncle followed, but without designating by any signs the points of transition of one class to another, contenting himself to unite families that conform in their essential characters: his divisions are taken from Nature; one would ask only that they were more numerous because each secondary class would then be less extensive, and the determination of genera and species would become easier; but he was hindered by the limited number of general characters that could be taken only from the essential parts of the fructification: the characters already mentioned are the only ones essential by themselves; they sometimes transmit their quality to others, as one has seen for the insertion of the corolla, but this quality is always dependent. Some others receive only a portion of their quality from these fundamental characters; they become general, but not essential, because they permit exceptions. Such are those that result from the following observations:[51*] the corolla which is a single part ordinarily bears stamens; when it is made up of several parts, the stamens practically never adhere to it, they [185] then have a common origin with it; whence it follows that the number of parts of the corolla and its insertion, considered together, suffice to determine with adequate precision the attachment of the stamens.

This covering can thus provide general distinctions, and sometimes shares with the seeds and sexual organs the exclusive privilege of providing the fundamental characters of the natural order. It will now be explained why the method of M. Tournefort, based on the corolla, is more natural than the system of M. Linnaeus, based on the stamens; [Linnaeus] in an essential part, has chosen considerations that are not;

the former, in distinguishing corollas that have a single piece from those that have several, has unknowingly followed one of the divisions taken from the attachment of the stamens, and in that come closer to the natural order. The method has another advantage over the system,[52] it is simpler, its characters are more apparent; if the Author had understood the functions of all the parts, it is probable that the relationship between the stamens and the corolla would not have escaped him, and he would have neglected the shape of the latter, having other, more reliable characters; we can make up for what that illustrious man left out; his work indicates to us a very simple way of multiplying the classes of the Trianon without deviating from the order which is seen there, without ceasing to take the insertion of the stamens as the basis for the secondary divisions.

This way consists of distinguishing the insertion, based collectively on two principles, the one *direct*, the other *indirect*; the first occurs every time the stamens attach directly to the pistil, to the support, or to the calyx; the second when the corolla bearing the stamens leaves at a point intermediate between them and the other parts; this distinction does not clash with the natural order although the two insertions are sometimes mixed in the one family because this mixture is rare and results in only a few exceptions. It is true that in the Leguminosae, characterized by a direct insertion of the calyx, two or [186] three plants have indirect insertion, where the stamens are attached to the corolla which is itself connected to the calyx, but this exception occurs only because the corolla, polypetalous in all the Leguminosae, becomes monopetalous in the plants in question; the fusion of the petals is a necessary condition for a change in the insertion of the stamens, and when this fusion occurs, one is almost sure to see the stamens borne on the corolla so as to conform to the proposition noted above: *When the corolla is of a single part, it usually bears stamens*; or its inverse, equally founded on observation: *The corolla that bears stamens usually consists of a single part*; this harmony ["sympathie"], striking in two characters, allows them to be combined, and [us] to decide, with some exceptions only, that one suffices to predict the other, from which it is natural to conclude that the character of indirect insertion can generally be expressed or designated by the term, monopetalous corolla.

This insertion always assumes [the presence of] a corolla; direct insertion, on the contrary, does not require the presence of this covering at all; entire families, like the Gramineae, lack it; others, such as the Cruciferae, Umbelliferae, and Leguminosae, have one; we will also add that it is rare in a family, that has for one of its characters the presence of a corolla, to find plants that lack it. *Cardamine impatiens* L., of the Cruciferae, and the *carob* [*Ceratonia siliqua*], in the Leguminosae, are examples; but as there are only a few exceptions it is possible, despite conceding them, to separate families that have a corolla from those that do not; this separation aims to increase still more the number of secondary divisions, an increase the utility of which is recognized; one will succeed in increasing them by guiding oneself always by the insertion of the stamens, by distinguishing only an *essentially direct* insertion from that which is *simply direct*.

When the corolla does not exist, the stamens have essentially a direct insertion to three points of attachment,[53] [187] because they cannot have an intermediary support; if on the contrary the corolla is present, this insertion is simply direct, because the stamens cannot then essentially adhere to the three points of attachment. The proximity of a part that elsewhere has the function of support can make its insertion change, and this change is so much the more possible as, in direct insertion, the corolla usually arises at the same place as the stamens; this common origin helps the fortuitous joining of the two parts: in such a case, the two bases being confluent, the corolla appears to bear the stamens, although in fact the filament of the stamen is prolonged on the limb of the petal up to the point of insertion; this remark can also be applied to indirect insertion and serves to confirm the primary general propositions on the insertion of the stamens.

Naturally apetalous plants never bear a corolla, those which, on the contrary, originally[54] have a corolla, sometimes lose it; from this comes the following two propositions: the corolla never exists in the essentially direct insertion; in that in which it is simply direct, it usually exists, but may sometimes be absent, provided that the possibility of its existence is recognized. We have previously remarked[55] that this corolla would be of several parts in its assumed insertion, and this makes its suppression less difficult; polypetalous corollas are liable to abort or become monopetalous. On the contrary, there is not a single example of a corolla of a single part ever having become suppressed: in addition, it is rare to see it divided into several petals, and it does not even become subject to this last variation unless it already deviates from the first rule in not bearing stamens; it is necessary to except from this proposition, the genus *Statice* which, with *Plumbago*, will merit special discussion.[56]

We will linger no more over these different observations, that which has been said suffices to conclude that the essentially direct insertion necessitates the absolute suppression of the corolla; that the simply direct insertion supposes [188] the actual, or at least possible, existence of a corolla of several parts; from which it follows that by the term "apetalous plants" we can designate the first of these insertions, and the second by that of "polypetalous plants": these consequences, firmly established and linked to that deduced from indirect insertion, will facilitate the clear comprehension of the method we propose for the school at the Jardin Royal.

The plants will be distinguished there, as at the Trianon, into acotyledons, monocotyledons, and dicotyledons.

The first class will remain undivided until the organization of the plants that make it up is better known: most have either invisible or barely evident pistil and stamens, and these most often separated in different flowers; this makes insertion difficult to observe, and consequently secondary divisions impossible.

The second class includes plants that do not have a corolla, but only a calyx, colored in most Liliaceae; the insertion of the stamens thus never being indirect, the class will be divided only into three; for this, characters will be taken from the direct insertion of the stamens to the support, to the calyx, and to the pistil.

The dicotyledons, which are much more extensive, will have, with the help of the corolla, more divisions; and it is here that one begins to deviate a little from the arrangement of the Trianon. To multiply the classes, it is proper first to distinguish indirect, essentially direct, and simply direct insertions, or, what amounts to the same thing, [to distinguish] monopetalous, apetalous, and polypetalous plants; each of these three secondary classes will be subdivided according to the insertion of the stamens.

The apetalous plants, being less complex ["composées"], will be placed first, and will immediately follow the monocotyledons, with which they have a character in common, that is, the absence of a corolla. One observes only two insertions of the stamens in these plants, namely, on the support and on the calyx; this provides two divisions. We

[SYNOPSIS OF JUSSIEU'S CLASSES]

ACOTYLEDONES			1
MONOCOTYLEDONES:			
	Stamens attached to the support		2
	to the calyx		3
	on the pistil		4
DICOTYLEDONES:			
Apetalous:	Stamens attached to the calyx		5
	to the support		6
Monopetalous	Corolla attached to the support		7
	to the calyx		8
	on the pistil	anthers joined	9
		anthers separate	10
Polypetalous	Stamens and corolla attached to the pistil		11
	to the support		12
	to the calyx		13
Irregular, stamens separated from the pistil			14

do not know [189] of a dicotyledonous, apetalous plant that has stamens borne on the pistil, unless the Aristolochiaceae are dicotyledons.

In monopetalous plants which follow, the insertion of the stamens being uniform, one uses the insertion of the corolla itself to the support, to the calyx, on the pistil, for making three other classes, of which the last may again be subdivided, in distinguishing plants that have united anthers from those that have them separate; in this manner, one thus obtains one more class, and the family of the Compositae is separated from those that also have a monopetalous corolla borne on the pistil.

The polypetalous plants also separate into three classes characterized by the insertion of the stamens on the pistil, to the support, to the calyx, or, if one prefers, by the attachment of the corolla to the same parts; the two are equivalent since the polypetalous corolla is normally held at the same place as the stamens.

To these nine classes of dicotyledonous plants it is necessary to add a tenth, which includes plants with irregular [flowers], exempt by their nature from following the laws of insertion; such are flowers in a catkin, those which by the arrangement of their parts and their particular conformation have the stamens essentially separated from the pistils in different flowers: these flowers are male or female, depending on the organ with which they are provided. The male flowers have stamens attached to the calyx, or to a scale that takes its place, or to a special pivot which is raised from the base of the calyx. One assigns no significance to these different attachments because insertion in general becomes important only when the essential character, the relative position of the stamens and pistil, is determined by its help; the separation of these organs makes this determination impossible in irregular plants; they occur last here, as at the Trianon, partly because of this irregularity which makes them differ from all others, partly because they normally have the stamens attached to the calyx, and approach by this last character the third [190] class of polypetalous plants. Plants in which the separation of the sexual organs is caused by a simple abortion are not placed here.

Such is the distribution of classes, fourteen in number, which are placed here together [in] the Table. [See the accompanying table.]

Each of these classes [in the table] contains several orders which are as many as the families of which the distinctive feature is based, not on a single character, as in the divisions of the méthodistes, but on the combination of several; this concurrence of characters becomes indispensable, because a family is nothing other than the assembly of plants that resemble each other in many characters: moreover, the result is that the number of characters is indefinite, and that they can be drawn indifferently from all parts of the plant. Some are essential and invariable; the others can vary independently; the first being more general, common to several families, serve to join them together, to characterize the class, and from their several combination[s] the entire general order results; the [191] second are less constant, having no importance except in their combination, and are employed collectively to distinguish the families one from another. Thus an embryo with two lobes, a monopetalous corolla bearing stamens and attached to the support, are the fundamental characters common to the Solanaceae, Boraginaceae, Labiatae, etc., but this latter family differs from all the others by a tubular calyx, an irregular corolla with two lips, four unequal stamens, a pistil made up of four ovaries which become so many naked seeds,[57] a style terminated by two stigmas, and opposed leaves and flowers: each of these characters can occur separately in another family, but their combination is found only in the Labiatae: each can also vary; the shape of the corolla changes, abortion suppresses the stamens or the ovaries, an irregularity doubles the style, and the seeds are sometimes covered by a pulpy substance: these accidental and individual differences do not remove from the Labiatae a plant that otherwise has all the characters.

In thus combining several characters to make an order, one is never liable to separate congeneric plants; in choosing as the basis of the method the most general characters, recognized as being uniform and constant in families, one is further assured of bringing together analogous families: these two advantages, which the methods published do not yet have, are unique to the arrangement at the Trianon; they are also found in the arrangement at the Jardin Royal, arranged on the same principles, and based equally on the number of lobes of the embryo and the relative position of the sexual organs; this arrangement is less perfect than the preceding because it has a defect of which no method is exempt, that of being unable to exist without allowing exceptions; although they may be few, one has to take account of them, to recognize that the insertion of stamens on the corolla is sometimes mixed in a family with one of the other three insertions, that the corolla itself is liable to vary in the number of its parts.

[192] It will always be impossible to avoid exceptions when one wants to make exactly methodical classes, and to increase [these classes] to aid in teaching. This inconvenience will cease if one ignores the simply general characters in the fundamental divisions, so to use only those that are the most essential and invariable; but because the latter are few in number, that of the classes in the natural order is not very great. The animal kingdom has only seven; one cannot find more in the plant kingdom when following the divisions of the Trianon. It is necessary to note again that the classes and families in these two kingdoms are not separated with the precision that is regarded as a merit in the artificial methods. Nature follows another path: analogous genera are not always equally distinct; they differ by very slight nuances in genera that are very close, more markedly in those that are different, and considerably especially in the two points in the one order that are most separated; these nuances seem to form in each family a kind of degeneration which, extending from the center to the extremities, establishes communication by these latter with neighboring families; the latter are joined with others, and so on successively. The classes resulting from their assemblage are also connected by intermediary points, of the kind that when reascending from the particular to the general, one always recognizes an analogy among the genera, the families, and the classes;

one attains the desired point, which is to find the chain of beings, to ascend by an insensible gradation from the least perfect animal to the most, from the smallest herb to the largest tree. This way of proceeding does not allow divisions that are too methodical; the Author of the arrangement at the Trianon [Bernard de Jussieu] has also avoided them; this modest man has not published his work because he believed himself too little advanced in the science. He first wished to reduce the number of gaps caused by the absence of unknown families, and waited for new discoveries to put underway the reform of [193] doubtful links. Regarding this work as suitable for Botanists alone, as a preliminary step toward perfection, and as a simple indication of the way which must lead there, he wanted, so as to satisfy the present goal, which is public instruction, to work more for the Students than the adepts; so that without departing from the true principles one sought to establish a method with more numerous and more precise classes that are consequently easier to grasp. He thought that it was still necessary to countenance as much as possible the received prejudice, which regards that method which is based on parts that are the most obvious and easy to observe as preferable. It was thought that this double aim might be achieved by combining with essential, but sometimes hardly apparent, characters, accessory characters that indicate the existence of the former and are constant and always visible; through associating the corolla with the stamens so as to designate classes. In a word, it appeared that under present circumstances a method which approached the order of the Trianon, and simultaneously combined the particular advantages of the other most valued methods, must be preferred. That of the Jardin Royal satisfied all these points; one finds there all the families and the fundamental divisions of the Trianon; [yet] apetalous, monopetalous and polypetalous plants are distinguished as in the method of M. Tournefort. Furthermore, being based in part on the stamens it is similar to the system of M. Linnaeus. The genera of this [latter] Author, although too detailed, are, however, the best that have so far been made; the species are the best ascertained; his nomenclature aids the memory by the trivial names, his descriptive phrases give an idea of the plant described. [Of all his works, these are] the best and most useful; having nothing better to propose, we think we must adopt his genera, species, and nomenclature. If sometimes we should come to make slight changes, to conserve old and generally accepted names and to substitute these for the new ones he wished to introduce; this liberty will not [194] harm the science. M. Linnaeus was perhaps wrong to wish to innovate so much; in not approving him on this point, we are inclined to render him the justice he merits on all others. Botany owes him a part of her progress; his system will always remain first-rate among the number of artificial methods and will help the study of plants. He did not restrict himself to this work; persuaded, like all Botanists, of the existence of the natural order, he tried to bring together in fragments the genera agreeing in the greatest number of their characters.[58] These fragments correspond to the families of the Trianon, but, formed by what appear to be different principles, they sometimes include plants which ought to be separated, and are themselves arranged without any order. Since M. Linnaeus added neither proof nor explanation, we cannot make other than simple conjectures as to the principles on which he based his work. One might be led to believe that in giving all characters almost equal value, he confused essential with variable [characters], and did not pay attention [to the fact that] in their enumeration, one of the former is equivalent to several of the second: in this, he approached M. Adanson, who allowed no essential and fundamental character in his families, but only some characters more general than others, and common to a larger number of plants. This Author [Adanson] goes still further than M. Linnaeus, because he recognized no single part in plants that was essential ["essentiellement existant"].[59] Whence it follows that, like M. Linnaeus, he had to ignore the basic divisions, that his general arrangement necessarily departs from that of the Trianon, and that his families must likewise differ in several points.

To decide which of these authors approaches most the Order of Nature, the following questions must successively be resolved: Do essential parts occur in plants? Do these parts provide fundamental and invariable characters that are uniform in families that are known? In regarding the family as an assemblage of plants that [195] resemble each other in the greatest number of their characters, is a single essential or general character equivalent to several particular or variable ones? Moreover, can one decide that families agreeing in these principal characters must form part of a single class? Simple replies will clarify these different questions; (1) one is acquainted with the generally received axiom which is difficult to dispose of:[60*] each organized being originates from a fertilized egg; this fertilization is brought about by the cooperation of the sexual organs: the egg and the organs are consequently necessarily preexistent; the seed, the stamens, and the pistil are hence the essential parts of plants:[61] (2) the number of lobes of the embryo contained in the seed, the relative positions of the pistil and stamens provide general characters which are the only ones that are uniform[62*] and invariable in known families. They will also be so in all other families if it be true that they must be formed on the model of the first; these assertions, the first founded on constant facts, the other on a strong probability, allow one to conclude at least that all other characters taken from plants are more or less variable in one or some families, and consequently they cannot ever be essential by themselves. Those that vary a great deal remain in the category of individual characters; those that vary little become simply general, they are usually connected with one of the essential characters; this affinity is the real cause that reduces the number of variations; it provides at the same time accessory signs that announce the existence of true characters, and it has been seen [196] in the proposed method what help these signs might be when used with care: (3) there is no need of a long argument to appreciate the relative value of all characters. A different arrangement in the plant embryo, during the development and organization of the plant, causes remarkable differences which make as many characters: these differences, being dependent on that of the embryo, the characters that they give depend equally on the single one that causes them; from which it follows that the character taken from the embryo should have a value equal to that of all the others combined: (4) it further results that this character equivalent to several must always be uniform in families, and if it is the same in two, they ought to be brought together. Such is the basis of the fundamental divisions that seem to have been shown in the natural order, and [which are] based on essential characters taken from the main parts of the plant. This principle, if it is never contested or destroyed, assures for the arrangement of the Trianon its preeminence over those of MM. Adanson and Linnaeus; it unmasks a new agreement between animals and plants [which are] arranged on the same plan.

The Sciences that deal with the two kingdoms have both their advantages and disadvantages. Botany has more species to describe, but the organs on which it bases its characters are fewer, better known, almost all external, and each has a fixed rank in the economy of the plant. Zoology, on the contrary, has fewer species and more characters to examine; digestion, respiration, movement, and sensibility, which have no place in plants, necessitate many individual organs which in different classes of animals must vary in number, shape and structure, and in the degree of importance of their functions; it is not always easy to determine which are most essential, and because they are all internal, it is necessary to have recourse to anatomy to recognize them, with external signs to indicate them; [197] he who contents himself with these secondary signs without establishing their affinity with internal parts will not have other than an imperfect idea of the true similarities that occur among animals. Systems are good for giving a first idea of species and their obvious differences; the natural order alone can extend and guarantee our knowledge. Botanists and Zoologists must therefore leave systems aside so as to work together in the search for this order, whose existence is demonstrated and which is based on fixed principles.

ON THE RELATIONSHIPS EXISTING BETWEEN THE CHARACTERS OF PLANTS, AND THEIR VIRTUES

SUR LES RAPPORTS EXISTANS ENTRE LES CARACTÈRES DES PLANTES, ET LEURS VERTUS.

Hist. Soc. Roy. Méd. 1786: 188–193.

Translated by Susan Kelley
Arnold Arboretum of Harvard University

[188] The materia medica, one of the most essential parts of medicine, merits particular attention from physicians who propose to do useful research in the diverse kingdoms of nature. The virtues of plants were regarded, justifiably, by the ancients, as an essential objective of botany, and they [the ancients] have left us a number of good observations on this point. The more valuable their discoveries, the more we desire to extend them, by a general study of the properties of all vegetation. The

botanist, who would occupy himself only with theory and systematic arrangement, would supply only half of the views that he should propose. Besides, if he wants to follow the march of nature, which is simple and uniform in its combinations, he will not separate the study of plants from that of their properties, because they are intimately linked, and because one necessarily leads to the other.

[189] Two beings that are similar in their conformation, and consequently formed of the same elementary principles, cannot differ in their properties, if it is true that each particular principle has its inherent property, and that the mixture of a given number of these principles, in equal proportion, in diverse bodies, constitutes for each the same nature, and must give the same results in all points. Experience and observation come in support of this reasoning to prove this truth and to put it to use every day.

Botany, in classifying plants, attempted to bring together those that had the most relationships between them; but in admitting arbitrary principles and artificial methods, it was always diverted from its objective. This is perhaps the place to speak of the natural order, which alone can establish the true similarities, and to develop its principles and its advantages; but that question interests medicine only in an indirect manner; we content ourselves with discussing the points that are relative to it. Supposing then the natural order being known, along with the distinctions of species, genera, families, and classes, we will observe that relationships based on the assemblage of characters are certain and invariable, and that they indicate with precision the nature of each plant and the properties that derive from this conformity in organization.

The individuals of a particular species, which do not differ in [their] principles, have then absolutely the same property, as long as they were not altered by accidental causes dependent upon the earth, culture, and exposure. As each plant extracts only the juices that are appropriate to it, it must have more vigor and virtue, when it finds a terrain filled with the same juices. The plants of the mountains are more potent than the same gathered in valleys or cultivated in gardens, because their dominant principle is diluted by a smaller quantity of water in their vegetative parts. Therefore the chicory [*Cichorium*] [190] of the fields is preferable for use in medicine; that of gardens, more succulent, serves for the nourishment of men. Like individuals have thus the same principles and the same properties, that vary only according to the quality of fluid that circulates in their veins.

One finds the same conformities in the virtues of the species of the same genus. The mallows [*Malva*] are all emollients, the poppies [*Papaver*] narcotics, the gentians [*Gentiana*] febrifuges, the cucurbits [*Cucurbita*] refreshing, scurvy grasses [*Cochlearia*] antiscorbutics, rhubarbs [*Rheum*] purgatives, absinthes [*Artemisia*] vermifuges, the aconites [*Aconitum*] and the hellebores [*Helleborus*] more or less caustics. One species substituted for another, in medicinal usage, produces almost the same effect, provided that one proportions the dose. The root of the violet [*Viola*] of our fields, acts like that of the ipecacuanha, which is of the same genus.[63] The *Polygala* of Europe, employed for the same uses as the *Polygala seneka* of Virginia, produced, with a stronger dose, the same effects. The properties are therefore almost the same among congeneric species. One would object in vain that sometimes certain species seem to remove themselves from the common virtue of the genus; this removal always depends on some difference in the characters. The ranunculus known under the name of lesser celandine, *Ranunculus ficaria* Linn., is not caustic like other ranunculi, but one can distinguish it very well by its leaves, by its calyx, and by its corolla, which would be sufficient to make it a separate genus. It would be the same with other exceptions that one could allege; always the species discordant in its properties, varies also in its characters.

Genera brought together by nature offer, in their usages, the same relationships as neighboring species, for example borage [*Borago*] and bugloss [*Anchusa*], anemone and ranunculus, wild thyme [*Thymus*] and oregano [*Origanum*]. The well known virtues of one serve to determine those of the other. The *Lobelia siphilitica* was employed in America specifically [191] for certain illnesses; *Phyteuma*, a neighboring genus in the natural order,[64] was tried in Europe for the same ills, and the test was successful. It would be the same with the other genera that one would like to examine with care, and one will only succeed in substituting in medicine plants from foreign countries, by following the analogy of the characters.

This identity of virtues occurs not only in closely related genera, but also in the numerous assemblages known by the name *orders* or *families*, which unite all of the plants whose conformation presents many affinities and few differences. Their particular properties seem always to correspond to a principal property from which they are derived, almost as the union of all of the generic characters forms the general character of the family. We will make this proposition more evident by applying it to a large number of facts drawn from the families known to be the most natural.

The grasses, which nature seems to have united, almost always have the same structure. Their fistulous stems are full of a more or less sugary pith; their fibrous roots have a laxative quality; the seed is full of a more or less considerable farinaceous body, good for fermentation, when it is diluted in a certain amount of water; this farinaceous body is not only very nutritious, but applied externally, it becomes a good resolutive.[65] Although all of the grasses have their common properties, one does not substitute them indifferently one for the other. Medicine employs only the most common species, those that work well, that have the greatest number of roots, those of the couch-grass [*Agropyron*] in the shops. For nourishment one cultivates only wheat, rye, oats, corn, rice, in which the farinaceous body is the largest, the most nourishing, and has the best taste. Sugary material is [192] taken only from sugar cane, which furnishes a considerable amount of it.

The Liliaceae, another very natural family, can be subdivided into many sections, each of which has its own property. That of the asparagus has aperient[66] roots, like the asparagus [itself and] the butcher's broom [*Ruscus*], or else sudorific[67] like the sarsaparilla [*Smilax*]. The bulbs placed at the beginning of the roots in the lily section are softeners when applied externally, and taken internally, they strongly expel the wateriness in the abdomen; such are those of the lily, scilla, onion, garlic. This last quality takes on another nuance in the section of aloes, of which the juice is a potent purgative of a particular nature. These three sections, of which one could make three distinct families, have therefore some affinities in their virtues.

The Labiatae which are joined in nearly all of the methods, and which have so many affinities between them in their outward characters, have no fewer between their stomachic and cordial quality. It seems to be a family composed of a bitter and an aromatic property, which are themselves a mixture of other simpler properties. But how to arrive at separating them? One knows that bitter alone is harsh and coagulative, that pure aromatic is too irritating, and that these two substances together correct themselves and make a beneficial mixture. It is all formed by nature in the Labiatae, like sage [*Salvia*], mint [*Mentha*], balm [*Melittis]*, rosemary [*Rosmarinus*], germander [*Teucrium*], lavender [*Lavendula*]. Nevertheless the proportions are not the same. Aromatic dominates in sage, bitterness in germander, which renders each of them appropriate to be used in certain cases. One remedies nervous disorders, and returns life and movement in the organs; the other, less active against these ills, is better at restoring the weakened tone in the fibers of the stomach, which would be irritated by the first. It is as if each remedy must be proportioned to the severity of the ill, and [193] that the most active, used in a less serious accident, becomes more pernicious than useful.

The family of the Compositae, the most extensive of all, does not have a well-defined general virtue; but each of the three orders that it contains admits some less marked ones. One recognizes in the Cichoraceae, like lettuce [*Lactuca*], endive [*Cichorium]*, milkweed ["laitron": ?*Sonchus*], dandelion [*Taraxacum*], viper's grass [*Scorzonera*], an incisive[68] quality, dependent probably upon the milky juice that they contain; some come out at the skin, and the others in the urinary conducting tissues. These virtues are found with a different nuance in the order of the Cinarocephalae which includes the sunflower [*Helianthus*], burdock [*Arctium*], carline [*Carlina*], artichoke [*Helianthus*] and the different species of thistle. The order of Corymbiferae, which is the third, is subdivided into several sections. All seem to enjoy a bitter and tonic property, but it manifests itself differently in the tansies [*Tanacetum*], the absinthes [*Artemisia*], the camomiles [*Anthemis*], the elecampanes [*Inula*]. Some convey their action to the skin, such as petasites [*Tussilago*] and the elecampane; others, like tussilago [*Tussilago*], *Arnica, Gnaphalium*, have a stronger action in the lungs; matricaria [*Matricaria*], artemisia [*Artemisia*], are used successfully to give tone to the fibers of the uterus. The different effects are produced by the same dominant principle, modified differently, which, following its combinations with other principles, determines the particular action of each of these plants.

The effect of the Umbelliferae on the human body is still not well determined. Most have hot and carminative seeds, or aperient or sudorific roots, such as smallage [*Apium*], fennel [*Foeniculum*], masterwort [*Astrantia*]; but why are some of them poisonous, and require so much precaution in their medical usage, while others, cultivated in kitchen gardens, become very suitable for nourishment? These apparently different qualities, perhaps depend only on the [194] different proportion of their principal constituents. Bitterness, less abundant in chervil [*Chaerophyllum*] and parsley [*Petroselinum*], is then only tonic and aperient; stronger and combined differently in hemlock [*Conium*] and water dropwort [*Cicuta*], it causes over-heating and is even dangerous; amended with an aromatic part, it is a cordial in masterwort and angelica [*Angelica*]. A change in the proportions could produce contrary effects. Too large a quantity of parsley could become dangerous, while a small dose of hemlock is beneficial. It is always the same principle that varies, not only in these diverse plants but even in the different parts of the same plant: reproduced in the leaves of the hemlock and in the root of the carrot [*Daucus*], but with a very different dose, it shows itself by the same good effects, when it is applied externally.

If we pass to the order of Cruciferae, we recognize there the existence of a common principle, on which depends the anti-scorbutic virtue, present in all of these plants, and which is strongest in cochlearia [*Cochlearia*], turnip [*Brassica napus*], cress [*Cardamine*], and pepperwort [*Lepidium*].

The order of Leguminosae, one of the most numerous, includes many sections that are distinguished by their characters, and thus by their properties. All, but especially the section of the beans, have more or less farinaceous and resolutive seeds: those of the acacias and the bladder-senna [*Colutea arborea*] have purgative seed, pods, and leaves, properties more marked in cassia and senna than in other analogous plants. The section of brooms [*Cystisus*] is aperient, and one again finds this virtue in the roots of some plants in neighboring sections, such as the false acacia, *Robinia* [*pseudacacia*], and licorice [*Glycyrrhiza glabra*].

From this enumeration of the known properties of families, one can conclude (1) that those of the plants of the same family are analogous; (2) that this analogy is in proportion to the affinities of the plants themselves. The most [195] natural families, like the Gramineae, the Labiatae, the Cruciferae, have more equal properties than the Liliaceae, the Compositae, and the Leguminosae, whose characters are less uniform. The same analogy is found in the Malvaceae, the Boraginaceae, the Myrtaceae, the Rosaceae, and other similarly natural orders.

If in some families there are genera that are, on the contrary, a little removed from the general character, their properties differ for the same reason. The fumitory [*Fumaria*], placed near the poppies, but differing in its stamens and corolla, shares few of their properties; valerian [*Valeriana*] in the Dipsacaceae, the peony [*Paeonia*] in the Ranunculaceae, offer the same distinctions. These differences are observed, not only in isolated genera, but also in entire sections. In support of the examples already cited, we will speak of the section of polemoniums [*Polemonium*], which, placed in the family of the Convolvulaceae (one can make of it a very distinct family[69]), does not have a purgative virtue; the water lily [*Nymphaea*] and the lotus [*Nelumbo*], deprived of the caustic property of the Ranunculaceae, yet they are related (more recently they have been removed, and returned to the Hydrocharitaceae, among the monocotyledons[70]). One can believe, with some foundation, that these genera and sections that differ like this, do not invariably belong to the family to which they have been assigned; most of these have already been separated, either to form distinct families, or to be reunited with others. Consequently, the discordant genera are to families what the discordant species were to the genera.

We will remark, however, that the properties of neighboring families always have an analogy conforming to that which exists between the same families, by the comparison of their characters. The cleansing quality of the aforementioned Umbelliferae, such as sanicle and pennywort [*Hydrocotyle*], is found in the Ranunculaceae that immediately follow. [196] The family of the Scrophulariaceae is connected to that of the Solanaceae, not only by its characters, but by its emollient and resolutive quality, by its ability to heal the infections of the skin. If one again compares the vervains [*Verbena*][71] and the Labiatae, the Plantaginaceae and the Amaranthaceae, the Crassulaceae and the Portulacaceae,[72] one will have new examples appropriate for confirming the analogy of families by their virtues.

We will forego many other facts and proofs that are not necessary to develop; the stated observations suffice to clarify and strengthen the following conclusions, which terminate this memoir:

1. The elementary principles of the bodies, each having their particular property, the general property must be the same in the plants composed of the same principles in the same proportion.

2. The analogy in the elementary composition, being characterized by external signs, the plants marked with the same signs are composed of the same principles, and are consequently endowed with the same properties.

3. The affinity of plants by their properties is thus proportional to the affinity of their characters or external signs. It is therefore strong in the individuals and species, less so in genera, slight in families, and almost none in the classes.

4. Observation demonstrates the truth of each of these propositions; it also proves that, if one species is removed a little from its genus, a genus or a section from its order, their properties also offer progressive differences.

5. There exists therefore a marked relationship ["rapport"] between the characters and the properties of plants, and the inspection of the one suffices to determine up to a certain point the other.

[197] Such is the relationship that it was proposed to establish and to prove in this memoir. If it is recognized and acknowledged, it will be less difficult to indicate the virtues of all plants which had been regarded up to the present as deprived of all properties ["propriétés"]. The study of the botanist will find itself intimately linked to that of a doctor, and his research will always serve to the benefit of humanity.

An Introduction to the History of Plants
[the Introduction to the *Genera plantarum*]

Introductio in historiam plantarum.
In *Genera plantarum* . . . ,
pp. i–lxxii. Paris: Hérissant et Barrois.

Translated by Susan Rosa
University of Wisconsin at Milwaukee

Note by Peter F. Stevens: The arrangement of the Introduction is not clear. Although there are sectional headings in the margins, the relationship of these sections to one another is not explained. There seem to be three main parts of the Introduction, with blank lines immediately before the discussion of "the true aim of the science" (p. xxxiv) and the brief conclusion (p. lx). The absence of clearly subordinated sections makes it difficult to find out exactly what Jussieu meant by his second way of attaining the natural method (see pp. xlii et seq.). The table of contents for the Introduction accompanies this translation; the pagination given there is that of Jussieu.

Table of Contents for the Introduction to the *Genera plantarum*

[Preliminary Discourse]

[i] The science that investigates and sets forth the nature of Animated and Vegetative Beings and Minerals is called natural history. Whether it examines individual bodies separately, or whether it includes all together, this science, ranging widely, has no fixed boundaries. Closely linked to the other sciences, it sometimes, proceeding in advance of them, offers light to the hesitant, and at other times it borrows from previous work the means to undertake useful investigations more profitably and bring them to a happy conclusion. It considers the varied forms of bodies, the remarkable connection among all the elements, or the texture of organs, and in the remote areas of the earth, traversing different regions, it recognizes everywhere the peculiar image ["typus"] and origin. Among individual species it admires the harmonious structure of parts and the happy collaboration of forces as it were in a contracted world; next, rising up in a more ambitious effort, it examines the entire sphere of the earth itself, whose daily and annual revolutions, as well as its uniform phenomena, make manifest

everywhere an unvarying harmony and a unique moving force moderated by an eternal law of the Divine Will. Without doubt, the various parts of bodies, as well as all bodies in the world, are driven by an incessant internal movement. The perpetual collision of attracting and opposing elements and the consequent loosening of old bonds agitate the great mass of minerals and bring together new combinations. Innumerable Plants, which adorn the earth in an inexhaustible variety, being subject to different alternations in their growth, are reborn or put forth new foliage every year. Animals exercise similar functions in the course of life, and, preserved by the reciprocal force of solids and fluids, and gradually renewed by nourishment, are propagated without variation by a certain successive transfusion of vital exhalation ["aura"].

Now this science, the contemplator of nature, which is concerned with the manifold investigation of forms and structures and actions and places, is not only magnificent but also excessively extended so that it cannot be grasped by human intelligence unless it is distributed into several branches. Nature, an ambiguous term which is to be understood in two senses, sometimes signifies a very general mass of beings or the universal condition of the earth, and [ii] sometimes refers to the structure of a single body and its special mode of existence. He who, too confident in himself, dares to undertake the study of nature in the first sense, will hold back inquiring minds and seduce them by a great preparation, and allure them by a splendid discourse and apt words; for the sincere investigator of natural events, however, a brief sample and incomplete knowledge of them will not suffice. He must first understand the nature of single bodies complete in all their parts, which is particularly determined by their elementary composition and their organic structure ["fabrica"].[73] This is a double science, very distinct, of elements and organs, the one part calculating the nature, number, proportion, and mutual affinities of the elements, the other measuring the structure, the number, the arrangement, and the pre-eminence and reciprocal action of organs. The old division of nature into three kingdoms—namely, Animated and Vegetative Beings and Mineral—is therefore to be rejected and more recently there has been more correctly substituted for it the division into the organic Kingdom and the inorganic Kingdom, the latter containing Minerals alone, and the former embracing Animated and Vegetative Beings.

Kingdoms of nature

The inorganic nature of minerals is determined by the joining of elements coming together in various mixtures, by the juxtaposition of which a mineral body arises, grows, and is brought to perfection; in its turn it is destroyed, the elements separate, then by a more recent bond came together in new bodies not coeval with the first but made from their detritus, very distinct from the organic being to be defined below. Mineralogy stops at the examination of the form and mutual position of Minerals; Chemistry proceeds further, separating by analysis the principles of bodies, discovering the number of these and considering the various affinities among them, and emulating nature itself through the fruitful art of Synthesis.[74]

Different from the preceding, the science which studies the nature of organs does not devote itself to discovering in them those principles which govern chemistry, but above all endeavors to gain [knowledge] of the form and connection ["contextum"] and the mode of life of organs. For the organic being is composed of solid and fluid parts which drive one another in reciprocal motion, of which some [solid parts] are very simple, *similar substances* ["similares"], others, *organic*, are made out of similar substances; the fluids flow within the vessels of the solids, or remain within their compartments ["cellulae"]. [Such a] being is born of a congeneric one, it grows in conformity with its predecessor, it gains strength, basic constituents of nutrition being incorporated ["intussusceptis"], from the beginning to the end it undergoes various transformations, it produces [iii] offspring like itself, persisting beyond it, and finally, when the usual time has elapsed, it dies. Hence its history[75] and nature are together expressed by the structure and action of the organic parts, and from thence the knowledge of organs and their functions produces the true science of the organic Kingdom.

A common structure ["apparatus"] sustains the lives of Animated and Vegetative Beings, although this is simpler in Vegetative Beings, which being principally deficient in nerves and muscles consequently neither feel nor exert spontaneous movement: consequently there is a science common to the two [groups] surveying their organization and vital functions ["vitam"], enumerating similarities and differences coming from a like or unlike mode [of life] of each, and thence establishing a natural series, in which similar [organisms] are close together, the dissimilar are separated, so that from this immutable arrangement of beings definite affinity and alienation[76] between them is perceived beforehand, and a few beings being known, knowledge of similar neighbors is more readily arrived at. From what has been said, the Animal is an organic body, living, feeling, and moving; a Plant, though, is an organic body capable of life, but lacking movement and sensation. Enough said of the analogy and difference of the two: it would not be right to dwell further on an enquiry into Animals, which is the proper concern of Zoologists; now we must completely turn our attention to the history of Plants, which is principally the reason for this undertaking.

Botany

The science of plants, called *Botany* or *Phytology*, is what inquires into the nature of Vegetative Beings, that is, it determines the number, texture, action, and value ["praestantiam"], the shared or unique position, the shape, proportion, and mutual differences of their organs, thence deducing the signs suitable for defining them and distinguishing them one from the other. This easier designation of plants is the highest achievement of that modern science which rests completely on the selection and arrangement of especially suitable and evident signs, but which is less concerned with other things. Unlike the preceding, however, the true science, devoted to a study of the entire organization, does not stop at a sample of signs, but is obliged to examine all aspects ["modos"] of all organs, an accurate exposition of their structure and action being preferred.

Parts and functions of plants

The simplest or similar parts of plants, called *fibers* and *vesicles*, the same in Animated Beings, coalesce in different ways to form vesicular webs, membranes, or vessels, swollen by nourishment, their own juice, or by air. From the interweaving of these are formed the central pith ["medulla"], the surrounding wood ["lignum"], and the outermost bark ["cortex"]. Larger and more perfect [iv] organs are formed from such conjunctions; some, like the root, the stem, and the leaf, fostering the life of the plant; others, like the flower and the fruit, are essential for the propagation of the species, all coming together in the general mechanisms ["organum generale"] of whole plant, whose [the organs'] structure, arrangement, and use it is profitable to investigate in the first instance.

The *root*, fixed immovably in the earth, extends upward into the *stem*:[77*] in both, the completely bladdery medulla makes up the medulla; the wood surrounds it in hardened[78] layers of longitudinal bundles of vessels and fibers connected by interspersed bundles with a utricular texture; the bark forms a covering made up of fibers and vesicles and separate series of utricles laid almost parallel, loosely interwoven, a network, so to speak, moist and succulent in the root, in a herbaceous stem it is soft and membrane-like, while in a tree-like [stem] it is dryer and often fissured. The root, in the fashion of milk-bearing veins, avidly sucks out its juice, divested of grosser particles like the chyle of animals, prepared in the earth, and sends it to the vessels of the stem, which carry it higher, most generously in the young plant, less so in the older one on account of the increased amount of solids and the decreased diameter of the vessels.[79] One part of this nourishment, conveyed between the cortex and the wood, condenses ["concrescit"] there on one side to become the *sapwood* ["alburnum"] or outer wood, and on the other the *liber*, or layer of the inner bark.[80] The other part, continuing along the bypath of the vessels, deposits stickier particles—now sweet or gummy, now resinous or bitter—in the utricles or glands, and then proceeding further it ejects more fluid substances chiefly through the skin and spiracles of the leaves. For the *leaves* possess a double series of vessels, both on their upper and lower surfaces, and these gape through fine passages, and both are covered by a paired membrane, which is easily separated after maceration. The pores located on the upper surface emit excess perspiration fluid ["humorem"] by a special force and the attraction of the Sun: therefore the upper surface of the leaves is always directed toward the free air, in such a manner that, even if the petiole has been turned upside-down, the leaf returns slowly toward the Sun. The pores hidden on the underside of the leaf and frequently crowded by shaggy hairs, inhale the morning and evening dew from the atmosphere, which once taken in, is driven all the way down to the root and serves as its nourishment. Thus [v] juice travels perpetually from root to leaves and from leaves to root, emulating the circulation of Animated Beings, and performing the same service for Vegetative Beings. Spiral tubes among the vessels, similar to the tracheae of the Vermes,[81] aid this alternate ascent and descent, filled with air entering through the passages of the roots and leaves which by reciprocal expansion and contraction elaborate the juice, as in animal respiration, and this is said to aid the circulation.

The aforesaid actions of the organs described suffice to renew the life of plants through nourishment, and support their vital activities ["motui"]; in addition, other parts and functions are dedicated particularly to their multiplication and propagation. To be mentioned here only in passing, however, are the arts of propagation best known to Gardeners, such as cuttings and layers taking root in the soil, and stolons sent out from the lowest part of the stem, and the setting of grafts, and the bulbils of *Allium* or *Narcissus*; more generally, propagation is carried out by the union of the sexes. For as long as the irrigating sap bathes the various parts of the Vegetative Being with nourishing dew, *flowers*, displaying sexual organs in a special covering, which surrounds them in double, or more rarely, in single fashion, emerge from buds and prepare the way for the next generation of offspring. The outer covering, called the *calyx*, more durable than the interior, the continuation of the skin ["cuticulae"] of the flower stalk, and consequently often the same greenish color, surrounds the other parts of the flower. Inside the calyx is the *corolla*, continuous with the liber of the flower stalk; [it is] another covering, sometimes abortive, often of a color other than green, undivided or consisting of numerous leaflets or *petals*, commonly identified with the flower on account of its great conspicuousness, [and it is] placed near the sexual organs, which are the surrounding male organs, or *Stamens*, and the central female organ, or *Pistil*. The stamens originate together with the corolla; having the same life span and sometimes connected to it, arising together from the same point, and shed together after fertilization, they bear the *anthers*, or folliculae, heavily laden with seminal *pollen* and often supported by a *filament*. The parts of the pistil are the gravid *germen* [= the ovary], with the rudiments of the seeds, like the womb; the *style*, similar to the vagina, originating most commonly at the top of the seed, and tubular or often hair-like, sometimes none; the *stigma* at the end of the style or placed upon the germen if the style is lacking, and often sticky and wet. Whether they share a covering, or whether they are in different places,[82] the stamens and pistil work together toward the same end. For as soon as they reach maturity, with elastic force the anthers break apart, and the emitted *pollen* clings to the moist stigma, washing the eggs with a fertile rain. [vi] This [rain] consists of innumerable vesicles, very small and of various shapes, which when they swim in fluid, they float about hither and thither, swell, and splitting along the side, emit gelatinous material which spreads like a cloud but does not mix with water; this must be considered to be the seminal breath ["aura"] which alone fertilizes. Relieved of their weight, the withered stamens soon fall off at the same time as the corolla, which in truth is an appendage of the stamens as well as a diverticulum for the fruit-bearing sap sent in advance until this sap, unmindful of its earlier path, flows directly into the young fruit where it will be more useful. Then the *fruit* or *pericarp* swells and often thickens, the utricles being turgid with copious sap. Inside this fruit, the conceived offspring secretly grow. The sap, weakened to a suitable nourishment for the tender young, enters the ovules, and ultimately brings them to their destined size. When its growth is complete, the lymph sap inside hardens and becomes solid, not all at once as in the viviparous class of Animated Beings, but through alternating periods of growth and incubation within the pericarp. Soon the fruit becomes dry, and generally splitting open, reveals the mature seeds affixed to an interior *receptacle* [= the placenta]; [the seed is] sometimes covered by a hard, bone-like or shelly *nut*, or by a membranaceous or leathery *aril*. The *seed* itself, very frequently covered by its own double membranes, is marked on the outside by the *hilum*, a navel or vestige of the funicle by which the seed was attached to the receptacle; inside it shows the *corculum* [= the embryo], or compendium of the plant, composed of the ascending *plumule*, the descending *radicle*, and the *cotyledon* or lateral lobe, which may be simple or double, [the embryo is] usually separate in the seed or provided with the *perisperm*,[83] a distinct body partly or wholly surrounding it and of different origin, useful perhaps for protecting the interior.

This is not all. The seed, master of itself, beginning a new life, is committed to the fertile bosom of the earth; then passages are opened, softened little membranes split, the lobes[84] separate, while underneath, the unimpeded radicle admits juices; within, the rising plumule grows to become the stem, to which the lobes supply food. Gradually, these lobes become *seed leaves*, which protect the little plant against all possible shocks; these provide and store lymph both taken from their own suction and borrowed from the more active radicle, without which it [the plant] would weaken and decline. The medulla is covered with a delicate cortex, fibers being almost absent. But new leaves develop and the seed leaves fall off; the suction of the root is increased by the beginning of perspiration; moisture ["humor"] flows in; a growing, cone-shaped bundle of fibers separates the medulla from the cortex[85] and forms the layer of wood at whose apex is [vii] the bud, the source of new shoots. Next, a kind of organic mucus, not soluble in water, is deposited between the bark and the wood, making the beginning of the fibers. Twisted in cylinders or spirals, these fibers create an unending series of small vessels; every year new bundles, new layers are built upon the old; the cortex becomes stronger as it extends; terminal buds are driven to the sides, to bring forth twigs; with the work of nutrition happily completed, the plant thrives and grows larger; sending out lush branches bearing beautiful foliage, it is adorned with flowers and produces fruit, the hope of new progeny. But gradually, by the mingling of the earthly principle, the soft vessels harden into more compact fibers; and, closed up, they deny access to the sap; now the solids rule over the fluids; the latter are halted in their course while the former grow hard; compressed, the medulla is altogether destroyed; the closing of the vessels in the bundles continues, the roots and branches perish and fall off. Air penetrates the interior of the mutilated body; by alternating moistness and dryness, it breaks up the utricles and the connections of the fibers; the earthly particles ["terreas moleculas"], deprived of their glue, separate, and, broken up, they crumble into dust: thus the Vegetative Being dies.

Such are the various changes of growth and decline in plants, always the same in each, and never distorted. The explanations of functions set forth here are only samples, gathered from the experiments of various selected and approved Authors and from their various systems of the nature, development, and action of the parts of the plant; several organs of lesser importance have been reserved until later to be mentioned in due course. Certain animal functions—such as the action of the senses and spontaneous movement, as well as preliminary digestion and the excretion of solids—are lacking in plants because of the absence of the [relevant] organs; nevertheless, plants can easily live without these things and still maintain vital functions ["motu"], thus showing obviously an essential sign of their clear difference. From what has been said, their organic nature may be reasonably concluded, as well as their affinity with Animated Beings, and it has become apparent that they both are subject to physical laws.

The differentiae of the parts

The parts enumerated, since they always fulfill the same function, differ in different plants according to their mutual location, number and form, interior and exterior structure, and duration, which principal differences will now be noted and defined one by one so that these being properly inspected, plants can be distinguished in turn easily and by an apt choice of terms.

[viii] The root, that part of the plant which descends into the earth and there draws out juices; with respect to position is *perpendicular*, descending straight, or *horizontal* or *repent*, stretched out sideways and crossways, and occasionally *stoloniferous*, emitting little shoots here and there. With respect to number, it is either *simple*, *branched*, or *fasciculate*, consisting of several [stems] bundled tightly together. With respect to structure and form, it is *fibrous*, or *capillaceous*, with fibrous rootlets very much like hairs; *tuberous*, compacted into a roundish and thick body, with the inside utricular and less fibrous; *turbinate*, stout and obconical, or *fusiform*, longish, and thick at the top and more slender at the bottom, or *praemorse*, longish and truncate below, or *nodose*, swelling, often with short interspaces. With respect to length of life, it is *annual* and vanishes every year, or *perennial*, surviving longer, *herbaceous*, soft, and *ligneous*, more solid. Certain roots are parasitic on trees, or implant themselves in walls; others are aquatic. The *bulb* is not to be considered a root; more accurately, it is the very lowest subterranean part of a scaly stem or [one] ensheathed by leaf bases, usually swollen, scaly or turbinate, and having roots at the bottom.

The stem, that part of the plant which emerges from the earth (more rarely, it is almost absent), and bringing forth the leaves, flowers, and fruit, is *bare* or *leafy*, *entire*, quite undivided, or *simple*, branching at the top, or *composite*, divided at the base and caespitose; with the divisions or branches *alternate*, emerging here and there, or *opposite* with paired shoots which grow from opposite sides of the stalk, or *verticillate*, with many shoots placed opposite in a circle, or *dichotomous*, dividing several times in pairs. In shape, it may be *terete*, cylindrical, or *ancipitous*, with two ridges opposite one another, or it may be *triquetous*, *triangular*, *quad-*

rangular . . . trigonous, tetragonous . . . polygonous from its number of angles or facets [ellipses in the original]. In surface, *striate*, furrowed with lines, or *sulcate*, hollowed by grooves, or *glabrous*, entirely smooth, or *tomentose*, like a cloth, or *villose*, with sparse, soft hairs, or *asperous*, with stiff hairs, or *scabrous*, roughened with points; or *aculeate*, with sharp points emerging from the cortex, or *spinose*, these protruding from the wood.[86] In orientation, *erect*, or *procumbent*, or *repent*; *flexuous*, [curved] at intervals, or *scandent* on neighboring supports, or *twining* spirally, *dextrorse*, ascending around them following the movement of the Sun, or *sinistrorse*, in opposition to the Sun's course, or *sarmentose*, woody, creeping, and almost bare. In *herbaceous* [plants], in duration a year, in woody *trees* most perennial, less in *shrubs*, least in *suffrutices*. In some [plants] the stem bears a different name, namely the *trunk* in a tree, [ix] the *scape* in certain herbs [where the stem] arising from the root bears flowers but not leaves, the *culm* in Grasses, hollow and nodose at intervals, the swollen *stipe* in the Fungi, and the *caudex* in *Dracaena*, *Yucca*, *Agave*, and the Palmae, where it is continuous with the root and as it were a cylinder in shape, bare beneath a simple tuft of leaves growing on top.

The leaves are the primary organs of perspiration and inhalation; they are almost always present, and rarely absent, sometimes *petiolate*, whether with a bare *petiole* or more rarely with wings, sometimes they lack it, and are *sessile*; the base may be bare, or *stipulate* or stipate, accompanied by membranes and scales which are given the name of *stipules*; they [leaves] differ primarily in location, position, orientation, and duration; they may be simple or compound in structure, and may also differ in foliation or early development.[87] Thence come many useful signs of these differences, not all to be rehearsed at length here, but only the most outstanding will be set out, as the others have been most accurately recorded and defined elsewhere. 1st. Leaves according to their place of attachment are *radical*, *cauline*, *branch*, and *floral*. 2nd. According to their position both individually or with respect to other leaves, they are *amplexicaul*, *perfoliate*, or *vaginant* when surrounding the stem altogether or in part, at the base like an embracing sheath, *decurrent*, winged on both sides at the base on the stem; *alternate*, emerging here and there without order, *opposite*, arising in pairs on opposite sides of the stem [and] with the bases occasionally *connate*, *verticillate* when three or more are opposite in a circle, *distichous* when disposed in two opposite rows, often *fasciculate* or *conferted* when close and more numerous, *imbricate* when covering each other. 3rd. According to orientation, they are *erect, horizontal,* and *pendulous, appressed* to the stem, or *patent, inflexed,* or *reflexed, straight* or *oblique.* 4th. According to their length of life, they are *deciduous, persistent,* and *evergreen.* 5th. According to structure, the majority are *simple,* or solitary and undivided above the petiole: of these the form is *linear, subulate, lanceolate, hastate, sagittate, deltoid* or triangular, *rhombic* or of four unequal angles, *cordate, obcordate, reniform,* or *orbiculate.* The margin is *entire, ciliate, spinose, dentate* with widely separated sharp teeth, *serrate* with these [teeth] close together, *crenate* when they are blunt, *sinuate, lobed,* the sinuses being distinguished by the lobes being either acute or obtuse, *palmate* with these coming together in one point, *pinnatifid,* oblong, with both sides deeply and frequently sinuate and lobed, [and] *runcinate,* almost the same, with pointed, retrorsely uncinate lobes. The apex is *acute, acuminate, obtuse, truncate, retuse,* [or] *emarginate.* The surfaces of the upper and lower sides [are] *shiny, smooth,* [x] *tomentose, lanate, green* or *hoary, hirsute* or *pilose* with distinct hairs, *hispid* with sparse stiff bristles, *scabrous* with sharp points, *punctate* with points not projecting and sometimes transparent, *aculeate* with a few prickles, *venose* with many conspicuous veins entwined in a loose network, *rugose* with veins standing out from intervening areas, *nervose* with nerves running undivided from the base to the apex, [and finally,] *avenous* or *enervate* with veins or nerves not at all apparent. The surface is *plane, convex, concave, plicate, undulate, crisped,* [or] *canaliculate.* The substance is *membranaceous* soft, *coriaceous* more solid, *terete* or cylindrical, *semiterete, fistulous* or tubulous, *thick, fleshy, triquetrous,* [or] thickened [and] *polymorphic,* as [in species] of *Mesembryanthemum.*[88] Compound leaves consist of several *leaflets* attached to the same petiole, namely, *conjugate,* with two leaflets, *ternate,* with three, *quinate* or *digitate,* with five or more emerging like fingers from the same point on the petiole, *pedate,* when the leaflets emerge like fingers, but are affixed internally to a divided petiole, *imparipinnate* when several leaflets arise on either side of a common petiole [and] with an unpaired terminal [leaflet]; *abruptly pinnate* [= paripinnate] when the petiole lacks its solitary leaflet; *bijugate, trijugate* . . . depending on whether two or three leaflets are present on each side of the *pinna. Decompound,* or twice compound [leaves], are, like the preceding, *bigeminate, biternate, bipinnate* with the pinnae pinnulate. *Supradecompound* [leaves], with three orders united in the same manner, are *tergeminate, triternate, tripinnate* with the pinnae bipinnulate. *Multipartite* [leaves] are made up of very numerous leaflets. 6th. According to their mode of foliation or early development [which may occur] inside or outside a bud, the folding is proper to each and every young leaf: *convolute*, with the edge of one side bent inward and surrounding the other in a spiral fashion, *involute* with each edge spirally incurved separately, *revolute* with each edge separately recurved, *circinate* with the apex spirally recurved toward the base, *reclinate* with the apex simply recurved toward the base, *conduplicate* with each side folded once and the edges are parallel to each other, [and] *plicate* corrugated with several folds. With respect to their mutual position, young leaves are *convolute* with the one on top rolled around the one below, *involute* or *revolute* with each rolled separately inward or outward, *equitant* with one conduplicate and including its like, *obvolute* when they are conduplicate with the alternate margins folded in one after the other, [and] *imbricate* when they are flat, lying against each other.

Now that the major differentiae in roots, stems, and leaves have been expounded in a summary fashion, we must inquire into the various dispositions and kinds of structure of the flower and fruit. In considering the *inflorescence*, or the [xi] disposition of the flowers, [we must give an account] of their location and insertion, orientation and number, and mutual position, as well as their added ["adjunctarum"] parts. 1st. By their place, flowers are *radical*, *cauline*, or *ramiflorous*, from the flower-bearing part, *terminal,* if at the apex of the branch or stem, *axillary*, if in the axil between the leaf and the stem, *supra-* or *extra-axillary* or arising without arrangement and *sparse*. 2nd. In insertion, they are *pedicellate* springing from a *pedicel* or individual support, or *sessile* lacking a pedicel. 3rd. In orientation, [they are] declinate, horizontal, or nodding, often prone, more rarely oblique or resupinate, certain of them follow the Sun, turning toward it unceasingly by a diurnal bending of the pedicel. 4th. In number, they are solitary or several together, *fasciculate* with distinct pedicels, or one-, two-, or many-flowered coming together in a common *peduncle*. 5th. In relative position, those crowded together and embracing the stem in a circle are *verticillate*; those [which are] sessile and clustered together on a common receptacle are *capitate*, and occasionally they are *aggregated*, grouped together inside a common covering; others, sessile or subsessile and growing here and there on a rhachis or common axiform peduncle, are called *spicate*, *distichous* or in a paired arrangement [alternating on opposite sides of the stalk], or *secund* or arranged on one side, or *spiral* when the spike is twisted into a spiral, or *amentaceous* in a denser, soft spike like a cat's tail, with unisexual, apetalous flowers, as is to be mentioned below; others, *racemose*, are similar to spicate [ones], but each [flower] emerges from a separate pedicel, with the pedicels subequal in length, one- or few-flowered; others, *corymbose*, differ from the racemose in that the lowest pedicels gradually become longer and erect, equalling the uppermost pedicels in height; others, *paniculate*, with several peduncles divided in various ways, growing unevenly, lax or congested; others, *umbellate*, with subequal pedicels joining to the same point on the one peduncle, like an umbrella; others, *cymose*, are racemes or corymbs joined at the base like an umbrella; certain ones become so-called *spadices*, chiefly in the Palmae and the Araceae, from the *spadix*, or undivided or branching peduncle, bearing flowers in a spicate or paniculate fashion inside a *spathe* or surrounding membrane.

In addition, flowers are sometimes naked, sometimes surrounded beneath the calyx or at the end of the pedicel with a *bract* or floral scale, single or many, lateral or clasping, colored or green, persistent or caducous; various in form and structure, some are assigned different names. In most Umbelliferae and some other plants, numerous leaves[89] placed in a verticillate manner around a peduncle are known as an *involucre*; [xii] the *glume* in the Gramineae is outside of and similar to the calyx and is often bivalved; the *spathe* in the Liliaceae and spadiceous flowers ensheaths the pedicel or peduncle at the base and surrounds the young flowers, and may be simple, split, or multipartite.

The parts of the flower mentioned above are the sexual organs and their coverings or Perianths, the latter being known by the names of calyx and corolla, and the former by those of stamens and pistil,[90*] which differ individually in their structure and arrangement, both of which we will now discuss in detail.

The calyx, the outer covering of the flower, or *Perianth* properly speaking, is a corticose extension of the floral peduncle, rarely lacking, placed as the protection of the sexual organs; it originates beneath the pistil, sometimes it grows partly or altogether upon that organ, covering it like a skin, *semisuperior* or *superior*[91*] when the upper limb is free, sometimes *inferior* when distinct and separate from the point of origin; *caducous* directly the flower opens, or *deciduous* later with the corolla, or *persisting* afterwards; *simple*, or *caliculate* with an often smaller *calyculus* on the outside; *polyphyllous* when it consists of several leaflets which are distinct, definite or more rarely indefinite [in number], and arranged in a single, double, or multiple order, or *monophyllous* and then *multipartite* with lobes joined only at the base, or *entire* if it is completely undivided, or undivided at the base, but divided at the limb. Its base [may be] short or longish or very long; [the calyx is] *ventricose* with a contracted throat, or *urceolate* with an open throat, or *tubular* [if] cylindrical, or *turbinate* [if] obconical, or if rather long, [it is] *infundibuliform*, or *patent* if entirely flat. The limb is connivent, or straight, or patent, or reflexed, undivided, [xiii] or crenate, or dentate, or sharply laciniate or *multifid*, or bluntly *lobed*; [the lobes] definite or indefinite [in number], of equal or unequal [length], and occasionally irregular. In the Musci, the peculiar calyx is very different on account of the calyptra, and is given another name, the *perichaetium*; in the Fungi, a membranaceous *volva*, like the calyx, often enfolds the young fungus; in certain Amentaceae *squamae* take the place of the missing calyx; in aggregate flowers there is a common calyx for many flowers.

The corolla, the inner covering of the flower, emergent from the liber of the peduncle, and very often colored, is present in most plants, but absent in a few;[92*] before fertilization it thrives; afterward it is very often *deciduous* along with the stamens, or more rarely *caducous* more transitory than these, or sometimes persisting and *marcescent*, and then partaking more of the nature of the calyx than of that of the petals; moreover it differs 1st. according to its location, namely *perigynous* or attached to the calyx surrounding the young germen, or *epigynous* placed on the pistil, or *hypogynous* when placed below [xiv] it; [it may be] alternate with the parts of the calyx, or more rarely opposite to them. 2nd. According to the number of parts, namely, *polypetalous* when with a definite, or more rarely, indefinite [number of] petals composed of a limb and a blade, or *monopetalous* in the same manner as a monophyllous calyx, sometimes deeply multipartite, now with an undivided tube and a limb definitely or indefinitely lobed, crenate, dentate, ciliate, or more rarely entire. 3rd. According to its form, namely *regular* with parts opposite each other similar and equally distant from the center, like a bell, a funnel, a cylinder, a wheel, a bowl, a star, or a cross, etc.; or *irregular* with unequal spacing and structure of the opposing parts, *personate* with the image of a face, labiate, horn-shaped, papilionaceous, or polymorphous [and] anomalous. 4th. According to its proportions, namely large, medium, or small, equal in size to the calyx and stamens, or larger or smaller. 5th. According to its color, which often varies in many; in many it is constant. 6th. According to the addition of other parts, such as a small gland, fovea, spur, appendix, scale, bristle, etc.

The stamen, or male organ of the Vegetative Being, consists of the anther strictly speaking, follicular, full of pollen, sometimes sessile and sometimes supported by a filament, of which are to be considered 1st. the position or place of insertion of the entire stamen, *perigynous* when inserted on the calyx or *epipetalous* when on the corolla, now *epigynous* when inserted above the pistil or *hypogynous* when below it, now attached to a glandular disk emerging from between the base of the pistil and the lowest part of the calyx; sometimes [it is] opposite or alternate to the calyx or corolla, sometimes entirely separate from these. 2nd. The position of the anther, sometimes upright at the apex of the filament, or incumbent or versatile with each end free, or almost umbilicate, or it may be adnate to the side of the filament, sometimes projecting a little above the same. 3rd. The structure of the anther, externally round, ovate, oblong, sagittate, two-horned, bisetose, arcuate, reniform, didymous, appendiculate, etc.; internally [it is] often bilocular, more rarely uni- or multilocular, very frequently dehiscent along its length, sometimes porose at the apex, or with a single valve [opening] from the base to the apex, or transversely bivalved. 4th. The structure of the filament, often capillary, sometimes flat, cuneiform, dentate, hirsute, squamigerous, [or] glandular, etc. 5th. Sometimes the joining of parts occurs, such as when the anthers unite into a single, tubular body; the collection of the filaments into a single organ resembling a ring, a pitcher ["urceoli"], a tube, or a column; or the filaments forming double [xv] or multiple bundles, bearing two, three, or more anthers. 6th. The number of stamens may range from one to twelve, or to a definite number over twelve, or indefinite, whence plants are called *monandrae*, *diandrae*, . . . *dodecandrae*, *icosandrae*, or *polyandrae*. 7th. The proportions of the filaments, equal or unequal to each other, or equal to, or longer, *exserted*, or shorter than the surrounding parts. 8th. The abortion of some or all of the stamens. 9th. The pollen, or seminal vesicle containing the spirit,[93] in shape in each [species is] unchanging, spherical, oval, didymous, angular, echinate, rotate, etc., soon becoming deformed when the spirit has been expelled, and serving as a waxy material for Bees.

The pistil, or female part, central in the flower and placed upon the floral receptacle, consists of the germen, the style, and stigma. The germen is either simple, *monogynous*, or multiple, *polygynous*; sometimes *superior*, with the calyx below, sometimes *inferior*, fused inside the calyx with that organ above it, or occasionally, when it is partially joined to the base, it is called *semi-inferior*; sometimes [it is] attached to a glandular disk, either surrounded by the disk below, or covered by a similar gland above; very often [it is] sessile, but sometimes raised on a stipe; in certain flowers [it is] abortive. In *monostylous* flowers the style is simple, and either entire or divided at the top, or in *polystylous* [flowers] it is multiple, or occasionally none; very often, it emerges from the apex of the germen, or sometimes from the side, more rarely from the base; [it may be] deciduous or persistent; varying in length, orientation, thickness, and form; very rarely staminiferous. The stigma is never lacking in a fertile germen; sometimes it is sessile and placed upon it, more often it is situated on the top of the style, simple or multiple, deciduous or persistent, varying equally in length, thickness, and form.

The four parts listed above together make up a *complete* flower; if any one is lacking, it becomes *incomplete*. In this latter category are *apetalous* flowers, or *acalycate* [flowers] which lack a calyx, or unisexual, *female* [and lacking] stamens, or *male* and deprived of a pistil, or else separately bisexual, *hermaphrodite*, or *neuter* and eunuch-like. Most plants, however, are *hermaphrodite*, being completely bisexual in every flower; certain exceptions are the *androgynous* or *monoiceous* [monoecious], which bear separate male and female flowers on the same plant; some are *dioiceous* [dioecious], which bear a single sex, *male* or *female*; more rarely, *polygamous*, which bear hermaphrodite and male or female flowers mixed together; some are *cryptogams*, with the sex organs inconspicuous on account of their minuteness, and consequently less known.

[xvi] Certain flowers growing in luxuriant fashion because of their very careful cultivation and abundant sap, multiply their coverings; this happens only rarely to the calyx, but more often to the corolla, sometimes in monopetalous and most often in polypetalous flowers, [and they may be] either truly sterile with all their stamens aborted, *full* ["pleni"], or fertile, with a few stamens remaining, *multiplied*. Because of this, radiant varieties,[94] which grow in gardens and are much in demand among enthusiasts, are disdained by Botanists on account of their impure organization, and count for nothing in science unless to

confirm, by the frequent conversion of stamens into corolla, the very great analogy between these two organs.[95]

The fruit is formed from the mature germen and consists of the enveloping pericarp and the seeds it contains. The pericarp is commonly identified with the fruit as a whole, and is occasionally lacking (the seeds then [being] bare),[96] simple or multiple, superior or inferior or semi-inferior; it agrees with the germen in position and number. In these [things] besides, it varies: 1st. In form [it may be] spherical, ovate, truncate, turbinate, conoid, angulate, compressed, orbicular, bladder-like, alate, articulate, etc. 2nd. The surface [may be] terete, sinuous, striate, sulcate, smooth, prickly, tomentose, villose, hispid, echinate, spinose, etc. 3rd. The substance [may be] membranaceous, coriaceous, bony, fleshy, pulpy, or succulent. 4th. Dehiscence [may be] none, the fruit being undivided; [it is] sometimes partial, occurring at the apex, side, or base, with teeth or pores; [it is] sometimes complete and across the fruit, *circumscissile*, the upper valve being simple and operculiform; sometimes [it proceeds in] a completely perpendicular direction, [and is then] bivalved, trivalved, . . . multivalved, sometimes elastic. 5th. The internal arrangement of the fruit [may be] unilocular, bilocular, . . . multilocular, with a *dissepiment*, variously arranged, which separates the loculi; [this is] very often perpendicular, and only rarely transverse, opposite or parallel to the valves, continuous with or attached to them, entire or divided; a *columella*, a central axis-like structure distinct from the dissepiment, sometimes being present, but more often lacking; a seed-bearing *receptacle* [may be] simple or multiple, attached to the center, or to the side, or walls, and free or applied to the dissepiment. 6th. The numbers of seeds contained in each loculus [may be] one or many, the fruit or loculus [being] *monospermous*, *dispermous*, . . . *oligospermous* with few seeds, *polyspermous* when they are many. 7th. Because of the preceding conditions, various names have been given to the fruit, namely *capsule* [if it is] membranaceous or coriaceous or crustaceous; *siliqua* [if it is] bivalved and capsular, with a seminiferous suture on each side; *legume* [if it is] similar to the siliqua, but with only one seminiferous suture; *nut* [xvii] [if it is] bony; *berry* [if it is] juicy, or pulpy, or fleshy, and full of separate seeds; *pome* [if it is] fleshy, solid, sustaining a capsule;[97] a *drupe* [if it is] fleshy and surrounds a nut; *strobilus* or *cone* [if it is] like the arrangement of amentaceous flowers, with many seeds or nuts interspersed with scales all crowded densely together on a small head or cone. 8th. Moreover, the floral involucre, peduncle, calyx, and the receptacle of the pistil sometimes swell with sap, and counterfeit a fruit, as they are commonly so called, and also [become] truly edible in *Ficus*, *Cassuvium* [*Anacardium*], *Morus*, *Fragaria*, etc.

The seed, or egg of the plant, the abridgement of the future plant, is sometimes naked and placed on the receptacle of the pistil; more often it is enclosed in the pericarp and attached to its receptacle in its interior, sessile or umbilicate, with a funicle, or the connection being broken, it nestles in the pulp; it contains the corculum and its special coverings; outside, it differs 1st. in position upright, inclining downward, or inverted; 2nd. in form and size and surface often variable; 3rd. with a terminal, basal, or lateral, broader or narrower hilum; 4th. with an aril, or easily loosened outermost skin,[98] present or not, monospermous or more rarely di- or trispermous; 5th. with special coverings, very often paired, of which the inner one is thin, and the outer one membranaceous or coriaceous, and cartilaginous or callous, alate or marginate.

The inner part of the seed is of primary and essential importance, it is a true little plant, called the *corculum*, always having ["gaudet"] a single radicle and plumule, but it is provided with a lobe or cotyledon, most often paired and opposite each other, occasionally only one, lateral, or more rarely, none at all visible: whence the distinction among seeds—*acotyledonous*, *monocotyledonous*, *dicotyledonous*.[99*] The radicle projects a little beyond the lobes, and is upright or bent back into the lobes, sometimes ascending and tending toward the apex of the fruit, on other occasions descending, turned toward the peduncle. The lobes may be slender or thickened, flat, or in part or altogether curved, or convolute, or corrugated. The whole corculum is solitary inside its coverings (very rarely [there may be] three, as in *Aurantium* [*Citrus*]);[100] it is either surrounded by the *perisperm*, a thickened body with a different origin, whose substance may be farinaceous, or fleshy, or horny, or woody, and which [xviii] may either cover the corculum entirely, or surround it in part, or be merely joined to it; or else [the corculum] may conceal itself in a small pit inside the larger perisperm.

To the parts of the plant listed above we should now briefly add certain others which are less essential: such are the floral bracts, the stipules of the leaves, and the projections, spinous or prickly, already mentioned; in addition, we might note the buds of flowers and leaves, considered separately or together, and characteristic of most trees, but denied to herbs;[101] or the spiral tendrils, undivided or branched, which are useful in binding climbing plants to supports; hairs functioning as excretory vessels; the colored saps of plants, milky, yellow, blood-red, etc.; the glands suited for the secretion of saps. Neither must we fail to mention the unequal rate of germination of seeds, faster or slower; in some plants, this process is very easily accomplished long after maturation, in others, however, it must begin earlier, perhaps because of the condition of the corculum; sometimes large and thriving in the absence of a perisperm, sometimes, very delicate, soon withering inside a perisperm deprived of sap. Nor must we pass over the fact that the periods of flowering, fruiting, and foliation are not the same in different plants; in some they are brief and circumscribed, in others more protracted. We must also remember the daily opening of certain flowers, which takes place at predetermined hours in the manner of a Floral clock;[102] and the morning opening of leaves in certain plants and their closing at evening, very much like sleep; and we should not forget the irritability of certain parts in other plants, which is to be elucidated and verified by further experiments. Nevertheless, it will not help to discuss further parts and functions that are less well-known, and which have, to a degree, been neglected in the practice of botany.

Characters

Whereas certain external signs of distinction may be deduced from the many above-mentioned mostly external differentiae of parts, only those expressing the mode of internal structure, which are known by the name of *characters*, provide sure indications of the organization of Vegetative Beings and thus of their nature. These characters, moreover, the true ends of botanical discourse, being necessary for the proper diagnosis and definition of plants, are distinguished according to number, value, and mutual affinity. First, the simplest characters considered by themselves come together with others to form composites of others joined together, or further unite to form general [characters] from composites, or finally into the universal [character], embracing all parts and modes of the Vegetative Being, which [the universal character] determines the habit unique to each plant, and, consequently, its external, [xix] manifest nature; whence, plants most similar in habit will agree in character and nature and vice versa. Now in the series of functions, some are superior to others, advancing the processes of growth and reproduction;[103] therefore, the organs serving these processes are considered primary, and their special characters are rightly set before all others. When, nevertheless, in the primary organs, the number and form and proportion of parts, as well as other similar causes, contribute little to the exercise of the chief function, or if they change while the [organ] itself nevertheless remains, then the signs deduced from these modes are less to be valued in proportion to their instability. Characters unequal in importance are therefore arranged in order according to the value of the organ under discussion and the extent of their variations, some being changeable and varied, others more constant, and others very constant or essential, and in comparing plants, they are thus not discussed indiscriminately, but in [that] order. Moreover, many of these characters are not at all or very little related ["cognati"] to one another, [even] in the same plant, and they are more or less separate from one another, as though each were made according to its own law; others, on the contrary, chiefly in flowers and fruit, show affinity mutually, and taken from the series of more constant or essential [characters], they work together or are connected to one another in such a way that the existence of one always or very often announces that of the other: thus, a monophyllous calyx is always borne upon ["adnascitur"] an inferior germen, and stamens are very frequently inserted on a monopetalous corolla and then are definite in number. Many examples will confirm these agreements in the future, and it will often become clear from a large number of indications, the utility of the more important of these being perceived from their dissimilarity.[104]

Species and Varieties

The characters known are of value for determining plants, and they indicate the similarities or differences, as well as the greater or lesser degree of agreement among all. Just so many plants agreeing in all their parts, or being consistent in their universal character, and born from and giving birth to those of like nature, are the individuals together constituting a *species* properly speaking; [a term] used wrongly in the past, now more correctly defined as the perennial succession of like individuals, successively reborn by continued generation. This assemblage and series of beings, immutable and perpetual in kind, is occasionally subverted for a while by chance or human industry; that is to say, some individuals may vary one from another on account of location or climate or disease or cultivation; some may differ from the first of their kind ["primigenio"], their flowers being multiplied [semi-double], full, mutilate, or proliferous, with leaves variously more luxuriant or deformed, though with the color remaining unaltered, or infiltrated [xx] by rust or smut, or with enlarged organs owing to a copious influx of juices.[105] But these *varieties*, obeying the law of nature, and committed to a new germination of seed; they return to the primordial species,[106] their character restored, if other factors do not interfere. Now the species produced by nature ["nativa"], limited with certainty by accurate definition, stands as the true basis of the science of botany, which completely depends on it, so it clearly distinguishes all species, their universal character being first understood, in turn bringing together relatives ["affines"] and separating those which are different, and from this special knowledge and general comparison, a complete knowledge of nature and the coherence of all things may be gained.

Genera

This complete knowledge of species cannot be put together until, as an aid to memory, they are gathered first in small bundles[107] according to their affinities, and thence in bundles composed of the smaller ones. The small bundles, called *genera*, embrace some species which agree in many signs, differing in a few, and which therefore conform to one another not in their universal but only in the general character. Their generic association of course demands a selection among specific characters, not arbitrary, but based on solid principles, such that [as a result] each and every genus contains species that are truly congeneric, disparate ones never being mixed up.

In early times, there were no principles for this generic constitution, and because the habit of plants only somewhat similar was not examined, plants from far and wide were ranked under the same name. The ancients, being mostly more concerned with medicine than with the science of botany, exhibited the virtue and not the character of plants, very often in works with a great show of learning,[108] but they are alien from today's concerns, and it would be superfluous to review their merit here, because men of a great but different erudition are not at all or only slightly of use to the science now to be elucidated. For him who nevertheless burns to know them, the best work to consult is the "Isagoge" of Tournefort[109] where he enumerates the Authors one by one and assigns to each his deserved share of praise. In the same work, he also commemorates the successors of these men for their conscientious and steadfast labor in describing plants, but because of a lack of principles, and an ignorance of certain primary organs, they produced descriptions which were very often imperfect, and formulated distributions [that were] always incorrect based on location, qualities, structure, or badly understood habit; they are to be highly commended, however, for their useful indications and histories [xxi] of known species, as well as for [their] synonymies of the Ancients.

The science, uncertain [and] restrained by no laws, meanwhile grew silent, until Botanists understood that its business was in the first place with the investigation of characters, the generic connection of species, and a further ordering of genera, and in the constituting of genera or orders, certain characters were to be considered before others. Nevertheless, this doctrine did not immediately prevail. Gesner first proclaimed that characteristic signs were by rights to be deduced chiefly from the fructification, and this was later agreed to by Cesalpino and Columna. Morison restored this principle when it had been neglected for a while, and Ray, Rivinus, and their contemporaries allowed it; more recently, Tournefort expressly explained it.[110] For, having examined the various parts of plants, he asserted that generic signs could be based with certainty on the flower and fruit; from thence he built up all his genera, namely those in the first rank marked by fructification alone, and designating others in the second rank, to which is added structure other than similarity of fructification. With few exceptions, the large[111] cohort of Botanists following him has accepted the law that was promulgated, and the stability of the numerous natural genera.

In his enumeration of the floral characters and organs, Tournefort disdained the parts of the pistil and stamens, taking them for excretory vessels,[112*] even though Camerarius, Cesalpino, Grew, and Malpighi had published something on the sex of plants, already presaged by the ancients.[113] At the beginning of the present age (1702 [see Burckhard 1750]), Burckhard explained at length in a letter to Leibniz the structure and function of the sex organs; next, Geoffroy outlined the various shapes of the male pollen; Vaillant, ignorant of Burckhard's opinion, published openly a description of the action of the sexes; [xxii] finally, the dehiscence of the pollen and the emission of the seminal aura became known through the observations of Needham and B. Jussieu. From that time, the pistil and stamens being more highly thought of, [were found to] exhibit better features ["notas"] which Linnaeus aptly put into his generic [characters]. Having greatly increased the number of generic characters inside the flower, he decreed that all were to be drawn from the fructification, and none from any other part, thus limiting the law set down by earlier authors. Rejecting therefore the secondary genera of Tournefort,[114] and reforging all the rest, he added useful features of the stamens and pistil, and in addition selected some up till then overlooked from the calyx, corolla, and fruit, and he arranged the parts according to number, form, proportion, and position, and in so doing created a uniform and enlarged series of genera, sometimes new and sometimes transformed, which has now been gratefully accepted by almost all Botanists.

This rule of the constitution of genera, while very useful in the present state of science, nevertheless seems arbitrary, and indefinite besides, and therefore not entirely adequate for the demonstration of truly natural genera, in contradiction to Linnaeus, who always judged his breaks between genera to be natural, and easily accommodated to any arrangement whatever. For there are parts and relations in the fructification, some of which are more valuable than others, and consequently characters unequal in value, some being primary [and] essential, others only secondary and general, and others, the lowest, for the most part varying, certainly not all equally suitable for the designation of genera. Certain others are scattered outside the fructification, more valuable than the lowest [characters] mentioned above, and to be set far above them; sometimes equal to the general [characters], [and] in part restoring the secondary genera of Tournefort; often conforming to others that are truly natural, they are sometimes helpful in the definition of orders themselves, and are therefore of more than moderate use to the science of botany. The Linnaean rules, therefore, are artificial, and do not express the inequality among floral characters, or the importance ["vim"] of certain extra-floral ones; the more valuable genera of Tournefort, however, show little discrepancy from natural genera; indeed, the most part can be admitted forever, fewer are to be revised, and taken together they provide a useful basis for science.

Nomenclature and description

When the species have been identified on the basis of characters and grouped into genera, there is another work for the Botanist to begin; that is, he must designate the genera and species with a suitable name, and define them by means of a sound description, in such a manner that by the name and description they can be distinguished from all others. This part of the science, while neither primary nor truly constituting the science [xxiii], must nevertheless be advanced, and is therefore not to be despised, but rather elaborated more accurately, so that by the facilitation of discourse Botanists together may more easily instruct [each other] by mutual observation and thought. Nothing useful in this respect is to be drawn from our early Writers, who very often offered neither apt descriptions nor unvarying names, and for this reason many plants that they accepted for medical use can in no way be found out on account of the lack of an accurate description. To these authors succeeded men more skilled in botanical matters, but who did not yet rely on a nomenclature with clearly defined principles, and who, not knowing how to reach a happy mean, were either too brief or too excessively diffuse in their descriptions of plants. Neither did those promote the science who, joining name and description together, designated plants by a definition attaining the length of an entire phrase which, too prolix for the naming of species, turned out unequal to describing [a plant]: whence the faulty nomenclature of the preceding century, although accepted by Tournefort, has posed a hindrance to the progress of science.[115*] To this difficulty the work of Linnaeus offered the greatest help, for having freed plants from the chains of botanical phrases,[116] he furnished[117] each with a double name, one part generic and substantive, the other adjectival or specific, then setting out the characters in an adequate description, with meaningful words, much being conveyed by few [words] in a technical style. Moreover, he transmitted these rules of nomenclature and description, [which are] for the most part excellent and ought to be generally accepted, [but] some are perhaps invalidated [by his own practice] and [are] not sacred in themselves.[118*] According to his wise arrangement, the *generic* name ought [xxiv] to be immutable,[119*] simple, not two words or composed of two parts in an improper fashion;[120*] it should not be sesquipedalian, or barbarous, or unpleasant to the ear;[121*] it should name[122*] rather than signify,[123*] and be unique to the genus it designates and not held in common with any other body;[124*] the mere addition or removal of one syllable [xxv] should not render it exactly similar or comparative.[125*] The species name must be both simple and easy, but it must also signify; in particular, it must be drawn from a solid character, distinguishing the species from all its congeners and then truly *specific*,[126*] or when this character is lacking, [the species name] must in the mean time be selected on another basis, namely, from the soil, region, climate, length of life, color, taste, odor, use, etc., then, in pejorative fashion, being called *trivial*.[127*]

A description is partial or universal: the latter is appropriate for Monographs describing a single plant, or for travelers uncertain about a vegetative being of a hitherto unknown genus, each and every part being examined in detail and in a uniform order (lest any should be omitted): namely, root, stem, leaves, inflorescence, calyx, corolla, stamens, pistil, fruit, and seed, as well as their function,[128] position, number, connection, form and proportion, to all this being added the habit, native place, soil, vernacular name and use. A more useful partial description delineates generic or specific characters in the same order, now more lengthy in dissertations and monographs ["phytographia"], now shorter and giving only outstanding distinguishing ["dissimilia"] signs in general treatments, as in the very recent Linnaean system with its adumbration of genera and descriptive phrases added as specific names. Many of the compendious descriptions of Botanists are inadequate, however, because of an improper selection of characters, an omission of essential ones, a lack of comparable characters, [or] an insufficient indication,[129] and therefore the task is to set these faults right again and again. At this point it will not be unprofitable to take up the subject of *synonyms*, or different names bestowed chiefly by the primary Authors[130*] [xxvi] on the same plant, which function like Ariadne's thread, revealing the labors of those gone before, and exposing errors to be avoided; the merits [of our predecessors] becoming clear, each may be given his own. Many have also devoted themselves with zeal to Ichinography, or the representation of plants by means of copper engravings; when not pleasantly deceptive, but rather genuine and simple, it is more effective than descriptions in conveying habit. The images of Early [Authors] were for the most part bad, those of Recent ones are more accurate; from these it is possible to distinguish certain plants previously mixed up under the same name, or directly discover a new plant not yet delineated. Thus, by means of convenient nomenclature, accurate description, dependable synonymy, and clear images, the designation of plants can be achieved, and certain knowledge both specific and generic truly established.

Classes and systems

The principles ordained by Botanists that were of great value in grouping species together, also proved very useful in creating classes of genera; in this case, the rule was similar, but more restricted. For just as the fixing of genera evokes the outstanding characters of species, so that of classes gathers primary [characters] from the series of generic [characters]. Thus, specific characteristics ["notae"] are present in all parts, generic ones in some, those of classes in only a very few. In its progress, science makes use of the simpler at the same time as more outstanding signs: the entire fructification supplies the generic designation, only one part of it supplies the [designation] of the class, this according to a law anticipated by Cesalpino, partly followed out by Morison, established more firmly by Tournefort, confirmed by others, and finally promulgated more strictly by Linnaeus. This rule of distribution was hidden from early Authors, who based their arrangements less on the observation of organs than on habit, location, length of life, time of flowering, medical use, etc.[131*] When first [xxvii] the importance of the fructification became known to true Botanists, each one arbitrarily picked out a more obvious part more suitable for multiplying outstanding characters. The sexes being not yet known, some [botanists] gave preference to the fruit, others to the calyx, and still others to the corolla, each creating for himself a method in his own way, differently arranged, but still truly systematic.

The usefulness of a method is this, that the genera, joining similar species together, in turn grouped in sections designated according to secondary characteristics, and several sections being then further assembled into well-ordered classes defined with one primary and very simple sign, [so that] by the help of this arrangement plants can easily be distinguished by a given character, and called by their proper names without delay. In the same manner, any soldier in an army is soon found, once one knows his regiment, battalion, squadron, and platoon. But these methodical distributions of plants are valuable only insofar as they are natural, easy, and uniform: 1st. Botanists who diligently obey nature in joining individuals and species feel ashamed to diverge from her too much in their bringing together of genera, of which there are many exceedingly natural series most acceptable to all; by some called *natural orders*, by others *natural fragments*. By nature's tacit command, whatever method that preserves undivided any of these is the more perfect the more it admits. 2nd. Nevertheless, a desire to clarify the science and accomplish the study often diverts the Méthodistes from nature, preferring easy knowledge to that of a more universal kind, they emphasize mainly the naming of a plant, less the knowing it in its entirety: for this reason, they require that all divisions of a system must be neither confused nor ambiguous, but very distinct and clear and easily comprehensible to the intellect, and supported by obvious and unvarying signs. 3rd. Those also deserve some praise whose primary [divisions] in the system are all drawn from the same organ, and their secondary divisions from another, so that, circumscribed and firmly based, as if proceeding and flowing from a single theme, [their groupings] rarely or never admit exceptions.

[xxviii] These are the precepts of systematic composition, somewhat arbitrary indeed, and in part repugnant to nature, but useful to a science until now uncertain, they have been variously interpreted, sometimes rigidly executed, sometimes relaxed at will. Meanwhile, many other systems have been formulated, all worthy of praise, and all providing a clear indication of the genius and erudition of their Authors; not one of them, however, altogether conforms to the preceding rules, but departing from them more or less, on the one hand, they reveal diverse limits ["modos"], as well as many affinities among diverse parts of plants, and these are of great usefulness to science, on the other, however, they clearly demonstrate the weakness of all artificial science, as they are supported by unstable principles and arbitrary laws. The truth of the preceding assertion would be confirmed by judging them all individually, but it is not profitable to delay here in order to support or overthrow[132*] them one by one, the discussion produces little shortening, nor usefully being transformed into other meditations. Let it suffice therefore to examine the systems of two of the most excellent among all, namely, Tournefort and Linnaeus; through a faithful exposition and mutual comparison and careful consideration of the contributions of each, which being known completely the advantages and dis-

advantages of both will become apparent, and also the usefulness and worth of the entire systematic doctrine can be gauged with certainty.

THE METHOD OF TOURNEFORT. Tournefort, Professor at the botanic garden [xxix] of Paris from 1683 to 1708, who first formed more adequate genera, first showed there a correctly delineated general method,[133] for a long time more excellent and acceptable to outstanding Botanists. Concerned especially with clarifying the subject and therefore eager to accomplish a convenient and uniform arrangement of plants, he based his classification on the corolla (or, as he expressed it, on the flower), because [it is] clearer, more outstanding for laymen, and [has] many characters that are so easily and adequately observed; for example, [the corolla is] either clearly present or not, either simple, multiple, or composite inside a common calyx, monopetalous or polypetalous, regular or irregular. Not yet altogether diverging from the early Writers, however, he placed first the old distinction between herbs and trees, [and so his classification] was not strictly based on one part, and thence he devised seventeen classes of herbs and five of trees. The first division of both the former and the latter is of those with a corolla, or petaliferous plants, and those deprived of the same, the apetalae; second, the petaliferous plants are divided into simple and composite; third, the simple are then divided into monopetalae and polypetalae; fourth, both are divided into regular and irregular; the fifth division is derived from the different forms of them all. Another division, into Apetalae and Compositae, is noted below.

Of the herbaceous [plants] bearing a single petal, in the first class with a monopetalous corolla, therefore, appear the regular *Campaniformes*, in the shape of an open bell, whether imitating the form of a tube, or a globe; in the second class are the regular *Infundibuliformes*, with the corolla shaped like a funnel, a bowl-stand, or a wheel; in the third class are the irregular *Anomalae*, with the shape varying greatly, [the corolla] preceding a capsular fruit; in the fourth class are the irregular *Labiatae*, with the limb [of the corolla] as though bilabiate, and moreover with naked,[134] tetraspermous fruit; in the fifth, with a polypetalous corolla, are the regular *Cruciformes*, with tetrapetalous flowers in which the petals intersect in a cross; in the sixth class are the *Rosaceae*, most often with five petals, sometimes more or fewer, arranged like a Rose; in the seventh are the rosaceous and five-petalled

Umbelliferae, but with one-flowered peduncles joined together in the shape of an umbrella, and with naked, inferior, dispermous fruit; in the eighth are the polypetalous and regular *Caryophylleae*, with long, unguiculate petals emerging from the depth of a tubular calyx as in *Dianthus*, or *Caryophyllus* as it is more commonly called; in the ninth class are the *Liliaceae*, or lily-like, tri- or hexapetalous or merely sexafid, always with trilocular fruit, which may be superior or inferior; in the tenth are the polypetalous and irregular *Papilionaceae*, possessing dissimilar petals (standard, wings, and keel) which imitate a butterfly, and with superior fruit [that is] leguminous [two valves separate and twist, flinging out the seeds]; [xxx] in the eleventh are the *Anomalae*, not papilionaceous, but multiform and not otherwise to be defined. To the simple succeed [forms with] a composite corolla, monopetalous and aggregated in a common calyx, each with its own germen affixed on a thalamus or common receptacle; in the twelfth class are all the *Flosculosae*, or tubular [corolla], with a quinquefid limb; in the thirteenth are all the *Ligulatae*, or [with a] tubular [corolla] divided from the apex, and thence produced as a ligule; finally, in the fourteenth are the *Radiatae*, that is, with the central [flowers] flosculous [i.e., tubular and quinquefid] and the marginal ones ligulate and arranged radially. Following these are the Apetalae, of which some, the *Stamineae*, are in the fifteenth class, with a flower (calyx and stamens) and fruit; others, the *Flore carentes*, are in the sixteenth class; these do not lack a seed; still others, the *Flore et fructu carentes*, are in the seventeenth class. Trees combined with bushes, when simply apetalous, are in the eighteenth class, the *Stamineae*; in the nineteenth are the apetalous *Amentaceae* with spicate staminate flowers [arranged] upon an amentum or common axis like a cat's tail, in the twentieth the *Monopetalae*, in the twenty-first, the regular, polypetalous *Rosaceae*, in the twenty-second, the irregular, polypetalous *Papilionaceae*. Sections of classes are sometimes determined by the form of the corolla, or by the arrangement of flowers, or more rarely by the habit of the plant, and more often by whether the fruit is superior or inferior, simple or multiple, baccate or capsular, uni- or multilocular, mono- or polyspermous.

Thus is carried out a method based on simple principles and manifest signs, and leading therefore to an easy arrangement and study of plants; on some occasions [it is] entirely faithful to nature, in a number of its classes truly natural orders being allowed, of which the Labiatae, the Cruciformes, the Umbelliferae, the Papilionaceae, the Compositae, etc., stand out, along with many others also forming whole sections of other classes. But it also proceeds in a simpler fashion than others, and for this reason it is seriously weakened 1st. through an improper distinction between herbs and trees which, not being clearly separable, [when used] dissociates related ["affinia"] genera in the Rosaceae, the Papilionaceae, and the Compositae, and separates congeneric species in *Aralia*, *Bupleurum*, *Salix*, *Rubus*, *Datura*, *Senecio*, *Genista*, *Coronilla*, and very many other genera in part herbaceous, in part trees and shrubs; these, although deceived by [such] method, are to be placed undivided into trees or shrubs;[135] 2nd. by an arrangement which is sometimes vaguely defined and not always uniform, for example, one class among the Apetalae [xxxi] is called lacking a flower but not a seed;[136] the Caryophylleae and the Liliaceae are defined only by means of habit and a certain similarity, and certain Liliaceae (*Hyacinthus*, *Narcissus*, *Gladiolus*), while clearly not polypetalous, are mixed in among the related ["cognatis"] Polypetalae in a manner countenanced by natural law, but contrary to the system; 3rd. by an improper choice of poorer characters of the corolla [yet] the most outstanding neglected, for example, the unvarying position of the corolla is rarely employed in defining sections, while its form is used to define various classes in the Campanulatae, Infundibuliformes, Anomalae, etc., relatives ["affines"] are repeatedly separated—*Campanula* and *Lobelia*, *Rubia* and *Rubeola*,[137] *Cerinthe* and *Pulmonaria*, etc.—and dissimilar [plants] are joined—*Aralia* and *Asparagus*, *Lobelia* and *Aristolochia*, *Butomus* and *Helleborus*, *Viola* and the Orchideae, *Paris* and the Cruciferae, etc.; 4th. by the enumeration of only a very few characters, with only those mentioned in a systematic fashion which at most serve to define and name species, and all others omitted; 5th. by the erroneous inclusion of the floral covering, which is clearly secondary, among the primary organs, as being equal to the characters of classes, while the truly primary sexual organs, namely the stamens and the style, are altogether omitted, and not even employed in the definition of sections.

THE SYSTEM OF LINNAEUS. As mentioned previously (p. xxi), Tournefort, ignorant of the function of the sex organs, disdained the stamens; if their manifest value had then been known, as it was later, they certainly would have been of use to botanists in the constitution of more perfect genera, and would have been employed in the creation of classes. Burckhard[138] noticed that the characters of the stamens were useful in delineating classes and concentrated his research upon them. Linnaeus, Professor at Uppsala from 1741 to 1778, went further, and confirming the value of the sexual organs, he constructed what he called in his own words a sexual system, in which the characters of the sections were drawn very often from the pistil, and those of a class always from the stamens. Of these latter he examined the number and proportion and mutual connection and fusion with the pistil, and their concealment and sometimes their position: whence arose twenty-four classes, simple or multipartite by reason of number of the germens or styles.[139]

Thus the first eleven classes with a definite number of stamens (from one to twelve, eleven never existing), and with appellations to correspond in the graeco-roman manner, are called *Monandria*, *Diandria*, *Triandria*, *Pentandria*, *Hexandria*, *Heptandria*, *Octandria*, *Enneandria*, *Decandria*, *Dodecandria*. To the *Icosandria* following them were then [xxxii] removed [those plants with] many stamens, often twenty, inserted on the calyx; to the adjoining *Polyandria* those with a large and indefinite number of stamens not inserted on the calyx. Next considered is the characteristic of the proportion of the stamens; tetrandrous and hexandrous plants, called the *Didynamia* and *Tetradynamia*, two or four of whose stamens are longer than the other two, are placed separately into the fourteenth and fifteenth classes. After this the connection of the stamens is considered, with the filaments being grouped in one, two, or several connate bodies [bundles], providing three classes, namely *Monadelphia*, *Diadelphia*, and *Polyadelphia*; from [those] with the anthers growing together in a cylinder a single [group], called the *Syngenesia*, is formed. Next follow the *Gynandria*, in which the stamens grow together [fuse] with the pistil or are affixed to it. In the classes following this series of hermaphroditic or bisexual flowers are [those plants] which produce unisexual male or female flowers, which may be mixed on the same plant as in the *Monoecia*, or they may be borne by different plants in the *Dioecia*, or, occasionally, some [are] promiscuous and have hermaphrodite [flowers], as in the *Polygamia*. Finally, assembled together in the *Cryptogamia* are the plants whose flowers and therefore whose stamens, on account of their great delicacy, are not at all or very little known, and whose sexual functions are consequently obscured. The different sections of the classes are for the most part defined according to the number of germens or styles, by common designations of both the former and the latter categories called *Monogynia, Digynia, Trigynia, Tetragynia, Pentagynia ... Polygynia.* Sometimes they are distinguished according to the various kinds of polygamous flowers; by seeds which are either *gymnospermous*, naked, or *angiospermous*, covered by a pericarp; by the fruit, siliquose or siliculose; the number of stamens itself, because it designates the class, is not used.

Of this system, like the preceding ones, certain parts are to be praised, and others are to be refuted. More accurate and consistent and devoted to one principle than the others, it draws all the characters that determine membership in a class from one part, and thus so clearly defined and circumscribed that they can be perceived with little effort, and they collect all plants in an orderly distribution and [so] to be recognized easily without the aid of an instructor. Moreover, certain of the classes, such as the Tetradynamia, the Monadelphia, the Diadelphia, and the Syngenesia, are almost entirely natural. But 1st. the system sometimes depends on the most delicate organs so very difficult to observe even with a lens[140] and dissecting needle, and causing vexation and fear of error if such a fault is overlooked.[141] 2nd. Besides this, [there is] arbitrariness in the manner of a system, the classes being multiplied, an attempt is made to derive all their definitions from one [xxxiii] part; then, on account of a lack of sound characters, to the essential ones he adds carelessly those more inconstant, and as these latter are more numerous, he very frequently uses them, the former generally being neglected. Therefore, less stable primary features being deduced from the number and proportion of the stamens, and disparate plants grouped together in the same sections: in the Diandria *Piper* and *Jasminum*, in Triandria *Tamarindus* and *Iris*, in Tetrandria *Rubia* and *Euonymus*, in Pentandria *Primula* and *Coffea, Ulmus* and *Cicuta*, in Hexandria *Berberis* and *Hycinthus*, in Octandria *Tropaeolum* and *Erica*, in Decandria *Malpighia* and *Alsine*, in Polyandria *Chelidonium* and *Tilia*, etc. In other places he separates didynamous plants from [similar] tetrandrous ones: *Chlora* [*Blackstonia*] from *Gentiana, Blitum*[142] from *Chenopodium, Erica* from *Andromeda, Centunculus* from *Anagallis, Canarina* from *Campanula, Myosurus* from *Ranunculus, Sibbaldia* from *Potentilla, Sempervivum* from *Sedum, Cassine* from *Ilex*, diandrous Labiatae from didynamous ones, triandrous Gramineae from hexandrous ones, etc. 3rd. Thus, not only are the most closely related ["cognatissima"] genera, which differ in the number of stamens, set apart by the system, but even congeneric species are also often unknowingly separated from one another, [as] in *Convallaria, Polygonum, Rivinia, Verbena, Gentiana, Valeriana, Rubia, Lepidium, Geranium, Cerastium, Linum, Tillaea,* etc., of which the very close bonds are not accounted for by the system, which is disfigured and rendered less uniform by these exceptions. 4th. As a parallel case, although much rarer, a difference in [stamen] number may characterize different flowers of the same plant, as in *Monotropa, Chrysosplenium, Adoxa, Trientalis, Disandra,*[143] *Ruta,* and others, thence being very difficult to classify, and this confirms the instability of this character of the stamens. 5th. In certain plants, such as *Bignonia, Gratiola, Chelone,* etc., the stamens are sterile or are suppressed by a chronic abortion, which changes the arrangement of these plants. 6th. The Gynandria differ very little from the classes mentioned above (especially from the Monandria), and are best restored to them, and besides, they include very few plants that are truly gynandrous. 7th. Neither ought the main classes [based on] the separation of the sexes to be kept, since many monoecious and dioecious and all polygamous plants are considered as such solely on account of the abortion of parts, and ought more correctly to be associated with the true hermaphrodites.

THE LIMITED UTILITY OF SYSTEMS If a method that is easy and most faithful to Nature, is by that to be judged as superior, the Tournefortian excels the others because it is less destructive of natural affinities, in its use of more obvious signs, and in [xxxiv] being quicker to run through. The arrangement of the Linnaean system is sometimes laborious, the designations are sometimes difficult, and connections are very frequently made between dissimilar plants. But if this system were more circumscribed, more surely defined, and dependent on signs sedulously worked out and drawn from one primary part, it could boast of its extensive army of Vegetative Beings, drawn up in orderly fashion under the authority of these signs, rarely deviating [from them]; enriched by superior description and nomenclature, both generic and specific, and possessing abundant advantages, it would compel the assent of almost all Botanists. But the systems constructed in an arbitrary fashion exhibit a false, unnatural science, and are devoted not to the complete knowing of plants, but to a great degree to compendious defining and more certain naming. They ought therefore to be regarded as botanical preludes, or well-arranged catalogues or non-alphabetical indexes, some more convenient than others, in which, following signs allowed by Botanists in order to make their labor in their own investigations easier and to communicate with one another, plants are tentatively arranged in an agreed order until that happier day when, through continued meditation, they can be arranged in a truly natural series.

The True Aim of the Science

From the preceding, the parts and functions of plants, and the differentiae of parts and the characters derived from them, have become known. Designations and appellations of species and genera have also been discussed, as well as more fitting ways of perfecting both. On the one hand, nobody is oblivious to the necessity for an arrangement of plants, and on the other [all are dissatisfied] with the arbitrary and imperfect systematic arrangement as it has been carried out until now, artificial not true, by no means suited to the revelation of true affinities, and useful only insofar as it provides a distant view of the truer science. [The science] will now be approached from a new angle, it will be subjected to those laws, not factitious but immutable, which Nature has clearly engraved on plants, and which are apparent to any careful observer. For a long time, the definition[144*] of botanical science has been incorrect, deceptive, or at the very least, vague; [xxxv] it has concentrated only on designating and naming plants; it has selected only the characters needed for this designation and omitted the rest, and then in the determination and naming of genera and species it has been very often uncertain and contradictory, being seen by laymen as a science of pure nomenclature, more especially suitable to the exercise of the memory than to that of the mind. The sincere contemplator of nature, who seeks to know her in plants, must, however, examine and disclose their entire organization; to that end, he must investigate all characters, and neglecting very few by remembering the method, he must search in such a fashion that omitting nothing he discovers the mutual affinities of all plants, and comes to possess in abundance a complete knowledge of them.

The natural method

This long-desired arrangement,[145*] far superior to all others, alone truly uniform and simple, always in conformity with the laws of affinities, is the so-called *natural method*, which links all kinds of plants by an unbroken bond, and proceeds step by step from simple to composite, from the smallest to the largest in a continuous series, as a chain whose links represent so many species or groups of species, or like a geographical map[146*] on which species, like districts, are distributed by territories and provinces and kingdoms. On one hand, permitting any given link to be closely connected with only two neighboring ones, it circumscribes the affinities of plants; on the other, it extends these, and shows that individual species, as well as the groupings of genera and classes, are equally neighboring ["conterminas"] with many surrounding them. However this may be, the leaders of Botany both past and present have recognized [xxxvi] implicitly and explicitly the existence and value of the genuine and outstanding connection among plants of this method, and either their labors have in view the enquiry into this, or [their] systematic classes imitate it, or often, however, by an unfortunate departure [from it], indicate the difficulty of the work undertaken.

Among the various Authors who have labored successfully in the investigation of a natural method, the principal ones are Linnaeus, Bernard de Jussieu, and Adanson. In the year 1738, Linnaeus first published his natural fragments (in *Classes Plantarum*[147]) and revised it many times, adding, however, no [new] principles to its basic structure or order ["dispositionis"]. In 1759, [Bernard de] Jussieu arranged the royal gardens of the Trianon according to a system of natural orders which he had worked out earlier but had never made known,[148*] except in private conversations, because he had little faith in it; even later, he did not attempt to publish it since he was not yet satisfied that [the orders] had been sufficiently confirmed. In 1763, Adanson brought out his plant Families built upon principles set down in a preface.[149] Each of these Authors ought to be given his deserved share of praise, although at present it is not profitable to explain or judge their contributions here, because once the laws governing the natural method have been laid down, examined in a proper manner, and duly recorded for posterity, the labors of these men will be more equitably judged, and consequently, their merit will emerge more clearly.

Obstacles to the natural method

The natural method, the truest and best aim of science, has been and still is hidden in part from Botanists on account of a lack of knowledge of all the plants to be arranged, and of the principles of arrangement. 1st. A continuous series of beings cannot really be established unless all have first become known. Botanical journeys produce discoveries daily, but there are probably many more exotic beings remaining to be discovered than have been found; an absolute knowledge of all existing beings is never ["nec unquam"][150] to be hoped for. But, even though all regions of the earth have not yet been explored, the rest [i.e., unknown areas] have nonetheless been noted by Geographers and carefully arranged in maps, each in its own location. Again, although numerous links of a chain may be missing or scattered at a distance, many are easily connected at different points in the chain, and are joined at various later times by new intermediate links, and gradually come to be separated by fewer intervening spaces. [xxxvii] Therefore, the method ought to be worked out so that it imitates a chain or map, despite the lack of exotic genera. 2nd. Neither does the objection drawn from the difficulty of the investigation of the natural principles have any validity: indeed, these [principles] are inherent in plants, and are readily apparent to the observer, and can be explained by a dual means which should be pursued in turn so that one as it were functions as the confirmation of the other.

The first means of attaining the same

The first and most common means is a progression from things simple and known to things composite and unknown, and from a uniform and concise series of observations and propositions, the true principles of the natural method may be elicited without serious effort.

The sure knowledge of species

Thus, the species must first be known, and defined by its proper signs: [it is] a collection ["adhaesio"] of beings that are alike in the highest degree, never to be divided, but simple by unanimous consent [and] simple by the first and clearest law of Nature, which decrees that *in one species are to be assembled all vegetative beings or individuals that are alike in the highest degree in all their parts, and that are always similar ["conformia"] over a continued series of generations*, so that any individual whatever is the true image of the whole species, past, present, and future.

The generic connection of related species

Yet the same law that gathers individuals completely alike into one species also determines, with equal right, the conjunction of similar species. For equally, and without anyone objecting, [it is stated] *those species are to be associated which conform to one another in a very large number of characters*, and therefore, those ought to be separated which differ in many signs. This previous law ["ratio"] of connections shows the first judgment of affinities, according to which all species are examined and thence are to be assembled in the same group.[151] A genus is thence constituted, a congeries of similar species, of which nevertheless there is taught no accurate definition, nor clear precept of construction, especially in the natural method. If there is any such, it differs very much from the systematic, it does not assign strict limits to genera, but through preserving the affinities connects all species by an imperceptible nexus, which it then for the sake of ordering the science, joins in distinct genera, with almost arbitrary cut-off points, enriched with few or many species at one's pleasure, and called truly natural when they do not separate related species or disturb the entire ["universalem"] series.

Moreover, many traces of Nature are not hidden, but imprinted here and there; these do not elude the observer, and are of the utmost usefulness to the rest following who would make manifest the progression of the whole. Without doubt, [xxxviii] there exist genera natural in the highest degree to which no one objects, such as *Aconitum*, *Delphinium*, *Ranunculus*, *Scabiosa*, *Geranium*, *Gentiana*, *Valeriana*, *Rosa*, etc., and when each of these is investigated, the true mode of generic construction will appear, whence others can be constructed with certainty following their example. Thus *Rosa* is a genus of shrubs with alternate and stipulate leaves, an urceolate, quinquefid calyx, pentapetalous and polyandrous, with the petals and stamens attached to the throat of the calyx, polygynous with the germens also hidden in the calyx and each with a lateral style and simple stigma, the seed wholly covered by a baccate calyx, the corculum of the seed erect, and no perisperm. The species conforming to the preceding designation [belong in] a single genus; they may differ among themselves in that the calyx may be ovate or globular, smooth or echinate, with its divisions partly appendiculate and partly destitute [of them], or very rarely altogether naked, [and they differ] in the shape and color and size of the petals, in the shape and number of the seeds, in the stem tall or prostrate, unarmed or thorny, the leaves pinnate or very rarely simple, in number and shape of the leaflets, in the solitary or sub-corymbose flowers, etc. *Valeriana* [is] another genus, herbaceous, with opposite leaves, flowers forming a corymb, a superior calyx, an epigynous, monopetalous, tubular corolla which has a quinquefid limb, epipetalous stamens that are definite [in number], a simple, inferior, monostylous germen, sub-monospermous fruit, a straight, ascending corculum, and no perisperm. Its specific differentiae are constituted by a prostrate or erect stem, the leaves shaped variously, simple or pinnate, the corymb of flowers dense or lax, the calyx entire or dentate, the corolla spurred or gibbous at the base, the limb regular or irregular, the stamens and stigmas various in number, and the fruit sometimes pappose and monospermous, or locular [and] bi- or trispermous and dentate at the apex. This genus may be divided in two on account of the fruit, or it may remain undivided; its different species cannot be separated one from another in the natural series.

From this enumeration of the characters of *Rosa* and *Valeriana*, it can be concluded with certainty, 1st. that the same characters are not always generic, and that some, such as the location of parts, are extremely constant in genera, while others, such as shape and number, etc., are sometimes stable, sometimes varying; 2nd. many characters drawn from the fructification as well as others from outside of it may be similar in species from the same genus; 3rd. most [characters] other than those of the fructification, and some taken from it, may vary in these [species]. [xxxix] Not all the characters of fructification, therefore, ought rightly to be valued above the others: for example, in *Rosa*, the location of leaves dominates over the shape of the calyx, while in *Valeriana* the same is superior to the number of stamens. 4th. Although some characters are very often mutable, others are alternately stable or varying, and others more consistently stable; therefore, they ought to be ranked ["ordinandi"] differently in proportion to their constancy, and when reckoned up, they ought not to be counted, but weighed in such a manner that each is given its own rank and *one constant character becomes equal or superior to many inconstant ones.*

The natural laws of affinities

To this maxim, confirmed by observation and to be placed among the truly natural laws, must be added the earlier law which states that *those plants ought to be grouped together which are related ["affines"] in a very large number of characters*; and from these two statements, suitably explained and amplified, depends the complete rule for a real ["genuina"] distribution of plants. If one character is equal in value to many, that character is always uniform in genera; if equal to fewer, it occasionally prevails; if equal to a very few, it is only to be employed in the defining of species: thus, the position of the leaves, to be ranked among the semi-stable characters, sometimes appears in generic designations by equal right with the semi-stable signs of the fructification. This triple division of characters, however, is not sufficient; [a proper] one allows several grades, but they are hardly to be defined; therefore, the most important work of the Botanist who follows Nature eagerly is and will be to ponder the weight of each and every character, to give each one its immutable position [in the hierarchy of characters]; but the graduated arrangement of characters will be established only by assiduous meditation on the truly natural genera and their mutual association.

The natural orders [families]

After adjacent ["confines"] species have been brought together in their natural genus, which may be simple or divided, the related ["cognata"] genera are to be connected and assembled in orders in a similar manner and by the same law. Examples of this kind of association are obviously put together by the favor of nature, namely, the truly natural and much-tested orders, [which are] often undivided in the methods of Systematizers; of such are the *Gramineae*, *Liliaceae*, *Labiatae*, *Compositae*, *Umbelliferae*, *Cruciferae*, *Leguminosae*. Just as only the outstanding characters of species are to be employed in constituting genera, so not all generic signs, but only the more constant, should deservedly first be prepared for designating orders. Indeed, the more general [i.e., the higher] is the ranking ["ordinatio"] of plants, the fewer the defining signs used; in accordance with this immutable [xl] law, those are taken to preponderate which, [being] outstanding and very important, equal in value many minor and more trivial ones. As this axiom is supported by reasoning, so it is confirmed by repeated observation, and particularly by carefully studying the orders set forth above.

Among the various characters already mentioned in the course of this work and to be discussed hereafter, occur an obvious difference and multiple value.

1st. Some [characters] in each of these orders are primary and always uniform, or essential, being drawn from essential organs such as the insertion of the stamens, or the mutual arrangement of these and the pistil, the location of the staminiferous corolla, or the number of lobes in the corculum. The latter is always one-lobed in Gramineae and Liliaceae, while in others it is bilobed. The stamens are uniformly hypogynous in Gramineae and Cruciferae, epigynous in Umbelliferae, perigynous in Liliaceae and Leguminosae, epipetalous in Labiatae and Compositae, of which the former are provided with a hypogynous corolla, and the latter an epigynous one.

2nd. Others are secondary and sub-uniform or generally present and varying only as an exception; [they are] therefore drawn from nonessential organs which may occasionally be lacking, such as the existence or lack of a perisperm, and a calyx, and a non-staminiferous corolla, the structure of this corolla, which may be monopetalous or polypetalous, the mutual position of the calyx and pistil, the nature of any perisperm there is. Thus, the corolla is very often formed similarly in the same order; in the Gramineae and Liliaceae [it is] non-existent, in the Labiatae and Compositae monopetalous, polypetalous in the Umbelliferae, Cruciferae, and Leguminosae; in the Leguminosae, it can rarely be monopetalous, and in these and the Cruciferae, it is very infrequently lacking. Thus the calyx exists in most plants, but is clearly absent in *Caltha* and a very few others; it is superior to the germen in the Umbelliferae and Compositae, inferior in the Gramineae, Labiatae, Cruciferae, and Leguminosae; in the Liliaceae it varies, being sometimes inferior and sometimes superior. In the Gramineae, the perisperm of the seed is farinaceous; in the Liliaceae it is horny, in Umbelliferae, woody, and in the Labiatae, Compositae, and Cruciferae absent; in the Leguminosae, it is sometimes absent, sometimes more rarely partially existing, fleshy, and imitating a thickened inner membrane of the seed itself; it is therefore a doubtful sign, and in the investigation of new orders fluctuates between primary and secondary [in value]. The same may perhaps also be said of the location of the corculum inside the seed, and of that of the seed inside the fruit, signs which are very often stable, but occasionally dubious and to be accurately examined.

[xli] 3rd. Other characters [are] tertiary and semi-uniform, now constant in the orders, now varying, sometimes drawn from essential organs, sometimes from others: for example, from the mono- or polyphyllous calyx, the simple or multiple germen, the number, proportion and connection of the stamens, the dehiscence of the fruit and the number of loculi, the site of the leaves and flowers, the arborescent or herbaceous stem, and from other similar or more trivial signs, which, suited separately for distinguishing genera, are employed together in the definition of orders. The necessity remains for a gradation to be instituted, as some are either much or only a little more constant than others. But such a valuation cannot proceed further which perhaps is now illustrated by an investigation of the few best known orders.

All this, however, only helps to conclude work already at hand; without doubt there exist distinct cohorts of characters, each of which possesses its own general degree of importance; [these groups], however, are not clearly circumscribed since their ends cling to each other, and are, so to speak, folded over each other ["complicantur"], and in the future, these will be again divided into several and more suitably defined when the number of truly natural orders has been increased, the field of observation widened, the modes of testing multiplied, and thence a more accurate valuation established.

If we admit meanwhile the threefold partition of generic characters, the dissimilar affinity of genera is derived from them, that is to say, none [i.e., no affinities can be derived] from the mere semi-uniform characters of little note, medium from the more constant and to some degree sub-uniform characters, and the greatest from the uniform characters.[152] And in instituting truly natural orders a different reason is to be had for each division: genera in them must agree in the primary characters, they must in general agree in the secondary characters, they must frequently agree in the more constant tertiary characters, and occasionally in those of the latter which are more unstable. Thus, in the order of the Cruciferae, the primary uniform characters are the hypogynous stamens and the two-lobed corculum. The secondary ones are the lack of a perisperm, the existence of a calyx below the germen, a hypogynous and polypetalous corolla, the descending radicle of the corculum, seeds in two lateral ranks opposite on the receptacle. The tertiary [characters] are a tetraphyllous, deciduous calyx, four petals alternate to the calyx, six tetradynamous stamens, a simple germen, a siliquose, bilocular, bivalved fruit, alternate leaves, non-axillary flowers; nonetheless, very rarely the calyx persists, the flowers by abortion [are] diandrous or tetrandrous, the fruit uni- or trilocular and indehiscent, the leaves opposite [xlii] and the flowers axillary; whence it may be seen that the tertiary characters in an order are less constant than others; nevertheless, if they vary individually and not together, they do not pervert the general character of the order. To these may be added a few inferior, as it were quaternary characters, which are common to certain genera in an order and only suitable for forming sections [groups of genera]: for example, in some the fruit may be long and siliquose and lacking a style, or it may be short and siliculose, possessing a single style. Through such a distribution of characters, the designation of the Cruciferae is easily elicited. He who investigates other known orders in a similar manner, will perceive the same progression among the characters, and when he turns his attention to delineating new orders, he will respect the established law of their inequality, and always value more highly the more general signs, while reckoning the strength of each. This is the only rule governing the construction of orders, the only means of laying bare affinities, to whose investigation the sincere Botanist devotes all his care.

The natural classes

Just as species come together in genera by a natural bond ["nexu"], and genera are assembled in orders, so related ["cognati"] orders continuing the series are grouped together in the same class. For Nature, not diverging from her established law, always assembles various small groups of plants into larger ones, which, by repeated connection, are themselves gradually enlarged, until at last she brings all together in one embrace. Since in a general distribution the superior characters ought to be placed first, it follows that in defining classes, the secondary, and certainly the primary characters, which are universal and uniform in all plants, ought rightly to take precedence over the tertiary ones, as these do not adequately support a firm definition ["designationi"]. But the secondary characters sometimes vary, while the primary ones always remain constant; therefore the value of the former and the latter is unequal, and to the primary characters must go the duty of defining primary divisions ["ordinationes"]. The primary characters are drawn from the corculum of the seed, from the stamens and the pistil, and from the staminiferous corolla. Do these possess the same strength and is there absolute equality among the four organs; or do the worth of the organs and [hence] the primary characters [drawn from them] differ?

Another means of attaining the natural method

In considering an easier solution to the questions raised above, it will be advantageous to proceed for a few moments in reverse order, moving not from simple to composite, but from composite to simpler, so that by tight reasoning and many accumulated observations, our previous statements may be confirmed, the inequality among characters appear more and more evident, and the more important ones [xliii] may stand out. The aforementioned natural orders, which are now admitted, contribute a great deal to this investigation; the botanical laws made known are tested against them, being carefully maintained as long as none contradicts any [natural] order, and unhesitatingly rejected if they offend even a single one.

1st. Characters that are mostly or sometimes unstable in orders are of absolutely no use in drawing up primary definitions, because a sign varying in a partial arrangement is certainly of no value for a more general one. Therefore, all parts, such as the root, the stem, and the leaves, that are occasionally or often dissimilar in genera of the same order ["coordinatis"] are laid aside. In the Liliaceae the root is sometimes tuberous and on other occasions capillaceous. The leaves of the Compositae are indiscriminately alternate or opposite, simple or pinnate. Some of the Leguminosae and Umbelliferae are herbaceous, while others are shrubby or arborescent. Therefore, among plants, herbs and trees are no more rightly to be separated than *Elephas* and *Mus* among four-footed Animated Beings, or *Trochilus* and *Struthio* among Birds. Moreover, the stem and the leaves are occasionally lacking, and cannot therefore be considered among the outstanding parts.

2nd. The primary characters, therefore, hide themselves in the parts of fructification, especially in the essential ones; consequently, the calyx and the corolla merit less attention, since both these coverings are less necessary, and either or both may be lacking without harm, the flower being none the less well adapted to generation, and therefore perfect. Moreover, their outstanding characters may also occasionally falter, with monopetalous and polypetalous corollas, [or] superior and inferior [or] monophyllous and polyphyllous caly[ces] being associated in a single order.

3rd. Excluding the calyx and the corolla, the sexual organs are preeminent by right, being necessary for the propagation of the species and therefore essential to it; if they are lacking in an individual because of abortion, nothing follows from this other than an imperfect specimen of the plant, diverging from nature and [counting for] nothing in the universal series. These true organs of the perfect flower work together for the generation of the seed or future little plant, which in germinating begins a new life, which will proceed until its destined time when it has produced similar progeny.

4th. With their work completed, the stamens fall, along with the styles, because they were useful only insofar as they helped bring about the fertilization of the seed hidden inside the fruit. This seed, or more correctly, the corculum hiding inside it, the first rudiment of a young plant, is that [xliv] especial and more general part of the Vegetative Being to the shaping, nutrition, and protection of which all the other organs of fructification contribute, drying up afterwards. The procreation of beings is the highest design of Nature, and her primary function. The apparatus for this, the importance of the sexual organs, and the usefulness of the surrounding flower, as well as the shedding of both of these latter after fertilization or maturation, prove the superiority of the seed so magnificently elaborated, and affirm that the primary divisions of Vegetative Beings are to be deduced from it or from the corculum, the essential part of the seed.

The primary classes [deduced] from the corculum

From the number of parts of the corculum (p. xvii) and from its mode of germination is elicited a threefold general partition, namely of Acotyledones, Monocotyledones, and Dicotyledones, already celebrated by some Authors, neglected by others, chiefly used in the garden of the Trianon. Immutable[153*] and natural, it is nevertheless absent from the "Fragmenta" of Linnaeus and also from Adanson's "Familles,"[154] both of which exhibit an arrangement in which Monocotyledones and Dicotyledones are occasionally intermingled. But then, is each and every order separately constituted by Nature without any regard for their mutual affinity? Or, on the contrary, are they all joined together by a common nexus of classes ["classicus"], so that they constitute an undivided series? The most approved orders, designated primarily by the structure [of the corculum], reveal at once the existence of classes, and the preeminence of the corculum itself; they also show, by the same sign, the evident relatedness ["cognatio"] of very many similar plants, and finally, an extraordinary analogy with [xlv] Animated Beings, whose incubation is similar to the germination of Vegetative Beings. The heart, the first living and moving part in the animal fetus, by its structure defines different classes of Animated Beings, each of which includes many orders. In Quadrupeds, Cetaceae, and Aves, it is bilocular and biauriculate, in Pisces and Reptilia, unilocular and uniauriculate, and in Vermes and Insectae, unilocular and nonauriculate; in each it never varies, and by its development and unceasing motion places those differences in a clearer light, predicting adjunct and secondary exterior signs so that each and every class easily comes to be recognized by its own general appearance ["habitu"].[155] In the same way, the corculum, the first living part of the plant, designates its most important rankings ["ordinationes"], and emulates the auricles of the heart by its two lobes, single lobe, or none, and once germination is accomplished, outer signs supplement its ephemeral character, and through these is constituted the general appearance proper to each class. These two primary organs, which constitute the entire being of both Animated and Vegetative Beings from the earliest moments after fertilization, embrace their entire organization, then compressed, but later developing through vital motion, and therefore, as if the characters are gathered all together into one, before dividing and being dispersed into many as the fetus grows.[156] Therefore the orders, similar in terms of the heart or of the corculum, truly conform to one another with many signs thus collected into one, and thus the first division of Animated and Vegetative Beings, drawn from the first-formed organ, is confirmed; thence the general distinction of plants into Acotyledones, Monocotyledones, and Dicotyledones remains correct and unshaken.

First occur the Acotyledones, simple in structure and very few in the entire series. They are followed, in a natural progression, by the Monocotyledones, more numerous and less simple. Finally come the Dicotyledones, exceedingly numerous, and having a more complex organization. The classes will not be strictly circumscribed in a systematic fashion, because they are united to one another without an interval at the boundaries by ambiguous genera, according to the rule of Nature. And this is not all: the same [classes], too large, are of necessity divided into secondary ones, similarly connected, the important characters of which must, by equal right, be uniform and are to be drawn from their essential organs.

The secondary classes

Does the corculum, that primary part of Vegetative Beings which defines the major classes according to the number of its lobes, reveal any other substantial signs [xlvi] by which the second[ary] classes can equally well be designated? In any plant whatever, its germination is always similar; in related plants it is often similar, while in less related ["diversis"] plants it occurs in diverse ways, examples of which are to be seen chiefly in the Monocotyledones, the general character of which is given below (p. 21 [in the body of the *Genera plantarum*]). But while some Monocotyledones are scarcely dissimilar in fructification, they differ in germination, as in the order Asphodeli (p. 51); moreover, within the limits of the Dicotyledones, germination is not clearly uniform; thus it is not possible to deduce substantial signs ["notae"] from it unless it is first closely examined in each [group]. Other chief characters of the same organ stand out: namely, the location of the corculum in the seed and of the seed in the fruit.[157*] The corculum sometimes lacks a perisperm and on other occasions is provided with it, [and then is] central or lateral to it, small, medium-sized, or large; it may send forth a radicle in an upward direction at the hilum, or push it out in the opposite direction; it often puts forth straight lobes, more rarely inflexed, or plicate, or corrugated. The seed is attached inside the fruit to a central or lateral receptacle, raised or much hanging down, sessile or strap-shaped. Each of these signs truly conform to one another in certain orders; in others, however, they have not been satisfactorily explored, or they may even seem at first glance to be opposed. Therefore, until the mode of germination of all, as well as the location of the corculum and the insertion of the seed have become better known by repeated and certain observations, it is impossible to deduce any stable character from them.

[DRAWN] FROM THE MUTUAL LOCATION OF THE STAMEN AND PISTIL, OR FROM THE INSERTION OF THE STAMENS. With the innermost part of the seed exhausted, the first characters of the inferior classes are then drawn by right from those outstanding and essential parts [that come] after it,[158] namely, from the sexual organs, which, acting in concert in the work of procreation, are necessarily preexistent to it, and are highly valued in the economy of Vegetative Beings. Just as, however, each of these [characters] is of no importance in itself [xlvii], and is only to be considered in the role it plays along with others in bringing about the generation of future progeny, so in the same way, taken separately, they produce no substantial characters, which together they reveal immediately. From what has been said before (p. xxiii), within the same order the number and proportion and connection of stamens are considered unstable; neither are absolutely unvarying signs to be drawn from whether the germen is superior or inferior, simple or multiple, uni- or multilocular, etc. (pp. xl, xli). And so we neglect these, but an essential character remains to be discovered, common to both organs, which indeed stands out, unvarying and truly primary, and clearly uniform in the tested orders; [this is] the mutual position of the stamens and the pistil, or the insertion of the stamens relative to the pistil, a sign omitted by many Authors. Linnaeus, for example, who concentrated on the more trivial features of the stamens, examined this stronger sign too little on those rare occasions when he added it to the generic definition; Adanson, however, repeatedly noted it in his *Familles*.

THE DIFFERENT [MODES OF] INSERTION OF THE STAMENS. The mutual location of stamens and pistil is threefold; for [either] one may lie beneath the other, or they may be placed next to one another: hence this excellent triple insertion of stamens, *epigynous* when they are above the pistil, *hypogynous* when they are under it, and *perigynous* when they are attached to a part surrounding it, or to the calyx.[159*] Each one of these is distinct in the highest degree from the others, and is never mixed with them in the same order: epigyny of stamens always defines the Umbelliferae, hypogyny the Cruciferae and Gramineae, and perigyny the Leguminosae and Liliaceae.

To the preceding ought to be added a fourth type of insertion; namely, *epipetalous*, or on the corolla, which sometimes defines whole orders, such as the Labiatae and Compositae, and sometimes is associated with the three aforementioned and outstanding types in the same order, genus, and even in the same flower. Thus, in the Leguminosae, the stamens are perigynous, but in *Mimosa* and *Trifolium*, genera in that order, they are occasionally epipetalous; and in the flower of *Dianthus* and the neighboring genera are observed five epipetalous and five hypogynous stamens. [xlviii] The great affinity of the corolla with the stamens explains this discrepancy because, as has been said (pp. v, vi, xvi), it is their true appendage and as it were luxuriant part,[160] arising from the same receptacle. When the corolla does not adhere to the stamens, it is of little importance; when, on the contrary, it coalesces below with them, it then becomes, so to speak, staminiferous, and thence existent by necessity, a sort of intermediate bed; its own insertion taking precedence over that of the stamens, which is completely assimilated to it, it becomes an essential character, agreeing in itself with the three preceding types of insertion. Thence *stamens joined to a corolla have the inserted part corolliferous, as it were*; thence also the very different triple insertion of the staminiferous corolla itself appears, namely, hypogynous, epigynous, or perigynous, and these three types of insertion are never found mixed together in the same order, hypogyny always [occurring] in the Labiatae, epigyny in the Compositae.

Therefore, a general double mode of insertion of the corolla ought to be carefully taken note of: the one *immediate* when the stamens are attached immediately at the aforementioned three points, the other *mediate* when they adhere to a mediating corolla, which then itself is staminiferous. Both modes are occasionally mixed in the same genus and even in the same flower, and are therefore judged to be clearly related; however,[161] they are the same both in their origin and their point of insertion. Thus, in *Trifolium*, which is distinguished by immediate perigyny, certain species exhibit the mediate mode of insertion; that is, they possess a staminiferous corolla which is itself perigynous; in the flower of *Dianthus* are observed five stamens immediately hypogynous, and five petals alternating with these, simultaneously staminiferous and hypogynous. These therefore are the three outstanding modes of insertion, differing completely one from the other, and never mixed in the [same] orders; when single, they are rarely double or a mixture; they are nevertheless characterized by a mutual relationship ["affinem"].[162]

THE FIRST DISTRIBUTION OF THE CLASSES ACCORDING TO THE LOBES OF THE CORCULUM AND THE INSERTION OF THE STAMENS. With the insertion of stamens examined, and its laws laid down, it is easy to draw up the first distribution of the aforementioned classes. In the Acotyledones, the sex organs are scarcely visible, and the location of stamens is obscure, therefore, they do not offer signs characteristic for a class and for the moment must remain undivided. The Monocotyledones are divided in tripartite fashion according to the mode of insertion—epigynous, hypogynous, or perigynous—despite the fact that in them the corolla is always absent. Similarly, a tripartite partition of Dicotyledones is drawn from their triple mode of insertion of the stamens, with each division comprehending either stamens inserted at one point of the flower, or the staminiferous corolla affixed at the same point. [xlix] Thus are produced seven secondary classes, defined by uniform signs of essential organs.

This latest arrangement of Vegetative Beings is confirmed by a similarly sevenfold division of Animated Beings (p. xlv) based in just the same manner on the most outstanding characters drawn from the first-appearing parts. For Quadrupeds and Aves, similar in their bilocular and biauriculate heart, differ in the structure of the sex organs determining generation, which in the latter is oviparous, and in the former viviparous. These organs are also clearly dissimilar in Reptilia and Pisces, which resemble each other in their unilocular and uniauriculate heart. Very suitably, therefore, just as branches grow from the same trunk, the two parts of the organic Realm agree in their plan and progression, which are simple and constant and certainly co-equal, and so the principles assigned to the one also suit the other, and in both, from the important organs and similarly, the functions, comparable divisions both primary and secondary may be drawn. In neither of these two parts of Nature are inferior organs, or inferior modes of superior organs, adequate to the defining of classes, since at best only tertiary characters can be derived from these; this will lead astray from nature's path any Méthodiste who depends on them, unless by some happy chance the features he selects are associated with the essential [organs], [so] when accepting an inferior sign, the primary [character] will be grasped unknowingly.

This inequality of characters, the worthlessness of the unstable and the value of the unvarying, did not escape the notice of the distinguished Author of the Trianon arrangement, Bernard de Jussieu, nor did the importance of the corculum and of the sexual organs, or the affinity of similar genera and orders agreeing in these signs or primary parts. The orders that he brought together are, generally speaking, truly natural and in accordance with principles noted earlier; they are arranged in the sevenfold division of classes described above, making use of a computation ["ratione"] of the structure of the corculum and the insertion of stamens and staminiferous corolla. In fact, he indicated no breaks between classes, as though he admitted no classes but only orders, but in private correspondence he always stated the aforesaid characters, and besides, reflection on his proposed series granted easy and abundant proof that this was on his mind, to connect his orders at a common nexus. Simpler than all other methods, and closer to nature, this distribution for the most part preserves the orders whole, but having in view an undivided series and on that account acknowledged by the Author to be neither strictly methodical nor [l] divided into sharply circumscribed classes, it eases the task of Botanists less. Therefore, a larger number of more limited classes is highly desirable so that the investigation of genera and species may proceed more promptly and expeditiously.

The lack of essential characters, however, is an obstacle to the multiplication of classes. For those outstanding characters, each of which alone is equivalent to all others taken together, and which separately define classes more than adequately designated by one such sign, are clearly to be considered natural. When, on the contrary, such primary signs are lacking, other subsidiary signs are employed, then many must be collected and their strengths combined so that the greater force of the primary one may be approached more nearly. [B]ut first their right value must be determined, and this is the more difficult the more numerous they are that seem equal at first glance; and in the meanwhile, all further division into classes ceases, inasmuch as it cannot be completed by these simple signs.

ANOTHER, MORE NUMEROUS, DIVISION OF THE CLASSES, BY MEANS OF THE COROLLA. As, nevertheless, Botanists eagerly await more numerous classes, their sound constitution must now be worked out, employing secondary characters (p. xl), preserving as much as possible the law of nature, and also preserving those genuine orders which cannot be broken up without detriment to science. Characters noted above, for example, the existence and insertion of the staminiferous corolla, [that are] very closely related to ["cognatissimi"] other essential ones, become in every way partakers of their immutability; others, such as the division of the corolla or its integrity, and its location when non-staminiferous, are called general (p. xl), and while related ["affines"] to the primary ones, are only half-sharers of their immutability, and vary, albeit infrequently: thus, it has been concluded from observations that a monopetalous corolla is very often staminiferous, but that when polypetalous it is very frequently non-staminiferous, and often emerges with the stamens from the same point: hence the insertion of the stamens themselves may be inferred with few exceptions from the insertion and number of parts of the corolla.

The corolla, therefore, because of its affinity with the stamens, can adduce auxiliary signs, by which new classes, or new divisions of previous classes, which are not foreign to nature, may be designated. On this account, the Linnaean system is less natural than that of Tournefort: for Linnaeus drew only tertiary signs from an essential organ; Tournefort selected secondary signs from a secondary one, and, in distinguishing monopetalous flowers from apetalous and polypetalous ones, conformed in part, though unknowingly, to the insertion of stamens. Because the sexes were not yet clearly known, he had overlooked the staminal organ [li] and its concord with the corolla, from which he might have elicited stronger characters. Those which he omitted must now be employed, the time being more opportune, and following his lead we must seek a simple means of multiplying classes, based on the corolla, without any subversion of orders and natural laws.

To this end, we must recall first the two insertions of stamens mentioned above (p. xlviii), the one *immediate*, the other *mediate*, which it is useful to distinguish carefully from one another, and which ought not to be mingled as in the sevenfold arrangement. For examples of these associated in the same order or in the same flower are rare (p. xlvii), and moreover very often when the insertion changes, the structure of the corolla is altered as well: thus, the corolla of *Fraxinus*, very closely related ["cognatissimae"] to monopetalous plants, shedding the stamens, becomes polypetalous or vanishes altogether; thus in the order Leguminosae, the corolla is polypetalous and non-staminiferous, [yet] certain leguminous *Trifolium* [species] bring forth a corolla both staminiferous and monopetalous, the fusion of the petals preceding, as it were, the insertion of stamens on the petals. Hence the following assertion can easily be made: *with few exceptions, a monopetalous corolla is also staminiferous,*[163*] or, to put it another way, *a staminiferous corolla is very often also monopetalous.* The relationship ["cognatio"] between the two characters is so great that one almost announces the other: generally speaking, therefore, a monopetalous corolla designates a mediate insertion.

The corolla is essential in this form of insertion, but some apetalous orders such as the Gramineae and Liliaceae, and some petaliferous orders such as the Umbelliferae, Cruciferae, and Leguminosae have immediate insertion. Only rarely are apetalous plants intermingled with the corollate orders: e.g., *Ceratonia* in the Leguminosae and *Cardamine impatiens* in the Cruciferae; neither generally do there exist orders partly petaliferous and partly lacking a corolla. These rarest of exceptions, therefore, do not prevent the separation of the corollate orders from the apetalous ones for the purpose of the multiplication of classes, provided that the insertion of stamens, which is then either *absolutely*, or *simply immediate*, is not neglected. For the lack of a corolla, or intermediate support, [lii] in the Apetalae, clearly and absolutely determines immediate insertion. In petaliferous plants, however, insertion may be simply immediate, or on certain occasions changeable into mediate, because from the beginning the stamens, born together with, but nevertheless distinct from, an existing corolla, sometimes coalesce with it at the base,[164*] as though placed upon it.

Plants by nature apetalous do not become petaliferous; petaliferous plants, on the contrary, sometimes cast off the corolla, or on rare occasions become monopetalous from polypetalous, as in *Trifolium*, *Saponaria*, *Cotyledon*, etc., very infrequently polypetalous from monopetalous, as in *Fraxinus*. Therefore, in an insertion that is absolutely immediate, no corolla is ever found; in simply immediate insertion it may perhaps be absent, but is more often present, and is then very often observed to be polypetalous.[165*] Therefore, as a group, such characters as a monopetalous corolla with mediate insertion, a polypetalous corolla with simply immediate insertion, and the absence of a corolla with absolutely immediate insertion, are considered analogous, as it were, with the ones announcing the others; consequently, for this insertion with awkwardly named and more hidden signs, the more obvious character and more accepted names of monopetalous, polypetalous, and apetalous plants can conveniently and safely be substituted. Thus, another method, differing little from the sevenfold classification of the Trianon, is easily delineated, including many classes, preserving the integrity of the natural orders, displaying the important Tournefortian characters of the corolla, and in addition based on the real ["solido"] sign of the insertion of stamens. [liii]

The method thence deduced, and [as] cultivated in the Garden of Paris

1st. The first division into Acotyledones, Monocotyledones, and Dicotyledones remains unchanged.

2nd. The Acotyledones stay undivided until their organization is better known. Their sexual parts are invisible, or often diclinous; for this reason, it is difficult to observe insertion, and so a division into classes is not yet to be attempted.

3rd. Because the Monocotyledones, in the absence of a corolla, are characterized by a single mode of insertion, namely absolutely immediate, their threefold division ["distributio"] cannot be further amplified, producing in three classes, [plants with] hypogynous, perigynous, or epigynous stamens; neither do these classes, laden with few orders, require further division.

4th. The exceedingly numerous Dicotyledones, whose number is almost ten times greater than that of the preceding, very truly require a multiplication of classes, and this can easily be perfected by means of the corolla; consequently, a recent change has been made in the Trianon arrangement according to the different modes of insertion, with apetalous, monopetalous, and polypetalous plants distinguished in turn.

5th. The Apetalae, inasmuch as they are simpler or lacking an organ, are placed first following the similarly apetalous Monocotyledones, and thence divided equally into three groups with the stamens immediately epigynous, perigynous, or hypogynous.

6th. In the Monopetalae, which follow, the stamens are very often epipetalous, and therefore scarcely vary in their proper insertion, for which has been substituted the triple insertion of the staminiferous corolla itself, hypogynous, perigynous, or epigynous. In addition, when the insertion is epigynous, connate anthers, which correctly define a numerous series of Compositae,[166] are distinguished from those that are separate: thus a fourfold class of Monopetalous plants is easily produced.

7th. The Polypetalae are divided by the threefold plan of the epigynous, hypogynous, and perigynous insertion of the stamens, or by the insertion of the corolla itself which is very often non-staminiferous, and generally emerges together with the stamens from the same point.

8th. A tenth[167] class may be added to the Dicotyledones, consisting of irregular, diclinous plants, not subject to the rules of insertion, namely, those which have the sexes separated in different flowers. In the male flowers, the stamens are attached to a calyx, or to a stipe located in the center of it, or, when the calyx is absent, to the scales that make up for it. These insertions, which do not determine the mutual position of the separated sexes,[168] are of little importance. [M]eanwhile, the diclinous plants complete our arrangement and that at the Trianon; [liv] these are either very different from the other classes, or when the stamens are inserted on a calyx or calycine part, they are very close to the polypetalous, perigynous [group], although they generally lack a corolla. With these true diclinous plants are never to be confused those spurious ones which are such only by abortion, and which can all be added singly to the related hermaphrodites in the series of aforementioned orders.

Evaluation of this method

This method,[169*] by multiplying classes, promotes the study of science, and only in this multiplication is it preferable to that of the Trianon, with which it agrees for the rest in the primary division according to the corculum, the undivided Acotyledones, the triple break of the Monocotyledones, the very close observation of the insertion of stamens, and the preservation of the natural orders in their entirety. While, however, for the sake of clarifying the science, it increases the three classes of Dicotyledones to eleven by a further division; while because of a lack of essential characters, it employs weak general ones liable to exceptions, it thus, participating in their inconstancy, correspondingly is forced to include exceptions which the Trianon arrangement evaded. Thus, some very closely related monopetalous or apetalous plants creep into the polypetalous orders; *Mimosa*, *Trifolium*, and *Ceratonia* in the Leguminosae; *Cotyledon* in the Sempervivae; *Trianthemum* in the Portulaceae; and *Tetragonia* in the Ficoideae; thus very rarely apetalous or polypetalous species are mingled in the monopetalous orders, for example, *Fraxinus* in the Jasmineae; or monopetalous plants are moved to apetalous [groups], for example, *Plantago* [is placed next to] the Amaranthi. This defect of the proposed method, its dependency on somewhat varying characters, is unavoidable; but it is much rarer here than in those [lv] systems from which the present method distinguishes itself, inasmuch as being utterly alien to obvious contradictions, for the most part it does not disturb the true affinities of genera. But its own disposition of dicotyledonous orders, employing primary characters drawn from the corolla—monopetalous, polypetalous, or completely lacking—and secondary ones drawn from the insertion of stamens, may perhaps on this account diverge from the natural series, both by preferring a sign that is merely general to one that is essential, and by separating generally related insertions, namely, those that emerge from the same point in mediate or immediate, absolute or simple fashion. If the earlier threefold division of the Dicotyledones based on the insertion of parts had been restored, and apetalous, monopetalous, and polypetalous plants examined in turn and each group placed in one of these three classes, these three divisions could have been much more intermingled in an inverse ordering of classes so that the Onagrae would have joined the Campanulaceae, the Amaranthi the Caryophylleae, etc., as in the Trianon distribution. The arrangement would be of greater value insofar as it multiplies classes and at the same time obeys natural law, it would have given preference consistently to primary rather than to secondary signs. However, the useful omission of the exceptions noted would not have followed from this, nor from a more perfect constitution of classes, which in both systems would have remained the same, and merely have been arranged differently. Little utility is therefore seen in changing this series, which, moreover, would disturb the Tournefortian divisions primarily based on the corolla, which are currently accepted by most Botanists.

The disposition of orders in classes and of genera in orders

Whatever disposition of these same classes may continue to be accepted, many orders are assembled in each joined together by a shared character of a class, essential or general; in like manner, many genera are brought together in any given order by means of inferior signs, secondary or tertiary, and hence are shown not singly, but collectively and strengthened by the conjunction. Thus, longer or shorter definitions of orders are produced, according to the number of common signs in the grouped ["coordinatis"] genera. Insofar as [the definitions] are long, the quick perception of orders is more difficult, and that of genera is easier, and vice versa: for when many common characters are brought together in an order, to that extent only a very few dissimilar ones are left over for calling in mind ["commemorandi"] the genera and suitable for distinguishing them. Related genera being grouped in sections that have been instituted in each and every order, there follows from this a more accurate distribution of genera, determined by the character proper [lvi] to a given section, which being segregated at the same time, a shorter and simpler generic designation stands out. In addition, among the characters common to orders and sections and genera, are listed not only those that the fructification displays, but many others, drawn from the inflorescence (p. x), the leaves, and the stem, [that] are by no means to be overlooked since they are observed to be common to related genera. Tournefort admitted some of these in his secondary genera;[170] Linnaeus strictly excluded them all from generic definition, in this most certainly overturning the law of Nature, which has established closely connected characters for the designation of orders, which besides, *a fortiori*, are to be applied to genera.[171]

The great conveniences of the natural method

Moreover, from what has been said above, the following advantages of the natural method are justly concluded, namely, 1st. Its definite arrangement using many signs is more difficult than a systematic one employing fewer, but the systematic one, while quicker, is greatly confused, while the natural [method], though slower, leads to a more than moderate degree of knowledge.

2nd. On the other hand, the natural designation of genera is easier than the systematic, because it uses fewer characters, the other, stronger ones having been placed first in the class or section: in this manner, the

Gramineae (p. 28) are defined at length, and sections being added, a bivalved glume, a univalved calyx with an entire apex, and spicate flowers complete the definition of *Alopecurus* (p. 29).

3rd. Since the natural designation of genera either expresses any common signs whatever, not only from the fructification but also sometimes from parts external to it, or follows on from those already expressed in the order, it is fuller and more perfect than the systematic one which, although far-reaching, often produces defective definitions: thus, for example, in the lengthy [descriptions of] Linnaean genera,[172] the insertion of stamens, and the internal structure of the fruit and the seed, as well as any number of extra-floral characters, are frequently omitted.

4th. In accordance with natural law, dissimilar plants are never mixed, or closely related ones separated, as happens sometimes in systems that sometimes proclaim as congeneric species [those] which are to be placed in very different orders. A majority of mixed genera of this sort is found in the writings of Early Authors, and also some in the works of certain more Recent Authors;[173*] [lvii] even Linnaeus himself, complying excessively with an arbitrary law, does not succeed in avoiding these; however, [his work] generally shows the best generic assemblages ["constitutiones"].

5th. An unknown plant cannot be arranged in a system unless it exhibits a character of a class, and most importantly, unless it has a complete flower. In the natural method, on the contrary, many signs are admitted that are often auxiliary to the primary ones, so while some of these are evanescent, others, persisting and making up the habit of the plant, are sufficient for the determination of its order and genus: thus, opposite leaves on a shrub, joined to one another by a single stipule in the middle, very often designate the character of the Rubiaceae (p. 196); alternate leaves, sheathing at the base, and revolute below when young, indicate the Polygoneae (p. 82).

6th. In addition to these auxiliary [characteristics] of lesser note, other more important ones, very much related to one another ["cognatissimi"], are observed in the same plant; they are frequently or always grouped together, as has already been stated (p. xix), and will be carefully set out below in the individual observations of the classes. The more there are thus assembled, the more they appear constant, because a change in one often determines a change in the other, and an alteration of many is more difficult than that of few. Thus, a monopetalous corolla requires a monophyllous calyx as well as, for the most part, epipetalous stamens that are definite in number, which, being attached to it, are not multiplied indefinitely unless they are first united at the base.[174] Therefore, the change from a non-staminiferous, polypetalous corolla to a monopetalous, staminiferous one is easier in the Leguminosae than in the Papaveraceae because the former already have a monophyllous calyx and stamens that are definite in number and joined at the base, whereas the latter have a polyphyllous calyx and stamens which are indefinite in number and separately inserted. Similar difficulties can be very satisfactorily explained if the affinities of characters are observed, and the law that many related ones change only with difficulty is borne in mind.

7th. Not only does the natural method convey a complete knowledge of characters and affinities, but it also, at the same time, indicates the virtues of plants, to the great advantage of the medicinal and alimentary arts. Because every small part of any Vegetative Being [lviii] possesses its own simple virtue, the joining together of those small parts in organs produces a sort of mixed power, which is always the same in organs that are most similar; it therefore does not differ, as experience testifies, in plants of the same species, or furnished with a completely similar organization, unless the native soil and the various proportions of the juices between mountains and meadows, gardens and forests, has produced a slight change, whether external or internal. Plants of the same genus, which are almost alike in characters or organs, scarcely differ in virtue:[175*] thus [species of] *Salvia* are all good for the head, *Anchusa*, for the chest, *Cochlearia* is anti-scorbutic, *Euphorbia*, cathartic, and *Rubia*, diuretic and tinctorial. A further connection is gradually revealed by the bringing together of contiguous genera, which are related ["cognata"] in many important organs, those somewhat similar to one another in virtue are conjoined, and observation confirms that this virtue is indeed generally uniform[176*] in truly natural orders, or in genuine sections of orders. Thus, generally the Gramineae are nutritious and farinaceous; Cruciferae variously fight against scurvy; the Labiatae are aromatic and bitter, such that the bitterness of some is of value in stomach disorders, [and] the aroma of others in diseases of the head. The outstanding virtue of the Liliaceae, Leguminosae, Compositae, and Umbelliferae is less well defined, or in accordance with a changed proportion of elements becomes unlike itself, as it were, and manifests itself differently in every section, all the while, however, retaining the same basic ["primigeniae"] nature. If, however, certain species are wrongly united together in a genus, or [lix] if genera themselves are grouped ["coordinata"] incorrectly, there follows a similar discrepancy in virtues among them: thus, the harmless *Ficaria* (p. 233) differs from the caustic *Ranunculus*, and *Fumaria* (p. 237), which regulates bitterness of humor, from the narcotic Papaveraceae. Therefore, there exists the closest affinity between characters and virtues, so close, indeed, that the ones are harbingers, so to speak, of the others, and from the known virtues of particular Vegetative Beings, we can almost certainly infer the unknown virtues of related ones. In this manner, the natural science indeed increases our knowledge of appropriate remedies and healthy food and promotes trade ["artium"], and being carefully improved it far surpasses the less fruitful systematic or artificial science.

From what has been said, the enormous importance of all efforts that contribute to the investigation of the natural method is clear; no one has undertaken these labors in vain who has assembled many species in true genera, and [in turn] brought these genera correctly together in orders, or who, undertaking an even greater task, has delineated a partial or universal series of orders. For any true affinities, either already detected, or remaining to be discovered in the future, will remain unchanged without any step backward on the part of science: thus many Tournefortian and even more Linnaean genera will never be overturned; thus some especially natural groupings ["connexiones"] of genera in the Linnaean system and many more in the Tournefortian method can never be separated; thus also in several systems there exist here and there excellent consociations of plants which ought on no account to be disturbed. Linnaeus deserves the thanks of the science for assembling his "Fragmenta," in which a far greater concurrence of related [plants] ["affinium"] is made known than in any system. However, the laws governing their construction and distribution have remained unknown until now on account of a lack of clear demonstration, and neither ought they to be extracted from it by contemplation of this arrangement, since it proffers untrustworthy[177*] signs opposed to whatever law. For instance, he equates many characters, indiscriminately mingling primary, secondary, and tertiary ones, scarcely admitting any difference among them. Adanson, the author of the *Familles*, [lx] more accurately worked out orders that are deformed by fewer uncertain characters and improved by many fixed groupings of genera, but since he admitted only certain general characters and no essential ones, he established no law for distinguishing them and therefore did not divide his almost arbitrary distribution of orders into classes, so that the general design of his system, like that of the Linnaean system, is difficult to understand. Nevertheless, essential and important organs [do] appear in plants (p. xliii), of which the primary (p. xl, xliv, xlvii) characters are assembled into orders that are always uniform; moreover, one primary character is of equal value to many secondary and tertiary ones (p. xxxix), and thus similar orders are brought together by a good character[178] (p. xliv); the arrangement of the Trianon garden is superior to the aforementioned "Fragmenta" and *Familles*;[179*] generally following the laws described above, [this arrangement] often reveals improved orders, and henceforth, these being established, properly allows their division into classes or is in full conformity with that to be established later.

Conclusion

Now that a sure definition of natural History, and of organic science and Botany has been established, and the parts and functions and differentiae and characters[180] of plants explained, species defined and more appropriate rules proposed for naming and describing them, arbitrary precepts in the construction of genera and classes exposed by injunctions, and some systematical arrangements of plants, especially the most prominent among them, delineated and duly examined, it is clear for all systems whatever that, while useful in the naming of plants, but less conforming with Nature, from their usurpation of the truer science, have deviated and taken a step toward an inferior status, [and have become] a prelude to Botany or a methodical index.[181*] [lxi] In place of these will soon be substituted a genuine science, which inquires not only into the names but also the nature of Vegetative Beings, surveying their entire organization[182*] and all their characters; supported by clear and simple principles, it reveals the natural series, or method, arranged according to affinities, so that related Vegetative Beings are associated by a very large number of characters, and in the enumeration of the characters, only the more important ones are considered. Moreover, certain very well known examples of truly natural genera and orders are adduced according to these laws of affinities, these confirming them, and proving useful for constituting new genera and orders by the same plan, and necessary for the establishment of a more accurate computation of characters. Clearly useful to these tasks also is the distinction among characters, which are now either uniform, subuniform, semiuniform, or never uniform but only specific. In addition, the mutual valuation of uniform [characters] has now been more certainly established; the primary divisions drawn from the corculum have been confirmed by repeated test; and the value of the sexual organs has been affirmed, as well as the arrangement arising from their mutual location. The great affinity of the stamens and the corolla has been proven, along with the special rule of insertion drawn from it, and this constitutes a source of new divisions. A class that includes orders is defined by a few primary characters; an order connecting similar genera is designated by a large or indefinite number of semiuniform signs, sometimes constant, sometimes varying, and therefore making their mutual valuation more difficult. That [sign], however, that remains constant among many orders, will be considered the more valuable. Thus, in the first place the number of orders deemed to be truly natural by unanimous consent [lxii] can be increased, and after this the computation of characters is instituted, with each receiving its deserved level of value according to in how many orders it is constant: in this manner, with sixty orders constituted, the structure of the corculum, in its highest degree of constancy, surpasses sixty times that of the leaf blade,[183] often varying and remaining scarcely uniform from one to another [order]. Public praise will be showered on those botanical laborers who framed this convenient multiplication of orders and extended computation of characters, although this happy event may not occur soon. The science, forswearing arbitrary laws and yielding to natural ones, now taking its first steps on a path not yet trodden, advances at first with a slow but firm pace, never looking back, and grows stronger more swiftly as Botanists join forces. The arrangements that have been examined are not yet absolute in all their parts, and neither will they be completed within a short time, all finally being connected in one. Bernard de Jussieu judged his own to be imperfect, inasmuch as it was frequently broken by empty spaces, because it admitted certain connections [yet] to be revealed, and because uncertain parts needed to be clarified by repeated observation. Our illustrious uncle, in whom the upright morals of early times seemed fitly joined to the learning of future ones, did not think that he had discovered a natural method, but only that he had taken a step toward this goal, many times to be purified by Botanists before it was revealed to beginners. Meanwhile, he much wanted to exhibit another distribution in the public garden, one which, while somewhat systematic, and not subverting the natural orders, would accommodate easy study by means of established and multiplied classes. Therefore, this method was carried out at the school of Paris; in it the orders and the primary characters of the seed and the sexual organs are preserved, and in addition the convenient primary divisions of Tournefort. Happy the author who, on the one hand, mindful of the work of true Botanists, has made clear the natural laws of affinities, and transmitted to posterity the accurate definition of orders and genera, and, on the other, for the sake of beginners has arranged orders and genera according to a method sometimes with the most certain of signs, sometimes supported by others that are general and obvious together,[184] and who has published a work welcome to both these [tyros] as well as to those [advanced in science], and has provided posterity with a memorial to all his own effort as well as created a monument to an uncle, [this book is dedicated] with a grateful and respectful heart. [lxiii]

BERNARD DE JUSSIEU
NATURAL ORDERS

ARRANGED IN THE GARDEN OF THE TRIANON OF LOUIS XV, ANNO 1759

[In the original, genera are also listed. The sequence of families is as follows: Fungi, Algae, Musci, Naïades, Aristolochiae, Filices, Orchides, Cannae, Musae, Irides, Narcissi, Lilia, Junci, Palmae, Aroideae, Gramineae, Cichoraceae, Cinarocephalae, Corymbiferae, Dipsaceae, Rubiaceae, Umbelliferae, Lysimachiae, Veronicae, Scrophulariae, Solaneae, Orobancheae, Jasmina, Verbenae, Acanthi, Gentianae, Sapotae, Apocina, Convolvuli, Borragineae, Labiatae, Cruciferae, Papaveraceae, Capparides, Ranunculi, Lauri, Rutae, Gerania, Tiliae, Anonae, Caryophylleae, Jalapae, Salsolae, Thymeleae, Polygoneae, Sempervivae, Myrtilli, Malvae, Leguminosae, Campanulae, Onagrae, Cucurbitaceae, Salicariae, Myrti, Rhamni, Rosaceae, Terebinti, Amentaceae, Euphorbiae, Coniferae.]

INDEX TO THE METHOD EMBRACING THE NATURAL ORDERS

				CLASS
ACOTYLEDONES				I
MONOCOTYLEDONES		Stamens hypogynous		II
		perigynous		III
		epigynous		IV
DICOTYLEDONES	Apetalae	Stamens epigynous		V
		perigynous		VI
		hypogynous		VII
	Monopetalae	Corolla hypogynous		VIII
		perigynous		IX
		epigynous	Anthers connate	X
			Anthers not connate	XI
	Polypetalae	Stamens epigynous		XII
		hypogynous		XIII
		perigynous		XIV
	Diclines irregulars			XV

[Note: This index appeared on p. lxxi of the *Genera plantarum.*]

THE SERIES OF NATURAL ORDERS

CLASS I

1 Fungi
2 Algae
3 Hepaticae
4 Musci
5 Filices
6 Naïades

CLASS II

7 Aroïdeae
8 Typhae
9 Cyperoïdeae
10 Gramineae

CLASS III

11 Palmae
12 Asparagi
13 Junci
14 Lilia
15 Bromeliae
16 Asphodeli

33 Plumbagines

CLASS VIII

34 Lysimachiae
35 Pediculares
36 Acanthi
37 Jasmineae
38 Vitices
39 Labiatae
40 Scrophulariae
41 Solaneae
42 Borragineae
43 Convolvuli
44 Polemonia
45 Bignoniae
46 Gentianae
47 Apocineae
48 Sapotae

CLASS IX

49 Guaïacanae

66 Acera
67 Malpgihiae
68 Hyperica
69 Guttiferae
70 Aurantia
71 Meliae
72 Vites
73 Gerania
74 Malvaceae
75 Magnoliae
76 Anonae
77 Menisperma
78 Berberides
79 Tiliaceae
80 Cisti
81 Rutaceae
82 Caryophylleae

CLASS XIV

83 Sempervivae

17 Narcissi
18 Irides

CLASS IV

19 Musae
20 Cannae
21 Orchides
22 Hydrocharides

CLASS V

23 Aristolochiae

CLASS VI

24 Elaeagni
25 Thymeleae
26 Proteae
27 Lauri
28 Polygoneae
29 Atriplices

CLASS VII

30 Amaranthi
31 Plantagines
32 Nyctagines

50 Rhododendra
51 Ericae
52 Campanulaceae

CLASS X

53 Cichoraceae
54 Cinarocephalae
55 Corymbiferae

CLASS XI

56 Dipsaceae
57 Rubiaceae
58 Caprifolia

CLASS XII

59 Araliae
60 Umbelliferae

CLASS XIII

61 Ranunculaceae
62 Papaveraceae
63 Cruciferae
64 Capparides
65 Sapindi

84 Saxifragae
85 Cacti
86 Portulaceae
87 Ficoïdeae
88 Onagrae
89 Myrti
90 Melastomae
91 Salicariae
92 Rosaceae
93 Leguminosae
94 Terebintaceae
95 Rhamni

CLASS XV

96 Euphorbiae
97 Cucurbitaceae
98 Urticae
99 Amentaceae
100 Coniferae

[Note: This list appeared on p. lxxii of the *Genera plantarum*.]

COMMENTARY BY PETER F. STEVENS ON JUSSIEU'S REVISED INTRODUCTION TO THE *GENERA PLANTARUM* (PUBLISHED POSTHUMOUSLY, 1837)

Antoine-Laurent de Jussieu made substantial changes to the Introduction to the *Genera plantarum*, and a revised version was published posthumously in 1837.[185] The major changes owe to the addition of new information on pollen and in particular on the morphology of seeds and fruits.[186] The revised Introduction also contains much more detail about the physiology and anatomy of plants, and here the tone is decidedly less Linnaean than in the earlier version. Finally, Jussieu replaced Linnaeus's comparison of the natural system to a map with that of bundles of sticks. Here I indicate briefly where the most important changes occur and what they entail; translations of some passages are given. Arabic page numbers refer to the posthumous Introduction; roman numerals to the original version.

In the posthumous version of the Introduction new material is introduced at the very beginning in which Jussieu discussed the nature of and differences between plants, animals, and minerals (98–99, 102; cf. ii–iii). The breakdown of the sciences was also given in greater detail (100–101, 103; cf. iii), and plant science was divided into vegetable physiology and botany proper. The former deals with the functions and external and internal structure of organs, the latter includes classification, and involves studies of the whole organization of plants.

Jussieu made extensive changes to the section in which the parts of the plant were discussed from a functional point of view (103–127; cf. iii–vii). The section on botany as a whole, and the physiology and anatomy of plants, was considerably extended (103–108; cf. iii–v). The correlation of different organs of the flower with different tissues of the stem is no longer evident (109; cf. v).[187] He added a long section on pollen and pollination (110–114), and he also made major additions to the section on the morphology of the ovary, fruit, and seed and that on germination (114–122; cf. vi). On pp. 119–120 Jussieu discussed the distinction between exogens and endogens made by Augustin-Pyramus de Candolle (among others; see chapter 5), although questioning it in a footnote. Pp. 122–123 consist of an outline of the general properties of plants, and pp. 123–125 deal with the metamorphoses of plant parts—teratological variation, variation within a plant, etc.; Jussieu connected such metamorphoses with the level of nutrition of the plant, as was common in the eighteenth century.[188] On pp. 125–127 Jussieu briefly discussed plant distribution, making the same distinction between habitations (biogeography) and stations (ecology) as made by Candolle.[189]

Jussieu introduced few changes to the lengthy section in the original Introduction in which the different parts of the plant are defined and discussed; there is generally good agreement between the two editions (as in 127–145 and vii–xviii). However, on pp. 127–128 (cf. viii) he added a section in which he categorized the different kinds of underground parts. On p. 131 (xi) he added a brief section on the times during the day when a flower might be open. The mention of the kind of tissue from which the corolla originated was omitted (134; cf. xiii; the last part of fn. 1, in which unusual perianth types were discussed, is new), as was the section on doubled flowers on p. xvi (but cf. 123–125). Jussieu added four ways in which the corolla might vary (135; cf. xiv) and two categories of variation in the androecium (136–137; cf. xiv–xv). The section on the pistil and germen was considerably enlarged (137–138; cf. xv) as was the last part of the section, on the fruit and seed (139–144; cf. xvii). A section on less essential parts of the plant such as glands, scales of various sorts, and stipules was added (144–145).

The sections on the systematic value of characters, on species and varieties, nomenclature and description, genera, classes and systems, and so on were largely unchanged (145–160, 193; cf. xviii–xxxv). The subheadings differ slightly, and on p. 150 (xxiii) one finds that fn. 2 and the last sentence of fn. 3 are new. On p. 152 (xxvi) two short sections were added, one on synonymy and one on illustrations.

Jussieu made an important change at the beginning of his discussion of the natural method (as system or arrangement: 194–195; cf. xxxv). Although the analogy between the natural method and a chain with numerous links was kept, Linnaeus's analogy of the map was replaced by that between the natural method and bundles of sticks, and bundles of bundles of sticks; the footnote (fn. b) in the original Introduction in which Linnaeus's analogy was quoted verbatim was also removed.[190] A similar change was made on p. 195 (cf. xxxvi–xxxvii) where the problem of plants that had not yet been discovered was discussed. A reference to Pierre Magnol as being the first to develop the natural method was also added (194).[191]

The middle part of the section on how related species were grouped into genera was considerably expanded (196–198; cf. xxxvii–xxxviii). Jussieu listed twenty genera and showed how the characters of cotyledon number and stamen insertion were constant in all of them.[192] Other characters, however, although constant in most genera might vary in one or two, or the extremes of the character could not be sharply distinguished. All but the last paragraph of the section on families ("ordines") was extensively changed (200–203; cf. xl–xli). As in the original Introduction, he took seven natural families and discussed the variation of different characters within them. He added a sentence in the introductory paragraph (200, xxxix) to the effect that genera only counterfeited an undivided being ("singula tunc ens simplex indivisum mentienta"), and that genera were grouped into families. He also observed that families did not have unique characters, rather, the characters of families were the characters of their included genera (and those of genera, their included species).

When dismissing characters that were variable as being of no use in drawing up basic definitions, Jussieu observed that such characters were more likely to be external than internal (205; cf. xliii); this is in line with his suggestion in the original Introduction that characters (in the strict sense) were necessarily based on internal differences.[193] Although the substance of the subsequent discussion on the characters used to distinguish primary and secondary classes was not changed, it was extensively reworked (206–209; cf. xliv–xlvi). Jussieu again mentioned the distinction of plants into exogens and endogens, but rejected it because he thought that there were problems in observing this anatomical character. In this part he also outlined the whole *scala naturae*, although in the earlier Introduction he had more simply compared the characters distinguishing the classes of animals and plants (he omitted a paragraph on p. xlix in which there is a similar discussion of these characters). This *scala* proceeded from complex to simple animals, then came organisms that were of uncertain affiliation, then simple plants, and finally more complex plants.

In the section in which the use of the type of corolla for subdividing overlarge classes was justified Jussieu added many examples of exceptions (214–215 and footnotes; cf. li–lii). Nevertheless, he continued to put considerable weight on the fact that groups overwhelmingly had petals or lacked them; exceptions were few, and some exceptions were perhaps not as important as they seemed. Although the Campanulaceae and Ericaceae were sympetalous and might be expected to bear stamens on their corollas, the stamens were basically free. However, since their corollas were marcescent, that is, more or less persistent, they partook somewhat of the nature of the calyx.[194] The rest of this section was considerably expanded (215–218; cf. lii–liv), and it included a new paragraph (216) in which the subdivision of the three natural classes (i.e., those based on the insertion of the calyx relative to the germen) of the dicotyledons were discussed. He also provided short phrases that could be used as names for the classes (218, fn. 2). There are few changes other than that of phrasing in the remainder of the Introduction, although the footnote promising a key to all genera recognized in the *Genera plantarum* was omitted (cf. lx, fn. b).

The main change in the last part is the addition of an extensive footnote on p. 220, which effectively serves as a new introduction to the section entitled "The arrangement of orders in classes and genera in orders." This footnote is of considerable interest, and I have translated the greater part of it here:

> The arrangement of orders into classes is not arbitrary, but subject to the laws of affinity, whence more related beings ["cognatiores"] are brought together one after another. Moreover, individual species of one genus generally resemble adjacent species, not two [species] as in the image of a link of a chain, but several almost equally, each in the center of its circle, as twigs collected into a bundle, and also as the regions ["partes"] and habitations in a geographical map of a diverse region, not, however, to be compared absolutely, since the affinities of plants are not strictly to be measured as in geographical space. The same agreement that has been observed for species is apparent for genera in orders and orders in classes, these fasciculi being collected into fasces. Thence forward fasciculi, fasces, and bundles ["glomeres," properly "balls"] more correctly portray the relationships of plants than chains and links.
>
> What, however, from the preceding is true and natural, the arrangement of species, genera and orders in fasciculi or as a map, cannot be allowed in printed works, in which memoranda of single parts alternate in a serial fashion, none being omitted. Therefore, no matter which part of two is brought forward necessarily alone, it precedes the other, the other follows, both being seen as more closely related ["affinior"] than the rest, not however by unanimous consensus . . . There being doubt in printed works, a similar situation occurs in botanical gardens.

APPENDIX 2

Candolle, Jussieu, and the Théorie élémentaire

Augustin-Pyramus de Candolle (1778–1841) made a number of important generalizations about taxonomic practice in one of his articles on the Compositae that was published in 1810.[1] On first reading, I found them difficult to understand, as it seemed that Candolle was making generalizations that directly contradicted the methodological prescriptions enunciated by Antoine-Laurent de Jussieu in the Introduction to the *Genera plantarum*. The article on the Compositae and other articles associated with it are based on a paper that Candolle read in Paris at the Institut des Sciences on 1 February 1808. I shall suggest that the first article in particular can be understood only from the circumstances surrounding its delivery, in particular, the relationship between Candolle and Jussieu.

When the articles finally appeared in print, Candolle had left Paris and had taken up residence at Montpellier. Candolle had been nominated to a position as professor in the Faculty of Medicine at Montpellier on 15 January 1808—shortly before he presented at the Institute the paper on the Compositae. By March he had made the move. Soon after, he failed to be elected to the Institut in his last active attempt to gain a seat. In 1810, partly at the instigation of Cuvier, Candolle was given the chair of botany at Montpellier,[2] and in 1813 his *Théorie élé-*

mentaire de la botanique was published. Social and political unrest, however, drove him to Geneva in 1816, and there he stayed for the rest of his life; his activities made Geneva one of the most important centers for taxonomy in the whole of Europe.

Candolle had been up for election at the Institut des Sciences three times before 1808. Candolle's letters to his father, Augustin de Candolle (nearly all of which are preserved), provide the most interesting, if necessarily one-sided, background to these events, and they largely confirm the briefer account in his autobiography.[3] We now enter the world of scientific politics in Paris, in particular, politics surrounding the always highly charged elections to the Institut. To be elected a member of the Institut signified that one had reached the pinnacle of the French scientific establishment, and, equally important in those days when salaried posts in science outside universities were few, membership was associated with a salary.[4] The first time Candolle had been considered for election to one of the vacancies in the Institut was in 1799, when Jules Arman Guillaume Boucher de Crèvecour (1757–1844), Jean-Louis-Marie Poiret (1755–1834), and Antoine-Nicolas Duchesne (1747–1827) were the candidates. His name came up again in 1800, when Jacques-Julien Houttou de Labillardière (1775–1834) was elected. Candolle seems not to have actively sought election on these occasions—indeed, he was only 22 in 1800—but there was another election in 1806, and he was more actively involved then.

Candolle wrote regularly to his father, who was a very prominent citizen of Geneva, holding among other offices that of Syndic de la Garde. Augustin was to have become the Premier Syndic when revolution forced the family into exile in Champagne, and it was to the elder Candolle in Champagne that the younger directed most of his correspondence. From these letters, we can see that Candolle thought that some of his ideas on systematics would lead to fundamental changes in the discipline. We can also see that Candolle was wary of offending his potential supporters at the Institut. The letters are particularly important because they help us understand the article of 1810 and also give us a tantalizing glimpse of the ideas that Candolle considered to be central to the *Théorie élémentaire* and to his systematic work in general. They also clarify the discussion noted in a paper by Jussieu,[5] and they provide details of the personal and professional relationships between Candolle and Jussieu and other scientists. Such relationships are rarely simple, and although it would be rash to suggest that scientific politics in Paris in the period 1790–1815 were any more convoluted than elsewhere, they were certainly complex. It is worth bearing in mind that in the early 1800s Jussieu, who became Candolle's "enemy," had close ties with Cuvier, Candolle's "friend."[6]

Perhaps most importantly, the letters reveal that scientific papers during this period express, just like papers today, only a part of an argument. One may only be able to understand the contents of papers, or even to realize in the first place that there were important matters at stake, by noticing key words or turns of phrase, or knowing which papers were being read at meetings. In what follows, Augustin-Pyramus de Candolle's desire to become a member of the Institut can be shown to have an important effect in shaping his scientific discourse, both in terms of its content and the timing of various publications, and even whether a paper was put out for publication at all.

Candolle's paper on the Compositae is one of these (at first sight) enigmatic papers—how could he have said what he did? When the paper is put in context, however, we see its importance in an ongoing debate about the shape of nature. Even nonpublication, as with Candolle's work on the Dipsacaceae in 1802, may signal an event in the life of a scientist that is as important as any publication.

Candolle and the Théorie élémentaire

Augustin-Pyramus was a botanist of obvious talents, as early became evident in his association with the Société d'Arcueil.[7] By 1800 he seems to have hit on some of the central ideas of the *Théorie élémentaire*, indeed of his botanical work as a whole. When he was in Paris in 1797 he attended, and enjoyed, Cuvier's lectures on anatomy. But, as he wrote to his father, this was because he realized that anatomy had a major role to play in the new approach to classification even then apparent—Cuvier's approach, not that of Jussieu. "This anatomy perplexes my mother, it seems to me, and I assure her that it is not for pleasure that I am studying this disgusting science, but it is because the rest of natural history is based on it."[8] His mother cannot have been calmed by knowing that he went to the anatomy lessons three quarters of an hour early so that he could be close to the human corpse!

In the spring of 1800 we find Augustin-Pyramus explaining to his father his decision not to take the position that had apparently been offered to him, that of botanist on an expedition to be led by Nicolas Baudin.[9] He had two reasons for declining. One was that he wanted to extend the observations on plant physiology that he had made in France; he had a strong interest in physiological problems, and his work at the Société d'Arcueil was largely of this nature. As he said of the second reason,

> [It] is of a kind I have not yet dared to acknowledge except to myself because it has the appearance of bragging. It is however necessary to tell you in confidence. I have in mind a big plan of reform in Botany which if I can carry it out properly must quite ["autant"] change the face of the science as that of Linnaeus did. Up to now plants have been arranged in a series in such a way that each of them is next to only two others; I maintain and I prove that each plant has a more or less pronounced relationship with 3, 4, 5 or 6 others, that as a result it is necessary to make a geographical map of the plant kingdom in which each plant will be in the middle of those that resemble it. There is the heart of my theory; you will understand that to carry it out it is necessary to own a considerable herbarium so as to be able to compare different plants thousands and thousands of times the one with the others. And how to acquire this herbarium? If I do not acquire it, it will be necessary either to give up this idea which I think is important or that I spend several years in towns where herbaria are.[10]

Candolle had clearly rejected the idea of a simple, linear scala, but it is unfortunately not clear from this letter whether he thought groups were discrete. Nevertheless, even at this stage he probably inclined to the former view, partly because the analogy between a classification and a map was not new. As we have seen, it had been used by both Jussieu and Linnaeus, to mention but two authors who Candolle is certain to have read; for both Linnaeus and Jussieu the analogy suggested continuity.[11] One even wonders if Candolle's later enthusiasm for the analogy has anything to do with Jussieu's later dropping it.[12] And the matter of the herbarium was soon solved. On 16 August 1800, Charles-Louis L'Heritier de Brutelle was murdered, and Candolle acquired part of his herbarium, so, as he put it, becoming a botanist.[13]

Aside from these issues, it is clear that some of the ideas that appeared in the *Théorie élémentaire*[14] thirteen years later had already been sketched out by the time he wrote this letter. He may even have developed some of these ideas by 1798; Candolle mentions outlining a plan for an elementary book in botany at around this time.[15] As he wrote when the *Théorie élémentaire* finally appeared in 1813,

> Apart from this [his catalogue of the plants growing in the botanic garden in Montpellier[16]] I have taken it into my head to publish before the opening of my course my Élémens de Botanique which it is true have been ready for a long time but which I am giving the final polish. Of everything I have published it is this on which I place most importance and this on which I base my fame if I should have any, as under this simple title I have really given a new theory of the science. I chose this form so as not too much to startle the jealousy which to my cost I have come to know.[17]

The "Élémens de Botanique" appeared with the title, *Theorie élémentaire*, and it is clear that Candolle was concerned about the reception of his book. Soon afterward he wrote again to his father, again observing that "It was [this] work on which I had placed the greatest importance in my entire life. I have thought about it for 15 years."[18] Largely confirming the account in his autobiography, he noted that some parts of the *Théorie élémentaire* had been written in his father's house in Champagne, the Maison Bertrand, and that he had deferred publishing the book until after his hoped-for election to the Institut, because he thought that Georges Cuvier and René Louiche Desfontaines, two of his powerful supporters, might have been alienated by the ideas in it, and so would not have voted for him[19]—he implies that he had made this decision well before he failed to be elected to the Institut in 1806.

Relations Between Candolle and Jussieu

To the extent that the version of Candolle's paper on the Compositae that came out in 1810 reflects the version that he read at the Institut in 1808, some of the ideas that distinguished his approach to classification from that of Jussieu (and allied him more to Cuvier) were clear before the election: there were distinct groups of organisms; anatomy was important; there was a hierarchy of characters at the level of flowering plants as a whole. The *Théorie élémentaire* differed from the ideas of Cuvier mostly in its less-developed idea of a functional connection between characters, in its rather rigid hierarchy of characters, and in its emphasis on the desirability of understanding the types of taxonomic groups, or their symmetry, and the laws that explained change of plant form. As we have seen, attention to this latter point would increase in Candolle's later publications.[20]

From Candolle's letters to his father one finds that it was perhaps Jussieu in particular with whom he did not see eye to eye, and that the ambiguous paper on the Compositae was the occasion of one of their conflicts. Even as early as 1802 there are signs that the young and obviously rather brash Candolle (he was then 24) did not always get on with Jussieu, who was then 54. Friction occurred when Candolle discovered what he thought were the real affinities of the Dipsacaceae. He described how for no particular reason he took a scabious flower from a vase in his wife's room, happened to look at it carefully, and discovered that its relationships were not what had been previously thought.

> Last Saturday I was fooling about in Fanny's room. I took a scabious, I went to dissect it and I saw something which so far had escaped all botanists and which turned its classification upside down; the next day I went right away to show my observation to Desfontaines who found it interesting and advised me to read it immediately at the Institut; at that very moment stopping everything I put myself to prepare a little memoir and I went yesterday to present it; as I thought that I was to read in front of the greatest botanists an observation that contradicted them all, I was obviously uneasy; my memoir was indeed disputed quite vigorously and I was obliged to keep up a long discussion with M. de Jussieu; fortunately I almost brought him round to my opinion; but despite that [the long discussion] I was indeed amused by the discussion.[21]

The tone is good-natured, even breathless; Jussieu comes across as something of an old fogey who had to be humored. Unfortunately, however, Candolle's enthusiasm over the importance of his findings was premature. Jussieu and Desfontaines were assigned to prepare a report on the paper, but no report was ever published. Candolle himself never published any paper proposing realignments of the relationships of the Dipsacaceae, but in his unpublished manuscripts that were examined after his death by Alphonse de Candolle we find listed "145. Sur les rapports naturels des Dipsacées avec les Nyctaginées. Ce mémoire est faux; l'observation inexacte. Il ne doit pas être imprimé. [The memoir is wrong; the observation incorrect. It should not be printed.]"[22] And in a paper on the Nyctaginaceae that appeared the next year Jussieu devoted considerable space to discounting any relationship between the Dipsacaceae and the Nyctaginaceae, a relationship he mentioned that Candolle had suggested. He agreed that the two families had similar inflorescences, but noted that they differed in several important floral details and in seed anatomy.[23] Candolle was mistaken, and one can only speculate as to whether Candolle's misinterpretation suggested to Jussieu, and perhaps others, that Candolle was not a sound botanist.

Jussieu had long been a force to be reckoned with when it came to elections at the Institut and scientific politics in Paris in general,[24] and unfortunately for Candolle relationships between the two seem to have soured. A few years later in 1806 we find Candolle having discussions with Jussieu, whom he described as "his sanctimonious and crafty antagonist."[25] There was soon to be another election to the botanical section of the Institut, but Candolle again failed to be elected. This time he was, however, more actively involved in his candidacy. The botanical section presented two candidates, Candolle and Ambroise-Marie-François-Joseph Palisot de Beauvois (1752–1820), and the latter won by three votes. Jussieu, Étienne Pierre Ventenat, and the chemist and politician Antoine-François de Fourcroy (1755–1809) promoted Palisot's candidacy, while René Louiche Desfontaines, Labillardière, and initially Lamarck were Candolle's main botanical supporters. But Fourcroy seems to have been able to detach Lamarck from Candolle's camp. As Candolle tells the story, Lamarck admitted that Candolle's publications were far superior to those of Palisot de Beauvois, but insisted that Palisot de Beauvois was senior and so should be elected.[26] Candolle described those against him as the "vieux savans" who were mediocre even in their time—although that seems a little unfair given Palisot's trials and tribulations in the 1790s—and who were now no longer abreast of the science. He mentioned that Cuvier, Jean-Antoine-Claude Chaptal (1756–1832; an applied chemist and politician, who was apparently Fourcroy's enemy), and the mathematician Sylvestre François Lacroix (1745–1843) were also for him.[27] Of Ventenat, Candolle later had little good to say, observing rather sourly that he would use any means to get elected to the Institut;[28] Candolle also made some less than enthusiastic comments about Palisot.[29]

At about the time of the election in 1806 Candolle described a new genus in the Rubiaceae, *Cuviera*, named after Cuvier. Candolle considered this work on the Rubiaceae, which was, however, never published in its entirety, to be more "philosophic" than were his previous writings;[30] this paper and one on some parasitic fungi[31] were read at the Institut in 1806 to further his attempts to get elected.[32] As Candolle observed in his dedication of the new genus, Cuvier had shown that all branches of natural history were part of the same science, and that advances in the theory of classification in one kingdom would benefit the other;[33] Candolle then proceeded to divide the Rubiaceae into four tribes.[34] Cuvier's ideas of the four embranchements may already have been in the air; although Cuvier did not finally publish what may be considered his mature ideas on classification until 1812, ideas that depended on this quaternary division,[35] he seems to have had the idea that nature was arranged in this way from 1807, if not before.[36] However, there is no obvious evidence of groupings of four being particularly common in Candolle's other work; for example, he did not divide the Compositae into four main groups.[37]

Early in 1808 Candolle was in the process of leaving Paris for Montpellier; there seemed to be no hope of immediate advancement in the capital. On 1 February he read a paper on the Compositae at the Institut (it had been officially minuted at the previous meeting on 18 January). He described to his father what happened:

> Fifteen [sic] days ago I read a memoir at the Institut that I consider to be the best thing I have done so far, but as it included new ideas, de Jussieu and Ventenat didn't let a sentence go by without making puerile objections to me; happily they had disguised their hateful jealously badly; to each objection they were obliged to admit that they were of my way of thinking. . . . What shocked me the most in Jussieu's behavior is that before reading the memoir I had been discussing it with him and he had approved everything in detail. But I will revenge myself only by obliging him if [I can] to make the report and consequently to retract in writing what [he] said verbally.[38]

Jussieu, along with Desfontaines and Lamarck, made up the committee of three that reported on 7 March on the paper, which was on the classification of the Compositae in an extended sense;[39] this was presumably the series of three that Candolle began publishing two years later. There is, however, little mention of general ideas about classification in the report, which is largely restricted to an outline of the particular classification proposed by Candolle. The tone of the report is hardly enthusiastic. Jussieu and his colleagues observed that the Compositae in an extended sense was a family so natural that it was kept intact in artificial systems—which of course both implies that Candolle's classification was artificial and skirts the issue of what Jussieu or Candolle thought should be done with the group in a natural classification. The committee criticized Candolle's new grouping of the Labiatiflores, which included genera that had previously been placed in two families, Jussieu's Cinarocephalae (e.g., *Jungia*) and Corymbiferae (e.g., *Mutisia*); genera with scaly or naked receptacles were placed together. On the other hand, Candolle's subdivision of the Cinarocéphales (no longer a family, but a subgroup of the Tubuleuses) using the attachment of the fruit ("seed") to the receptacle was more positively received. The previous division had been based on whether or not the involucre was spiny, a trivial character; the new division was based on the seed, that is, on one of the most important organs of the plant.

Ventenat died on 13 August 1808, and there was an election on 31 October for the seat he had held. The second candidate this time was Charles-François Brisseau de Mirbel (1776–1854).[40] Of the botanists in the Institut, both Jussieu and Palisot de Beauvois were for Mirbel, or as Candolle observed in his autobiography, "ou plutôt, contre moi"; Desfontaines and Labillardière again supported Candolle, and Lamarck was again undecided. Candolle thought that the Jussiaean party was campaigning against him in private; to compound his problems, it seems that the Queen of Holland, Hortense Beauharnais, wrote letters in support of Mirbel.[41] The botanical section presented two candidates, and Candolle lost by six votes. Candolle had wanted to keep his position at Montpellier for the time being on the principle that a bird in the hand was worth two in the bush, but his prudence seems to have been presented in another light—an attempt to increase his emoluments—by his opponents, thereby alienating nine mathematicians who were in the Bureau des Longitudes.[42] However, Mirbel was a formidable opponent of an altogether greater caliber than Palisot de Beauvois had been.

After the vote, Candolle told his father that he would be revenged on the Institut through his publications; they would be so good that it would become clear that a grievous mistake had been made in not electing him.[43] Although he had failed in his attempts to be elected to the Institut while he lived in France, he considered it to be a great honor when in 1826 he was finally elected an "Associé étranger" of the Académie des Sciences (as the Institut was then called). It was, as his son Alphonse remarked, rather like being given the baton of a field marshal.[44]

Notes

Letters mentioned in the notes are held by the following institutions:

GH. Gray Herbarium at Harvard University, Cambridge, Massachusetts. Letters filed under senders' names.
G-FAC. Fondation Augustin de Candolle, Geneva. Letters filed in numbered bundles.
G-JB. Conservatoire et jardin Botaniques de la Ville de Genève. Letters filed under senders' names.
K. Royal Botanic Gardens, Kew. Letters and manuscripts filed in bound volumes:
Bentham *Corr.* (G. Bentham, Correspondence).
Bentham *Bot. Mss.* (G. Bentham, Botanical Manuscripts).
Bentham *Letters* (Bentham's letters to Sir J. D. Hooker, 1842–1884).
Bentham *Misc. Mss.* (G. Bentham, Miscellanous Manuscripts).
Bentham *Diary* (Diary of G. Bentham)
[Various Senders] *Dir. Corr.* (Directors' Correspondence).
Gray *Corr.* (Kew Correspondence, Asa Gray, 1839–1873).
W. Hooker *Letters* (Sir W. J. Hooker's Letters to Mr Bentham, 1842–1862).
P. Muséum national d'Histoire naturelle. Letters filed under senders' names.

Chapter One

1. Brongniart 1837:19; Hamy 1893:15; Hahn 1971; Stafleu 1964.
2. Jussieu's book is the starting date for conserved family-level names in plant nomenclature (Stafleu 1964:[xxiv]).
3. Morton 1981:311.
4. Cuvier 1845:298; see also Cuvier 1810:231. All translations are mine unless otherwise indicated.
5. Stearn 1961:xvi; see also Morton 1981:311. Sir J. E. Smith (1817:416–418) provides a jaundiced appreciation of Jussieu, but Smith never shifted from his early allegiance to the Linnaean system. Patrick Tort (1989:214) simply dismisses Jussieu, but he is a partisan of Michel Adanson. It is the body of theory and practice incorporated in the *Genera plantarum* that is so important over the ensuing century, and if I mention "Jussieu's" ideas, it is only to express this commonly-accepted (and in this case largely accurate) filiation. I do not compare the principles of Jussieu and those of his predecessors such as John Ray, Joseph Pitton de Tournefort, and Andrea Cesalpino; there are important connections to be made, but these are usually not of relevance to my thesis.
6. E.g., Brongniart 1837:16; Cuvier 1843:106; Flourens 1838; Sachs 1875:125–126; Stafleu 1964:[v]–[xiii], 1971:321–332; Jahn 1985:278; Williams 1988:30. See also chapter 3.
7. E.g., Outram 1986:336–337.
8. Mayr 1982a:194–195. See Sneath 1964:486–489; Guédès 1967; Burtt 1966; Nelson 1979:16–21; Bernier 1984:6–8, for sometimes contradictory analyses of Adanson's procedures.
9. Mayr 1982a:204, also p. 196.
10. Synthesis: "the putting together of parts or elements so as to make up a complex whole" (Oxford English Dictionary). Synthesis was frequently used in the sense of the gradual building up of the natural system (Bicheno 1827:492; Wallace 1855:188 [cf. Strickland 1841:189]; Bentham 1857:33; Baillon 1864:179; Sprague 1925: 9–10). There are other meanings of synthesis current in systematic literature; one is the use of all available data in constructing the natural system.
11. See Daudin 1926a:209; Stafleu 1971:326–328; Bernier 1984.
12. Larson 1971:42; Jahn 1985, and references. Naturalists who commented on the continuity of nature include John Ray (1682:[4]), Joseph Pitton de Tournefort (1694, 1: cf. pp. 2 and 13—he compared a classification as a whole to a map, individual genera to bouquets of flowers), Albrecht von Haller (1742:14) and Pyotr Simon Pallas (1766:23).

13. Lamarck 1778, 1:i–cxix, reprinted unchanged in Lamarck and A.-P. de Candolle 1805, 1:1–60.
14. Jahn 1985:282.
15. Lovejoy 1936: chap. 2. For some recent comments on Lovejoy's work, see Bynum (1975), Gram and Martin (1980), and Kuntz and Kuntz (1987).
16. Lovejoy 1936:52, emphasis in original.
17. Daudin 1926a, 1926b. He did not look at the classifications made then in any detail.
18. The word "taxon" (plural, taxa) refers to groupings of organisms at all levels of the taxonomic hierarchy—a subspecies of maples and the family Orchidaceae, with some 20,000 species, are both taxa.
19. See Anderson 1982:41–42, for a discussion.
20. E.g., Daudin 1926a:176–188.
21. Donati 1750:xix–xxi; Linnaeus 1751:27; Pallas 1766:23; Rüling 1766:chart; J. Hermann 1777; Giseke 1792:623–627 and chart facing p. 623. Diagrams of these early systems of relationships are quite frequently reproduced (Barsanti 1988:figs.3–24 is a particularly good source).
22. Cuvier 1812a and 1816, 1:xx–xxi; Mirbel 1802, 1:7; Candolle 1813b:198–203. See chapter 4.
23. Cuvier 1805, 1:63; Candolle 1813b:156.
24. Wallace 1856:195.
25. An analytic classification might recapture the natural relationships existing between the organisms and their essences if the naturalist was dealing with a classification of analyzed entities (Cain 1958).
26. There has been widespread misunderstanding of the goals and practice of early proponents of the natural method. Buffon in his anti-Linnaean polemics claimed that it was a "metaphysical error" if relationships in indivisible nature, which could be recognized only by looking at numerous characters, were suggested by emphasizing one character at the expense of the rest (Buffon 1749:21, 22). Here Buffon implies that Linnaeus's avowedly largely artificial sexual system was his best attempt at a natural series (see Daudin 1926a:135, 136; Roger 1970:577–581; Farber 1982a:21, 24 cf. Malesherbes 1798, 1:59–60, 74–78; Gusdorf 1972:292; Larson 1980:74; Lyon and Sloan 1981:22; Anderson 1982:37–38; Barsanti 1984:90–91; Stevens and Cullen 1990). Another long-standing controversy in the literature concerns the work of Michel Adanson. Cuvier suggested that Adanson had produced a series of tables in which groupings of plants held together by individual characters were evident, and had then recognized as families plants that were most often associated in these tables; this view of Adanson's practice came to be widely held

(Adanson 1764, 1:ccxxv–cccxii; cf. Cuvier 1819, 1:281–283; also Jussieu 1789:xxviii; A.-P. de Candolle 1813b:26–28, 70–72. See note 8). However, Condorcet's reaction to Adanson's work was very "neo-Adansonian," down to suggestions that characters could be tabulated by an assistant and their combinations analyzed by mechanical means (Baker 1962 and 1975:123–124). Adanson may not have put organisms that were adjacent in his numerous individual tabulations of characters in the same group, and he cannot have analyzed in his head all the characters that he tabulated, but at least some of his contemporaries caught the spirit of his proposals.

27. E.g., Conry 1987:9–61.
28. Guédès 1982:95.
29. C. C. Gillispie (1960:170) thought that taxonomy did not attract the historian of scientific ideas: "The problems were formal and practical, but the question whether classifications were artifical or natural did not ultimately prove interesting." William Stearn (1959:95–96) suggested that the concepts behind the search by Linnaeus and others for the natural system prior to 1859 were difficult to formulate, but he was inclined to think—and here he referred to A. J. Cain (1958)—that it was the search for essences.

    It is often suggested that there is a "natural history" stage in the development of scientific disciplines; this is largely descriptive and has empirical generalizations. It is the subsequent theoretical stage on which historians of science have focused (Northrop 1947: chaps. 4 and 5; Hempel 1965:119–120, see also Hesse 1974:73). Since the making of classifications is often seen as a largely empirical discipline, it is not surprising that so little attention has been paid to it. One of my theses is that the making of classifications, particularly after 1789, was not really empirical in this sense.
30. Greene 1971:9.
31. Farber 1982b:149–150, fn.11.
32. A.-L. de Jussieu 1837:230, quoting Linnaeus, e.g., 1751:27. Linnaeus never changed his mind over continuity, although John Ramsbottom (1938:217) noted Linnaeus had removed the phrase from his own copy of the *Philosophia botanica*. But Ramsbottom mistook the removal of an inadvertently repeated phrase for a fundamental change in opinion.
33. A.-L. de Jussieu 1777, 1778, and 1788.
34. As far as I know, this is the first time translations into any language have been published. An abridged version of the Introduction appeared (in French) in the *Dictionnaire des Sciences Naturelles* in 1824 (see also Flourens 1838:liii). A reprint of the *Genera plantarum*, arranged by Paul Usteri, appeared in Zurich in 1791. Usteri added a few signed com-

ments of his own and changed the order of the prefatory material; there are otherwise no differences from the original version. Jussieu worked on a second edition of the *Genera plantarum*, and his revisions of families or groups of families in the nineteenth century were its precursors, but this second edition was never published. A revised version of the Introduction with changes made by Jussieu was published posthumously in Latin (A.-L. de Jussieu 1837). Although his opinions did not shift fundamentally from those he expressed in 1789, some changes are outlined in Appendix 1.

35. But see Bentham 1857 and 1861b:147. Ernst Mayr (1982a:207–208) and others have recently emphasized this distinction.
36. Simpson 1961:115, also 114–117.
37. I discuss these distinctions further in chapter 7.
38. See also Joly 1986:53–54.
39. Simpson 1961:7.
40. Ibid., 9–11. The term "taxonomie" was coined by Augustin-Pyramus de Candolle (1813:19, 24); he used it to refer to the theory of classification.
41. Stafleu 1969a:31.
42. Griffiths 1974:90; Queiroz 1988:241.
43. A.-L. de Jussieu 1789:xxvi–xxxiv.
44. Linnaeus 1738. There is similar ambiguity in the work of Hermann Boerhaave (Sloan 1972:51, fn.137). I cannot see any difference between what Linnaeus called systems and what he called methods. He seems to have been more concerned to distinguish a synopsis, in which objects were contrasted in pairs, from a system, in which they were placed several together—the latter was more convenient (Linnaeus 1751:97–99). However, in the *Classes plantarum* (1738:441–442, cf. 485–486, see also 1751:97) he designated his admittedly artificial division of plants using stamen and pistil number "the sexual system," while at the same time he proposed "a fragment of the natural method." Buffon mentioned system and method extensively in the Premier Discours of his great *Histoire naturelle*, but did not oppose the two clearly. When he inveighed against Linnaeus and the use of single characters in circumscribing groups, he referred to system (for a translation, see Lyon and Sloan 1981:100–108). Georg Christian Oeder (1764:103) distinguished between method "per partitionem and per compositionem," but not between system and method.

    It is not easy to follow Adanson's distinction between artificial methods and systems, since the latter were also necessarily artificial; the main difference is that the principles of systems are absolute, while those of methods were "variable," i.e., characters had no fixed or absolute importance (Adanson 1764, 1:xciii–cii). He provided a "natural method" (p. clv) while at the same time he was working toward a new "system of nature" (p. cc). He also distinguished between a partial system, based on single characters (Newton was mentioned here) and universal systems, using all characters (p. cliv). Nevertheless, he thought that "the march of nature" was represented by the natural method (p. clxvi), and the "unique, universal or general" method would result only from the examination of all parts of the plant (p. clv, xcvii. Patrick Tort [1989:211–262] has emphasized the importance of what he calls the principle of exhaustion—the study of all attributes of all plants—in Adanson's approach to the study of nature; this was theoretically possible, even if not achieved in practice [ibid., 236]). In the index to volume 1 of his *Familles des plantes*, Adanson (1764:iii–xcviii, cf. 180–181) listed all the classifications of his predecessors as systems, but referred to some of them in the text as methods. In index entries to individual characters, Adanson (1763, 2:171, cf. 177) noted when they had been used by earlier botanists, so under "corole" he lists 27 systems (including that of Tournefort, although he often refers to Tournefort's method in the text), while under the entry "grandeur des plantes" we find listed 24 methods. Lamarck was also unclear. The *Flore françoise* is an example of his "method," which was both analytical and artificial; he barely mentioned either method or system when discussing his proposals for detecting the natural order (Lamarck 1778, 1:lix). Giulio Barsanti (1992:160–214) discusses systems and methods in some detail.

Confusion, or at least lack of clarity, persisted (e.g., É. L. Geoffroy 1762:ix–xvii; J. E. Smith 1798:49–162; Bicheno 1827:488–495). Yet another meaning of method that may have been in the backs of naturalists' minds in the middle of the eighteenth century are the methods of Pierre de la Ramée and Ramon Lull, methods that might help to organize thoughts (as in the Ramean Tree) or more generally, serve as an aid to memory (Yates 1966:231–242, 368–389).

45. A.-L. de Jussieu 1777:215, 219, 221. "Méthodistes" might form groups that were quite unconnected with any general series of nature.
46. Jussieu 1778:197.
47. Cournot 1872, 2:26–29; Callot 1965a:47; Morton 1981:303, fn.32. Roselyne Rey (1981) outlines the general problem of the fluid nature of the meaning of many classificatory terms at this time.
48. Potez 1909:143; see Lyon and Sloan 1981:38.
49. Descartes's immensely influential *Discours de la méthode* provided the support for the position of Fontenelle and others; Étienne Bonnot de

Condillac's *Traité des sistèmes* of 1749 ensured that "l'esprit de système" remained suspect. I return to this important issue at the end of chapter 3.

50. Anonymous 1807, 8: (Classification); C. A. Agardh 1819:65; Fries 1821:ix; Kirby and Spence 1826, 4:356; W. J. Hooker and Arnott 1855:v–vii.
51. Duhamel du Monceau 1755:v, ix and 1758, 1:xliv; Adanson 1764, 1:clxvi. For Diderot, see Callot 1965b: 295. Linnaeus (e.g., 1751:12–13) usually mentioned systematists in a negative fashion, while Jussieu (1789:xxxix) later alluded to "Systematicorum methodis"—the methods of systematists—in a neutral sense. Malesherbes (1798:39, 45) mentioned naturalists who were both "systématiques" and "méthodiques," but he did not clearly distinguish the two.
52. The brief entry "Systematists" in Rees's *Cyclopaedia* of 1817 defined the word as follows: "in Botany, those authors whose works in that science are principally employed about the arranging plants into certain orders, classes, or genera" (Anonymous 1817, 34: (Systematists). The word "systematic" and its derivatives were quite frequently used in English books (J. E. Smith 1798:95, 98 [used as an adjective, the sense not entirely clear, but not negative]; Shaw 1800, 1:tp., ix [used as an adjective, "systematic natural history"]; Whewell 1847, 1:486 [systematics]). When Kirby and Spence (1826, 4:356) used the term, they also referred to methodists, but they mentioned the sense in which they were using them—opposite to common French usage. I know little of later French usages. Mirbel (1810a:113, cf. 120, fn) used the adjective "systématique" in both a positive and negative sense. Pierre-Étienne-Simon Duchartre (1867:760) considered systematic botany and the art of description of plants to be synonymous; the word "floriste" was also used as a term of opprobrium for those who confused system with method (Le Maout 1852:xxii).

    Réjane Bernier (1975:174) suggested that the natural method became the natural system only when the idea of evolution became securely founded, thus providing the laws that she thought Adanson (for example) needed for his system, the "invariable idea" attached to its principles, but naturalists were grappling with other issues at the end of the eighteenth century.
53. Whewell 1847, 2:557; Cuvier 1816, 1:vi. Note that Cuvier here was inclining to an analytical approach to constructing the order of nature.
54. A glance at the Oxford English Dictionary (for example) shows that there were other usages of both "systematist" and "methodist" current in the late eighteenth and early nineteenth centuries, but none was common then.
55. See chapter 9.

Chapter Two

1. For comments on the circumstances surrounding the writing of the *Flore françoise*, see Burkhardt 1977:23–27. It seems unlikely that Daubenton wrote the preface (cf. Barthélemy-Madaule 1982:6).
2. The work of Lamarck has been extensively analyzed (Daudin 1926a:190–198 and 1926b, 2: chap. 10; Guédès 1971; Laurent 1975; Winsor 1976b:16–20; Burkhardt 1977:46–58, 124–126; Jordanova 1984:16–22; Williams 1988:26–29; Le Guyader 1988:37–40; Corsi 1988:40–47).
3. Lamarck 1778, 1:lxxxvii–lxxxviii, see also 1786:30. Notice that Lamarck sometimes refers to *a* natural order, not *the* natural order.
4. Ibid., lxxix–xc.
5. Ibid., cxi.
6. See also Lamarck 1788a:632. Lamarck is adopting a very Buffonian position; Buffon had also described the Linnaean sexual system as representing a metaphysical error; single characters could not give natural groups (Buffon 1749:21, 22). But Linnaeus himself did not think that the sexual system was natural (e.g., letter to Albrecht von Haller, 3 April 1737; see J. E. Smith 1821, 2:229), although he was not always clear about this; even the most natural groups in the fragment of his "natural method" did not have single, diagnostic characters (e.g., Giseke 1792:xv–xvi; also note 24, following).
7. Lamarck 1778, 1:v.
8. Ibid., xcii, emphasis in original; see also Poiret 1797:133.
9. Lamarck 1778, 1:xciii–xciv.
10. See Burkhardt 1977:63–65, for Lamarck's ideas on irritability.
11. Lamarck 1786:30.
12. Ibid., 29–30.
13. Lamarck 1778; cf. 1786:33. See Daudin 1926a:202; Legée and Guédès 1981:25. Denis Lamy (in litt.) notes that the relationships Lamarck recognized between different mosses made a "réseau touffu," a great, tangled net.
14. See Burkhardt 1977:55–58 for a discussion; Lamarck 1809: chap. 5.
15. Lamarck 1786:33.
16. As Réjane Bernier (1984:7, fn.1) observed, this is perhaps the earliest example of a mathematical calculus of affinities used by a naturalist.
17. Lamarck 1778, 1:xciv–cx.
18. See also Lamy 1992.
19. Lamarck 1778, 1:cxiii.
20. Beckner 1959:27; Sneath 1962:291. In discussions of polythetic groups

there is usually no mention of characters common to all members of a group that also occur elsewhere, albeit perhaps only rarely (e.g., Needham 1976:350–357). Borderline cases are an essential aspect of "polytypic classes" (sic), and theoretically adequate definitions of such classes are impossible (Beckner 1959:24–25), although for believers in continuity, definition and continuity would be antithetical, and any attempt at definition in this sense would not be worth making.

21. Lamarck 1785:442 and 1788a:631. See Burkhardt 1977:53.
22. Lamarck 1788a:632.
23. Lamarck 1791:xiv–xv.
24. Lamarck 1792d:470. Lamarck (1778:cx–cxvii) thought that his way of proceeding was superior to that of Linnaeus who gave names to the larger groups in his own natural arrangement, but did not describe them. This is indeed largely true, although in the *Musa Cliffortiana* Linnaeus (1736a) did briefly characterize families ("ordines") in his class Palmae. However, it seemed to Linnaeus (1751:100) that as more genera were discovered, the delimitation of classes, at least, would become more difficult. Paul Dietrich Giseke recorded a conversation that he (along with Martin Vahl and two other people of less importance for our story, Edinger and Tislef) had when a student of Linnaeus in Uppsala in 1771. In a socratic exchange, Linnaeus showed that the characters Gisecke thought defined one of the most natural of families, the Umbelliferae, either occurred elsewhere—for example, its two-seeded fruits—or did not occur in all its members—for example, the umbellate inflorescence itself (Giseke 1792:xv–xvi). Thus any simple definition of the Umbelliferae was problematical. (Hedge [1967:74–78] provides a brief account of Giseke.)
25. Poiret 1808:1. *Trifolium* had leaves with three, almost sessile leaflets, while in *Melilotus* the leaflets were petiolulate, having little stalks, the two lateral leaflets thus being a little below the apex; the inflorescence of *Trifolium* was capitate, that of *Melilotus* much more diffuse.
26. Lamarck 1778:lix–lxxxii, 1783:142–143; see also Lecomte and Leandri 1930:32–35; Voss 1952:8–10.
27. Westfall, Glen, and Panagos 1986:66. See Osborne 1963:149–152 for a mathematical analysis; Voss 1952 and Leenhouts 1966 for a history of the use of keys in biology.
28. Lamarck 1783 and 1792a.
29. Lamarck 1785:442.
30. Lamarck 1792b:362–364 and 1809: chap. 5.
31. It is difficult to evaluate the influence of the ideas Lamarck discussed so clearly in his *Flore françoise*. In the third edition that was prepared by Augustin-Pyramus de Candolle (Lamarck and Candolle 1805) the theo-

retical Discours Préliminaire was retained unchanged, despite the numerous changes and additions that Candolle made elsewhere; it is quite possible that the young Candolle was not allowed to make changes here. There were 5,000 copies, each costing 2 louis, of the third edition (A.-P. de Candolle 1862:161), and other printings in 1815 and 1829; it was a copy of the former that helped George Bentham, still a teen-ager, along the road to becoming one of the greatest botanists of the nineteenth century (Jackson 1906:16); I shall discuss Bentham's approach to classification later (see especially chapter 5).

32. E.g., A.-H.-L. de Jussieu 1843b:515–516; Maury 1864:275.

Chapter Three

1. It was not finished until after Candolle's *Théorie élémentaire* was published in 1813, appearing in installments between 1783 and 1817 (see also Guédès 1971:63–64). Lamarck's collaborators were J.-L.-M. Poiret, L. A. J. Desrosseaux, M.-J.-C.-L. de Savigny, and A.-P. de Candolle.
2. Adanson 1763, 1764. A. J. G. C. Batsch (e.g., 1786, 1787) provided descriptions for the natural families that Linnaeus had named, but had not described, that grew in the Jena area.
3. Lamarck 1783:xxxvi–xxxvii.
4. See also chapter 1. There are several excellent summaries of Adanson's life and work (Stafleu 1963, 1964, 1966a; Guédès 1967; Tort 1989:211–262); Chevalier (1934:110–122) summarizes Adanson's defense of his own originality, while Gidon (1934) explores the relationship between the work of the two Jussieus and Adanson. Bernard de Jussieu, a great teacher (Condorcet 1780:104–106), had outlined his natural arrangement of plants in April 1759 to Claude Richarde, Adanson himself, and a M. de Bombarde, a "célèbre curieux" and amateur natural historian (Stafleu 1964:[vii]), and this was before Adanson had published his book. It could thus be implied that Adanson's work was really Jussiaean (Condorcet 1780:97). Adanson had apparently written to Bernard de Jussieu from Senegal in 1750 about his new way of grouping plants; at that stage he claimed that he had read neither the work of Pierre Magnol (1689), who had been perhaps the first to propose the possibility of grouping plants and animals into families, nor that of Bernard de Jussieu (1741; see also Condorcet 1780:108–109) who had also suggested a way of arranging plants naturally. The details of Bernard's arrangement seem to have been developed as early as 1747 (Stafleu 1971:277, 321–332).
5. Milne 1770: article on Method.
6. Batsch 1787, 1:297.

7. J. E. Smith 1817:418 and 1793:2 (for his belief in continuity).
8. Jussieu himself had early realized the importance of characters taken from the fruit (A.-L. de Jussieu 1787), and the significance of Gaertner's *De fructibus et seminibus plantarum* (J. Gaertner 1788–1792; see also C. F. von Gaertner 1805–1807), appearing as the *Genera plantarum* was being printed, was immediately apparent to Jussieu, although he could only note its appearance in the appendix (Jussieu 1789:453).
9. Cuvier 1810:235 and 1845:298.
10. E.g., Jussieu 1770 and 1837:102.
11. Jussieu 1770. Antoine de Jussieu had defended a thesis with an almost identical title in 1721 (Davy de Virville 1954:66), while somewhat later, René Louiche Desfontaines also defended a thesis with a similar title ("An inter oeconomiam animalem et vegetalem: differentia, analogia"; see Laissus 1967:148). The general topic seems to have been popular at this time.
12. Laissus 1965, 1967.
13. Laissus (1967) describes this episode.
14. Jussieu was not elected to the Royal Society until 1829, over forty years after he had first come up for election (Stafleu 1964:[xii]). Sir J. E. Smith, who must have been in part responsible for this belated recognition of Jussieu's merits, castigated him as having religious scruples (unspecified) and as defending the idea of animal magnetism (Smith 1817:416). In a minority report of a commission reporting on this phenomenon, Jussieu (1784) had been open-minded as to the reality of the phenomena attributed to it (Hamy 1909; Gillispie 1980: 279–284), and Adrien de Jussieu observed that the report had made his father many enemies (a note on the copy of Jussieu's report in the library of the New York Botanical Garden). Indeed, Joseph Banks had written to Jussieu in 1788 telling him that opposition to ideas of animal magnetism among members of the Royal Society was responsible for Jussieu's failure to be elected then (O'Brian 1987:309-310). Banks died in 1820, and Smith in 1828.
15. Brongniart 1837:15, cf. Stafleu 1964:[vi]. Pascal Duris (1993) has recently shown that opposition to the Jussiaean method persisted for a surprisingly long time in France istelf, more particularly in the provinces.
16. See also Foucault 1970:70–71. François Dagognet (1970:6) started his historical study by assuming that naturalists' groupings were not conventional, but conformed to the nature of beings.
17. E.g., Jussieu 1789:xxxviii and 1824:39.
18. Daudin 1926a:209; Bernier 1984:10.
19. Buffon 1749:12–13; Bonnet 1764, 1:28–30; Pallas 1766:5, 11, 13–24. For Linnaeus, see Stevens and Cullen 1990; for Lamarck, see chapter 2;

for Treviranus, see Hoppe 1971:217–218, 224–225. A belief in continuity was commonplace at the time; as Joseph Schiller (1971:89–93) observed, continuity seemed evident in part because of the numerous larval forms that had not been linked to the adults into which they would change. Note that continuity might well be branching; thus Lenoir (1981:184) observed that Treviranus's vision of nature was of a reticulating order—a chain could be formed only by using a single character. I shall return to this problem in chapters 4 and 7.

20. Jussieu 1837:220, fn.1. See appendix 1.
21. Jussieu 1778:184.
22. Ibid., 192, 193.
23. Jussieu was one of the few members of the Académie who also belonged to the Société Royale de Médécine (Gillispie 1980:195, fn.23).
24. Hallé, in Jussieu 1789:(12).
25. See chapter 8 for the array of meanings given to this important word.
26. Jussieu 1789:xxxvi–xxxvii. See appendix 1 for an annotated translation by Susan Rosa; note that "ordinatio" is translated as "arrangement" rather than "classification."
27. I have not seen any mention of fossils in Jussieu's work, still less any discussion of the issue of extinction. In the 1780s many people still thought that organisms represented by fossils would eventually be found living somewhere on the earth, as had Antoine de Jussieu (1719:289–296), Jussieu's uncle, at the beginning of the century. Antoine-Laurent's belief in continuity made his acceptance of the idea that fossils represented extinct species unlikely; gaps in the continuity of nature would fatally compromise the search for the natural order as he saw it.
28. A.-L. de Jussieu 1789:xxxv (italics in original). Julius von Sachs (1875:131) seems unaware that Jussieu accepted this metaphor.
29. Jussieu 1777:221.
30. Buffon 1749:21–22; see Burkhardt 1977:49. Mention of the search for the philosopher's stone continued to be made (Rafinesque 1814:20; see Cain 1991:93 and Dubois-Reymond 1877:393.)
31. Jussieu 1778:178; see also Stafleu 1964:[xx].
32. Jussieu 1778:190 and 1789:xxxv.
33. Jussieu 1777:215 and 1789:xxx–xxxiii, p. lvii for identification. At around this time, other naturalists like Adanson (1764, 1:clv) and Johann Friedrich Blumenbach (see Lenoir 1981:131–132) made similar comments.
34. Jussieu 1777:217, 230.
35. Jussieu 1789:xviii.
36. Jussieu 1788:196.
37. Jussieu 1789:xlv.

38. Ibid., xxxiv. The "differentiae" of plants (p. vii et seq.) included external and internal structures.
39. Jussieu 1778:196, 197. He had just suggested that all characters were dependent on differences in the embryo! Jussieu's attitude to anatomy in animals is like that of Louis-Jean-Marie Daubenton, Buffon's collaborator (Farber 1974:67).
40. Jussieu 1778:183, 196.
41. See Jussieu 1787; for Jussieu and Gaertner, see note 8.
42. Jussieu 1789:iv–vi, xiii and fn.a, xvi, xvii. In his thesis "An oeconomiam animalem inter et vegetalem analogia?" Jussieu based his analysis of plant structure on that of Henri Louis Duhamel du Monceau, although Jussieu placed greater emphasis on the role of the medulla than did Duhamel; Linnaeus was, however, not mentioned there (Jussieu 1770:3–4; cf. Duhamel du Monceau 1758, 2:14–64).
43. Stevens and Cullen (1990) provide a detailed treatment of Linnaeus's cortex-medulla theory and its ramifications.
44. Later in his life Jussieu became less obviously "Linnaean" in his interpretation of plant construction, cf. Jussieu 1837:134 (see appendix 1).
45. Jussieu 1778:178.
46. Jussieu 1789:xxxvii–xlii.
47. Jussieu 1789:xxxix, xlii and 1824:28.
48. Jussieu 1789:xxxvii.
49. Jussieu 1777 and 1778:179.
50. Jussieu 1778:179 (italics in original); see also Condorcet 1777a:35–36.
51. Jussieu 1777:238.
52. Jussieu 1778:184, cf. 1789:xviii.
53. E.g., Jussieu 1789:xxxvii–xlii. Jussieu is apparently still discussing the first way of forming the natural series, that of synthesis.
54. Note that I am restricting the rank of class to the fifteen family groupings Jussieu recognized, not the other higher order groupings. Jussieu himself used the terms primary, secondary, and tertiary classes for the three highest levels of grouping (Jussieu 1789:xlviii–xlix).
55. Well, almost; see note 73.
56. Jussieu 1789:xli and 1824:30, 31. Jussieu's perisperm includes both perisperm in the strict sense (maternal tissue) and endosperm (tissue formed from cells with both paternal and maternal genes).
57. Jussieu 1789:xxvi–xxvii.
58. Ibid., xxxviii.
59. Jussieu 1778:196 and 1789:l. Jussieu discussed the use of several characters of lower rank together in the context of the recognition of taxa at lower levels of the hierarchy (ibid., lv).

60. E.g., Jussieu 1789:xli.
61. See chapter 1, notes 8 and 26, for various interpretations of Adanson's work.
62. Jussieu 1789:xxxix (his emphasis). I suggested earlier (Stevens 1984b:173) that many characters of low weight could prevail over a few high weight characters; this was not Jussieu's approach.
63. Or "gradus" (Jussieu 1789:xxxix).
64. Ibid., xxxix. This is the basis for the development of taxonomic instinct—see chapter 10.
65. Jussieu 1777:217 and 1789:l.
66. See chapter 5.
67. "Essentiellement existent"—Jussieu 1778:194.
68. Jussieu 1777 and 1778:179.
69. The characters of Jussieu's taxa were nominally those of their included members.
70. Jussieu 1778:178, 179; also 1789:xxxix and 1824:25–30.
71. Jussieu 1777:218–219. This suggests the much-quoted distinction later drawn by Charles-François Brisseau de Mirbel (1810a:128–129) between "familles en groupe" and "familles par enchaînement," the former having well circumscribed limits and indistinct genera, and the latter the reverse. See chapter 4.
72. Jussieu 1789:xlii–lii. Only pp. xlii–xliv specifically deal with this approach, but subsequent sections, which include discussion of characters that are progressively less important when characterizing groupings of plants, appear also to be an integral part of it.
73. Jussieu did not make this argument about lobed corolla tubes. The single cotyledons of plants like *R. glacialis* are derived from ancestors which had two cotyledons.
74. Jussieu 1789:xliii–xlvii.
75. Jussieu 1778:179, 180, 183.
76. The corculum is strictly speaking (i.e., classically) that part of the seed where the plumule, or shoot, and radicle, or root, join; Jussieu (e.g., 1789:xvii) used it to refer to the whole embryo.
77. Jussieu 1789:xliv.
78. Ibid., liv.
79. Jussieu 1777:221 and 1789:xlv.
80. Jussieu 1777:221 and 1778:180. Again, this is not a strict comparison. The whole embryo is not comparable with the heart, although the corculum *sensu classico* might be.
81. Jussieu 1778:196–197.
82. Jussieu 1789:xliv-l.

83. Jussieu called this the "support."
84. Jussieu 1777:230, 1778:185–189, and 1789:l–liii.
85. Jussieu 1789:lxii; he did not specify the aspect of the leaf blade about which he was thinking. See also appendix 1, note 183.
86. See chapter 2.
87. Jussieu 1777:216.
88. Ibid., 221; 1778:180 and 1789:xlv.
89. Jussieu 1777:230, 1778:195, 197, and 1789:xlv.
90. Jussieu 1777:217, 226, 228 and 1778:175. Another term used for describing relationships between groups was "analogie," which might refer to relationships between plants and animals as a whole (Jussieu 1777:220), as well as to those between smaller groups; the term "rapport" was also used for relationships within and between groups (Jussieu 1777:216, 225 and 1778:175, 177). But there were other usages of the word "analogie." It might specify relationships of similarity between apparently different organs ("homology" to some now), as in Jussieu's comparison of the petals of *Myosurus* with the spur-like structures of *Aquilegia* and *Helleborus*. Since they were not really different, but were analogous, variation in the petal-like organs in the Ranunculaceae at most provided generic characters (Jussieu 1777:226). "Analogie" also rarely denoted the interdependence of different parts of the flower (ibid., 231), and this returns us to one of the usages of "affinité."
91. Jussieu 1777:232–233 and 1778:186.
92. Jussieu 1789:lvii.
93. Jussieu 1777:225.
94. E.g., Jussieu 1777:238–239, 1778:175, 1788:195, 196, and 1789:lviii, lix.
95. Jussieu 1789:l.
96. Jussieu 1778:193.
97. Jussieu 1789:xxxii, xlv.
98. See Jussieu 1789:lx, fn.b, 416.
99. It has been argued that if Aristotle's species are parts of a continuum, being separated only by the more or the less, they still may have essences, the unique relationships between the organism and the environment (Lennox 1980). A. J. Cain (in litt.) observes that the angles and relative length of the sides of different triangles are continuous, yet the genus triangle has species (equilateral, etc.) with essences (see Cain 1958).
100. J. E. Smith 1813.
101. And whatever is written, there is still the question of what an author actually saw. For example, did George Bentham see, consciously or unconsciously, variation in how the leaves of *Cassia* are borne on the plant? He never mentioned this feature in his revision, but it neverthe-

less correlates very well with his classification (Dormer 1953:315, cf. Bentham 1875b).

102. A way of approaching this problem is to look at size distributions of taxa (e.g., Minelli, Fusco, and Sartori 1991; Cronk 1990; Cuerrier, Barabé, and Brouillet 1992; see also C. B. Williams 1964, chap. 6). Thus E. W. Holman (1985:37–38) compared size distributions of genera, and classifications of organisms, minerals, and diseases, in the work of Linnaeus and of François Boissier de Sauvages. Despite differences in the numbers of monotypic taxa recognized by the two authors, Holman thought that their "intuitive" classifications contained valuable scientific information. All too often such studies are carried out without knowing how the authors whose works are being analyzed interpreted their own classifications or even without much regard for the contents of the groups. I return to the issue of the statistical properties of classifications in note 134 and in chapter 8.
103. Jussieu 1778:188 and 1789:liii.
104. There is a circular element in the whole arrangement; see note 149.
105. Linnaeus (1751:37, 139) suggested that any such distinction was "lubricious."
106. Jussieu 1777:221, 237.
107. Jussieu 1837:206–209.
108. Jussieu 1778:192.
109. See note 174. Naturalists very often thought of such transitions as being degenerations (see Benichou 1983 for a summary); Lamarck (1809: chap. 6; see also Barthélemy-Madaule 1982:46–52) provides a particularly noteworthy example.
110. Jussieu 1789:xxxvii, xxxviii and 1807b:310.
111. Jussieu 1777:236.
112. Jussieu 1778:192.
113. Jussieu 1804d:20.
114. Jussieu 1789:lvi. This consideration was advanced much earlier in 1583 by Andrea Cesalpino. (For a translation, see Greene 1983, 2:815–817; note that the translation was inadvertently set up as part of the general text. See also Adanson 1764, 1:clviii).
115. Jussieu 1789:151. These genera form the core of what is now the Loganiaceae, but *Theophrasta*, now placed in the Theophrastaceae, was also placed here.
116. Jussieu 1803b:276.
117. Jussieu 1815:442.
118. Cf. Jussieu 1778:190 and 1789:lxxi. This order was a relatively late

addition, and was not mentioned in the original arrangement of plants provided by Jussieu in 1774.

119. See Walters 1961:75–77 for a discussion on this point, also Broberg 1990:69.
120. Linnaeus (1751:28–29) recognized a small family, the Nucamentaceae, which included genera like *Iva* and *Xanthium*, although most genera were included in his much larger Compositi; the two families were placed near each other in the system, but not immediately adjacent.
121. This subdivision is not original with Jussieu; for example, Sebastien Vaillant (1719) and Tournefort (1694) had recognized three groups, and John Ray recognized three (1682:70–71) or four (1703:27–34) groups.
122. Jussieu 1806a:379 and 1806b:183–184. Note that when François Joseph Lestiboudois later listed the families native to or cultivated in Belgium, he was still able to keep the total number at 100. Although he did not include some of Jussieu's families, he made seven families of the Compositae (Lestiboudois 1804: chart facing p. xi), and because of his subdivision of families like the Umbelliferae, Gramineae, and Compositae, his families range in size from 4 to 49 genera, and 90 families have fewer than 34 genera.
123. E.g., Jussieu 1824:50.
124. Jussieu 1788:193.
125. Jussieu 1807b:332. He implied that if the genera were fewer, but included more species, he would be able to keep the family within the theoretically acceptable size bound of 100 genera.
126. Jussieu 1820.
127. Jussieu 1805a:105. See also 1804a:159 and 1804d:20.
128. Jussieu 1805b:199, 208–209.
129. Jussieu 1789:xxxvii.
130. Jussieu 1807a:251–252.
131. Jussieu 1789:xlv, liv.
132. Ibid., l, liii. See also Stafleu 1964:[xxi]–[xxiii].
133. E.g., Jussieu 1824:45.
134. McNeill 1979:491–495. Jussieu's arrangement functioned as a memory system. A fixed and often small number of species per genus, etc., was a feature of such systems (Knowlson 1975:81–82); as we shall see in chapter 7, such small numbers are connected with constraints of the human mind. The Natural Method itself was something that was not "greatly fatiguing" to the imagination (Descartes 1962:14; see also Tournefort 1694:4 and Atran 1990:165).
135. Jussieu 1789:415. Bernard de Jussieu may have been inclining toward

the recognition of a fourth major grouping (see Chevalier 1934:113). The question of what Antoine-Laurent would have done had large numbers of plants with several cotyledons been discovered early in in the nineteenth century is unfortunately moot.

136. Jussieu 1824:28.
137. Ibid., 28 and 1837:200–203.
138. Note, however, that as continuous nature is subdivided, it becomes easier to find characters for groups; we saw in chapter 2 that Lamarck suggested as much. Some of these characters may by chance be coterminous with a group.
139. Jussieu 1789:lv. He discussed classes and families independently in a rather confusing fashion, and elsewhere (ibid., 228–231) returned to the problem of what the overall sequence meant.
140. Ibid., 87, 299.
141. Jussieu 1804c:423–425. The Dipsaceae he mentioned were included in the Valerianaceae by Augustin-Pyramus de Candolle, while *Opercularia* itself was later placed in the Rubiaceae by Jussieu himself. This style of discussion of relationships, with its emphasis on intermediates and linking taxa, remained a constant in his work (Jussieu 1789:237, 1807b:313, 1809a:134–135, and 1819a:233, 240–241).
142. Jussieu 1804d:23 and 1824:49.
143. See appendix 2 for the context of Jussieu's comments on the relationships of the Nyctagines.
144. The Nyctagines (Nyctaginaceae) are predominantly tropical.
145. Jussieu 1803b:278–279.
146. Cf. also the idea of chaos in Linnaeus's writings (Stevens and Cullen 1990:202–204).
147. Jussieu 1789:2.
148. *Taxus*, another conifer, was the only genus now included in the Coniferae in Jussieu's preceding, but unnamed group, a group which also included *Casuarina*.
149. Bentham (*Misc. Mss.*, p. 7, verso. K) also noted this pattern of relationships: see chapter 8 for circular arrangements in the nineteenth century. Lamarck may have alluded to a similar pattern of relationships when he suggested that general arrangements of plants often started *and* finished with organisms in which the parts of the fructification were indistinct or apparently incomplete (Lamarck 1786:30).
150. Groups three and four of the Filices, the family preceding the Naiadaceae, have both anthers and pistils, as do the Naiadaceae themselves; no other Acotyledones were recorded as having these organs (the situation in the fifth and last group of the Filices is unclear). The fourth

group includes *Pilularia* and *Lemma* (that is, *Marsilea*), both now included in the Pteridophyta, but in which anthers and pistils were carefully described by Bernard de Jussieu (1741, 1742).

151. A calyx was described as being present in all genera of the penultimate family, the Amentiferae, with the exception of *Salix* and *Myrica*, which Antoine-Laurent de Jussieu placed in the middle of the family; he was silent about its condition in *Platanus*, the last genus in the family.
152. A.-L. de Jussieu 1789:xiii.
153. Ibid., 384, 392. Jussieu also suggested that some other families of the Apetalae had structures like petals. Thus he mentioned a pseudopolypetalous corolla as occurring in the Thymelaeae, the third of the ten families in the Apetalae, and a marcescent, deciduous structure mimicking a corolla tube in the Plantagines, the antepenultimate family of the Apetalae. This latter structure he also described as being a corolla tube (ibid., lii, fn.b, and 89). Scattered genera or groups of genera elsewhere in the Dicotyledones were also described as lacking petals.
154. Jussieu's definitions of hypogyny, perigyny, and epigyny are confusing. This is partly because of his distinctive interpretation of the calyx and corolla; partly because he used different terms when he discussed the insertion of the calyx with respect to the ovary and the position of the ovary with respect to the insertion of the calyx (cf. Jussieu 1789:xii, xv); and partly because of the emphasis he placed on the insertion of the stamens on the corolla (ibid., xlvii–xlviii). This latter character was treated as being almost coordinate with the three taken from the insertion of the stamens on the ovary, although it was connected with a character of lower rank, whether or not the corolla was single or of several parts. It was nevertheless used at a very high level, and Jussieu implied that it was very important.
155. Willem Frederik Reinier Suringar (1895:14) noted the regular alternation, hypogynous/perigynous/epigynous.
156. Too much should not be made of this particular overlap, since Jussieu (1789:lv) suggested that the family adjacent to the Campanulaceae in the natural system, i.e., before the classes had been artificially increased in number, was the Onagrae (this family also has an inferior ovary). Elsewhere he noted that the Bicornes (that is, Ericaceae) were an exception to the correlation that he had observed between a monopetalous corolla and stamens being inserted on the corolla (ibid., li, fn.a). Interestingly, another exception to the correlation of fused corolla and epipetalous stamens occurs in *Statice*, the second (and last) genus of the Plumbagines, the last family of the Apetalae. The petals were usually free, but the stamens were inserted on the petals. *Statice*

was placed between *Plumbago*, which had free petals and stamens free from the corolla, and the Lysimachiae, which is a typical member of the Monopetalae, being both monopetalous and with epipetalous stamens. These plants had long been noted as being problematic (e.g., Burckhard 1750:135–136).

157. Jussieu 1805c:308.
158. *Nephelium* was included here because its spiny fruit seemed like the burrs of *Xanthium*. Jussieu did not know the plant well, and it is in fact a member of an unrelated family, the Sapindaceae.
159. Jussieu 1815:441–442.
160. Bureau 1864:10–13.
161. Jussieu 1803b:274.
162. Jussieu 1789:256–260. The core of the Aurantia is today included in the Rutaceae.
163. Genera in the first group are placed in the Elaeocarpaceae, Dipterocarpaceae and Sapindaceae; those in the second group are placed in the Olacaceae. None of these families is particularly close to the Guttiferae or Rutaceae (Aurantia), although in many family sequences the Elaeocarpaceae and the Dipterocarpaceae are placed near the Guttiferae.
164. Jussieu 1808:231–235.
165. Jussieu 1809b:411.
166. Jussieu 1823:109–110. Candolle provided incorrect citations, "Juss. Ann. mus. 17. p. 397" (Candolle 1813b:214; [properly 14. p. 397]) and "Juss. ann. mus." (Candolle 1819:244; see Jussieu 1819b:247), but Jacques Denis Choisy had first described the family.
167. E.g., Jussieu 1804b:316.
168. Jussieu 1777:236.
169. For the relative unimportance of seed number see Jussieu 1777:235. Note that he considered the single seeds of the third class to be naked, although this is now considered to be an incorrect interpretation.
170. Jussieu 1802:54.
171. Jussieu 1789:242–245, 376–383. Alphonse de Candolle (1880:86) long ago drew attention to this element of Jussieu's practice. Along the same lines, Duchesne (1766:41–47) had shown to his satisfaction how in roses and their relatives, genera with fleshy fruits differed only in the part that was fleshy; the flowers and the calyx were basically the same. But he did not extend his comparison to genera with dry fruits.
172. Given the importance Jussieu placed on the linking genera in establishing the natural series, one cannot assume that he thought these *genera affinia* were really out of place, and so did not include them in his descriptions.

173. Atran 1990:205 and fn.22.
174. For Jussieu and nomenclatural types, see chapter 6; Jussieu's general approach is compatible with modern prototype theory, also discussed in chapter 6 (see also note 99 in this chapter).
175. Jussieu 1778:187, 192 and 1789:liv.
176. I have not found any suggestions in Jussieu's writings that he thought that individual characters might gradually shade away and disappear, although Pierre Turpin (1820:12) later took this position. Jussieu (1789:xli) thought that *groups* of characters, not individual characters, shaded away. For Denis Diderot, continuity, and types, see Callot 1965b:305–308.
177. Foucault 1970:81.
178. Roscoe 1813:74–75. This was Roscoe's "Congreve rocket against the French botanists" (Lady P. Smith 1832, 2:344).
179. Jussieu 1803a:135.
180. George Bentham to Asa Gray, 9 December 1872, GH. This is in line with the Linnaean aphorism, "The character does not make the genus, but the genus the character." (Linnaeus 1751:119).
181. Cassirer 1951: chaps. 1 and 2; Knowlson 1975: chaps. 5–7.
182. For a translation, see Descartes 1962:19.
183. Stafleu 1964:[xxi]–[xxii] and 1971:327.
184. See, for example, Haldane and Ross 1931, 1:237.
185. E.g., Cassirer 1951:7–14; Paul 1980:28–37.
186. Knowlson 1975:164–166.
187. James Knowlson (1975:162–209) and Robin E. Rider (1990) discuss the general relationships between language, thought, and progress in science in late eighteenth century France; Pietro Corsi (1983) focuses more on botanists and zoologists.
188. See especially Knight (1968:53–78) and Rider (1990:114–120). Analysis, algebra, and the language of science are intimately connected in Condillac's work.
189. Gillispie 1971:382.
190. Fontenelle's "geometric spirit" had lost its popularity (Cassirer 1951:16). John E. Lesch (1990:74–82) gives a good treatment of the "geometric spirit" in classification, emphasizing particularly the work of Linnaeus and Adanson. In the Port Royal Logic, written in 1662 and many times revised and reprinted over the next two hundred years and more, mention of geometry and synthesis vastly preponderate over analysis (Arnauld 1964:309–355; cf. Dickoff and James 1964:xliii–xliv). The situation changed, but although Jean Le Rond D'Alembert (1757:637) favored analysis over synthesis, he also talked about the

"esprit géométrique," the spirit of "méthode et de justesse" (ibid., 628; see also Daubenton 1765b).

191. Condillac 1749:440–449. See also Gillispie 1971:382; Baker 1975:11–34; Knowlson 1975:165–174. Leibniz had little influence in France, with the exception of a few people such as the Marquise du Châtelet (Barber 1955:149–156; Callot 1965b:257; Lyon and Sloan 1981:20–27). Jean Baptiste Payer (1844:33) is exceptional in his mention of Leibniz's "De methodo botanico," although he dismissed it as being purely theoretical.

192. See also Cournot 1872, 2:28–29; Roger 1971:465–468; Dickoff and James 1964:xliii.

193. Jussieu 1789:ii. Adanson (1764, 1:xciii–xciv) also made some of these distinctions.

194. Cuvier 1795c:149–151; see also Mirbel 1815, 1:833. Cuvier confused the issue by comparing the latter approach (which I call analytic in chapter 5), or at least the exposition of its results, with induction. Toby Appel (1987:47) briefly relates Cuvier's work to Condillac's proposals for science; Cuvier, although accepting Condillac's general ideas, thought that mathematics was not an appropriate tool for the study of organisms (see also note 199).

195. Rider 1990:116; Lesch 1990:83–84.

196. The chemist Antoine-François Fourcroy (1794, 1:3–8) observed that chemistry was the science of synthesis, not analysis, even the analysis that yielded the elements out of which compounds were formed; he allowed, however, that both synthesis and analysis occured in nature—he was discussing synthesis in the second sense. Early in the nineteenth century there were related arguments over the relative value of induction and deduction among French mathematicians (Dhombres and Dhombres 1989:466–477); conflict between proponents of algebra and those of geometry are again involved.

197. The names given to plant groups were also discussed from this point of view (Jussieu 1777:218).

198. Linnaeus 1751:97.

199. Diderot had argued against the use of mathematics in the description of nature (Cassirer 1951:74–77).

200. See Cassirer 1951:25.

201. Cf. Le Guyader 1986:82. John Heilbron (1990:2) embraces classificatory studies in general in his concept of "the quantifying spirit" on the grounds that they were involved in ordering and systematizing. The latter is true, but there was in fact precious little real quantification involved. Ilse Jahn (1985:283) links a combinatorial approach to nature

with a belief in continuity. The former is particularly evident in the writings of Marie-Jean-Antoine-Nicolas de Caritat, Marquis de Condorcet, secretary to the Académie in succession to Fontenelle, although Linnaeus's approach to characters used to distinguish genera is also combinatorial (e.g., Linnaeus 1751:116–117; for Condorcet, see Baker 1962 and 1975:109–124). Condorcet himself had observed that only 13 of the possible 45 classes that could be derived from the characters Jussieu used at that level had yet been found (Condorcet 1778:28). Jussieu had found only some associations of characters and as a result there were gaps in the grand combinatorial of all characters (see also Malesherbes 1798, 1:22–23). Such gaps might not, however, be evidence for the existence of discrete groups, but simply a demonstration that nature could not be visualized as a completely continuous Linnaean landscape—and we see that Jussieu later moved away from this analogy. Condorcet also noted that a rigorously comparative approach would rationalize description; for each group there would be a framework ("quadre") ready to be filled (Condorcet 1777b:38; see Adanson 1777). Pietro Corsi (1988:23–39) describes a possible relationship between the thought of Condorcet and developments in comparative anatomy and crystallography.

Chapter Four

1. Ventenat 1799; Jaume Saint-Hilaire 1805. See R. L. Williams 1988:233–271.
2. Frans Stafleu (1969b:218–220) suggests that Gaertner was inspired by Adanson as well as J. G. Koelreuter.
3. R. Brown 1810b; see Mabberley 1985, chap. 9.
4. See Bentham 1866:5–7; Stafleu 1969b:218 and 1971, chaps. 6–9; Guédès 1973; Williams 1988.
5. See Green 1914:329.
6. Fries 1849:vi, vii; F. G. Planchon 1860; Marchand 1867:7–22, 48–73.
7. Merriam 1893:353. See Coleman (1964:46–48) for developments in zoological anatomy at the end of the eighteenth century.
8. Mounted lenses were commonly used by botanists after the 1830s. Alphonse de Candolle makes some interesting remarks about the use of such lenses early in the nineteenth century in a letter to George Bentham: Candolle to Bentham (24 Janv. 1879, Bentham, *Corr.* 3: 1024. K). Perhaps surprisingly, mycologists in the first quarter of the nineteenth century often paid scant attention to the microscopic details of the organisms they studied (Ainsworth 1976:68).

9. Daudin 1926b, vol. 2, chap. 8; Coleman 1964; Bourdier 1971; Jacob 1976:108–112; Outram 1984; Appel 1987; Roger 1990; Stemerding 1991:107–109, chap. 5.
10. Geoffroy Saint-Hilaire was initially Cuvier's closest associate when the latter came to Paris, and they collaborated on a number of papers (for the young Geoffroy Saint-Hilaire, see Appel 1987:27–29).
11. E.g., Outram 1984:120–121; Appel 1987:50; Bernier 1988:95.
12. The discussion of the work of the young Cuvier has focused either on how similar his approach was to that of Geoffroy Saint-Hilaire (Appel 1987:32) or whether or not he was a believer in the *scala naturae* (Bourdier 1971:522–523). But Cuvier's knowledge of Jussieu's works antedated his close contact with Geoffroy Saint-Hilaire.
13. The first use of the phrase I have come across is by Lamarck (1778, 1:xci [= cxi]); he claimed that subdivison of groups by the "pretended subordination of characters" only broke relationships.
14. Cuvier 1795a:388, 394–396, also 1795c. Only two years before, Lamarck (1792c:271) had taken the opposite tack: there were too few of the animals known to base a classification on them. Also, if the classifications of molluscs were to be based on the animals inside the shells, the collections of the shells themselves that were the pride and joy of many cabinets would become useless from the point of view of the study of natural history. The lecture Cuvier gave at the beginning of his course on comparative anatomy (Cuvier 1795c) also throws interesting light on the approach he adopted; he was quite aware of its importance (Cuvier 1805, 1:64). Lamarck himself was not really opposed to anatomical studies (e.g., Barthélemy-Madaule 1982:52).
15. In the late 1790s the two were close and visualized nature in a similar way—the papers that Geoffroy Saint-Hilaire published by himself are similar in this respect to those he published with Cuvier.
16. É. Geoffroy Saint-Hilaire and Cuvier 1795a:167–168, 171–173.
17. Cuvier 1798:4–5.
18. Ibid., 14–17; 1805, 1:45–58; 1819 (read in 1807), 1:280. See Daudin 1926b, 2:13–25, 48–68; Outram 1986:334–335.
19. Geoffroy Saint-Hilaire and Cuvier 1795a:168–169.
20. Cuvier 1795a:386–390, 1795b:434, and 1798:20. Cuvier referred to Jussieu in the second article. Hardly surprisingly, the relationship between importance and constancy tended to go in both directions, and the functional importance of characters could be deduced from this taxonomic subordination (Outram 1986:337). This leads to an inherent tension in Cuvier's work. Along the same lines, it has been suggested that Cuvier's claim that he could deduce the structure of a poorly known fos-

sil from his understanding of the functional relationships of animal parts was poorly founded; he was in fact first identifying the fossil fragment as a member of a particular group and then predicting the other features of the whole organism represented by the fossil based on the taxonomic co-occurrence of characters in that group (e.g., Coleman 1964:102–104). A similar problem bedeviled Augustin-Pyramus de Candolle.

21. Cuvier 1795a:390, 1798:17, 1816, 1:xvi, and 1819, 1:281.
22. Geoffroy Saint-Hilaire and Cuvier 1795a:166–167.
23. Cuvier 1805, 1:62–63 and 1816, 1:xiv; see Outram 1986:332.
24. Geoffroy Saint-Hilaire and Cuvier 1795a:169; Cuvier 1816, 1:vi. Bernard-Germain-Étienne de Lacepède (1795:450–451, 456) adopted a similar position. He favored the analytic method "of Bacon, Locke and Condillac" over the synthetic approach, although conceding that both were necessary.
25. Michel Foucault (1970:81) allowed that the two were not necessarily connected, but he suggested that there was no passage, no gradation, and no unity of series in Cuvier's work; I would agree only with the last proposition, at least as regards the early Cuvier.
26. Marchant 1858:246, 247. See Outram 1986:336–337. It was probably Linnaeus's bust that Cuvier had on his desk in Normandy, not that of Jussieu. Toby Appel (1987:29–34) and Dirk Stemerding (1991:149–152) provide summaries of Cuvier's early work, but do not link it to Jussieu's theory and practice. Bourdier (1971:522–523) suggests that Cuvier may have encouraged Lamarck's use of the idea of the chain of being, but he provides no documentation.
27. Cuvier to Pfaff, August 1790; see Marchant 1858:178–179.
28. Ibid., 247 (October 1791), 261 (11 March 1792).
29. E.g., Piveteau 1981:191–193. The general literature on Cuvier emphasizes this rejection of the unbranched *scala naturae* as being a landmark in his thought.
30. Cuvier 1825:266. See also Appel 1980:299; Yeo 1986:274.
31. Marchant 1858:214–215.
32. Cuvier 1805, 1:59–60.
33. Cuvier 1792:18, 19. "Squilles" are shrimps, etc. "Écrevisses" are crayfish and their ilk (A. J. Cain, pers. comm.).
34. Ibid., 25.
35. É. Geoffroy Saint-Hilaire 1796:20, 21. See also Appel 1987:28. Geoffroy Saint-Hilaire and Cuvier (1795a:164) observed that forms, particular combinations of organs, were repeated in nature.
36. Geoffroy Saint-Hilaire and Cuvier 1795b:452–454 and 1798:185–186. Geoffroy Saint-Hilaire and Cuvier link the fecundity of the tropics, the

ability of species to cross, and the diversity of both apes and parrots, alike imitators of humans. Note that Geoffroy Saint-Hilaire (1794, see Appel 1987:27) had also rejected a linear continuity, and this rejection, in an unpublished preface to a paper describing the aye-aye, *Daubentonia*, was apparently quite emphatic. He also thought that relationships between groups were multiple. There were often no major differences between groups, since nature usually avoided all abrupt jumps; exceptions were striking, as he noted in his discussion of Pallas's genus *Galeopithecus*, because then nature could escape from its own plan (Geoffroy Saint-Hilaire 1794:208). Again, the language was like that of Cuvier's, with mention of nuances and the chain of being (ibid., 200 and 1798:343).

37. Cuvier 1805, 1:60; see Stemerding 1991:151.
38. Cuvier and Geoffroy Saint-Hilaire 1795:147. Cuvier (1795a:394) even discussed the progressive nature of variation, noting that the six groups into which he divided "worms" (including crustaceans, molluscs, hydroids) could be arranged according to their "different degrees of organic perfection."
39. Cuvier 1799:440.
40. Christian Gottfried Ehrenberg singled out for criticism this kind of language in Cuvier's work (Winsor 1976b:33). Cuvier clearly had not broken with all aspects of continuity, even if he had disowned the scala (see also Appel 1980:299, fn.31).
41. Cuvier and Geoffroy Saint-Hilaire 1795:154. See chapter 5, note 218, for more on the level of study and type of procedure used.
42. Cuvier 1796:310. Cuvier collected his early work on fossil animals in his *Recherches sur les ossemens fossiles des quadrupèdes* (Cuvier 1812b). However, his attitude to extinction (and creation) was not without its ambivalences, as William Coleman (1964:136) shows. For Cuvier on fossils, see also Rudwick 1976:101–163; Outram 1984:146–160; Appel 1987:43–44: Laurent 1987:14–27 stresses Cuvier's catastrophist ideas.
43. Some of the fossils he found in Montmartre were the "intermediate" Artiodactyls (A. J. Cain, pers. comm.).
44. Cuvier 1804:223–224.
45. Cuvier 1812a:76, 79; for an early use of the word "embranchement," see Geoffroy Saint-Hilaire and Cuvier 1795a:174. Cuvier (e.g., 1816, 1:xx–xxi) believed that species were usually distinct.
46. Cuvier 1812a:83–84.
47. Cuvier 1825:264–265. See also Foucault 1970:81; Piveteau 1981:195. There were also obvious gaps in the order of nature that collecting had not filled.
48. The positive characters of Cuvier's four embranchements might be con-

sidered to represent the types of these groups. In the work on fishes that he and his collaborator, Achille Valenciennes, were carrying out toward the end of Cuvier's life, there is a well-developed typological approach. In each group, the first species or first genus is described in detail, all others being considered variations on the theme of the description, perhaps simply changes in proportion of the parts (for Cuvier on types, see Coleman 1964:98–102). The language Cuvier used is similar to that of both Buffon and Diderot (Lovejoy 1936:278; Barsanti 1992b), and it is interesting to see that these types were used in the Acanthopterygian fish in which groups were barely recognizable.

49. Lamarck 1778, 1:xci[=cxi].
50. Mirbel 1809c; see also Mirbel 1802, 1815. René Louiche Desfontaines (1807) provides an interesting analysis of Mirbel's early work.
51. Mirbel and Jolyclerc 1802–1806; Lamarck and Mirbel 1803. Mirbel wrote fourteen of the eighteen volumes of the first work and thirteen of the fifteen volumes of the second. For Mirbel's bibliography, see Payen 1858.
52. Payen 1858:48; Payer 1854:v.
53. See Mirbel 1802.
54. Mirbel 1810a, 1810b.
55. Mirbel 1810a:137–138; see Cuvier 1810:231.
56. Mirbel 1801:213 and 1802, 1:29–34. Apparently a friend and collaborator of Mirbel named Massey (about whom I have been unable to find anything) checked his results (Mirbel 1802, 1:50).
57. Mirbel 1802, 1:7. There are perhaps echoes of Adanson (1764, 2:clxiv) here: "It is in the differences shaded ["nuancées"] more or less perceptibly, and of which the totality is more obvious, that make up the gaps or spaces that one observes between beings, the lines of separation of which the number or total sum continue and are preserved constantly in the totality or ensemble of beings, although they may perhaps be mutable and changing with regard to each individual being." When he was teaching, Mirbel presented groups in a single series, the members of which were all joined by these nuances, but he explicitly adopted this approach for didactic purposes (Mirbel 1814:35).
58. Mirbel 1815, 1:484–485. He recognized only Agamous plants, Cryptogams, and Phénogams later in the same volume (ibid., 380). He was even not so emphatic that groups were discrete (ibid., 380–381).
59. Mirbel 1802, 1:15, 19, 25 and 2:21–31, 42–44. Here Mirbel seems to have reverted to Jussieu's position, but in the context of organography, not the establishment of relationships: when looking at organs, one should adopt the same approach as Nature had followed in life, pro-

gressing gradually and looking at simple structures first, only later examining more complex forms (ibid., 29).

60. Mirbel 1810a:141.
61. Ibid., 140.
62. Mirbel 1810b:214.
63. Mirbel 1810a:138, 139. Mirbel did not define the word, nor did he use it elsewhere.
64. Ibid., 136, 137. The issue of the legitimate use of analogy was even then a vexing one. Augustin-Pyramus de Candolle (1804:8, 13–34) emphasized the use of "natural analogy"—a mixture of theory, observation, and experiment—in establishing the relationships between the medicinal properties of plants and their external form. The use of analogy in a similar situation had been discussed earlier, for example by Jussieu himself (1788:196).
65. Mirbel 1810a:130–134, 137; Mirbel may have been referring to *Samolus*, a genus of the Primulaceae, which is unusual in this respect. The decision that the unnamed genus really was hypogynous is an example of typological thought.
66. Mirbel 1810b:233, 237. He apparently considered that Labiatae with two stamens actually had four stamens; for other examples of how he established the "typical" structure of the family, see pp. 232, 235, 236; chapter 6 for a discussion on types.
67. Mirbel 1810b:213.
68. Mirbel 1810a:136, 137. Mirbel, like Jussieu, included what are now called endosperm and perisperm in his perisperm.
69. Ibid., 137. Mirbel (1809b:336, 347) also made critical comments on the nature of the calyx.
70. Mirbel 1810b:231; for mention of Theophrastus, see Mirbel 1810a:111, 113, 122, 141. There is perhaps a parallel here with Cuvier's use of Aristotle as the founding father of animal anatomy (Coleman 1964:60–61). Mirbel noted that in the seventeenth and eighteenth centuries people were either anatomists or systematists, but the disciples of Jussieu must sooner or later overcome the lack of interest of the Linnaean school in anatomy (Mirbel 1810a:113, fn.; also 1809b:346–349 and 1809c:57–58 [on the use of microscopes]. See Stevens and Cullen 1990:210 for Linnaeus's attitude toward anatomy).
71. Mirbel 1810a:110, 111.
72. Mirbel 1810b:251, 252.
73. Mirbel 1815, 1:475; the book is appropriately dedicated to Cuvier. For Cassini's review, see Cassini 1826, 2:310, first printed in 1816 and reprinted in 1817.

74. Mirbel 1810b:214.
75. Ibid., 240–244; see note 66 above.
76. Mirbel 1810a:128–129. Lamarck (1783:xxxvii) suggested that there might be two kinds of families.
77. Mirbel 1815, 1:482–485. Louis Édouard Bureau (1856:27–36) provides an application of this idea, and his work is discussed in more detail in chapter 6.
78. Mirbel 1815, 1:482–483.
79. Ibid., 475–476.
80. Ibid., 833. Note that he transposed analysis and synthesis.
81. Augustin-Pyramus de Candolle lacks a good biography and a critical study of his work. Integral to both will necessarily be the remarkable series of letters he wrote to his father from the late 1790s onward; these, as well as much other important material, are preserved by the Fondation Augustin de Candolle in Geneva. Excerpts of some of these letters as they relate to the relationship between Candolle and Antoine-Laurent de Jussieu are included in appendix 2. Candolle's autobiography, published in 1862 with a few editorial notes by his son, Alphonse, is, as far as I can see, and as far as the autobiography itself goes, quite reliable, although a little edited when compared with the original manuscript (Jean-Marc Drouin, pers. comm.). Additional details of his life and work are scattered elsewhere (e.g., Sachs 1875:136–150; Miège 1979; Stevens 1984a; Le Guyader 1988:261–268; Drouin 1994). Roger de Candolle (1966) has published part of the correspondence between Candolle and Madame de Circourt.
82. For the theory and practice of classification, see Candolle 1813b, 1819, 1844; plant geography, Candolle 1820 (see Nelson 1978, for a useful analysis); morphology, Candolle 1827; and physiology, Candolle 1832. The *Prodromus systematis naturalis regni vegetabilis*, which he began after his unfinished *Regni vegetabilis systema naturale* (2 volumes, 1817–1821), was itself too ambitious an undertaking. Although 17 volumes appeared between 1824 and 1873, it never included the monocotyledons.
83. A.-P. de Candolle 1813b:198–203 and 1819 (2nd edition), pp. 228–233. References will usually be given to the first edition only; a third edition was published by Alphonse de Candolle in 1844, the few additions Augustin-Pyramus had made to the second edition being simply footnoted. For the same general argument about gaps, see also Lamarck 1809, 1:145–146 and Cuvier 1825:261–268. Jacob Gruber (1991) provides an account of the history of the classification of the duck-billed platypus.
84. This was an original, or, at the very least, decidedly unusual interpretation,

although the use of analogy itself had become almost hackneyed by this time.

85. Candolle 1804:23.
86. E.g., Candolle 1828a (chart), 1828b:82–85, pl.2, and 1829b, pl.1. Elsewhere Candolle (1826, pl.27) may have thought about relationships in a rather different way—see chapter 7.
87. Candolle 1810a:140, 141. For an early example, see R. Brown 1814:539. This approach was rejected by Mirbel.
88. Candolle 1813b:189, 194 and 1828b:14.
89. Candolle 1819:116–117 and 1827, 2:239–241.
90. Candolle 1810a, 1810b, and 1812. Some background to these papers is provided in appendix 2.
91. Candolle 1810a:141; see Léman 1810.
92. Candolle 1813b:213–220; cf. Jussieu 1789.
93. Candolle 1813b:188, quoted by C. G. D. Nees von Esenbeck 1831:58.
94. Candolle 1813b:186.
95. Atran 1985:309 and 1990:38, fn.19. See Lamarck and Candolle 1805, 4:404–408. Although the young Candolle might have suggested subdividing a narrowly circumscribed Cactaceae into several genera as a concession to Jussieu, he seems to have been an independent person; see appendix 2.
96. Candolle 1813b:189.
97. Candolle 1829b:22–24. His work on the Valerianaceae and Polygalaceae may provide additional examples.
98. Candolle 1813b:67. For an early use of "tâtonnement" in this context, see Cuvier 1798:21.
99. Candolle recognized three major divisions within the Compositae: the Cichoracées, the Labiatiflores, and the Tubuleuses. He divided the latter into the Cinarocéphales, Corimbifères, and Helianthées; these groupings were based largely on the variation shown by the corolla in disc flowers of the capitulum (Candolle 1810a:141, 142–146 and 1813b:186). Again Léman (1810) picked up the important point—Candolle's equal valuation of characters.
100. See also Candolle 1810a:141 and 1828a:14.
101. Candolle 1813b:86–90.
102. See Tesi 1982. Dorinda Outram (1986:365) rightly noted that Candolle found physiological characters were often of little importance in classification; little was known about plant physiology then.
103. Candolle 1813b:144.
104. Candolle 1827, 2:240–241.
105. For Candolle's concept of fusion, see Barabé 1978.

106. For Candolle on symmetry, see Stevens 1984a:60–68. Candolle's use of an analogy between René-Just Haüy's laws governing crystal form and plant symmetry would emphasize the discreteness of plant groups (e.g., Conry 1987:126–127); there could be no intermediate crystal shapes. The comparison between botany and crystallography, which pervades the extract quoted above, was integral to Candolle's thinking; he was using Haüy's findings on crystallography to inform his research program in botany (Stevens 1984a).
107. Candolle 1819:177–178. There is an extensive quotation from Cassini here, but I have not been able to find where it came from. (For other early references to symmetry, see Linnaeus 1738:487; Adanson 1764, 1:clv.) Candolle's ideas on symmetry are connected with those of Corrêa da Serra (1805:376–377); the latter emphasized characters that held a group together and considered similarities more important than differences.
108. Candolle 1819:116–117 and 1828a:14.
109. Candolle 1827, 2:239–241.
110. Terata were not to be ignored. There might be a theory of monstrosities along the lines of that provided by Isidore Geoffroy Saint-Hilaire for animals, but the study of terata would also illuminate the normal course of plant development (A.-P. de Candolle and A. L. de Candolle 1841:1–2)—see also chapter 6.
111. Stevens 1984a:64–66; Bernier 1984:9. For symmetry as parallelism, see chapter 8, note 19.
112. In 1800 he was working two hours a day at the microscope in his study of stomata in leaves, and he also used a microscope in his work on rusts. In his general work, however, a mounted lens was the optical equipment of choice (Candolle 1806; see also 1862:92 and 1807a:58).
113. Candolle 1810a:142.
114. Candolle 1813b:199.
115. Ibid., 192.
116. Ibid., 152.
117. Stevens 1984a:65, 77; Bernier 1984:9.
118. Candolle 1827, 1:xii, 7–8.
119. Lamarck and Candolle 1805, 1:64.
120. Candolle 1804:25; see Jussieu 1788:195.
121. Candolle 1813b:155. Karl Fuhlrott (1829:95) rightly emphasized the importance of this aspect of Candolle's work.
122. Candolle 1813b:156; see the next chapter for a discussion on analysis versus synthesis.
123. Ibid., 202.

124. A.-L. de Candolle 1880:324.
125. A.-P. de Candolle 1813b:203.
126. A.-P. de Candolle 1844:195, fn.1. He is referring to Candolle 1828a and 1828b.
127. Candolle 1827, 1:vi–viii, 2:242–244, and 1832:xiii.
128. Candolle 1813b:84.
129. See appendix 2.
130. Cuvier 1812a:73. His summary is not entirely accurate.
131. Candolle 1828a:3–4.
132. Ibid., 7, 9, 11.
133. The sequence of families as outlined in the *Théorie élémentaire* (Candolle 1813b:213–220) is notably lacking in structure; i.e., the higher level taxa provide few suggestions of internal relationships because they are little subdivided.
134. Outram 1984; Appel 1987.
135. Cuvier (1804:223–224) adopted a similar position.
136. This issue needs further study; to what extent is our interpretation of Jussieu constrained by his use of Latin?
137. A.-L. de Jussieu 1778:196–197.
138. M. P. Winsor (1976b:3) alludes to the importance of Candolle's different vision of nature.
139. Fuhlrott 1829:8, 9, 12, [85], 97–106. Fuhlrott himself rejected the idea of a continuous Chain of Being (ibid., 7); gaps between organisms might become obliterated, but new gaps appeared. He suggested that groups showed multiple relationships, and in this context he used Linnaeus' map analogy (ibid., 66).

Chapter Five

1. Appel 1980:314–316. Jussieu's contemporaries are another matter. Guédès (1973:215–216) discussed the work of two of these, one of whom, J. P. Buisson (1779), apparently believed in continuity, while the other, J. F. Durande (1781), did not. Aubin-Louis Millin de Grandmaison (1792:I), one of the two secretaries of the Société d'Histoire Naturelle de Paris and a member of the short-lived Société Linnéene de Paris, may have believed in continuity, while Antoine-Nicolas Duchesne (1766:219–220) held to some kind of a reticulum, possibly also continuous.
2. Nägeli 1847:96 (emphasis in original). See Kützing 1843:xii–xiii; Schleiden 1842, 1:263–266. For further details of this controversy, which involved polemics between followers of Schelling (including Kützing himself) and those of Hegel, see Kützing 1844; Schleiden 1842 1:7 and

1844:79–81. Zaunick (1960:17–18) and Esposito (1977) provide recent commentary, the first on Kützing himself and the second on Schelling's philosophy of nature. C. G. D. Nees von Esenbeck was also involved in the controversy.

3. Macleay 1819:161. See chapter 8 for a discussion of such systems.
4. Brongniart 1868:4; see R. Brown 1814:539.
5. Lenoir 1978:90–97. Lorenz Oken's combinatorial nature (see Mägdefrau 1992:84) might yield discrete and bounded groupings, but the pattern would depend on how he ranked his characters.
6. Willdenow 1792:148.
7. Jacquin 1785:15–18. Frans Stafleu (1971:183–191) discusses Jacquin's work, calling him a nominalist.
8. Cassel 1817:98–100.
9. A.-P. de Candolle and Sprengel 1821, 1:104–109; see also chapter 10.
10. Voigt 1806.
11. Fuhlrott 1829.
12. Kurt Sprengel (1817:294) observed that the divisions in Jussieu's system corresponded with those in nature, while Heinrich Gottfried Ludwig Reichenbach (1837:109) noted that Jussieu's work had emphasized the analogies that brought organisms together; future work would stress differences. (I am grateful to Uschi Haufstingl for translation of this passage.)
13. Jaume Saint-Hilaire 1805:lix; for a similar statement by a zoologist, see Duméril 1806:39–40.
14. Adanson 1764, 1:clviii and 1777.
15. Adanson 1764, 1:cliv. He was ambivalent as to what caused gaps in nature, or even whether or not there were gaps at all. Maybe future discoveries would fill the gaps; maybe some kind of evolution would obliterate them; or maybe there had been extinctions, with monstrous quadrupeds, fishes, plants, and a prodigious number of shelled organisms all dying out (cf. Adanson 1764, 1:clxv; see Bernier 1975:172–173).
16. See also Morton 1981:307. Only seven times in Adanson's accounts of families did he suggest that one family was related, or similar, to more than two other families, or related to families that were not immediately adjacent in the series (Adanson 1763, vol. 2).
17. Adanson 1764, 1:cxcix.
18. Ibid., clxxxviii; see also chapter 7. For a possible connection between Cartesian thought and Adanson's method, see Stafleu (1966b:[xi]).
19. Jaume Saint-Hilaire 1805:xlii–xliv. Auguste-François Saint-Hilaire 1825b:[iii] also thought that the limits of higher taxa were arbitrary.

20. Ibid., lvii. Jaume Saint-Hilaire equated "genera afinia" (sic) with "genres analogues."
21. The "spherical nettle," i.e., the nettle with spherical inflorescences, may be *Urtica pilulifera* or a similar species; an example of a nettle with panicles is *Urtica dioica.* Elsewhere Jaume Saint-Hilaire (ibid. 2:313–314) gave a rather different sequence of genera, with *Gunnera* and *Gnetum* and some other genera linking the Urticaceae and Amentiferae. His ideas are taken from Jussieu (1789:407).
22. Ibid., xvii, xxiv.
23. Ventenat was a staunch supporter of Jussieu, Stafleu and Cowan (1986, 7:701) calling his book an edited French translation of the *Genera plantarum.* His book is the companion volume to Cuvier's *Tableau élémentaire de l'histoire naturelle des animaux* (Cuvier 1798). Ventenat (1795:306–308) strongly defended Jussieu's interpretation of the calyx and corolla.
24. Ventenat 1799, 1:lviii.
25. Ibid. 1:iv.
26. Ibid. 1:xxii–lxi, 158, 218–219, and 2:223.
27. Ventenat 1794/5:191.
28. Ventenat 1799, 1:xxii, 170, 312.
29. Ventenat 1798:74.
30. The list could be lengthened. Louis-Marie Aubert Aubert du Petit-Thouars (1811:33, fn.9) clearly believed in reticulated relationships, if not a complete continuum.
31. Richard 1826:496.
32. Ibid., 501.
33. Ibid., 502 and 1834:84–86.
34. Richard 1826:501.
35. There were nine editions by 1864, and translations into English, German, and Italian. Henri Milne-Edwards (1851) wrote its zoological companion.
36. A. -H. -L. de Jussieu, 1843b:503.
37. Ibid., 529, 537–538; also 1825 (pl.29) and 1843a (pl.23.)
38. Jussieu 1843b:538.
39. Ibid., 517, 527.
40. Ibid., 528–529.
41. Ibid., 517.
42. Ibid., 529.
43. Jussieu 1823:351–352. Grady Webster (1987) provides a good treatment of the systematic history of the Euphorbiaceae; Jussieu's monograph is discussed on p. 24. Jussieu started off with six quite distinct genera that were either large or well known, and then added other genera.

44. Jussieu 1825:444 and pl.29; see chapter 7.
45. Jussieu 1843b:538.
46. Jussieu 1848:47–48. Asa Gray (1842:192) also envisioned relationships in a similar fashion; see also chapter 6. Emmanuel Le Maout (1852:58–59) agreed strongly with Jussieu's ideas, although he also used the old analogy between natural relationships and a reticulum (Le Maout 1844:165 and 1852:xxviii).
47. Jussieu 1848:48. There are several mentions of genealogy.
48. Brongniart in Brongniart, Duméril, Decaisne, and Milne-Edwards 1853:3; Decaisne 1854:390–391.
49. Jussieu 1843a.
50. For the use of floral diagrams and their implications, see chapter 6.
51. Thus relatively little of interest for our purposes can be gleaned from David Mabberley's (1985) careful and exhaustive account of Robert Brown.
52. W. J. Hooker's ideas of relationships in ferns have recently been analyzed and compared with those of some of his contemporaries and successors (Paris and Barrington 1990).
53. William Stearn (1990) has recently provided a useful account of Lindley's life; Richard Drayton (1993:207–211) comments on Lindley's work. Marion Filipiuk is editing Bentham's autobiography, while I am studying his botanical work.
54. See A.-P. de Candolle 1833a:147; Bromhead 1836:254; W. J. Hooker 1835:x; Gray 1836:xi; Wight 1840, 1:xi. This was the first time the natural system had been discussed in detail there, because in the earlier edition J. E. Smith had been responsible for the article.
55. J. E. Smith 1793:2.
56. R. Brown 1810b.
57. His important and elaborate revision of the Proteaceae (Brown 1810a) also helped to make his reputation.
58. Brown 1810b:v. See also Turrill 1963:116.
59. Smith 1811:369–370, see Mabberley 1985:157. Brown's comments would not have disturbed Smith, who seems to have believed in a continuous nature. In the *Prodromus florae novae hollandiae* Brown included *Brunonia* in his extended Goodeniaceae (Brown 1810b:589–590).
60. Brown 1811:13; see also Smith (1817:418) on the "clogging" of descriptions by exceptions. Like A. -L de Jussieu, Brown may have tended to exclude the characters of some of the "peripheral" genera of the family in his family descriptions (see A.-L. de Candolle 1880:86).
61. Brown 1811:13.
62. Brown 1810b:534–535. It may be significant that Brown described few

families that had only a single genus. Although he included only a single genus, *Pandanus*, in the Pandaneae (ibid., 340–341), he knew of the existence of a second, later described as *Freycinetia*. However, the very distinct Casuarineae (Brown 1814:571) was monogeneric.

63. Brown 1818a:429.
64. Brown (1818b:82) criticized statements that appeared to be of fact, but to him were really hypothetical, for example, that a monopetalous corolla was made up of several petals, although later in the same paper he ventured his idea of what a typical flower "complete in all its parts" might be (ibid., 89–90).
65. Bennett and Brown 1839:17, fn. Rather characteristically, the footnote occupies parts of seven large quarto pages, and itself has footnotes.
66. Brown 1822:211.
67. Ibid., 210. Brown not only circumscribed families in a manner that his successors have found satisfactory, but he also recognized relationships between them—as between the Capparidaceae, Cruciferae, Resedaceae, and Papaveraceae (Brown 1826:220)—that have stood the test of time. (But see chapter 10 for the doubtful value of this test.) Brown's recognition of groups of families is considered to be important in its marking botanists' developing ideas of relationships, although some of these groups simply reflected inequalities in the taxonomic hierarchy (Brown 1814:539).
68. Bentham 1863–1878.
69. Bentham and Hooker 1862–1883.
70. In his diaries he records at some length hearing about continuity from Benedict Prevost at Montauban in 1818; globules from the leaf of a purslane plant produced animalcules when placed in water, so plants and animals were not utterly distinct (Bentham, *Diary*, June 24th, 1818. K).
71. Bentham 1823, table 1, fn.
72. Bentham 1836a:xlvii.
73. Bentham 1857:31.
74. A.-P. de Candolle 1844:255.
75. Bentham 1845:232. Lindley (1853:xxxi) also disagreed with Candolle's ideas, calling them "utopian."
76. Bentham 1856:54. Bentham's work was cited with approval by J. D. Hooker (1856b:183, fn). Hooker himself thought that some families were linked by several genera, and even when there was only a single genus osculating between two families, that genus might be large. Hooker gives the impression that family limits were decidedly vague, even if families were not part of a complete continuum (ibid., 182–183). See also chapter 11, note 44.

77. Bentham also clearly distinguished between grouping and ranking, and he realized that taxa at the same hierarchical rank were not necessarily equivalent in any other sense (e.g., Bentham 1857). His guidelines for grouping in the *Genera plantarum*—no more than 200 families to be recognized, and all groups with at most twelve members at the next lower rank (Bentham, *Bot. Mss.* 161–168. K; see also Hooker 1859:xxxii, fn)—are compatible with both continuity and discreteness.
78. Bentham 1857:33.
79. Bentham 1858:2, 1861a:iv, and 1866:7.
80. Bentham (1847:477–478) did suggest that botanists should study outward form and relate it to anatomical structure and physiological differences, but this he thought would be a consequence of arranging plants following the natural order. This might imply that Bentham considered anatomy and external morphology to be linked, so obviating the need for a naturalist to study anatomy.
81. Lindley 1834:50–53; Gray 1861:125.
82. E.g., Lindley 1830:xvi and 1853:xxx–xxxi.
83. Lindley 1835a:42.
84. Ibid., 40.
85. Ibid., 40.
86. Lindley 1830:xvi.
87. Agassiz 1859:8–10.
88. Lurie 1960:205–208.
89. Lindley 1835a:42.
90. E.g., Lindley 1853:xxviii.
91. Lindley 1832:318.
92. Lindley 1829b:4.
93. Ibid., 3.
94. Ibid., 12–13; see also 1830:xvi and 1835a:39.
95. Arnott 1831:31.
96. Ibid., 30.
97. Ibid., 59–60.
98. Ibid., 59.
99. Ibid., 92.
100. White 1836:217–218; see Arnott 1831:30. For White's idea of history, see Wolfsohl 1991.
101. W. J. Hooker and Arnott 1855:vi. In the 8th edition, published in 1860, this verse was omitted.
102. Haworth 1821:vi. Such ideas are not so evident in his later work (Haworth 1825a, 1825b). Haworth (1768–1832) was well known both as a botanist and as an entomologist; he was elected a Fellow of the Lin

nean Society in 1798 and was a member of the council in 1823 (Gina Douglas, pers. comm.).

103. Stafleu 1971:55–56; Andrews 1980:68; Browne 1983:91–94.
104. Browne 1983:94; see also Frankel 1981.
105. E.g., Rudwick 1976; Outram 1984:142–155. The work of people like Adolphe Brongniart (1822, 1828a, 1828b–37) and Kasper Maria von Sternberg (1820–38) made it reasonable for botanists to allow that at least some fossils represented extinct plants (for the situation in the eighteenth century, see Andrews 1980, chap. 3), Brongniart (1822:205; see also Stafleu 1966b) in particular being clear that there had been extinction.
106. For some zoologists a belief in extinction necessarily entailed extant groups being distinct (Wallace 1856:206). I have not seen comparable statements in the botanical literature.
107. Browne 1981:93–94; see A.-L. de Candolle 1834.
108. Brongniart 1822:208.
109. See A.-P. de Candolle 1813b:198–203. (See also chapter 3.)
110. Mabberley 1985:212, 258, 329, 332; J. D. Hooker 1888:66.
111. R. Brown 1851.
112. Lindley and Hutton 1831, 1:ix–xxii, 93–110. Later (ibid. 2:165–173) they reevaluated the plants they had called palms, removing *Flabellaria* and *Noggerathia*, but adding other examples.
113. Ibid. 1:xvii–xx.
114. Brongniart 1846a:1397–1399; see Laurent 1987:143–147, 279–281.
115. E.g., Hooker 1856b:252–254 and 1859:xxiii.
116. Göppert 1865:196; Lindley and Hutton 1831, 1:89–90, pl.28–29.
117. Corda 1845:7.
118. Stafford 1989:221 and 1990:81–83. Hans Sloane (1707:[viii]) observed that Jamaican ferns had been found in slates in carboniferous strata in England.
119. Martius 1822a, 1825.
120. Martius 1822b. See his figs. 6 and 7 for *Lychnophora pinaster* and *L. villosissima* (and 1825:49–56). *Lepidodendron* is somewhat related to the cryptogam *Selaginella*.
121. A.-L. de Candolle 1834:99.
122. Klotzsch 1855:127. The belief that fossils alone could establish relationships that were otherwise problematic became very persistent and prevalent.
123. Turpin 1820:12.
124. Laudan 1987:78–82. Tournefort (1694:2, 14) seemed sympathetic to anatomical studies, but he also suggested that a microscope should not

be needed to see generic characters. Buffon (1749:19, see Daudin 1926a:153–156, but cf. Lyon and Sloan 1981:92) had complained about the use of microscopes and lenses that was necessary in any attempt at identification in the Linnaean system. Linnaeus considered the microscope to be a necessary instrument, but himself had allowed, albeit reluctantly, that he had examined all plants with the naked eye (Linnaeus 1735:[8]; see Stemerding 1991:57–58; Stevens and Cullen 1990:210).

125. A.-L. de Jussieu 1778:196, 197. Jussieu (1770:8) had earlier noted the greater complexity of both structure and function in animals.
126. Jussieu 1789:xxxii. In his negative comments about the user of the Linnaean sexual system needing a lens, it is possible that Jussieu was thinking of the system as an identification tool.
127. Louis-Jean-Marie Daubenton (1797:675) emphasized the value of anatomy; William Roscoe (1813:76) suggested that the study of internal characters was essential, and that because Jussieu's work was not based on such characters it was unlikely to reflect natural relationships. In general, naturalists' attitudes to anatomical studies were mixed. About this time Jean Senebier commented that Jean Étienne Guettard had called for a study of the insides of plants, but he, Senebier, thought that the outsides were still poorly known (Senebier 1775, 1:66–67). On the other hand, John Hill (1759, 1:150) observed that anatomy (type unspecified) was the basis of the natural method, because the form of external parts (equivalent to Antoine-Laurent de Jussieu's differentiae) depended on internal structure (characters). (Hill's sentiments were echoed almost a century later by Lindley [1853:xxvi]; internal structures regulated the function, growth, and very existence of the organism.) Other naturalists allowed that examination of the inside of the organism would suggest the same relationships as that of the outside (Laudan 1987:78–80).
128. Fries 1825:25–26. Fries distinguished what he called habit from external apperance; there may be a connection here with the distinction that Jussieu made between characters (internal) and differentiae (external) and the links he tried to make between the two (see chapter 3). Some botanists who worked on cryptogams extolled the value of microscope-aided studies; Johannes Hedwig (1785:[v]) thought that study of the outside provided knowledge that was "nominal," while that of the inside yielded real knowledge.
129. A.-P. de Candolle 1813b:152.
130. A.-H.-L. de Jussieu 1843a:519; see also Candolle 1827, 1:xii.
131. A.-P.-C. de Candolle 1879.

132. G. Bentham to Casimir de Candolle, October 1879, G-JB.
133. Desfontaines 1798:497.
134. Ibid., 501; see Dagognet 1970:61–63.
135. Hans Solereder (1898:2) later remarked that Mirbel had looked mostly at stem anatomy, in which there was little variation in the Labiatae.
136. Brongniart 1829.
137. Brongniart 1828a:12.
138. Gooday 1991:319–341.
139. A.-P. de Candolle 1827, 1:7–8.
140. Mohl 1831, 1: pl. E.
141. Kieser 1814, III–IV. In a remarkable attempt to show the same object at different focal planes, Joseph Duval-Jouve (1864, pl.4, figs.1, 7) provided a series of drawings on superposed pieces of transparent paper.
142. See comments by Charles Gaudichaud-Beaupré in Mirbel 1843:1235. By this time there was much discussion over phyllotactic sequences (see Montgomery 1970), which was to continue for the rest of the century. There were also sometimes acrimonious arguments over what the basic units of plant construction might be (see Cusset 1982; Rutishauser and Sattler 1985:442–443 for references); anatomists tended to be involved in such disputes.
143. The issue of sampling persisted. See Chodat 1891:134.
144. Candolle 1819:239–240.
145. Desfontaines 1798:485–486, citing work of Louis-Jean-Marie Daubenton, implied that this distinction could be drawn. Daubenton (1797) suggested that palms did not grow in width throughout their life, and that the trunk was basically formed by the leaf bases. Jean-Marc Drouin (1990) discusses Desfontaines's anatomical work.
146. Mirbel 1843, 1844; also Mohl 1831, 1:XVI–XVII. The mistaken interpretation persisted in the literature for a decade or more (e.g., Lindley 1854:20–22).
147. A.-P. de Candolle 1821b:210, 211.
148. Trécul 1845:286.
149. Richard 1822:340.
150. Duméril 1806:x–xiii. See chapter 11, note 87.
151. See Chatin 1892:[vi]. Duval-Jouve (1871:472, 476) rather unfairly made A.-L. de Jussieu into a supporter of the use of anatomy and A.-P. de Candolle an opponent.
152. It should not be forgotten that his early work on plant vegetative anatomy was comparative: he tried to understand one species in detail, and then compared it with others (Mirbel 1802, 1:50).
153. Mirbel 1809b:332, 346–349.

154. Mirbel 1815, 1:475–476.
155. Desfontaines 1798:492–493.
156. Audouin, Brongniart, and Dumas 1824:vii, xi–xii.
157. Duméril 1806:x–xiii. See chapter 11, note 87.
158. Candolle 1813b:204–205.
159. Hooker maintained that there was little value in anatomy as a general comparative tool (correspondence between Hooker and Alphonse de Candolle in 1883: G-JB; *Dir. Corr.* 135. K).
160. J. D. Hooker and Thomson 1855:15–17. Hooker was quite familiar with the use of the microscope, as is shown by his meticulously documented study of the then recently discovered *Welwitschia*, one of the most bizarre-looking plants known, which entailed a detailed analysis of its anatomy and morphology (Hooker 1863).
161. Duval-Jouve 1864:144. Auguste-François Saint-Hilaire (1819) justified using developmental studies to detect relationships.
162. E.g., Mirbel 1802, 1815; Moldenhawer 1812; Corda 1842, pl.A; Mohl 1851. G. C. Ainsworth (1976) and John Farley (1982) discuss the state of cryptogamic botany at this time.
163. Brewster 1833:68–69.
164. Brongniart, Milne-Edwards, and Valenciennes 1854:3; Brongniart 1868:12, 111; Chatin 1866:2–3. See Payer 1854–1857a.
165. Bonnet 1745, 1: fig. facing p. xxxii. See Daudin 1926a:99–103; Foucault 1973:151–152; Gould 1977:17–28; Haller 1983:890; Barsanti 1988:65–66; and especially Anderson 1976 and 1982:34–58. Joseph Schiller (1971a:102) noted a connection between preformation, continuity, and anticlassificatory attitude in Bonnet's work. For Lamarck's table, see figure 1 in chapter 2.
166. E.g., Bather 1927:lxxvii; Appel 1980:310 and refs.
167. See Anderson 1976:46, 1982:40 and refs. Anderson questions whether Bonnet believed that the chain was real.
168. Bonnet 1764, 1:28. Bonnet's concern about the absence of intermediates may also be connected with his belief in plenitude, because in the absence of a particular intermediate, a different arrangement might result (ibid., 16).
169. Anderson (1982:51–52) suggested that sometimes, at least, Bonnet did look beneath the surface of the organism in an attempt to understand it.
170. Bonnet 1764, 1:30.
171. Ibid., 28–29, emphasis in the original.
172. Wallace 1856:195, emphasis in the original. See also Darwin 1859:432 on "arrangements." Wallace (1855:188) was more favorably disposed toward synthesis.

173. Wallace 1856:206; Darwin 1859:431–432, 489. For the relationship between extinction and groups, see also Huxley 1868b:358.
174. For a translation, see Lyon in Lyon and Sloan 1981:89–128.
175. Buffon 1749:11–12. The literature on Buffon is extensive, and his views on species and discontinuities changed during his life and as he worked on different groups of animals. Daudin (1926a:137–143), Roger (1970), Mayr (1982:180–182), Sloan (1987), and Barsanti (1992b) provide an entry into the literature. Lamarck, Adanson, and Jussieu all addressed issues raised by Buffon, using his distinctive phrases, or simply quoting him.
176. Adanson 1764, 1:cliv.
177. Bentham 1845:232.
178. A.-L. de Candolle 1837b:237. Candolle (1844:254–260) suggested that groups and characters were both subordinated. Like his father, although less strongly, Alphonse thought in terms of groups having a particular symmetry or type (Candolle 1837b:228–232, 427).
179. Asa Gray (1858b:117) considered that Lindley sometimes overemphasized the idea of degradation of types; Gray himself thought that there was a change in the type. This might suggest that Asa Gray believed that there were groups in nature, but I have not come across any statements of his to this effect.
180. Comments in the entry "natural history" in the *Encyclopaedia Britannica* are also interesting; continuity seems to be both accepted and rejected (Anonymous 1817, 14:629–633; 1842, 15:739–740). The *scala naturae* itself was rejected.
181. Saint-Hilaire 1840:652–653.
182. Mirbel 1810a:120, fn.1. Ventenat was easy game for Mirbel—he was dead, and Jussieu was not. Anyhow, Mirbel softened his criticism by saying that Ventenat had seen the error of his ways; it was just that he hadn't got around to putting this in print before he died.
183. Ventenat 1799:liv–lvi, and 1794/5:193–212. Flourens (1838:xxii) also gave the impression that Jussieu's hierarchy of characters was rigid.
184. A.-L. de Candolle 1837b:221–233. On pp. 235–236 he gave an example of a computation using these weighted characters. His confidence in this hierarchy was increased since he found that even if he produced it in different ways, the characters occupied the same relative positions.
185. A.-P. de Candolle 1813b:73–77.
186. Strickland 1845:217.
187. See Dickoff and James 1964:xxii, xliii–xliv.
188. Lindley 1830:iv; see Candolle (in Lamarck and Candolle 1805, 1:64) for the same comparison; see Layton 1973:63 for further references. Tact

was also a valuable commodity in such circumstances (De La Rive 1844:353; G. Planchon 1860:102); see chapter 10, note 34, for the related idea of instinct.

189. Henslow 1837:117–122.
190. A.-L. de Jussieu 1824:47–48. Jussieu (1789:lxi) earlier suggested that the determination of the correct weightings of characters depended on the numbers of families being increased, but when he discussed the matter again in 1824, he used only Bernard de Jussieu's families, not even the 100 families he himself had recognized in the *Genera plantarum*.
191. Hooker and Thomson 1855:12; see also appendix 2 for Augustin-Pyramus de Candolle's comments to his father about the need to have access to such a collection.
192. Gray 1858b:120.
193. Grady Webster (1987:24–34) discusses this in the context of the history of the classification of the Euphorbiaceae.
194. Buchenau 1890:[1]; Dubois-Reymond 1877:393 and references.
195. J. Müller 1875:254–255, emphasis in the original. See also Baillon 1862:288–290, 1863:28–30.
196. Lindley 1844; Brongniart 1826:vi. All of Lindley's genera and three of the five that Brongniart described were considered by J. C. Willis (1973) to be synonyms. Lindley (e.g., 1853:xxvi–xxviii) tried to establish a hierarchy of characters for flowering plants in general, but because his hierarchy was heavily based on functional and anatomical attributes, he was forced to fall back on the constancy or otherwise of characters. Palisot de Beauvois (1812:v–vi, lv–lxi) proposed a "new method" in classification. Genera of grasses were vague, yet he wanted to provide good characters for them; he used the same characters to divide different groups. Such an approach leads to the appearance of parallelisms between groups—see also chapter 8.
197. A.-P. de Candolle 1826:160–161. See also Strickland 1840:224–225; Miquel 1843:37; Duby 1844:418–422; Wallace 1856:193–194; Hooker 1859:xx, fn.
198. J. É. Planchon 1847a:27 and 1846:591. A similar comment was made by Johann Reinhold Forster in 1774, albeit in the context of anthropological studies (Hoare 1976:118).
199. The family number given by Jussieu (1789:lxxii).
200. J. E. Smith 1822:211–212.
201. Brongniart 1850:8. See also Payer 1844; Clos 1869. Lamarck (1792b) was early worried that relationships might seem arbitrary; *he* thought that they were real.
202. It will be clear from the discussion at the end of chapter 4 that both

analysis and synthesis have a variety of meanings, some contradictory (see also Bernier 1984:8–9).

203. For this "a prioristic" approach, compare with Stafleu 1964:[xxi]–[xxiii] and 1971:331.
204. Carl Sigismund Kunth (1831:183–186) emphasized this aspect of Jussieu's work.
205. A.-L. de Jussieu 1824:10, 47–48. Perhaps there is an echo of Cuvier here.
206. Stafleu 1964:[xxi]–[xxii].
207. A.-P. de Candolle 1813b:150–151; the whole passage is emphasized in the original.
208. Ibid., 124–150; cf. 78–88.
209. Ibid., 155.
210. Gray 1846:134–135. See also Duméril 1806:x–xiii; Audouin, Brongniart, and Dumas 1824:x; Lindley 1853:xxvi; Hooker and Thomson 1855:15; Bentham 1861b:147.
211. Lindley (1853:xxvi–xxix) provides a detailed discussion on this point; see also Cuvier 1810:234.
212. E.g., Cassini 1826:310; Bentham 1861b:147–149. See Daudin 1926b, 2:12–62; Cain 1959a:188–204 and 1959b:236–237; Stafleu 1964: [xxi]–[xxii]; Winsor 1976b:8–11 and 1985:67–68.
213. Jussieu 1789:xxxvii, xliii.
214. Cuvier 1816, 1:xiv; Cain 1959a:204. Cuvier (1816, 1:9–11) described the subordination of characters as "method." Daudin (1926b, 2:22–26) observed that in Cuvier's emphasis of the downward movement, he disregarded the differences between the importance of organs and the value of characters.
215. Henslow 1835:137 (see also p. 136). Henslow spent some time in establishing a functional hierarchy of characters (ibid., 139–143).
216. Linnaeus 1758, 1:8. This parallels the distinction drawn between the "demonstration" and "invention" of the natural arrangement (see Epilogue, note 14).
217. Although Linnaeus thought that his genera and species of plants were in nature, this was less true of the higher groupings he recognized, except, of course, those in his fragments of a natural system (see also Cain 1958; Stevens 1991:159).
218. Sneath (1962:297) made this suggestion.
219. Daudin 1926b, 1:iv–v; Stemerding 1991:166.
220. Cuvier 1805, 1:62–63.
221. R. Brown 1814:539; Ampère 1838, 1:113; Brongniart 1868:4.
222. Sprague 1925:9–10. One might question how the human mind syn-

thesizes natural relationships at the highest taxonomic levels, which was Sprague's ultimate goal (note that he was using "synthesize" both in the sense of building up the system and also using many characters).

223. Lindley 1829a:ix; Mirbel 1810a:137. Mirbel's apparently contradictory earlier comments (e.g., Mirbel 1802, 2:21–31) bore on the issue of understanding characters, not groups.
224. Lindley 1846:xxix.
225. Ventenat 1799, 1:lvii.
226. James Ebenezer Bicheno's observation that the fragments of the natural system that were then known were almost all "framed with a view to distinguish" (Bicheno 1827:489) may also reflect his perception that the approach adopted by Ventenat and others had a decidedly analytic cast to it.
227. Adanson 1764, 1:cxcii–cxciii, see also A.-L. de Jussieu 1789:lvii.
228. Milne 1771:31.
229. Brongniart (1837:18) thought it unfortunate that Jussieu used what were basically "key" characters to subdivide his family groupings in the otherwise largely natural arrangement of the dicotyledons.
230. Lestiboudois 1781. Identification was sometimes accomplished by what were called "analytical tables." These are not to be confused with "figures analytiques," drawings of the separate parts of the flower after dissection (Richard 1834:87).
231. Duméril, influenced by both Lamarck and Lestiboudois, but more particularly by Augustin-Pyramus de Candolle, provided a key to animal groups that allowed groups to be keyed out following the sequence of their natural relationships in his "Zoologie analytique" (Duméril 1806:xvi, xix), and Étienne Geoffroy Saint-Hilaire and Georges Cuvier (1795a:171–172) suggested a similar approach.
232. Gray 1836:311.
233. Rafinesque 1814:46; 1815:103. See Cain 1991:97, 158, 235. Rafinesque (1836:8–11 and 1814:41) wanted shorter generic definitions and absolute characters, but he was not in favor of looking even at the anatomy of the seed.
234. Bentham 1875a:34–35; see also Gray 1879:330, fn.2. Bentham may have emphasized this point partly because his collaborator on the great *Flora australiensis*, Ferdinand von Mueller, was trying to combine the two.
235. See Stevens 1980:344–345 for later references to this distinction.
236. An earlier article by William Roscoe (1813), "On artificial and natural arrangements of plants: and particularly on the systems of Linnaeus and Jussieu," was also reprinted.

237. Bicheno 1827:486–487.
238. Ibid., 489–490.
239. Macleay 1829:207–209.
240. Ibid., 208. Macleay (1830:436) allowed that Fleming's binary system was analytical.
241. Bicheno 1827:488–491. He also suggested that one advantage of the natural system was that it abridged the labor of reasoning, an idea equivalent to Jussieu's desire to be able to generalize the characters of a genus or family, another remark to which Macleay (1829:202, 210) took exception.
242. Macleay 1829:204.
243. Fleming 1829. Macleay (1830:56, 136bis–137bis) nevertheless talked about the "law of continuity," emphasizing that members of a group might be separated by numerous and small or fewer and large links; a hiatus, but not a saltus, might occur in his system. Thus, the Linnaean canon, "natura non facit saltus," remained inviolate. What he meant was that there were no *unanticipated* gaps in his system.
244. Colebrooke 1829.
245. Haworth 1821; cf. 1825a, 1825b.
246. Exchanges like those between Bicheno and Macleay need more careful analysis than I have been able to give them here (DiGregorio 1982:230–236 provides some more details). Even at a superficial level, the desire for Linnaeans like Bicheno to maintain the circumscription of Linnaean groups was clearly another element complicating the argument. Bicheno (1827:486–487, 494–495) argued for stasis in the circumscription of such groups, but then had to allow that taxonomic ranks were not equivalent. Macleay (1829:201–206) suggested that the subdivision of a group made its "internal construction" more evident, and he professed bewilderment as to what Bicheno meant about ranks not being equivalent.
247. Whewell 1823:xv, xix.
248. Ibid., xix.

Chapter Six

1. For general discussions, see Sokal 1962; Farber 1976; Van der Hammen 1981:4–28; Desmond 1982; Stevens 1984b:181; Winsor 1985:63–68.
2. E.g., Cannon 1978:17; Broberg 1990:69.
3. Farber 1976. C. G. Hempel (1965: chap. 7) gives another classification of type concepts.
4. Antione-Laurent de Jussieu used this type concept toward the end of his

career, as, for example, when he mentioned that *Ardisia* was the type of the Ardisiaceae (Jussieu 1819a:230, 1819b:247). For another early usage of this concept, see the mention by John Bellenden Ker (1817: pl. 183) of the "prototype" specimens of *Aster novae-angliae* in Clayton's herbarium.

5. Classification type concepts may be considered more a part of natural history and morphological type concepts a part of natural philosophy (see also Laudan 1982). G. E. Allen (1971:160–163) distinguished between idealistic morphology and functional morphology; in Germany, the idealistic plant morphology of the early 1800s was in part replaced by a more causal-analytic and physiological approach in the 1840s (Zimmermann 1930:8–12). Chapter 9 continues this theme.
6. Idealistic and descriptive approaches to morphology have been linked, as well as experimental and functional approaches (G. E. Allen 1981:160–163).
7. Hagemann 1982:46–47; see also Baron 1931:314 for comments on "morphology."
8. Goethe 1790. See Cusset 1982:24–42; Stevens and Cullen 1990. E. S. Russell 1916 (esp. chaps. 5–7) provides a useful summary of zoological thought of this kind.
9. E.g., A.-P. de Candolle 1827, 1:547–558, and 2:248–249. G. Cusset (1982:24–39) provides a perceptive discussion of the similarities and differences in Goethean and Candollean morphology, and the influence of the two approaches (see also Guédès 1972).
10. Saint-Hilaire 1837:5, fn.2. I do not know if there are any copies of Dunal's paper in existence; they seem to have been destroyed during disturbances in Montpellier in 1819. For Turpin's "Essai" see Turpin (1820); Turpin may have developed his ideas by 1804 (Bernier 1988:169–170). I do not know anything about J. M. Pelletier. See also Link 1798:141; Brown 1818b:89 and 1822:211, fn.
11. Saint-Hilaire 1838a:437.
12. Saint-Hilaire 1823:149–152.
13. See Wigand (1846) and Germain de Saint-Pierre (1855: esp. v–vii) for good contemporary reviews. Frederick B. Churchill (1991:21) noted that in zoology, too, in the first part of the nineteenth century development-based studies had as their primary goal the detection of rules of development.
14. F. G. Pictet (1839:318–319) distinguished between these two kinds of types, as did Carl Heinrich Schultz (1847:11–21, 26–27, 234–242)—cf. his "morphologischen" or "phytodermischen Wuchstypen" and his "Verwandtschaftstypus."
15. William Montgomery (1970) provides an account of their work; Karl

Mägdefrau (1992:174–220) summarizes German morphology in the nineteenth century, and includes a discussion of development and cell theory.

16. Most commentators saw Candolle's work bearing more on an understanding of plant form than on relationships. For Brongniart, Candolle's main philosophical work in botany was not that of showing that all the appendicular organs of the plant were simply modifications of a single structure—after all, Linnaeus and Goethe had done that—but demonstrating that "the main differences in each of the organ systems depended on abortion, on reduction or increased growth, on fusion or separation of parts; modifications which might more or less alter the basic symmetry which the mind, guided by laws, succeeds in finding in the disposition of the parts of the plant; an important discovery which, in the infinite variety of forms, showed the simplicity and hardly ("presque") the variety of the basic plan, that great principle of creation." (Brongniart 1846b:8; see also Dunal 1842:24, 37; Flourens 1842:xii–xvii; Martius 1843:225–226; Daubney 1843; Anonymous 1854; Guédès 1972:254; Drouin 1994).
17. For one of the earliest students of terata as terata, Georg Friedrich von Jäger, the major class of "Missbildungen" was described as being the result of metamorphosis (Jäger 1814:252–281). The work of Thomas Hopkirk (1817) is another early example of the revival of interest in terata.
18. Moquin-Tandon, in Léotard 1893:76–77; Cusset 1982:31, 35. There are parallels with Linnaeus's understanding of nature which came not only from the fragments of the natural system he worked out, but from the continuity of form evident in his ideas on morphology and development (Stevens and Cullen 1990:211–214), although the observations of use for the one would be quite useless for the other. A zoological author who deserves mention here, being mentioned by a number of botanists, is Isidore Geoffroy Saint-Hilaire (e.g., 1832b–1836 and 1841:200–201; see Laurent 1987:471–473).
19. A.-P. de Candolle 1827, 2:238–239.
20. Moquin-Tandon 1841. Typological thought was probably most important for him at the level of establishing family limits and relationships. But continuity still caused problems for groupings; as he wrote to Auguste-François Saint-Hilaire (7 April 1847) about a new genus in the Chenopodiaceae that he had just discovered, "Now here is a plant which shows seeds belonging not only to two different genera, but to *two different sections*! I have called the new genus *Diplocarpa natura non facit saltus*!!! the more one studies natural history, the more one looks with pity at our classifications! . . ." (Léotard 1893:203).

21. Reichenbach 1828.
22. Ibid., vii–ix.
23. Reichenbach 1837:2–5, see also 19–96 for an extended discussion of metamorphosis in plants.
24. Moquin-Tandon 1841:15–16; see also Dunal 1829:19. For Moquin-Tandon's friend, Saint-Hilaire (1840:650–651), "la symétrie veritable" and "la symétrie des familles" were different; the former could be truly symmetrical. Alexandre-Henri Gabriel de Cassini (1816:337–338) suggested that the primitive type was regular, the peloric mutation (as that of *Linaria)* being an accidental return to a basic ("primitive") type, and the peloric flower a regularized flower (see note 40, also Chavannes 1833:54–65 and Stevens 1984a:69 for references). Symmetry, regularity, and type were becoming thoroughly confused, so Jean Baptiste Payer (1857b:259–262) tried to distinguish between them; he noted that for some symmetry was a regularity in the disposition of parts of a flower, for others a floral diagram or type, or more specifically, the plane of symmetry.
25. Flourens 1842:xiii–xiv, xxix–xxxi.
26. Baillon 1864:179–181. These types were simple forms. For degeneration, see chapter 3, note 109.
27. G. C. D. Nees von Esenbeck 1817:25, 58, 1831:58 ("typi universalis"), and 1836:2; Cassini 1817:15, 21 (an important paper); C. A. Agardh 1819:58; also Reichenbach 1828:vii–ix. Heinrich Robert Göppert (1864:221) also used the idea of a prototype, which was for him an organism uniting groups with different organizations. Zoologists mentioned yet other types; for example, Louis Agassiz (1850) discussed progressive types, prophetic types, embryonic types, synthetic types, even progressive embryonic and prophetic embryonic types, and Alfred Russel Wallace (1855), type and antitype or ancestor.
28. Macleay 1825:58–59.
29. Turpin 1819:pl.30, 31, 1820:pl.43. Such diagrams were used earlier to show how leaves were folded and arranged in bud (Linnaeus 1751:305–306 and pl.10; Willdenow 1792:pl.8, figs. 250–264). Brent Elliott (1993) provides a useful and well-illustrated introduction to the development of floral diagrams. The use of floral diagrams spread quickly, the idea being picked up in 1821 by Augustin-Pyramus de Candolle (1821a:pl.6). In the German-speaking world, Alexander Braun (1831:pl.32) provides interesting examples that were inspired by his studies on phyllotaxis, in particular, the arrangement of scales on a pine cone; his floral diagrams were diagrams of types (Braun's diagrams of 1831 are obviously not the first—cf. Mägdefrau 1992:162, fn.173, where he includes references to other early examples of such diagrams). In Eng-

land, John Lindley used them in his *School Botany* (1839, but cf. 1854), while the first floral diagrams in the Transactions of the Linnean Society appeared in 1837. A book that confirmed the value of floral diagrams was August Wilhelm Eichler's important *Blüthendiagramme* (1875–1878).

Some characterizations were intermediate between diagrams and formulae (e.g., Saint-Hilaire 1838b: "comparative diagram" at the end, and 1840:645–650, 795; Masters 1877); they, too, could be used to describe either the type of a group or an individual flower.

30. As Turpin (1819:429) remarked when showing that the capitulum of the Compositae could be readily derived from the umbel of the Umbelliferae by imagining the pedicels of the former disappearing, abstractions were not necessary, since nature herself had provided all the intermediates between different forms. Turpin's contemporaries considered some of his ideas to be rather odd; he was "the lost child of botany" (see Guédès 1972:269).
31. A.-H.-L. de Jussieu 1825:pl.14–29; see also A.-L. de Candolle 1837a:pl.8, fig.9.
32. Bentham 1837:pl.17, cf. Meissner 1826:pl.2–4.
33. Dunal 1829:pl.3.
34. Kunth 1833:pl.2, fig.3.
35. Cassel 1820; the book is dedicated to Quetelet.
36. Saint Pierre 1784, 2:282–283. The idea came from Jean Jacques Rousseau.
37. Seringe and Guillard 1835. Along similar lines, Ferdinand Bauer (see Lindley 1832:429–431) developed standard abbreviations to be used in illustrations. This was another idea that had limited success, although such abbreviations were used by Adrien de Jussieu (see figure 7 in chapter 5).
38. Grisebach 1854:69, 106, 134 and 1848:339. For uses of comparable formulae in zoology, see Artedi (1792) for fish. Such formulae were quite commonly used in descriptions of fish, as well as in other groups, such as insects (Redtenbacher 1843).
39. Moquin-Tandon (1841:15) discusses this point.
40. Moquin-Tandon 1832:228, 246–247; Cassini 1816:337–338; for peloria, see Linnaeus 1744a.
41. Martius 1826:218. Types usually represented the perfect form of a group (e.g., Brongniart 1850:16), and this would entail their being radially symmetrical (Cassini 1817:15).
42. I. Geoffroy Saint-Hilaire 1827:119, fn.1.
43. Moquin-Tandon 1841:20. The study of terata was integral to his own work; as he put it, terata were simply outside custom, somewhat out of the ordinary, but they were not outside nature (ibid., 116).

44. Schultz 1832:134–135.
45. Payer 1857a:718–719.
46. Schleiden and Vogel 1838:77. Of course, not all those interested in types looked at plants in this way. Thus for Grisebach (1854:56–58) the features of the type were to be taken from gross morphology, not development; he was inclined to think that chemical analyses of saps and secretions might also tell one a lot about relationships.
47. Rousseau 1785:48; Mirbel 1810a:130–134, 137; A.-P. de Candolle 1819:177–178. Grisebach (1854:56–58) also inclined to this position.
48. Timothy Lenoir (1978) discusses this distinction in detail; Karl von Goebel (1926:93–97) mentions differences of opinion between Wilhelm Hofmeister and Alexander Braun that turn on a similar distinction.
49. Candolle 1819:177–178.
50. Bureau 1864:39. A similar approach was followed by Linnaeus (1751:131; see Pennell 1930:10–11 and Stearn 1957:37–38) and Adanson (1777).
51. Bureau 1856:36–37.
52. Ibid., 27–36.
53. For the distinction between families "en groupe" and "par enchaînement," see Mirbel 1815, 1:482–485.
54. Spring 1842:7–8. He was careful to point out that his description of types did not prejudice the issue of the transformation of species.
55. Grisebach 1854:56–58; also 124, 132.
56. Brongniart 1868:103–106.
57. Baillon 1861a:323–325.
58. Saint-Hilaire and Moquin-Tandon 1830a:305–320.
59. Saint-Hilaire and Moquin-Tandon 1830b; see Jussieu 1789:236–237, 246. The temperate bias of botanical knowledge is clear here; the papilionoid flower is common in but one of the three major subdivisions of the Leguminosae, a subdivision that is very common in Europe. The Cruciferae, too, are a distinct family especially in the context of Eurocentric knowledge; they are very common and of great economic importance in Europe.
60. J. É. Planchon 1851:409.
61. Ibid., 411–412.
62. It is not clear that the latter fact is more than coincidental; I suspect it was not.
63. Ibid., 403; see chapter 11 for a translation.
64. Giseke 1792:xvii–xix.
65. The general style of discussion in his monograph on the Linaceae is also similar, and it is clear that when it came to establishing relationships,

intermediates were essential (J. É. Planchon 1846:592–593 and 1847b:591–597).

66. Whewell 1847, 2:424. Even Macleay's quinarian system could be given a gloss that brings it into line with Whewell's ideas (Colebrooke 1829).
67. Ibid., 1:494. Whewell's approach is similar to that of modern prototype theory (e.g., Berlin 1992:24–25, chap. 3).
68. Whewell 1847, 2:370–372. Others adopted similar distinctions. Carl Schultz (1832:135) distinguished between "Typenverwandtschaft," which allowed one to establish the boundaries of groupings, and "Reihenverwandtschaft," which formed the transitions between them. There is a similarity between Whewell's way of thinking about types and the later idea of "common is primitive"—the condition that occurs commonly among members of a group is the primitive condition (see Crisci and Stuessy 1980:117–119 for general references). The species at the center of a group would tend to have characters that were common in a group.

    It may not be irrelevant to note that Planchon was fluent in English, and had spent 1844–1848 at Kew as W. J. Hooker's assistant. The occasion for his departure seems to have been that he got what was considered above himself and contemplated marriage with one of John Lindley's daughters (W. J. Hooker to G. Bentham, 4 Sept. 1848, W. Hooker, *Letters*, 495–496. K; G. Bentham to W. J. Hooker, 7 Sept. 1848, *Dir. Corr.* 36:38–41. K). Unfortunately, I do not know if Planchon had philosophical as well as matrimonial inclinations when in England; his obituaries do not suggest any.
69. Whewell 1847, 1:495.
70. Sensu Farber 1976:93–95.
71. Augier 1801:v–vi; see Stevens 1983. Cuvier (1812a:81) and Adrien de Jussieu (1823:351–352) made observations that are compatible with the approach which Augier claims to have followed when producing his tree.
72. Strickland 1845:219.
73. Strickland 1844:69.
74. Farber recognized that they might also play this role, although in the context of more overtly idealistic systems of classification like the quinarian system (Farber 1976:117).
75. E.g., Gray 1850.
76. Gray 1836:311; see also A.-P. de Candolle 1828c:11.
77. Gray 1842:102, 107.
78. Gray 1842:193 and 1858b:117.
79. E.g., Kunth 1815:62–64; Reichenbach 1828:viii; Martius 1835:IV; Brongniart 1850:11–19.

80. Hooker 1856a:26 and 1853:xvi. In a letter to Charles Darwin in 1844 he expressed similar reservations (Burkhardt and Smith 1987:26–27). Alexandre Carl Heinrich Braun (1868:790–791) wondered just how a "Hauptform" or "Centrum" of a species could be recognized.
81. Hooker 1859:xxiv, fn. This position might well lead him to believe that relationships of some plants, at least, would be impossible to ascertain; Adrien de Jussieu (1848:48–49) makes much the same point.
82. What Hooker seems to be saying here is that his types—types of individual organs, rather than of organisms—did not operate across wide gaps; they depended on the establishment of continuity of form when a relationship between perhaps similar structures on different plants was being contemplated. Having seen similarity in structure through a continuum in form, he could relate the two ends, calling one a type and the other a retrogression from it. He does not seem to have thought of his type as much more than a convenience in discussion, an assertion that one end of the continuum was "up"; he does not often mention types in his general work.
83. Bentham 1871:6. If the morphology of the type was uncertain (e.g., H. L. H. Müller 1881:348–349), little would seem to be gained by using the term.
84. Schleiden and Vogel 1838:77.
85. Saint-Hilaire and Moquin-Tandon 1830b; for dédoublement, see Moquin-Tandon 1826.
86. Saint-Hilaire 1838b ("comparative diagram" at the end).
87. Farber 1976:100–113. Denis Barabé and Jacques Vieth (1979) discuss how difficult it is in general to use morphological concepts.

Chapter Seven

1. Linnaeus's belief in continuity is evident in his classificatory sequence. Studies have so far concentrated on the last major group of the Animalia, the Vermes, that Linnaeus recognized in the tenth edition of his *Systema naturae*, and within both the Insecta and the Testacea, now called Mollusca (Linnaeus's Mollusca were soft-bodied animals: Winsor 1976a; Cain 1981:153–154; Stevens and Cullen 1990:199–202, table 2). These studies show that the distribution of characters is not coterminous with the limits of groups; characters tend to "lap over" into adjacent groups, the result being a catena of variation. Thus features that might be considered both restricted to and characteristic of plants alone, features such as the possession of bark and flowers, or primordia of five year's growth contained within a single bud, are all to be found in animal groups

included in the Vermes. These groups are placed adjacent to plants in the general series.

2. E.g., A.-P. de Candolle 1810a:141; see also R. Brown 1814:539.
3. Yeo 1986:266.
4. This is another area in which we know little. There are few critical, comparative studies of systematic practice (but see Winsor 1979, where the focus is on species-level problems in animals in the first part of the nineteenth century, and 1991:105–118, where she treats generic-level problems later in the century).
5. Laurent 1975:360.
6. See also Yeo 1986:264. W. F. Bynum (1975:22) noted that the three principles involved in "the great chain of being" had no absolute or necessary connection, and the same is true of Jussieu's prescriptions.
7. Hooker 1856a:25.
8. Bureau 1864:91.
9. James Bicheno (1827:488–495) provides a particularly good discussion of this point.
10. Thus Adrien de Jussieu (1848:35, fn) noted that the systems of *Naturphilosophen* favored analysis over synthesis, hardly a mark in their favor.
11. Mirbel 1810a:128–129; Stevens 1984b:173–174, for references. J. D. Hooker (1859:vi, fn) noted that there was an ambiguity in the term "natural"—it could mean either groups of closely related genera that formed families that could be readily defined, or simply groups of genera more closely related to each other than to others. Hooker made a similar distinction between objective and subjective orders. Joseph Woods (1838:1–3) also noted an inverse relationship between the ease of recognition of genera and the naturalness of a family; Carl Schultz (1832:186) discussed similar differences.
12. Pujoulx 1803:19–20. Louis-Marie Aubert Aubert du Petit-Thouars (1811:35 and 1817:226) also considered that a linear series resulted from trying to place anomalous genera somewhere in the sequence; nature was not linear.
13. Vahl 1804, 1:xi.
14. Mirbel 1815, 1:482–485.
15. Ibid. 1:482–483.
16. Bureau 1856:37, emphasis in the original.
17. Turpin 1819:429.
18. Bentham 1845:232.
19. A.-H.-L. de Jussieu 1823:351; see Jussieu 1825:444, for "enchaînement" in the Rutaceae.
20. A.-P. de Candolle 1828a, diagram at the front; cf. De La Rive

1844:352–353; Candolle 1862:388–389. I cannot remember where I read about this—I think it was in a reminiscence by Alphonse. Most of the diagrams of relationships in Augustin-Pyramus de Candolle's papers imply continuity of relationship (see section 4), whether or not that was his intent.

21. J. É. Planchon 1851.
22. A. Milne-Edwards 1864:15.
23. Don 1832:341.
24. A.-L. de Candolle 1880:52–53. He noted, however, that a few botanists like Antoine-Laurent de Jussieu, Robert Brown, and Stephan Ladislaus Endlicher tended to exclude atypical genera from their descriptions; this was not the case, however, with the majority of botanists (ibid., 86–88; cf. Bentham, *Bot. Mss.*, p. 167. K). The elderly Bentham remembered that when he was young he had been told by Antoine-Laurent de Jussieu, himself then elderly, that good generic characters—i.e., a description—could be drawn up only when all its species had been studied (George Bentham to Asa Gray, 9 Dec. 1872, GH). Even Heinrich Friedrich Link (1824:435) emphasized the importance of taking the natural character of a genus from all its species and basing the natural genus on many characters, rather than making an essential character, using few features.
25. Arnott 1831:62.
26. E.g., Mirbel 1815, 1:499; A.-P. de Candolle 1813b:187; Cassini 1826:311; Bicheno 1827:484; Bureau 1856:37. See Linnaeus 1751:119.
27. E.g., Klotzsch 1855:127; Bentham 1871:14–16; Scott 1896:994.
28. Bentham 1873a:14.
29. Gould 1987b:195.
30. Macleay 1830:56, 136bis–137bis. The words "jump," "hiatus," and "gap" had very specific meanings (Canguilhelm, in Foucault 1970:83).
31. The paper by Lincoln Constance (1964) is a good example, and George Lawrence (1953:119) also invoked the idea.
32. A.-P. de Candolle 1826:160.
33. Schultz 1832:xxi, 135–136.
34. Bromhead 1836:247.
35. Ibid. Bromhead mentioned Whewell.
36. Drude 1879:672.
37. Earlier (Stevens 1984b:186) I emphasized the importance of a prior belief in relationships in forming groups. But if a character were considered to be of particular importance, it would largely determine group circumscription.
38. Maury 1864:108.

39. Hooker and Thomson 1855:17. This was despite Hooker's suspicion of taxonomic types. Asa Gray (1842:19) saw a similar role for anatomy, as is evident from the way in which he subdivided botany; for Robert Brown, see chapter 5.
40. Jean Baptiste Payer's *Traité d'organogénie comparée de la fleur* (1854–1857a) and Hugo von Mohl's *De palmarum structura* (1831) are classics from this period; the former in particular is still very much worth consulting. The remarkable studies on the life cycles of plants carried out by Wilhelm Hofmeister (e.g., 1851) also deserve mention here; they provided an understanding of the life cycles of land plants, and how these could be related.
41. E.g., Anonymous 1855:286; cf. Klotzsch 1855. Bicheno (1827:491) also thought that the size of a group was immaterial. The argument was a perennial one; in 1718 it was suggested that Ray's genera were too small and Tournefort's genera too big (Anonymous 1719:46).
42. Bentham, *Bot. Mss.*, pp. 161–168. K.
43. Bentham and Hooker 1862–1883. This is shown clearly by an analysis of the size classes of taxa in the *Genera plantarum* (unpublished data—I am grateful to Stuart Davies, Bill Hoffman, and Santiago Madriñán for carrying out this project). E. W. Holman (1992) has shown that in several fully worked-out classifications the mean taxon size is about 7, a number probably connected with the upper bound of 7 (plus or minus 2) of variables with a unidimensional stimulus that the human mind can deal with simultaneously (Miller 1956). The popularity of some idealistic systems based on repeating low numbers (to be discussed in the next chapter) was probably bolstered at an unconscious level simply because they were so memorable.
44. Hooker 1859:xxxii, fn. He and Bentham recognized 201 families in the *Genera plantarum* (Bentham and Hooker 1862–1883).
45. Bentham, *Bot. Mss.*, p. 162. K. There were of course other opinions on this question, and Joseph Decaisne thought that isolated groups would stay isolated (Joseph Decaisne to George Bentham, 15 February 1875, Bentham, *Corr.* 3:809. K; Bentham to Decaisne, 7 March 1875, P).
46. A.-L. de Candolle 1867a:8–10 and 1873a:313. His nomenclatural ideas may be connected with this belief. It has also been suggested that John Ray thought that most organisms had already been discovered (cf. Jenyns 1837:3–4), hence an inventory of them was feasible. Linnaeus took a similar position, perhaps because the collections from the tropics came from near centers of European expansion, and these latter had already acquired a pantropical "tramp" flora and fauna after 250 years of European voyages between them (e.g., Merrill 1954: chap. 5; Stearn 1958:193).

47. A.-L. de Candolle 1867b:301–304.
48. E.g., Brown 1810b:535–536; Don 1831:151; Klotzsch 1860:9–11.
49. E.g., Hooker 1859:xxxii, fn.
50. Bentham 1857:31.
51. This tradition was of course instituted by Jussieu himself for families, and by Linnaeus (e.g., 1737) for genera; it favored the persistence of concepts based on north temperate plants.
52. See A.-H.-L. de Jussieu 1843b:538.
53. R. Brown 1818a:429; Bicheno 1827:486–487, 494–495; Bentham 1857. Bentham (*Bot. Mss.*, pp. 171–172. K) provides a restatement of the issue in the context of the *Genera plantarum.*
54. Haworth 1819:40. For Haworth and genera, see also 1821:viii–x.
55. Haworth 1819:39–41. Linnaeus's famous admonition, "habitus occulte consulendus est" (Linnaeus 1751:117), should not be forgotten.
56. Paradoxically, in biogeographical studies carried out at this time very strong assumptions about the nature of the taxonomic hierarchy were made. Groups at the same hierarchical rank were nearly always treated as being equivalent; thus the diversity of an area could be measured by the number of genera and families that occurred in it. Janet Browne (1983, chap. 3, esp. 73–85) observed that much "botanical arithmetic," the tabulations and numerical comparisons of distributions so common in the period 1810–1855, was carried out in the absence of any ideas of how distribution patterns related to theory, and certainly these tabulations depended on taxa at the same hierarchical rank being equivalent. Only very rarely in the nineteenth century (and since) was the formal classification rejected in favor of informal groupings which differed in their circumscription, but which reflected relationships of a kind that were appropriate—as the authors saw it—for biogeographical studies. Again Bentham (1860:7–8 and 1875b:342–343) is an exception. Similarly, J. D. Hooker (1856b:182–185), in his review of Alphonse de Candolle's *Geographie botanique raisonnée*, observed that the absence of general agreement on family limits prevented any useful study of the distribution of these families (see also Hooker 1862:279). He concluded his review with an almost damning summary: "It remains to say a few words upon the general subject of Botanical Geography. It is no fault of M. de Candolle's work that we lay it down more impressed than ever with the vagueness of its principles, the inexactness of its methods, the puzzling complexity of its phenomena, and the purely speculative characters of those hypotheses upon which all inquirers base their efforts to explain its facts and develope its laws (ibid., 249)."

    The issue was not only that the study of plant distributions without a clear idea of what caused them was likely to be a thankless task, but that the very data used in a biogeographer's tabulations were suspect. Hence the "purely speculative" nature of much biogeographical explanation. It is possible that the size of the data sets used and the nature of the pattern being studied may have avoided some problems (see Sepkoski 1993 for an analogous claim).

57. Miers 1845:365–368.
58. Miers 1851:173, 176.
59. A.-H.-L. de Jussieu 1825:pl.29.
60. Roger 1989:381; Ritvo 1990:265, 271; see also Yeo 1991:44.
61. E.g., A.-P. de Candolle 1813b:213; Saint-Hilaire 1827:121; Lindley 1835a:4; Gray 1837:297; Arnott 1831:92.
62. See Le Grand (1988:260–262) for a brief discussion of this topic in the context of geology. Pietro Corsi (1983) describes models and analogies used in the reform of natural history at the end of the eighteenth century, emphasizing the importance of crystallography (see also Stevens 1984a). Several works contain illustrations of the diagrams used by naturalists to depict ideas of relationship (Nelson and Platnick 1981 [the diagrams are interpreted as branching diagrams of a particular sort; it is not clear that the authors of the diagrams would have been happy with the interpretations offered]; Stevens 1984b; Barsanti 1988, 1992a).
63. For the latter half of the nineteenth century, see chapter 11. The situation in botany is apparently different to that in ornithology where R. J. O'Hara (1988b and 1991:256) has recognized quinarian, map-making, and evolutionary periods, respectively 1819–1840, 1840–1859, and 1859 onward.
64. Brongniart 1868:5.
65. O'Hara 1991:270. Buffon used the metaphors of chain, map, *and* tree throughout his great *Histoire naturelle* (Barsanti 1992b: esp. fig. 25); I have not found quite this degree of catholicism in any of the people I mention here.
66. See also A.-L. de Candolle 1841.
67. See also A.-L. de Candolle 1880:324. It would be interesting to explore the relationship between the use of geographical analogies for the natural system and the possibility of transferring that geography to plantings in a garden. Desfontaines and Antoine-Laurent de Jussieu "planted out" their developing ideas of relationships in the Jardin de l'École at the Muséum in Paris (Pujoulx 1803:20; for contemporary discussion see Le Maout 1844:871; Brongniart 1850:7), and Bernard de Jussieu had earlier put gardens to a similar use (Gidon 1934). L. H. J. Smith (1852, chap. 13) discussed the possibility of planting arboreta following the arrangement of a natural system rather than a linear sequence.

68. Ventenat 1799, 1:iv; see also Petit-Thouars 1811:38.
69. Fries 1821:xvii.
70. Weddell 1856:40–41, also Baillon 1858:253–254, 279.
71. A.-L. de Jussieu 1824:39. This analogy is less prominent in the Introduction to the *Genera plantarum*, but in the posthumous edition it replaced the Linnaean landscape (Jussieu 1837:220, fn.1; see appendix 1).
72. Giseke 1792, chart facing p. 623; see Ventenat 1799, 1:lviii. Note that Giseke's diagram conveys both an idea of the size of plant families and the relationship between them, and in conveying both these aspects, it is distinctly unusual for its time (cf. Tufte 1983:44; I am grateful to David Hendler for drawing my attention to this reference).
73. A.-P. de Candolle 1828a:4–5, fig.1; 1828b:82–85, pl.2; and 1829b, pl.1.
74. Wallace 1855:187.
75. Delise 1822:24–25 and 1825 (figure at beginning); see also Cassini 1816–1817:pl. 83; Choisy 1833:101–105; Plée 1842 (fig. facing p. 52, Famille des Graminées:[5]); Gray 1842:196; Le Maout 1844:867; J. É. Planchon 1847a:31 and 1847b:597; Miers 1852:129. Thomas Castle (1833) provided a mobile, circular "planisphere" of plant groups; this was largely a tool to aid identification and comparison of different systems (Jim Reveal provided this information). Note that George Bentham did not accept circular arrangements (cf. Stevens 1984b:175; Barsanti 1992a:69, fn. 106).
76. It is unlikely that the circles form an element of a larger circle that represents the next higher rank in the hierarchy (cf. Macleay's quinarian system; see chapter 8).
77. Dunal 1817 (chart).
78. A.-H.-L. de Jussieu 1825:pl.29. Jussieu (1843a:pl.23) provided a more complex diagram, including information both on the distribution of individual characters and on geographic distribution.
79. Rüling 1793 (chart at end); see also Batsch 1802:vi–x, chart at end; Link 1824:428–431; Fenzl 1841:263.
80. Lindley 1846. This idea was apparently suggested to him by Hugh Strickland (Wallace 1856:206).
81. J. É. Planchon 1847b:591–592.
82. Colebrook 1829. See F. G. Planchon 1860:74–85; Stevens 1984b: 174–179 for references, also Petit-Thouars 1817:226; Saint-Hilaire 1827:121; A.-L. de Candolle 1837b:228; Gray 1837:297. Palisot de Beauvois's two-dimensional arrangement of mosses, which may represent their relationships, is still extant (Lamy 1990:figs.3, 4).
83. A.-L. de Candolle 1841:150–168, see A.-P. de Candolle 1813b:202 (for a translation from the third edition [Candolle 1844:193–195], see Nel-

son & Platnick 1981:102-103). Emmanuel Le Maout (1844:866–871) discussed natural relationships in detail; this analogy is prominent. Hugh Strickland (1844) proposed a similar idea for depicting relationships between birds. Gordon McOuat (1992:248–249) shows how such analogies were also used to help emphasize the distinction between species (island) and infraspecific forms (land separated by an isthmus).

84. A.-L. de Candolle 1837b:[281]; see also Bureau 1856:37.
85. A.-H.-L. de Jussieu 1843a:16, pl.23.
86. Weddell 1856:40–41. His idea was enthusiatically taken up by Henri Baillon (1858:252–254, 279–280) and Asa Gray (1858a:110); Baillon (1863:43) later seems to have become less happy with the idea.
87. Cf. Panchen 1992:48. Georg Uschmann (1967) discusses some of the early zoological literature.
88. A.-L. de Jussieu 1789:xlix, "just as branches grow from the same trunk, the two parts of the organic Realm agree in their plan and progression."
89. Turpin 1820:11–14.
90. Payer 1844:6–8.
91. Duchesne 1766:220–221. Both Buffon (1755: chart facing p. 228) and Duchesne (1766: 219–228, chart facing p. 228) give diagrammatic genealogies (of dogs and strawberries respectively); both diagrams are reticulating and both are inverted compared with most later botanical and zoological trees, the origin being at the top of the page as in human genealogies (Tassy 1991:24–31). Duchesne (1766:223–224) noted that relationships between his strawberries were like those between different branches of the same "house." Augustin Augier (1801 [chart]; see Stevens 1983), provided a particularly fine tree, although it is unusual both in its extent (it includes all plants) and time (it is the earliest botanical tree of relationships I know about).
92. See Seringe 1815 (chart at end); T. F. L. Nees von Esenbeck 1820 (chart); Stevens 1984b:178–179, fig.3. Alexander Andrejewitsch von Bunge (1862:12–15, pl.2) drew a tree for *Anabasis*, in the Chenopodiaceae; as in Seringe's tree, characters are drawn on it. Bunge also provided a table showing the differences between genera, equivalent to a twentieth-century data matrix, and also a diagram like a cross section through a tree, again indicating characters (ibid., 15 and pl. 3). For early representations of trees, see Barsanti (1992a:76–96); for literature on early trees in zoology in particular, see chapter 11, note 86.
93. Fleming 1829:311–312. Although Ramean trees seem preeminently "artificial," they were not always seen to be so. Jeremy Bentham (1817:254) provided careful instructions on "How to plant a Ramean Encyclopâedical tree, on any given part of the field of art and science,"

that is, by careful analysis providing proper categorisations of the branches of human knowledge. He was close to thinking that nature itself could be represented by a Ramean tree. There is a most magnificent tree of knowledge with trunk and leaves in the supplement to the great *Encyclopédie* (Anonymous 1780, chart at beginning); this shows the "geneaological" relationships between the different branches of knowledge.

94. Reichenbach 1837:105, 107. He also made the interesting comparison between the different parts of the plant life cycle and different groups of organisms.
95. Duchesne 1766:14bis–16bis. See also note 101.
96. A.-H.-L. de Jussieu 1843b:538.
97. Jussieu 1848:42; Haworth 1821:xi.
98. A.-P. de Candolle 1826:pl.27. The details of the shapes of the circles and of exactly where they touched were probably important to Candolle. The diagram is similar to that drawn rather later by Henri Milne-Edwards (1844, fig. facing p. 98); this is frequently mentioned in the zoological literature.
99. The relationships envisioned by Elias Magnus Fries (1821) can also be converted into a tree, although Fries described his system in terms of circles. Late in the eighteenth century Pyotr Simon Pallas (1766:22–23) saw relationships as forming a sort of shrub—there were branches from the base, as in the colonial zoophytes he was then studying. Erwin Stresemann (1975:188–189) noted that Pallas believed in continuity.
100. On the other hand, Adrian Desmond and James Moore (1991:419) suggest that by the 1850s the tree was the accepted metaphor among naturalists. Even if this is true of the zoologists they had in mind (but cf. O'Hara 1988b), one still has to bear in mind the issue of how the tree was interpreted.
101. Berlin 1992:18, 175. Ideas of genealogy, generation, and birth readily became attached to words like genus, family, and even nature (Magnol 1689 [see Stevens 1984b:174]; Duchesne 1766:14[bis]–16[bis]; L'Heritier de Brutelle 1795:30; Pujoulx 1803:19–20); Roselyne Rey (1981:58–59) discusses the general problem in its historical context. Jean-Baptiste Payer (1857b, 1:1, fn.1) was against the use of the term "family" for this very reason; the word "order" more correctly implied a group that had a certain number of characters in common—he alluded to the orders of nobility, clergy, Benedictines, and the like.
102. Rafinesque 1836:69–70, also 35; he observed that he had made a complete botanical map as early as 1815 (on pp. 41 and 93 he mentioned the possibility of evolution). Du Petit Thouars (1817:226) also mentioned an analogy between taxa and relationships and leaves and

branches in a general survey of analogies used by botanists to discuss relationships; he found the analogy, although perhaps accurate, to be unworkable.

103. F. G. Planchon 1860:87–102.
104. A.-L. de Candolle 1837b:238–240.
105. Le Maout 1852:57–58.
106. E.g., Conry 1987. Adrien de Jussieu (1843b:531–538) believed that a natural classification should represent the "different degrees of organization in their ascending progression." Thus the Monopetalae were "superior" to the Polypetalae, and so he ended his series with the former group; the result was an arrangement quite different from that of his father (see also Brongniart 1868:5), who had largely arranged dicotyledons according to the number of their petals. Pierre Turpin (1820, chart facing p. 23) provided a diagram in which relationships of animals and plants were shown. It was a V-shaped series, with plants and animals joined by plant-animals at the bottom of the V, and the series of plants and animals separately graduated and culminating with the "highest" forms of life, buttercups and human beings respectively (Rafinesque, 1815:38, capped *his* series with roses and man). Carl Adolph Agardh (1817:3–4) believed that relationships were reticulating, yet he used a similar vocabulary. His son, Jacob Georg Agardh, presented in 1858 yet another version of the natural system (this one heavily dependent on variation in ovules and ovary) in a book with the subtitle "in series naturales dispositio, secundum structurae normas et evolutionis gradus institutio." For other good examples, see Schultz (1832:137–148), Baillon (1858:289–290), and Duchartre (1867:808–809). J. D. Hooker (1859:xxiii–xxiv), in his important introductory essay to the *Flora tasmaniae*, in which he discussed the consequences of evolutionary thought for botany, evaluated evolution in terms of its ability to explain the progression of organisms. For the situation in geology, see Rudwick 1976:154–155. Note, however, that although there might be overall progression, most authors described relationships of the particular group of plants on which they were working in terms of degeneration from the locally perfect type (see also chapter 3, note 109).
107. Weddell 1856:40–41.
108. Gray 1858a:110. Gray elsewhere showed himself to be much interested in this question of highness and lowness (e.g., Gray 1858b:120).
109. E.g., Fuhlrott 1829:14; Bureau 1864:107; F. von Mueller 1882:vi. John Miers (1858a:353 and 1858b:36) believed in the pervasiveness of circular relationships in nature, but he emphasized linear arrangements, if only for practical reasons. Barthélemy Claude Joseph Dumortier

(1822:49–50) simply took exception to a single measure of highness and lowness; continuity itself was not, however, disputed.

110. Hence the use of such language in Augier's discussion of relationships in his tree (Augier 1801; Stevens 1983:209); trees viewed from the side readily lend themselves to ideas of highness and lowness.
111. E.g., R. Brown 1810b:590; Vriese 1854:2; Baillon 1855:25–26; J. Müller 1857:75. For a zoological example, see Wallace 1856:207; cf. figs. on pp. 215 and 205.
112. A.-P. de Candolle 1826:151. Note that he placed the Leguminosae between the Rosaceae and Anacardiaceae, with the Detarieae being intermediate between the Rosaceae and the Leguminosae (ibid., 141).
113. Wallace 1856:207.
114. J. D. Hooker to Alphonse de Candolle, April 1883, G; Alphonse de Candolle to J. D. Hooker, April 15 and May 15, 1883, *Dir. Corr.* 135:418–419. K.
115. Lawrence 1951:32; cf. Bentham and Hooker 1862–1883 and A.-L. de Candolle 1873a:307–311. Charles Bessey (1894:241–243) realized that Bentham and Hooker's groupings might not express relationships.
116. Three of the four major groupings of the dicotyledons in the Candollean sequence have different limits in Bentham and Hooker's work, and the numerous groupings recognized by the latter (the Candolles recognized very few) make the two very different when they are interpreted strictly hierarchically or when the branches are allowed to rotate at the nodes.
117. See also Hull 1970:28.
118. Slaughter 1982:42.
119. Lindley 1839:24; see also Suringar 1895:13–15.
120. Mirbel 1814:35; for Lamarck, see Burkhardt 1977:58. However, in the *Philosophie zoologique*, Lamarck (1809, chap. 5) thought that the almost linear arrangement reflected nature, but the hierarchical classification of groups was more useful for teaching. Julius von Sachs (1875:148) correctly observed that Augustin-Pyramus de Candolle adopted ideas of progression in organizing his series, while at the same time disavowing the idea that the series represented nature. Merxmüller (1972:319) notes that teaching still tends to be "linear."
121. Cf. Stevens 1984b:200; see also Roger 1989:381.
122. Bentham 1875a:27.
123. These terms are based in part on distinctions made by Michael Ghiselin (1967:53, 85, 97). It is fatally easy to brand a person as being a nominalist without realizing that the nominalist position as currently defined is distinctly less than monolithic (see also Winsor 1991:23; cf.

Corsi 1983:387, 396; Rieppel 1988:20–21). As to claims that a particular naturalist was an essentialist, it should not be forgotten that all naturalists had to describe groups, whatever their opinion of their ontological status. Since descriptions abridge and summarize variation, groups will seem to be more discrete than they really are, but this does not make all naturalists essentialists or typologists (Hull 1965:317–318; Mayr 1988:345–348; Greene 1992:269; but cf. Grant 1957:43 and Mayr 1982a:256–272 and 1987:304 for examples of confusing usage).

A comparable table can be made for naturalists' attitudes to grouping and ranking at the species level; the same naturalist may have different positions in the two tables. One can be a limital nominalist or a hierarchical nominalist as regards higher taxa and a realist when it comes to species—the early Lamarck and the early Bentham, respectively, are perhaps examples. Additional columns like "belief in evolution" (yes/no) can also be introduced.

124. The same ambiguity is evident in Alphonse de Candolle. Candolle accepted Bentham's suggestion that evolution meant that taxa at all levels of the hierarchy were equivalent, yet at the same time he thought that some ranks had particular properties; they would not interbreed (species) or they could not be cross-grafted (families) (Alphonse de Candolle to George Bentham, 27 February 1875, Bentham, *Corr.* 3:1001. K.
125. See also Gregg 1950; Mayr 1982a:200; Stevens 1986:329. Gusdorf (1972:298–354) provides a useful general discussion.
126. A.-L. de Candolle 1873b, chap. 7, and 1880:215–217; cf. Jean le Rond D'Alembert (1765) and other entries on nature in the *Encyclopédie.*
127. The word is only rarely qualified, although the term "Gilmour-natural" (Heywood 1988:52 and 1989:89–90) has been proposed to denote groups that are based on overall similarity and are maximally "useful" (see also Walters 1988:16).
128. Bather 1927:lxix–lxxi. Bather first gave a version of this talk in about 1890 to the London Scientific Society (ibid., lxiii, fn.1).
129. E.g., Davis and Heywood 1963:6; Sokal and Sneath 1963:19, 170. I have not looked at later nineteenth-century literature with the issue of predictivity in mind (but see Maroske and Cohn 1992:545 for Ferdinand von Mueller).
130. Stafleu 1964:[xxii] and 1971:330; see also Flourens 1838:xxx, lii.
131. E.g., Cesalpino 1583:6; Grew 1673:9, 13; Camerarius 1699; Linnaeus 1736b:33 and 1747:11–12; Bernard de Jussieu 1741:254–255; Duhamel du Monceau 1758, 1:v, xxix–xxxii; Haller 1983:1151. For David Sigis-

mund Büttner in 1760, see Lenoir 1981:125–126. Sebastien Vaillant (1718:32–36) gives an interesting early discussion on correlations between different floral features. I would expect to find ideas of predictivity in Aristotle.

132. See also Coleman 1964:120–121.
133. A.-P. de Candolle 1804:8 and 1807a:62.
134. Delile 1827:101.
135. E.g., Lindley 1829b:12–13.
136. Only four or five families remained to be discovered (62 or 63 were already known), and Adanson suggested that one would be placed between the Myrtaceae and Umbelliferae, one or two between these families and the Compositae, one between the Compositae and Campanulaceae, and one between the Campanulaceae and Cucurbitaceae. There were 400–600 genera to be discovered, mostly in 23 of the known families, and 25,000 species (2,000–2,200 genera and some 18,000 species were then known). But the discovery of these groups would not affect the shape of nature (Adanson 1764, 1:cccxiv–cccxxiii).

    According to A. G. Morton (1981:310), this kind of prediction came from Adanson's understanding in principle of the connection between classification and phylogenetic relationships, but it seems better interpreted as dependent on a modified *scala* or, even more mundanely, based on Adanson's general knowledge of collecting and collections. François Dagognet (1970:47) thought that Adanson's predictions depended on his understanding of the combinations of characters that were possible. Later, Alphonse de Candolle (1873a:313) suggested that most genera and families had been discovered, although many species were as yet undescribed; this prediction came directly out of his tabulations of the numbers of these taxa that had been described in successive volumes of the *Prodromus*.
137. The tendency to maintain well-established names and to favor observations made on the outside of the organism would only increase the problem; see also chapter 10.
138. See also Sachs 1875:148.
139. Hooker 1859:xxiii.
140. Hooker 1856b:182–183.
141. R. J. O'Hara (1988a) provides a good treatment of this point.
142. E.g., A.-L. de Jussieu 1778:186, 1788, and 1789:lvii–lix. See also chapter 3.
143. Lovejoy 1936:232.
144. In systems like the quinarian system there is a rather high level of predictability for both the nature and number of undiscovered groups;

there are only so many gaps in the system, and the organisms that fill them will have predictable features.

145. Given the constraints of tradition, there would be no equivalence within a rank; Robert Brown (1818a:429) early noted this point.
146. Bicheno 1827:487–488; cf. Macleay 1829:403–404.
147. See chapter 1.
148. Candolle 1827, 2:244, and 1828:15–16.

Chapter Eight

1. See Rehbock (1983) for a discussion, also chapter 9 here.
2. Along these lines, Vernon Pratt (1985:427) suggested that "systematists" in the eighteenth century were interested in the structure of nature in its entirety; the simple cataloging of new species was not their ultimate goal. Details of Pratt's thesis need emending in the context of the discussion here.
3. Stafleu 1964:[xxiv]; Gusdorf 1972:301–302; Ridley 1986:107.
4. Foucault 1973:145–147; Anderson 1982:38–39; Barthélemy-Madaule 1982:61–62; Barsanti 1992b:281.
5. Larson 1971:143.
6. Henderson 1917:45. Continuity and gradation were also generally compatible with Christian doctrine and social structure at the end of the eighteenth century (Bynum 1975:5–7; Ritvo 1990:265). Most people still considered social relationships in hierarchical terms, and the systematic and social hierarchies would resonate, the one confirming the other. The analogy Jussieu used, that of a classification to a map or continuous landscape, was commonly used at the end of the eighteenth century, and it also served to clearly establish the limits between the known and unknown (Yeo 1991:26–27), providing a sense of closure to the knowledge of nature or whatever branch of science was under consideration (see also Stafford 1989:206–207). The notion of unity in the *Encyclopédie*, a unity stemming from linkages among its parts (McRae 1961:119; d'Alembert 1965:19, 57–62), is similar to that of Jussieu's continuity. Jussieu's species were indivisible and more or less discrete entities, but they were also part of an indivisible whole, much like the individual entries in the *Encyclopédie* itself.
7. Stevens and Cullen 1990:210–214.
8. Baehni 1957:84.
9. Broberg 1990:59. These figures were higher than most estimates.
10. See also Linnaeus 1751:137; Lamarck 1788a:632.
11. Rüling 1766, chart; relationships discussed in the text (pp. 435–436, 459–462) are added.

12. Batsch 1802, chart. See Stevens 1986b (fig. 1) and Barsanti 1988 and 1992a (pp. 61–73) for reproductions and discussion.
13. Cassini 1822:576–581; for his network, see Cassini 1816–1817:pl. 83.
14. E.g., Lamarck 1786:33 and 1788b; A.-L. de Jussieu 1777:220–221, 1778:180, and 1789:xlv. See also Ventenat 1799, 1:304; Légée and Guédès 1981:24–28.
15. See DiGregorio 1982:231, 235.
16. Macleay 1830:56, 136bis–137bis. See Canguilhelm in Foucault 1970:83.
17. E.g., C. A. Agardh 1819:61; Macleay 1825:53.
18. Macleay 1819–1821. Gilbert Thomas Burnett (1830:371–372) thought that the number in which Macleay was really interested was 7, and that both 5 and 7 were imperfect approximations of the number 3, which was the number in which *he* was interested. As has often been noted, Macleay's ideas contributed substantially to discussions on the shape of nature in England in particular immediately prior to the publication of Darwin's ideas, and his diagrams have been much reproduced (e.g., Ospovat 1981:101–113; Rehbock 1983:26–30; Gould 1984, 1987b:195; DiGregorio 1984:22–26; Winsor 1976b:81–97; O'Hara 1991:256–259). Nevertheless the full extent of his influence is probably still underestimated. It would be interesting to see how many "circular" systems can trace their conceptual ancestry to Macleay; his ideas are specifically mentioned by Henri Milne-Edwards (1844:79) and Elias Magnus Fries (1825:21). Systems such as those of Macleay and Swainson were largely based on an examination of external characters (Swainson 1834:165–172; see Maroske and Cohn 1992:532), and tended to founder on closer examination of organisms.
19. For Fries's ideas as conveyed to an English-speaking audience, see Lindley 1826; Eriksson (1969) includes a general study of Fries's ideas. Fries (e.g., 1821) was definitely looking for patterns in nature. Organisms showed relationships rather like those evident in a Linnaean landscape; there were circles of relationships at the various levels of the hierarchy, with the perfect member of the group, the best expression of the type, forming the centrum. Candolle had a different goal in his study of the Cruciferae (A.-P. de Candolle 1821a, table facing p. 200). He isolated two sets of five characters, one from the fruits and the other from the seeds, and showed in a table how genera might be grouped using the two sets of characters. It was probably only coincidence that he used five characters; toward the end of his life he adopted Cuvier's idea of there being four embranchements for plants, too, but questioned the idea that there was any significance in the fact that both groups could be divided into *four* subgroups (A.-P. de Candolle 1833b). It is possible Candolle's table was

designed to show parallelisms (cf. Brongniart 1850:19). "Tâtonnement," groping, seems to have informed Candolle's initial construction of the table, but Candolle was struck by the regularities in it. Rather confusingly, he referred to them as exemplifying "the law of symmetry."

20. Miers 1852:129. See also Miers 1858b:36; Hooker 1856b:182. Similarly, Adrian Hardy Haworth (1823, 1825a, and 1825b:428) saw all nature as being continuous, and he emphasized both binary division *and* circular relationships. He had rejected simple continuity, preferring the reticulation allowed by the botanical map; he also toyed with the idea that a genealogical table might be useful (Haworth 1821:x–xi). See also chapter 7 for other botanists who described circular relationships.
21. Macleay 1825:63. Theodore Charles Hilgard (e.g., 1857:132–156, pl. 6–7) developed (independently?) a remarkable system with 5-armed stars; relationships were detected by the use of "immediate catholic affinities"; "mediating genera" also abounded.
22. Wallace 1856:212.
23. Strickland 1845:176–177.
24. Lindley 1832:3. Lindley (1832:324–348) provided an interesting commentary on them; as he observed, "it is possible that some useful ideas may be elicited from the wildest of such speculations" (ibid., 324).
25. Macleay 1825:52, fn.
26. Macleay 1825:52 and 1830:139bis. Affinity was later called homology by many (Roth 1991 provides a convenient entry into the literature). Useful early discussions include those by Pyotr Simon Pallas (1766:5, 11–12, 23), Fries (1821, vol. 1), Johann Heinrich Friedrich Link (1824:46–47), Henri Milne-Edwards (1851:77–82), and François Gustave Planchon (1860:60–68). The development of such distinctions has been discussed elsewhere, although largely using zoological examples (e.g., Winsor 1976b:82–87; Mayr 1982a:202–203; DiGregorio 1982:282–284; Rehbock 1983:27; Knight 1985:94; O'Hara 1991:258–259). Distinctions between analogy and affinity were initially made by proponents of more idealistic classifications, with apparently independent developments in both botany and zoology (e.g., Macleay 1825:63; A.-P. de Candolle 1826:158; Jenyns 1837:18–19).
27. Lindroth 1973:123; Knight 1985:86–88 (for the creator and the natural order); Desmond 1985:164; Desmond and Moore 1991:312 (for circular systems).
28. Wight 1845:ii–ix. Wight also believed very strongly that relationships in nature were circular.
29. Fries 1821, 1:xvi–xviii, xxii–xxiii, 19–20, and 1825:15. See also G. C. D. Nees von Esenbeck (1817, e.g. frontispiece, p. 126, fig. facing p. 270);

relationships were represented by intersecting circles, semicircles, and lines. Carl Friedrich Philipp von Martius (1826:276) mentioned how circular relationships in the Phytolaccaceae were formed by joining two series.

30. O'Hara 1991:265–266.
31. Gould 1987b:184–195.
32. Wight 1845:27fn. Lindley (1839) included numerous floral diagrams in the book, but in later editions there were only a few (e.g., Lindley 1854:17 and figs. 68, 69. I am indebted to Brent Elliott and David Mabberley for information on these two publications, which I have not seen).
33. O'Hara 1993:233.
34. Burnett 1830:368.
35. Augier 1801:238–239; see Stevens 1983:208–209, fig.1.
36. Martius 1825:274, emphasis in original. He used an analogy with music to heighten the point he was making.
37. Reichenbach 1837:108. He quoted Antoine-Laurent de Jussieu here, but I have not been able to locate the source of the quotation.
38. Allman 1835:[134]. David Hendler brought this diagram to my attention.
39. Bromhead 1836:250–253.
40. Ibid., 252–253.
41. He was, however, well connected with a number of important figures in British science (Hyman 1982), and was elected to the Royal Society in 1817 and the Linnean Society in 1844.
42. Bureau 1864:107; see also his schema of relationships on p. 109.
43. I. Geoffroy Saint-Hilaire 1842, 1845. Geoffroy Saint-Hilaire (1832a:378, fn.1) suggested a connection between his ideas of parallelism and the "loi de parité," the similar tendencies he thought were to be found in the symmetry of organs or organisms in different groups. Henri Milne-Edwards (1844:79–80) also emphasized parallelisms of "affinité collatérale."
44. Chatin 1857:161–162; Caspary 1858:507, "In solche Weise kann Man aus Alles Allen machen."
45. Weddell 1856:40–41; Baillon 1858:247–249, 279–280 and 1861b: 354–355. Baillon did not refer to Chatin's work. Adolphe Brongniart (1850:19) was also receptive to the idea of parallelism, likening the pattern of relationships it showed to a double entry table (see note 19). Somewhat later, Joseph Duval-Jouve (1865) described the extensive parallelisms found between different species in the same genus. At this level, however, it was not so much an interesting pattern reflecting the fundamental operation of nature as evidence that narrowly circumscribed species were of little value.

46. Adrian Hardy Haworth (1825a) called the features that gave the high-level characters in his system, analogies; he called features showing close relationships, affinities.
47. E. M. Fries 1821:xvii and 1825:6.
48. According to M. P. Winsor (1976b:92), Macleay considered that the relationships between adjacent members within his circles of affinity were not different in kind from those between organisms in different circles (analogies), and he specifically contradicted Thomas Henry Huxley who had suggested that analogies and affinities were different. See also Knight 1985:74; O'Hara 1991:258.
49. Dumortier 1864:169; see also Dumortier 1822:42, 45; Turpin 1820, fig. facing p. [23].
50. Strickland 1841:185–192 and 1845:172, fn.
51. Strickland 1840:225–226.
52. A.-L. de Candolle 1837b:237–238.
53. E.g., Grisebach 1838:18–25.
54. A.-L. de Candolle 1830:64. He saw no symmetry in nature; fossils certainly could not provide it (Candolle 1834:99).
55. A.-P. de Candolle 1826:158–162, pl.27. Parallelisms occurred in features that were not subordinated by a dominating character.
56. Hooker 1856b:120.
57. Agassiz 1859. E. S. Russell (1916) describes the work of other zoologists that resulted in the invocation of various kinds of parallelisms, particularly between development and position in classification.
58. Reichenbach 1837:118. Lorenz Oken produced a system with sixteen classes each based on single characters like wood, flower, foliage, berry-like fruit, and with each class successively subdivided three times into sixteen subgroups, the result being a plant universe with 65,356 species (Mägdefrau 1992:84).
59. See note 19 for Candolle and his division of plants into four major groups.
60. Duchesne 1795:289–290. See also Marchant 1858:214–215 (for Cuvier); Decaisne 1842:357; Barsanti 1982a:64–65, fig. 38 (for discussion of Duchesne). One wonders how Duchesne's contemporaries reacted to his paper. To show that man also had the properties of plants, he mentioned the insensitivity of a paralytic; on the face of it, hardly convincing evidence. Duchesne (1766:219–220) had earlier suggested that there was no straight line of relationships; as with races of the same species, so with higher taxa, relationships were multiple.
61. L'Héritier de Brutelle 1795:28–29.
62. A. Smith 1980, II:12; see Skinner 1979:22. Chaining thoughts or obser-

vations thus seems to be a "natural" way of thinking (see Hays 1976 for its application in folk taxonomy).

63. A.-L. de Candolle 1844:195, fn.1.
64. A.-P. de Candolle 1829a:53–54. See also Audouin, Brongniart, and Dumas 1824:viii–ix; Lindley 1829b:3.
65. Desmond 1985:163. See also note 45.
66. Broberg 1990:69.
67. Baillon 1863:41–43.
68. A.-L. de Candolle 1880:323–328; see also Comte 1851, 1:657. Even by 1817 it seemed to some authors that botanical relationships were labyrinthine in nature (Petit-Thouars 1817:226).
69. The analogy of the web used in a social context described the complexity of relationships in the new industrial world, complexities so great that they were barely above chaos and any meaningful pattern (Marcus 1974:57–71).
70. See Balan 1975:291–295; Jacob 1976, chap. 2; Schiller 1978; Russo 1981.
71. Bonnet 1766, 1:44, emphasis in original.
72. Schiller 1980, chap. 7.
73. A.-L. de Jussieu 1789:iii, xvi, lviii; also Hallé in Jussieu 1789:(12).
74. Jussieu 1778:180, 183, 188 and 1789:lx. Jussieu (1778:188) once used the term "composées" in this sense, as when comparing the less complex apetalous plants with the rest of the dicotyledons.
75. Jussieu 1789:xxxv, xlv, lxi.
76. Jussieu 1778:196.
77. The ambiguities persisted. Jussieu (1824:24) thought that the goal of botany was "not to name a plant, but to understand its nature and total organization, since it will suffice to know what its family is to appreciate immediately the totality ["ensemble"] of its principal characters." He also thought that "by comparing all the characters of single individuals of different species one could comprehend the organization and nature of each, and hence decide on their relative affinities." (ibid., 15 and 1806b:183–184). This takes him close to meaning the fundamental constitution of the plant (see also Jussieu 1837:102, 105).
78. Adanson 1763, 2:32–36.
79. Lamarck 1785:440.
80. Poiret 1804:76.
81. Poiret 1816.
82. Jaume Saint-Hilaire 1805, 1:iii, xxi.
83. Ventenat 1794/5:14–47.
84. Ventenat 1799, 1:419–420.

85. Richard 1828:388. This was also a hope expressed by Jussieu and Elias Magnus Fries, among others.
86. Richard 1822:340, cf. 1826:501. Cuvier (1798:14) discussed the term as if it were a place-holder, or perhaps something more, for life itself. For him, organization was "the [totality of] form and structure the result of a force transmitted from generation to generation, of which the origin goes back to that of organized bodies themselves." And comparing two linked papers on the invertebrates he published in 1794, we find the title of one beginning "le structure interne et externe, et sur les affinités . . . " and that of the other, "l'organisation et les rapports . . . " (cf. Cuvier 1795a and 1795b). For Mirbel (1802, 1:13), organization might mean the way that life itself was constituted. But he also frequently equated organization with an anatomy that had physiological overtones (Mirbel 1806, 1809a:313, 1809b, 1809c, and 1810a:118, 139), "organisation interne" might refer more specifically to anatomy (1810a:127 and 1810b:252), while "organisation" could refer both to characters that were visible only on dissection or close examination (1810a:131) and to all characters (ibid., 138).

    English-speaking authors used the word less often, but when they did, they tended to mean only general features of the organism. Thus John Lindley (1846:xxvi) observed, "That which really determines affinity is correspondence in structure. . . . But it will be obvious that an examination of all plants through every detail of their organization is impracticable; it has not in fact been accomplished in any one case. Experience must have shown us that the organs of vegetation are of very different value in determining resemblance in structure." Yet characters of high value were important physiologically or functionally, and/or had an anatomical basis (Lindley 1831:liii and 1846:xxvi).
87. E.g., Desfontaines 1798; Lamarck 1802b; Mirbel 1809c; Cuvier 1816.
88. E.g., Lamarck 1802a:8; Treviranus 1832:75.
89. Kieser 1814. His book, *Mémoire sur l'organisation des plantes*, was divided into two main parts, anatomical and functional-physiological.
90. Its contemporary usages extended beyond biology, and like the analogies used for the natural system itself, the word "organisation" also resonated with contemporary societal issues (Balan 1975; Figlio 1976:39–41).
91. Pascal Tassy (1991:111–130) provides a good account.

Chapter Nine

1. Bentham 1866:7.
2. G. E. Allen 1981:160–163.
3. See also Farber 1982b:152. When there is more than a single element in

a concept such as natural history, following the change in the concept over time is a particularly tricky exercise (for useful discussions, see Laudan 1987:103–107; Mayr 1982b; Hull 1988:200–207, 377–378; Shapere 1989).

4. Farber 1982a:124–125. Similarly, G. E. Allen (1981:160–166) distinguished between a largely descriptive natural history and a more experimental causal-analytical biology that followed it.
5. Browne 1981:286; see also Farber 1982b:145–149; Rehbock 1983:5.
6. Sloane 1707, Preface:[vi].
7. *The Compact Edition of the Oxford English Dictionary*, 1971, 1:1311. For comments on the ahistorical nature of early natural history, see also Martin Rudwick (1976:69) and Wolf Lepenies (1976:40–66).
8. This links with the antiquarianism that was evident early in archaeology; there a historical approach started to develop both in England and on the Continent in the eighteenth century (Trigger 1989:45–65; see also Ashworth 1990:321–322).
9. Nicolson 1987:169–170.
10. See also Linnaeus 1744b:2.
11. Linnaeus 1735:[1].
12. As with all attempts at an easy categorization, this too fails and misrepresents. Linnaeus was interested in much more than classification—for example, the cortex-medulla theory that underlies much of his thinking about organisms (Stevens and Cullen 1990) is a far cry from the sexual system.
13. A. J. Cain (1993, 1994) gives a good treatment of Linnaeus's practice; Vernon Pratt (1982:215–223; see also Foucault 1973:132) discusses the historical background. Marius Jacobs (1980) also described how the focus of those who studied living beings progressively narrowed after the Renaissance.
14. See W. B. Ashworth Jr. (1990) for an interesting treatment.
15. Baron 1966:9.
16. E.g., A.-P. de Candolle 1828c:16. See Foucault 1973:161; Burkhardt 1977:20–21; Lyon and Sloan 1981:2.
17. See Withers 1991:206–207 and 1992:300–301.
18. Outram 1984:178.
19. But see McMullin (1990:820–821) for what might be considered to be "Baconian." The idea that there is such a thing as "pure description" is scarcely tenable.
20. E.g., Valmont de Bomare 1768, 2:474–475.
21. Larson 1971:40. "Natural hierarchy" is a somewhat ambiguous concept.
22. A distinction made very clear in the entries "natural history" and "nat-

ural philosophy" in the *Encyclopaedia Britannica* (Anonymous 1817, 14:628–641). Elias Magnus Fries (1825:2–3, cf. the translation made by John Lindley 1826:82) distinguished between *natura naturans* (natural philosophy) and *natura naturata* (natural history). A similar distinction between "Naturbeschriebung," the study of outer form, and "Naturwissenschaft," the study of inner form, was made (Cassel 1817:1); this leads to the idea, particularly prominent in Germany, that certain kinds of botany are more "scientific" than others. Rachel Laudan (1982:7) noted that natural history and natural philosophy were also more or less independent in geology.

23. Daubenton 1765a:226–227 and 1751:341. Richard Burkhardt (1977:20–22) outlines Daubenton's attitude to natural history. Roselyne Rey (1981:48–50) briefly mentions differing ideas over the nature of natural history in the late eighteenth-century France.
24. Daubenton 1765a:228. He thought that the goal of comparative anatomy was to understand the "mechanique" of man in particular and of nature in general (ibid., 226; see also Anonymous 1842:738). François Dagognet (1970:122) noted ambiguity in Daubenton's use of anatomy; Paul Farber (1975) discusses Daubenton's work in some detail.
25. Anonymous 1805, vol. 5 (Botany); the author was probably William Wood (E. Rudolph, pers. comm.). However in a later volume the article "Natural history" did not suggest such an all-embracing compass, being characterized as "a description of the natural products of the earth, water, or air, either on a systematic or geographical basis"—see Anonymous 1813, vol. 24 (Natural History). Gusdorf (1972:246–247) suggested that there were a variety of meanings of natural history in the eighteenth century; the one I outlined earlier was certainly common then and on into the nineteenth century.
26. Anonymous 1765:712–717. There were two entries, "Plante," the entry just mentioned, and "Plantes (bot. méth.)," which largely devoted itself to a discussion of Tournefort and Linnaeus.
27. J. E. Smith 1798:118 (emphasis added). This remained a not uncommon circumscription, being that adopted by Le Maout and Decaisne (1868:1), although there the emphasis was on the functions of organs and the whole topic was dealt with in a section only twelve pages long.
28. E.g., Sloan 1979:129–131.
29. Lyon and Sloan 1981:3.
30. Ibid., 20–27.
31. Buffon 1753:356; see Sloan 1987:124. Buffon's attitude to natural history has been discussed in detail (Roger 1989; Sloan 1990).
32. Buffon did not consider anatomical observations to be an integral part of

natural history (but see Lyon and Sloan 1981:92), although Daubenton, his collaborator for the first parts of the *Histoire naturelle*, was more interested in gross anatomy (Daudin 1926a:133; see note 24), and he encouraged its use in botany (Daubenton 1797). The detailed and extensive descriptions of individual species that Buffon included in the earlier parts of his *Histoire naturelle* in particular approach Baron's "Wissenswerte" as much as his "Systemkunde" (Baron 1966).

33. Cassirer 1951:79–80. Little was known about fossils in the middle of the eighteenth century—see chapter 5.
34. Another problem Buffon faced came from the way in which he went about discussing nature. His approach was that of "l'esprit de système," and so was opposed to late eighteenth-century thought in both England and France. J. E. Smith (1798:143fn) described Buffon as "a painter and colorist, rather than delineator, of Nature" and complained about his "capricious and precarious" hypotheses. The Marquis de Condorcet, in his capacity as permanent secretary of the Académie, delivered an éloge of Buffon that made similar points (Condorcet 1791:esp. 66–67); Buffon was more interesting than instructive (ibid., 83).
35. See Baron and Sticker 1963:20–26 for Blumenbach's attitude to history.
36. Nicolson 1987:175; see also Lenoir 1981:170–174.
37. Whewell 1847, 1:512.
38. Ibid., 480; cf. 535–541. Even Cuvier (1810) included anatomy with physiology.
39. Whewell 1848, 3:414.
40. Hoppe 1971:230; Farber 1982a:145. There are good treatments of biology in the early nineteenth century (e.g., Baron 1966:1–4; Schiller 1971b:240; Caron 1988). From the point of view of definitions, note that it is the new field that takes a new name. Ironically, botany (classification) was innovative in the late eighteenth century, and this is in part why it was sometimes *excluded* from natural history then (Daudin 1926a:146).
41. See Lenoir 1982:1.
42. E.g., Treviranus 1832:75. See Hoppe 1971; Lenoir 1981:175–190 and 1982. Treviranus had a younger brother, Ludolf Christian, who studied the anatomy and physiology of plants.
43. Lamarck 1802b:443. Earlier he had suggested that "physique végétale," which developed from the intimate knowledge of the organization of plants, the nature and form of their parts, their development and modes of reproduction, the "qualités" and movement of sap, etc., was the first part of botany (Lamarck 1785:439–448, also 1778, 1:4. See Burkhardt 1977:46–48; Outram 1986:336).

44. Candolle, in Lamarck and Candolle 1805, 1:63–64. Although Candolle apparently rewrote the *Flore française* without Lamarck's help, it would be interesting to know if Lamarck had any input into this edition; the inclusion of Lamarck's original introduction without any change suggests that there was some. Candolle's compartmentalization of the botanical endeavor was common in the nineteenth century (e.g., Kunth 1831:[3]–6; Lincoln 1832:32; Richard 1834:82; Duchartre 1867:761; Bellynck 1869:36–37). The split he described had been evident toward the middle of the eighteenth century, Étienne Louis Geoffroy (1762:1–2) noting that classificatory studies were rather different from other studies of organisms. Nevertheless, Candolle, as well as John Lindley, Asa Gray, Julius Sachs, Charles Bessey, Eduard Strasburger, and others, included all aspects of botany in their textbooks; however, as the century wore on, it became obvious that the vast amount of information, not to mention the growing difficulty of one person staying current in all areas, made it difficult to include the whole subject in a single volume.
45. Rehbock 1983:7–11.
46. Lindley 1834. Bentham was to report on the rest of botany, i.e., classification studies, but his paper was not given until 1874 (Bentham 1875a).
47. Caron 1988:230–238. For Auguste Comte, see p. 234.
48. Daudin 1926a:144–149. Buffon (1750:10-15) thought that there were many more individuals of plant than of animal species, although the latter were more distinct; in general, animals did not look as much like each other as did plants. This attitude perhaps comes from his experience with European plant communities, which are species-poor but individual-rich; Condorcet (1780:97) made similar comments, although comparing botany with "histoire naturelle" (see note 40).
49. Baron 1966:5.
50. D. E. Allen 1976a; Merrill 1989.
51. See appendix 2. Similar decisions were made by Linnaeus (Blunt 1971:117–118), Cuvier (Outram 1984:61–64), and Asa Gray (Dupree 1959:57–69), although Cuvier suggested that tropical explorers who stayed in one place for a number of years could make valuable contributions (Cuvier 1829:7). Jules-César Savigny, partly at Cuvier's urging, went on Napoleon's ill-fated expedition to Egypt and made extensive collections there (Winsor 1976b:25). Of course, a disinclination to travel may not represent a rejection of field studies. Linnaeus had a long-suffering fiancée on his mind, and Gray could not wait for preparations for the United States Exploring Expedition to be finalized. Cuvier utilized his access to living animals in zoological gardens (Stemerding 1993:217), but this does not represent his endorsement of field studies.

52. A.-L. de Candolle 1862:214, 216. He wrote this paper on species limits because he had so little in the way of taxonomic novelties to describe (ibid., 212). Some authors did call for a broader approach to the study of nature—John Fleming (1829:305–309) in this respect taking a Buffonian position, although he was an anti-Buffonian. It has been suggested that field-based knowledge was becoming more important in American universities by the 1830s (Kohlstedt 1988:426), but there was no obvious effect on taxonomic practice then.

Although botanical collectors might make quite detailed field notes, such notes were often considered of little importance once the specimens arrived in the herbaria (Bentham 1861b:135; see Shaw 1987:2–5; Howard 1988:66–87; Madriñàn 1994). It is not clear how the field observations of botanical explorers like Robert Brown, J. D. Hooker, and Carl Friedrich Philipp von Martius affected their classificatory work; one can only suppose that their field observations guided their collecting and hence what they later described. Brown (1814, 2:547) commented that the "native inhabitants" in the Port Jackson area used field characters to distinguish between species more readily than botanists could. In discussions with systematists I have been unable to get a clear idea of whether Brown or George Bentham, whose work on the flora of Australia was based on herbarium material—albeit with Ferdinand von Mueller's (sometimes unappreciated) observations based on *his* field experience—delimited species in a more satisfactory fashion. Hooker certainly campaigned against the use of field characters in delimiting species.

It is interesting that the accurate observations on the growth patterns of some trees in Amazonia made by Richard Spruce, a great field botanist, although published in the *Journal of the Proceedings of the Linnean Society*, were not cited in the early literature on tree architecture, a subject to which they directly pertain (Spruce 1861, cf. Hallé, Oldeman, and Tomlinson 1978). Spruce (ibid., 7–8) thought that the different branching patterns he had found in different species of the genus *Diospyros* might, if coupled with floral differences, be grounds for the recognition of distinct genera. However, William Philip Hiern, who monographed the genus soon afterward, was unable to confirm Spruce's suggestion, and even found that within *D. paralea* there were specimens collected by Spruce that apparently had different branching patterns (Hiern 1873:240–241). Growth characters were of no use in taxonomic studies, and there was no other body of botanical knowledge to which they could then be related. (Santiago Madriñàn [1994] has recently studied Spruce's work on tree architecture.)

Similar attitudes prevailed in zoology. Field studies were no part of the ornithological tradition as seen by the leaders of the discipline in the middle of the nineteenth century (Farber 1982a, chaps. 6 and 7), nor of other areas of vertebrate zoology (Outram 1984:61–64). As the ornithologist Jules Verraux noted in 1845, "I hope very much if my voyage ends as I desire, to be able to spread this love [of the habits of birds] . . . My only fear, however, resides in the great difficulties that I think I will encounter among our scientists, especially those at the top of this science." (Farber 1982a:94). Gordon McOuat (1992:210–211) notes that somewhat later in the century field studies were considered to be of little importance at the British Museum. Although Hugh Strickland thought that more knowledge about habits, geographic distribution, and oology were desiderata in ornithology, this was to ensure that better collections were made. It was on these collections that a classification of birds based on their anatomical features could be securely founded (Strickland 1845:219, cf. Tutt 1898:25).

53. Merrill 1989:80–82.
54. Bentham 1871:7; see also A.-P. de Candolle 1862:216. Cuvier, commenting on a work by Humboldt, observed "it is only in the cabinet that one can traverse the universe in every sense" (Outram 1984:63. Outram did not identify the work by Humboldt).
55. Lowe 1976:518. It is interesting in this context to note that much of the literature on natural history was written by women who were presumably not professional botanists in any sense of the term (Barber 1980, especially chaps. 2 and 9. See also note 80, following). Ann Blum has recently (1993) shown how style in nineteenth-century American zoological illustration also illustrates the split between field- and collections-oriented naturalists.
56. Keeney 1992:23.
57. Merrill 1989:118–122; D. E. Allen 1976b:509.
58. Barber 1980:40–44.
59. In the 1830s both Darwin and Henslow used "naturalist" or "finished naturalist" to refer to somebody who could make classifications (Burkhardt and Smith 1985:128–129, 512; for T. H. Huxley, see L. Huxley 1901, 1:7).
60. Hooker 1853:xii–xvii, xxvi and Hooker and Thomson 1855:24–36. See also Bentham 1871:7.
61. It was the species described by botanists like Alexis Jordan, narrowly circumscribed and based in part on field and cultivation experiments, that caused problems for herbarium botanists, and this remained true throughout the century. Toward its end more field- and experimental-

ly-oriented studies came to be seen as of greater general interest than herbarium studies.

62. Outram 1984:33, 61–63.
63. A.-P. de Candolle 1862:121; see also appendix 2.
64. Compare Linnaeus 1753 with Bentham and Hooker 1862–1883; the latter only estimated numbers of species. Baillon (1891:155) estimated that there were some 200,000 species of phanerogams. Similarly, the number of species of birds recognized in 1862 was twenty times greater than that recognized by Linnaeus in 1758, the number doubling in the last thirty years of that period (Barber 1980:65). Even Buffon had had to give up his plans for a detailed description of all animals when he moved from the relatively few mammals known to the much more numerous birds and fish. 1500–2000 species of birds alone were known (Anker 1938:32–33).
65. German-speaking botanists may be partial exceptions. Gottlieb Wilhelm Bischoff (1830, 1:8–9) included within pure botany both "Naturlehre" (comprising anatomy, physiology, etc.), and botany in the strict sense, "Naturgeschichte" (largely classificatory studies in which a historical component was minimal.) Plant geography, which did include historical elements, was a part of botany *sensu stricto*, although Bischoff thought that it might have to be a third major branch of pure botany. Indeed, Walter Baron (1966:9, fn.2) suggested that in Germany *special* botany was to be synonymized with systematics. However, Christian Gottfried Nees von Esenbeck (1820–1821), much under the influence of "Naturphilosophie," equated natural history with botany, including idealistic plant morphology in the purview of the latter.
66. E.g., Mirbel 1815. For Henslow (1837:122–123), "botanical physiology" included anatomy.
67. Bory de Saint-Vincent 1825:252.
68. Bentham 1865:69–76.
69. If one looks at the numerous publications listed by Stafleu and Cowan (1976–1988) one gains an impression of great productivity, but it was not by productivity alone that plant taxonomists were judged by their peers.
70. Lindley 1835b:vi; Fleming 1829:304. Lindley's comments were directed against remarks such as those of Sir J. E. Smith (1817:380) that the very words of Linnaeus were sacred, they were "a kind of public property, the current coin of the botanical realm." Brent Elliott (1992:474) discusses the influence of Linnaeus in English botany of the early nineteenth century; see also Mabberley (1985:179). Pascal Duris (1993:191–215) notes the perhaps surprisingly strong support of Linnaeus in early nineteenth-century botanical circles in France.

71. Gillispie 1980:192.
72. Monk 1818:435, 437. The occasion of this outburst was a pamphlet written by J. E. Smith (1818) describing his attempt to be appointed to the vacant chair of botany in Cambridge University (see Walker 1988:45–48). S. Parks (1821:55) made the same association of botany with horticulture, but also quoting Bacon, "it is the purest of human pleasures, and the greatest refreshment to the spirit of man." Perhaps not suprisingly, botany was a subject "that amused the fancy and exercised the memory" (White 1836:217, and earlier editions); nothing else.
73. Anonymous 1864:357; see also Ker 1823.
74. See Desmond 1985; Gage and Stearn 1988.
75. Anonymous 1864:355–357; D. E. Allen 1976a, chaps. 9 and 10. P. D. Lowe (1976:522) thought that a widespread disillusionment with traditional systematics began in the 1880s, but the malaise is of much longer standing. As Captain Cook observed of Joseph Banks's collections, "most of them [were] unknown in Europe and in that alone constituted their sole value," and mathematicians in the Royal Society claimed that Banks, president from 1778–1820, attempted "to amuse the fellows with frogs, fleas and grasshoppers" (Carter 1988:78; O'Brian 1987:209). But they refused to be amused.
76. Bentham 1836b:77 and 1865:52–55; Bentham to Alphonse de Candolle, le 29 Janvier 1849, le 23 Xbre 1850, and le 16 Fevr 1862, G-JB; Bentham to J. D. Hooker, 9 March 1852, *Dir. Corr.* 36:69. K.
77. Limoges (1980:225–240). He suggests that a similar decline occurred in both England and America (ibid., 213–214).
78. Outram 1984:181. Charles-François Brisseau de Mirbel (1833:570–571) also observed that Desfontaines's lectures were well attended and that they were the first lectures to be given in vegetable physiology; he thought they marked the beginning of a brilliant period in the history of botany. But the association that Mirbel is making is between plant physiology and botany.
79. Lepenies 1976:57.
80. Alic 1986:110–118, 176–178. The close association between natural history in general (and in particular botany, collecting, and Linnaean classification) and women, teaching, and children has been a matter of general comment (e.g., Rudolph 1973, 1984; D. E. Allen 1980: 241, 249–250; Shteir 1987; Kohlstedt 1988:422–423; Keeney 1992, chaps. 4 and 5), and it persisted throughout the century. J. Arthur Thompson (1899:449) observed that a "well known philosopher" considered natural history to be "one of those Kindergarten subjects," and many books on natural history were written to help mothers or other (female) guardians instruct

children. Pascal Duris (1993:103–105, 182–189) suggests that in France Jean Jacques Rousseau was in considerable part immediately responsible for the popularization, vulgarization, and "puerilisation" of botany.

81. Beal 1881.
82. Adams 1887.
83. Lonsdale 1970:49. I have not located any more information about this story. David Layton (1973:55–74) details John Stevens Henslow's efforts in botanical education; it is possible that Henslow's work is here being (mis)recorded.
84. Watson 1833:102. He suggested that in general the French were better descriptive botanists than were the English, and the English were better than than the Scots. Unfortunately, he did not make a phrenological comparison between English women and French men. (Gordon McOuat showed me this fascinating article).
85. Lindley 1835b:vi; see also Kirby and Spence 1815, 1:v.
86. Anonymous 1831:64–65. For a brief discussion on the Declinist controversy see Hyman 1982:74–87.
87. Only shortly before, Lindley (1834:25) himself had excluded "mere systematic botany" from the main questions in philosophical botany being debated in England, and he also admitted the harm that the association of (systematic) botany with women was causing the discipline (Lindley 1829b:17).
88. Anonymous 1864:356. Similar definitions are common. A good example is that of Jacques Nicolas Ernst Germain de Saint-Pierre (1870:135–153), who began the entry "botaniste" in his *Nouveau dictionnaire de botanique* with a one-line paragraph, "Le botaniste est un homme heureux," and he continued in the same rhapsodic vein throughout.
89. Lefébure 1835:103, 109; see Lefébure 1817 for his idiosyncratic views on classification and system (he was a Linnaean botanist [Duris 1993], and his comments must be read bearing in mind his allegiance.)
90. Lepenies 1976:53.
91. Jussieu 1789:xxviii.
92. Green 1914:329. Jacob Georg Agardh's *Theoria systematis plantarum* of 1858, in which the Orchidaceae were placed with the Orobanchaceae (parasitic plants related to foxgloves), was simply one among the dozens of systems that appeared in the nineteenth century, although it was rather extreme. Adrien de Jussieu (1848:62), observing the variety of arrangements of plants, commented weakly that light coming from many sides would illuminate a large number of points. Augustin-Pyramus de Candolle (1829a:53–54) made similar comments about the state of botany

(see also Audouin, Brongniart, and Dumas 1824:viii–ix; Lindley 1829b:3).

93. Clos 1869.
94. Hooker and Thomson 1855:13–14; see also Browne 1983:46. Similar complaints were voiced intermittently in the nineteenth century (e.g., Payer 1857b:4; Thistleton-Dyer 1895:838, 842–843).
95. D. E. Allen 1976b:509–510. Thus E. Hart Merriam (1893:354) considered that it was a negative aspect of the new, laboratory-based biology that it separated the professional and the amateur. The fluid boundaries between professional and amateur in nineteenth-century North American mycology have been discussed elsewhere (see papers in Haines and Salkin 1986). Elizabeth Keeney (1992:2–3, 97–98) suggested that in North America amateurs became divorced from "scientific" systematists during the course of the nineteenth century, although she thought this might not have been the case in Great Britain. She also compared this with the situation in ornithology and astronomy in the United States. Although she tends to conflate "biology" with professional systematics, with her thesis applying more to the former than to the latter, there may indeed be national differences; I would expect the line between professional and amateur to be less clear in Britain than in the United States. Richard Drayton (1993:172–211) shows how in Britain classificatory botany was affected by its close association with amateurs.
96. At the beginning of the twentieth century, the need to write Latin was seen as a way of keeping the hordes (of amateurs, species mongers, and the like) from the gate (Bessey 1908:222; Cowles 1908:270). But similar attitudes were already evident by the middle of the nineteenth century in both botany and zoology; Louis Agassiz asserted that only professionals could describe good species (Winsor 1991:89–90), while the attitude of some botanists that only workers at large institutions could correctly delimit species also drew limits around the ranks of true botanists—in this case excluding professors in German universities and benighted colonials alike (see note 60).
97. Farlow 1913:79.
98. See Stafleu and Cowan 1976, 1:909–911. Of Michel Gandoger's infraspecific taxa, 150,000 were ultimately ignored because he did not follow nomenclatural conventions.
99. Farlow 1887:310–313.
100. Pfister 1986:72–74; Edgar 1991:[1]–[3]. Farlow seems to have considered botany proper to consist of classificatory studies. W. J. Beal, writing in 1871, suggested that for twelve years after 1850 plant morphology was the leading idea in at least popular botany; after that studies of

growth and pollination came to the fore. Even at this level, systematics, associated with the learning of innumerable names, was pushed to the background.

101. Galton 1895.
102. Schleiden 1842; Schwann 1839.
103. Schleiden 1842, 1:9; he repudiated the excesses of the *Naturphilosophen*.
104. Cohn 1895:(11)–(12); Goebel 1925:5–6. Frans Stafleu (1966c:27) suggested that the rebirth of German botany after 1840 centered on Martius's monumental *Flora brasiliensis*. Although definitely providing a focus for systematic endeavors in Germany and elsewhere, and also a small injection of cash into the botanical economy (contributors were paid), it seems to have done little for the status of systematic versus "wissenschaftliche" botany.
105. Nägeli 1865:7, 92. He wrote the entire contents of the short-lived "Zeitschrift für wissenschaftliche Botanik" that appeared between 1844 and 1846; Schleiden was co-editor (Stafleu and Cowan 1981, 3:683).
106. Sachs 1868 (much reprinted, the second and third editions were in 1870 and 1873 respectively); Bessey 1880; Gray 1880:337. Farlow (1913) and Rodgers (1944:230–231) provide general background. For additional information, see Bentham 1873a:[3]; Bessey 1935; Pool 1935:235–240; Bower 1938:20–21, 36; Coleman 1971:2–3. The "New Botany" was laboratory botany, as Bower observed. Biology—anatomical, physiological, experimental—was not necessarily restricted to studies of particular kinds of organisms, but zoologists had largely taken over the right to use the word (Macmillan 1893; see Pauly 1984:392); this zoology was also laboratory-based (Martin 1877). Thompson (1899:437) suggested that in England natural history and zoology were synonyms, but by then natural history and biology (effectively equivalent to zoology) had become linked. (Note that recently Joseph Caron [1988:240, 255] has shown in detail how the term "biology" first became popular without being associated with a definite research program and how it then became banalized.)
107. E.g., Farber 1982b:146; Hagen 1986:198–202; Coleman 1986:189.
108. For contemporary comment in America, see Merriam 1893, and in England, see Thompson 1899; Oliver 1906; Bower 1938:26–27: for recent discussion, see Dupree (1959:384–403), Kohlstedt (1988:408–415), and especially Winsor (1991:175–197, 242–244). Camille Limoges (1980:232–233) suggests that in Paris the collections at the Muséum became further marginalized during the period 1904–1934 by a return to a more "naturalist" orientation, i.e., collections-based descriptive work was emphasized.

109. E.g., Nägeli 1847:4, 96.
110. Brongniart 1868; see also Zahlbruckner 1901, Wettstein 1901, and Strasburger 1895. Anatomy and physiology, especially when placed in an ecological context, were popular during this period (Wettstein 1901:199–203, 207–209).
111. For a good discussion of Hofmeister's work, including an outline of its conceptual background, see Farley 1982:82–100; for a biography, see Goebel 1926.
112. Lyon and Sloan 1981:2; cf. Sloan 1990:297.
113. See also Larson 1986:488.
114. Burkhardt 1977:46.
115. Cf. Browne 1981:286.
116. See also Farber 1982b:147–150 for ornithology.
117. Keeney 1992:21.
118. Farber 1982a:124.
119. Farber (1982a:124–125) considers these to be traditions within natural history in the eighteenth century.
120. Anderson 1982:54.
121. See notes 38 and 39. In zoology organ-level comparative anatomy was integrated with classification in a way that it could not be in plants. As Leonard Jenyns (1837:2–3, 13–15) observed, comparative anatomy and physiology were properly subordinate departments of that part of natural history called zoology, and the former in particular was essential for classification (see also Anonymous 1852:738, fn.1; Whewell 1848, 3:414). Even looking at the subdiscipline of conchology, a zoologist would be disabused of the idea that the external and internal parts of an organism were in simple agreement, and so both had to be studied (Anonymous 1838). Cuvier's incorporation of anatomy and physiology in his *Leçons d'anatomie comparée*, a work purportedly dealing with natural history, challenged old disciplinary boundaries (e.g., Burkhardt 1977:46; Outram 1984:62). Although it was conceded that botanists began to study monsters before zoologists did, monsters had proved to be less useful in botany; there was a similar story when it came to a study of positional relationships between organs, which became so important in determining what any particular structure "was" (Audouin, Brongniart, and Dumas 1824:viii–xi).
122. See Browne 1983, and especially Larson 1986. Humboldtian plant geography, with its emphasis on tabulation and descent, was also closely connected with the *ecological* relationships between organisms (Nicolson 1987:174–175, 186–187).
123. Geology was becoming profoundly historical in the early nineteenth

century (Rudwick 1976:135; Corsi 1988:54; Laudan 1987:88–97; Sloan 1990:299). The study of strata would seem almost inevitably to give a historical complexion to geology (Laudan 1982).

124. G. E. Allen 1981:165–166. Compare the natural history of Louis Agassiz (1863), largely descriptive as it was, with that of E. B. Wilson (1901), who, although a moderate in this context, clearly saw the discipline in a very different light. (Note that the nomenclature may get confusing; zoology and natural history were sometimes equated in North America and Britain [Keeney 1992:57; see also notes 65 and 106].)
125. For the circumscription of botany in the twentieth century, see Smocovitis 1992b.
126. Gray 1879:[1], fn. See also Bentham 1865:72, fn. Earlier Gray and Hooker had exchanged letters over the state of botany. Gray noted that botany was a science of observation, and although physiology and anatomy were not botany, they were taking up too much of the botanical curriculum (Asa Gray to J. D. Hooker, May 5th, 1857, Gray *Corr.* 143. K; J. D. Hooker to Asa Gray, June 21, [18]54, GH). See also Hooker and Thomson 1854:17–19, 23–24.
127. Keith Vernon (1993) describes the lowly status of systematics earlier this century, but the malaise that he describes in the discipline must be seen in the context of a general decline which stretches back almost two centuries. It is against this broader background that attempts by systematics to rejuvenate itself, or "botany" to redefine itself (Smocovitis 1992b), must be seen.

Chapter Ten

1. Flourens 1842:xii–xvii.
2. F. von Mueller 1882:vi.
3. Davy de Virville 1954:70. The three families into which Jussieu divided what are now known simply as the Compositae or Asteraceae are the most general exceptions. It is superfluous to cite more than a few of the papers which see post-1789 developments in higher-level classification as simply tinkering with the Jussiaean superstructure (see Saint-Hilaire 1827:125; Fuhlrott 1829; Martius 1843:218; A.-H.-L. de Jussieu 1848:62). As J. D. Hooker (1888:55) observed of Robert Brown's work on the Australian flora, "it is rightly regarded as the complement of the great work of de Jussieu in respect to the perfecting and extending the natural system of plants." On a more whimsical note, Pierre Flourens (1842:ii; see A.-P. de Candolle 1862:5) suggested how the mantle of greatness was passed between botanists: Augustin-Pyramus de Candolle (1771–1841) was born

one month after the death of Linnaeus (born in 1707), two months after the death of Albrecht von Haller (born in 1708), and only three months after the death of Bernard de Jussieu (born in 1699); their spirits must have surrounded the crib of the infant Augustin-Pyramus.

4. To my knowledge Réjane Bernier (1984:10) is the only author who has recently noticed this difference, although Michel Foucault (e.g., 1970) locates Jussieu's work firmly in the tradition of continuity, and Scott Atran (1986:129–130) has made the same connection. Of course, Henri Daudin's work is still relevant.
5. Bentham 1866:7.
6. Figlio 1976:37.
7. Sachs (1875:155–156) notes that 1825–1845 was the golden age of systems, with no fewer than 24 being proposed then; he did not include those of the *Naturphilosophen* (see Dumortier 1864 for a summary of several of the systems; Green 1914:329).
8. Julius von Sachs (1875:157–161) touches on some of the points I raise here.
9. Dean 1979:211; Winsor 1985:59–62, 69–70 and 1991:8, 99–100. Olivier Rieppel (1988:6) suggests that the science of comparative biology has maintained close links to philosophy; this is not true for the generality of the discipline.
10. Cullen 1968:175; Hull 1970:20; Reif 1983:195; Stevens 1984b:199.
11. See also Hyman 1982.
12. For the former, see Zimmermann 1930:8–12; for the latter, see chapters 1 and 3. As late as 1857 Jean-Baptiste Payer (1857b:6, fn.1) used a quotation from the Marquise du Châtelet to argue against system-building.
13. A. J. Cain (1993) suggests that for a naturalist in Linnaeus's time the word type may have had unfortunate medical connotations; this does not seem to have been an issue in the nineteenth century.
14. A.-P. de Candolle 1828c:12.
15. Candolle 1827, 1:242–244.
16. Ibid., x.
17. Sachs 1875:160–161.
18. Morton 1981:316, fn.47.
19. Marchant 1858:160. Unfortunately, the correspondence does not suggest the nature of this schism.
20. Candolle 1828c:11.
21. J. E. Smith 1817:389. See also Anonymous 1817:635–636, 1823:63; Bentham 1861b:133, fn. For Hooker, see Turrill 1963:212; Browne 1983:155–156.
22. Gillispie 1980:152.

23. In A.-P. de Candolle 1862:ix.
24. Cuvier 1832:iii–iv, quoted by Bentham 1862:6. That Cuvier's own approach to classification was hardly devoid of theory (Coleman 1964:73; Outram 1984; cf. Bourdier 1971) is to the point.
25. Hooker 1853:xxvi. Zoologists had similar attitudes (McOuat 1992:208–210).
26. Henslow 1837:115.
27. Winsor 1991:118; the quotation is from Darwin 1859:484.
28. In his analysis of the work of Andrea Cesalpino, C. E. B. Bremekamp (1953:581) suggested "Most of us will agree that in judging the scientific value of works written in the past we must leave the author's philosophical or theological orientation as much as possible out of consideration, as this value is, as a rule, but slightly influenced by these factors." And Max Walters (1988:18; see also Walters 1961:83 and 1986:529, 539, 544; Turrill 1963:200–201, 212–213; cf. Griffiths 1974:87) observed that philosophical hair-splitting usually seems of little importance in systematics. Bruce Trigger (1989:4–5) discusses early classifications in archaeology that were apparently theory-free.
29. See also Sachs 1875:161.
30. The popular idea of nature in late eighteenth- and early nineteenth-century thought had both shifting and emotive connotations; I do not do this important subject justice here (Charlton 1984 and Stafford 1984 should be consulted in this context).
31. See Condorcet (1780) on Bernard de Jussieu; Jussieu was able to study plants in the order nature gave to us.
32. Adanson 1764, 1:clvii–clvii.
33. Richard 1828:389; see also Ventenat 1799, 1:xxii. Rather earlier Jean Senebier (1775, 1:5; also 2–12, 55) had proposed how a naturalist should go about seeing:"The Observer is a person who looks at Nature like a book; in which he must carefully read the characters without thinking to imagine the meaning they must have." The observer was like "a lover who contemplates the face of his beloved, without wishing any feature changed."
34. See Stevens 1990:391–394 for a discussion. Augustin-Pyramus de Candolle (1827, 1:viii) called both Adanson and Jussieu instinctive systematists; Linnaeus (1751:97, 117), É. Geoffroy Saint-Hilaire and Cuvier (1795a:172), and Cuvier (1829:3) provide other early uses of the word. Pierre Turpin (1820:14) noted that what was "seen" by intuition could not be communicated, although a sound taxonomic intuition could to a certain extent be acquired by much observation. Christian Gottfried Nees von Esenbeck (1831:58) observed that the laws of nature were

written by instinct, not artificially; he opposed "instinctu" by "manu." Adrien de Jussieu (1848:5), although coming from a very different philosophical background, thought that the grouping of species into genera, and still more of individuals into species, was a natural function of the mind. Although Nees von Esenbeck's instinct is unlikely to be the same as Jussieu's natural function of the mind, in both cases the impression is given that things just happen. William Whewell (1848, 3:355–366) thought that an instinctive feeling about relationships was often found to be based on physiological differences—but we have seen that plants showed little in the way of obvious physiological variation (see also McOuat 1992:68–69). I am far from deriding instinct, but to invoke it tends to terminate discussion of the important subject of how a systematist judges relationships. Carlo Ginzburg (1983) suggests how such instinct might work in a nineteenth-century scientific context.

35. Bentham 1875a:27. See also Cain 1959b:243;Simpson 1961:[vii]; Winsor 1991. M. P. Winsor (1991:272) suggested that the apprenticeship system would lead to systematics being marginalized (the context is zoology in the later part of the nineteenth century).
36. In the 1930s L. M. Perry was told by B. L. Robinson, of the Gray Herbarium, to read Bentham's descriptions (in Bentham and Hooker 1862–1883) "if she wanted an example of good Latin." At the beginning of every family Perry subsequently took up for the "Plantae papuanae archboldianae," a work that runs for some 700 pages, she went to Bentham and Hooker's *Genera plantarum* to see how that family was treated (L. M. Perry, pers. comm.). Bentham himself remembered his conversations with Antoine-Laurent de Jussieu (George Bentham to Asa Gray, Dec. 9th, 1982, GH), so there are just three intellectual generations in some 200 years.
37. Mirbel 1815, 1:860.
38. Bentham 1827.
39. E.g., Bentham 1858, 1861b; the largely unpublished manuscript written at the time he gave a copy of his *Handbook of the British Flora* to the Linnean Society and the manuscript introduction to the *Genera plantarum* also bear evidence to this (Bentham, *Bot. Mss.* pp. 1–143, 157–192. K). Bentham later figures very prominently as both logician and systematist in Stanley Jevons's *The Principles of Science* (Jevons 1874: Bentham was the most-cited botanical systematist after Linnaeus), and Jevons in turn is closely connected with the work of J. S. L. Gilmour (e.g., Gilmour 1989. This indirect link between Bentham and Gilmour has been overlooked.) Indeed, Hooker preferred to leave discussions of the theoretical basis of higher taxa to Bentham (J. D. Hooker to Asa Gray, Feb. 20th, 1857,

GH). Bentham, however, was averse to what he called Hooker's "speculations" on evolutionary matters (George Bentham to J. D. Hooker, Aug. 1st, 1859, *Letters*. K.)

40. A.-P. de Candolle 1862:29–32. See also Dunal 1842:12; Martius 1843:219. Systems generated by those imbued with ideas of *Naturphilosophie* were obviously linked directly to philosophical concerns. Systematists also tended to be reluctant to think about the broader implications of their work, and J. D. Hooker was rather exceptional in his belief that the ability to theorize, and by that he meant the ability to think about larger issues like evolution, was necessary for anybody who aspired to be in the front rank of the discipline (J. D. Hooker to Asa Gray, 1858, GH). Yet we saw above that Hooker downplayed "theory" when it came to the description of new species. See also Darwin's comment about Robert Brown: "He never propounded to me any large scientific views in biology." (Mabberley 1985:328).
41. Bessey 1894:244.
42. If stability of classifications is to mean anything, it must be stability after the evaluation of new characters, or agreement of the classification with relationships suggested by completely different data—William Whewell's consilience of induction (Whewell 1840, 2:230). Although new data have been added to classifications over the years, given that there is neither any clear way of evaluating data nor of presenting the data on which a particular classification was based, relating classifications to observations is no simple task. Knowledge of character variation is not always cumulative; knowledge can also be lost. Kevin de Queiroz (1988:253) comments on stability; he suggests that because biological science predated recorded history, this would impart inertia to classifications (ibid., 241).
43. Linnaeus 1736b:485–486; see Stafleu 1971:126–127. For Ray, see Sloan 1972:32–39. Tournefort (1694, 1:20, 28) also claimed simply to be doing what people before him had done.
44. Saint-Hilaire 1827:125, and 1825a:107, 108.
45. For contemporary comments on taxon limits being reached by consensus, or being a matter of convention, see Whewell 1823:xix; A.-H.-L. de Jussieu 1843b:529; Robinson 1906:85–88.
46. See Sachs 1875:157.
47. E.g., Marchand 1867:22; Hutchinson 1923:67; Davy de Virville 1954:70; Stafleu 1963:204–207, 1964:[xxiv]–[xxv], and 1971:129–135. Tod Stuessy (1990:197) comments on the stability of generic concepts. Max Walters (e.g., 1961) sees much eighteenth-century classification as the formalization of groupings that were particulary apparent in Euro-

pean plants to the Europeans who were cataloging nature, these groups consequently had a strong Eurocentric bias. This may well be true, and I have commented elsewhere in this book about areas where such a bias is apparent, but what is interesting here is the stability of these groupings and the fact that many naturalists believed that they represented Nature.

Some of the hitherto most stable and accepted groupings at all levels of the hierarchy are now being questioned. Several of Jussieu's uncontestably natural families will have their limits changed as the limits of taxa are brought to conform with the phylogenetic principles that many think should inform systematics. The Umbelliferae will include the Araliaceae (as Bernard de Jussieu had suggested—see Jussieu 1789:lxvi) and the Cruciferae will go with the Capparidaceae (Judd, Sanders, and Donoghue 1994); the Liliaceae have been dismembered, and just where the consituent elements will appear in a phylogenetic system is unclear (Dahlgren, Clifford, and Yeo 1985 do keep most of the families into which they split the Liliaceae within the same higher grouping, but with the addition of families like the Orchidaceae); and the Labiatae and the Verbenaceae seem at this stage almost inextricably entangled (Judd, Sanders, and Donoghue 1994), and thus what appeared to Jussieu to be an easy separation is disappearing. Although the Compositae (which Jussieu of course did not really recognize), the Leguminosae, and the Gramineae are three groups that remain intact despite these differing views of nature, their ontological status has changed. John Ray (esp. 1703; Morton 1981:203–206 provides a good summary) early drew attention to the distinction between the monocotyledons and dicotyledons, but it is unlikely that the two will be maintained with their present circumscription (see Donoghue and Doyle 1989:37), which, after the excision of the "gymnosperms," goes back to Jussieu.

48. Biological nomenclature is supposed to deal with names alone, not with concepts, but historical examples show how far wrong this idea can be (La Vergata 1987; Stevens 1991). Gordon McOuat (1992) has recently shown that in England in the first part of the nineteenth century social arguments and scientific nomenclature were connected.
49. Adams 1887:117. Charles Withers (1992:300–301) suggests that a practical emphasis in the study of natural history may have been widespread at the end of the eighteenth century; names of organisms were needed for the numerous acclimatization experiments that were common then.
50. Smith 1817:389, often repeated.
51. Dagognet 1970:54; see A.-L. de Jussieu 1788:191. Geologists have a public that is very directly affected by their findings. In the nineteenth century, miners in particular and prospectors in general utilized geo-

logical knowledge, while in this century, oil companies have become ever more important. If a shaft is sunk because strata are wrongly named, the negative consequences are immediately apparent; moreover, such errors retard progress in geology itself. Mine-owner and geologist alike would seem to have common interests in the correct circumscription and nomenclature of strata. However, in botany changes of relationships seem to have been of little consequence to anyone other than systematists. The relationship between the use of classifications by nonexperts and their construction by experts needs more detailed study; ornithology is another area with some of the same pressures that occur in botany.

52. A.-P. de Candolle 1832, 3:1524. Success in grafting could be predicted from the position of the plant in the natural order, not from its general appearance. A few other remarks suggest a weak linkage between the natural system and the general utility of classifications based on it. The Museum of Economic Botany at Kew was arranged in part following natural affinities (Drayton 1993:323). W. T. Thistleton-Dyer (1880:14) observed that the natural system might help predict if a new seed had poisonous properties. Richard Drayton (1993) has recently provided a comprehensive discussion on the justification of nineteenth-century classifications in England by their utility. See also Farley 1982:50; Brockway 1989 is of general interest for this period.
53. McOuat 1992, chap. 8.
54. Thistleton-Dyer 1880:8.
55. A.-L. de Candolle 1880:100.
56. J. D. Hooker to C. Darwin, [25] Feb. 1858. See Burkhardt and Smith 1991:34; L. Huxley 1918, 1:454; Stevens 1990:403.
56. Elliott 1992:473–476. Although one might be disinclined to dimiss as rhetoric the suggestion made in 1831 (Anonymous 1831:65) that the splitting of species might be accommodated by the experts, but not by amateurs, such a statement should have made naturalists pause for thought. Who should tell them what their species—nominally the basic unit of their botanical currency—should or should not be?
58. H. K. Airy Shaw (1902–1985) and John Hutchinson (1884–1972; a self-taught botanist) were both at the Royal Botanic Gardens, Kew, and they disagreed over the relationships of many families. They moved some genera around the herbarium to their "proper" places, and these genera were then temporarily missing (B. Verdcourt, pers. comm.).
59. Engler 1893; see Drayton 1993:404–430.
60. See chapter 7.
61. D. G. Frodin (1984:25–29) discusses the utilitarian aspect of floras. He

notes that particularly since the 1930s, floras have taken a dominating position in the discipline. They have been important for far longer.

62. Sprague 1925:10. For later papers that raise some of these issues, see Camp 1940:381; Greenman 1940:371, cf. 373; Sherff 1940:376; Walters 1961:83–84.
63. Condorcet 1778:28–29.
64. Buffon 1749:14–15. The "dangerous" separation that Henri Daudin (1926a:189–195; see also Baehni 1957:84) saw developing between identification (class hierarchy) and the description of the natural order (continuity) would seem to be almost inevitable.
65. Mark Barrow (1992, chap. 4) discusses this issue in the context of nineteenth-century ornithology.
66. Gray 1836:311; see also A.-P. de Candolle 1828c:11.
67. A.-P. de Candolle 1810a. See chapter 4 and appendix 2.
68. Desvaux 1813:285.
69. Alfred Moquin-Tandon suggested that the disposition of Candolle's projects in his will was not well-advised (Léotard 1893:116–118). Moquin-Tandon thought that Alphonse de Candolle did not really understand Augustin-Pyramus's *Théorie élémentaire*, and it would have been better if Michel Félix Dunal, not Alphonse, had been entrusted with bringing out the third edition of that book.
70. A.-P. de Candolle 1862:216. See Candolle and Sprengel 1820, translated by Jameson in 1821 and reprinted in 1978 by the Arno Press, New York. Lincoln Constance (1955:407) relied on this version for his discussion of what he took to be Candolle's species concept (see A.-P. de Candolle and Sprengel 1821:95–97; cf. Candolle 1819a:193–215).
71. E.g., Voigt 1806; Fuhlrott 1829.
72. Duval-Jouve 1871:475–476.
73. Baron 1931:331. Baron was also reading Candolle in translation, but this time a more accurate translation by Johann Jacob Roemer of the first edition; on p. 244 of this translation, Roemer had it right (cf. Candolle 1813a:27 [the reference Baron picked up], 212, cf. 213).
74. Bonnet 1764, 1:30.
75. See A.-L. de Jussieu 1778:196–197. But compare with Roscoe 1813:76; Marchant 1858:214.
76. Stevens 1986:325–329; see also Merxmüller 1972:319.

Chapter Eleven

1. Morton 1981:311.
2. A.-L. de Jussieu 1837:239.

3. But cf. Schama 1989.
4. See Nuttall 1818, 1:vi (he compared the revolution in science with that in politics); A.-P. de Candolle 1828c:7, 11 and 1829a:34; Cuvier 1829:4; A.-H.-L. de Jussieu 1834:303 (a paper that was written to set the record straight over the genesis of the *Genera plantarum*); Gray 1836:311; Flourens 1838:xxvi, also Hamy 1893:13; Cohen 1985:529; Le Guyader 1988:45 et seq. But Linnaeus could equally be described as a revolutionary because of the nomenclature he devised and his ordering of species into genera (Condorcet 1781:72; Desfontaines 1795:517). Moreover, the revolution that Asa Gray and Augustin-Pyramus de Candolle mentioned was not so much carried out by Jussieu himself as facilitated by developments in plant morphology (for which Jussieu was not directly responsible) combined with Jussieu's approach to the problem of construction of the natural system.
5. Adanson's earlier, but ill-fated endeavors in this direction should not be forgotten.
6. A.-L. de Jussieu 1837:230. Remember that his championing of continuity may be connected with ideas current in late eighteenth-century Paris of what was proper procedure in science—see chapter 3. Also, "Jussieu's" ideas are simply ideas that he held, not necessarily ideas that were original to him.
7. The ambiguities in Jussieu's position over what needed to be observed should be borne in mind and also his distinction between characters (based on internal features) and differentiae (lacking this basis).
8. See also Gould 1987a:16–23.
9. Some plant morphologists have been more inclined to see a continuum of form, although this is largely because form is divorced from its systematic context.
10. Dobzhansky 1937:305; Turrill 1942:270; Davis and Heywood 1963:30; Blackwelder 1967:7–10; Sloan 1972:2; Patterson 1977:632; Dean 1979:216–217; Stevens 1984b:183–202 and 1986:326; Mabberley 1985:144–145; Winsor 1985:57–62 and 1991:115; Joly 1986:57. However, the period c. 1860–1900 has been called the "evolutionary period" (Alston and Turner 1963:39), with evolution supposedly profoundly affecting systematic thinking.
11. Greene 1971:22.
12. Darwin 1959:485; Eichler 1875:iv, 1–2; Naef 1919; see Coleman 1976; Desmond 1982; Di Gregorio 1984:80.
13. See Stevens 1992 for a brief summary. Again, botanists lagged behind zoologists in the adoption of the new approaches, and the changes barely affected herbarium systematists.
14. See Smocovitis (1992a) for an interesting account.

15. Delpino 1869a:268 and 1869b:302. The main axis of his diagrams was across the page; smaller genealogies were descending (Delpino 1890:5, 8, 19). The famous treelike diagrams which Ernst Haeckel (1866, 2:pl.1, pl.2) used to represent his idea of evolution of life are earlier (for Haeckel, see Oppenheimer 1987). Charles Victor Naudin (1852:106) also used the tree metaphor; although he believed in evolution, I do not know that he made any diagrams of relationships. For the interesting work of the paleozoologist Albert Gaudry (e.g., 1866:36) see the recent discussion by Pascal Tassy (1991:60); Darwin (1859, chart facing p. 117) also included a schematic tree in the *Origin*.
16. Delpino 1888:6. Peter Endress (1993:79–82) summarizes Delpino's evolutionary system.
17. Delpino 1889:32. He placed *Sciadopitys*, of the Coniferae, near a branching point to allow all its relationships to be satisfactorily depicted; his classification is not at all closely related to his tree.
18. Buchenau 1890:52–59.
19. Kusnezow 1896:79–84, figs. on pp. 81, 84. An evolutionary study by Thomas Henry Huxley (1888) on gentians came in for extensive comment (Kusnezow 1896:26–32). Another anatomist, Émile Constant Perrot (1899, fig. 29 on p. 285), also produced a diagram showing reticulating relationships in the Gentianaceae. Reticulating relationships are also shown by Ferdinand Albin Pax (1888, numerous little diagrams with reticulations from p. 164 onward—only a few of these diagrams depict hybridization). Yet another anatomist, Julien Vesque (1889a, pls. 3, 17, 26, 76, and 1889b, pl.39), depicted reticulations, both within and between sections; the complexity of the relationships shows was further increased by the distributions of characters that Vesque included in his diagrams.
20. Tchouproff 1895:552.
21. Bunge 1872:11, pl.2, see also p. 9.
22. Bentham 1873b, pl.11; cf. Cassini 1817:20–21 and 1816–17, pl.83.
23. Bennett 1888, fig. facing p. 60. Stevens 1984b:192–195 and Lam (1936) provide further references. Alston and Turner 1963:16–35 include a selection of diagrams from the later twentieth century. The tree that Huxley (1888, pl.2) produced in his work on the Gentianaceae is very much in the spirit of that of Bennett, but some characters are indicated on the tree. Hans Gottfried Hallier (1893:586) shows a tree drawn from above, so the focus is on the central "Urform" (see also Hallier 1903:19); groups are joined serially. Some of Hallier's diagrams show considerable symmetry (Hallier 1893:533).
24. O'Hara 1988a; Bowler 1988:197; Ruse 1988:99–104.

25. Thomas Junker (1989) discusses the reception of Darwinism by German botanists; the people he deals with include several systematists. At least Nathanael Pringsheim, Julius Sachs, and Carl Wilhelm von Nägeli toyed with ideas of orthogenesis; such ideas are often connected with those of parallelism (for which, see note 35: Junker does not discuss diagrammatic representations of nature). Peter Bowler (1983: chaps. 6 and 7) shows clearly how widespread ideas of orthogenesis had become on both sides of the Atlantic and among various biological subdisciplines at the end of the nineteenth century.
26. Botanists like Hallier and Buchenau (notes 18 and 23) continued to talk about types.
27. Strasburger 1872, vol. 1, fig. facing p. 264.
28. Engler 1874a:111.
29. Engler 1874b, pl.13, fig.28.
30. Engler 1881:398–401, pl.4. He noted the problems caused by convergence and the difficulty in distinguishing between primary (primitive) and secondary (advanced) characters.
31. Lankester 1873:321. This is not the happiest of comparisons, because family pedigrees are simple only because they focus on a subset of the descendants.
32. Bowler 1989:286–290.
33. Heckel 1894:PPP; Wettstein 1895:288.
34. But see Ferdinand Albin Pax (1884:416–420, fig. on p. 417) and Otto Warburg (1897:100–110, fig. on p. 102). Pax, closely associated with Engler throughout his career, tried to show parallelisms on his tree, and parallelisms were of interest to Warburg (see also Cusset 1982:37). Willem Frederik Reinier Suringar (1895:13–16, chart at end; see Anonymous 1895, for a French abridgement) produced a magnificent tree showing relationships between all plants, including fungi. Suringar was interested in series, and his tree is very different from those of Pax and Engler.
35. Engler (1897) listed 32 principles of "scientific (evolutionary) classification"; along with progressions, parallelism was of considerable interest (Constance 1955:409–410; Meeuse 1975:156–157 discusses Engler's ideas of progress in the angiosperm flower).
36. See Wettstein 1895:283–289 and 1896:368–379 (the latter work discusses relationships within European species of *Gentiana*).
37. Solms-Laubach 1887:1–2.
38. Göppert 1881:212. Bernard Renault (1879), in a paper cited by Solms-Laubach as one of those that established "genetic relationships" between plants, took exception to the idea of prototypes; they were not the linking groups they were supposed to be.

39. Tassy 1991:118. Adolphe Brongniart (1828c:255–258) gives a good example of the kinds of tabulations involved.
40. Tassy 1991:59–63, 70–72, 83–110.
41. Ibid., 96, 111–130.
42. A.-L. de Candolle 1880:323–328, esp. 327. Candolle discussed evolution in his published work and in his correspondence (Candolle 1862 and A.-L. de Candolle to G. Bentham, Feb. 27, 1875, Bentham, *Corr.* 2:1001. K); the possibility of widespread hybridization does not seem to have been an issue. José Triana (1872:3) provides a smaller-scale example of how difficult it was to see relationships.
43. J. D. Hooker 1873:993.
44. Ibid., 994. Relationships were reticulating (see Turrill 1963:116), directly between families, and ideas of highness and lowness were close to the surface (ibid., 81–85). In the famous debate with Samuel Wilberforce, Bishop of Oxford, at the British Association meeting in 1860, Hooker compared relationships within a group to the threads and knots of a spider's web; "at least" half of plants could be placed in groups (Burkhardt, Porter, Browne, and Richmond 1993:596)!
45. Darwin 1859:411. See note 57.
46. Cf. Bentham 1856:54–55 and 1873a:14–16. William Philip Hiern (1873:61–63, pl.1) includes a diagram of relationships that is consistent with Bentham's visualization of nature.
47. Hugh Strickland (1841:190) early noted the similarity between the analogy of the tree and that of a map.
48. See especially O'Hara 1988a, 1991; Craw 1992:68, refs. Bunzô Hayata (1921:104, 177) out-hookered Hooker in *his* visualization of relationships being comparable to a net of infinite extent with crystal beads, each one on a mesh of a different color and reflecting the images of all the other crystals. It is not surprising that T. G. Tutin (1952) despaired that little progress had been made in understanding the origin of any major group of angiosperms in the 92 years since the publication of *On the Origin of Species*.
49. Bessey 1915, fig.1. See Bessey 1894, fig. on p. 251, and 1897, figs. 1–3, for early versions which look more like Delpino's diagram. Alain Cuerrier and I have recently looked at Bessey's ideas of phylogeny, evolution, and classification in some detail; Bessey's classifications, rather like those of Jussieu, do not represent simple hierarchies.
50. Rodrigues C. 1950; see Stevens 1984b:191–193, 200–201 for discussion. One of Rodrigues C.'s redrawn horizontal Besseyan diagrams shows physical continuity between all groups (ibid., fig.1).

51. Rodrigues C. 1950, fig.1. Parallelisms are not considered "real" relationships in this context.
52. Bather 1927:lxxi.
53. Coleman 1976:168; see also Bather 1927:lxxxv. My earlier comments (Stevens 1984b:200–201) on this "vice" were wide of the mark; the issue is not reticulating relationships, but continuous relationships.
54. A. L. Panchen (1992:25) emphasizes the differences between minimal spanning trees, the technical term for trees discussed here, and minimal Steiner trees, which include genealogical trees in which terminal taxa are connected only by extinct ancestors.
55. English 1987:311–312. No way of integrating genealogy and amount of change has been suggested (Mayr and Colless 1991:123); in attempting to do this, Tod Stuessy (1990, chap. 9) shows how difficult it is.
56. Bessey 1894:238–239.
57. For example, Darwin (1859:419, 426) suggested that chained relationships, with organisms at the two ends of the chain showing no similarities, were one result of the evolutionary process; he also noted that adaptations caused problems in the detection of genealogy (ibid., 440). Although Darwin's ideas of genealogy and classification are complex (ibid., 414–428), it is overstating the case to suggest that "Darwin's systematics is the theory that taxa stand in relation one to another, as ancestors and descendants." (Nelson and Platnick 1984:146. See also Patterson 1982:64; cf. O'Hara 1993:233.)

    Zoologists like Edwin Ray Lankester were ambiguous about the relationships between genealogy and classification (Bowler 1989:286–290), and Thomas Henry Huxley was a particularly interesting case. He changed neither his methods of detecting relationships nor his way of producing classifications after 1859 (DiGregorio 1984, chap. 3; Hull 1988:97). His diagrams of relationship are not easy to interpret and will repay a close study; groups are joined to groups, and levels of advancement are prominent. As he observed, "*Classification by gradation*, and the formation of natural series, is another stage in the same progress, and must by no means be confounded, as it often is, with the ultimate result—though, in all probability, it represents the true genetic classification more nearly than any other arrangement can do" (Huxley 1868b:361, fig. on p. 362). Add Hooker to this group (note 44), and Darwin's major supporters hardly demonstrated a novel approach to classification or to understanding relationships (see also DiGregorio 1982:253–254). Insofar as there were glimmerings of a new approach to evaluating data, constructing trees, and producing classifications from those trees, they are to be found in the zoological literature; Fritz Müller is an important early example (Müller 1864:6–7; see Craw 1992:69–72).

58. Sloan 1979:132. A.-L. de Jussieu (1789:xxxix, xlii) had implied that there was a parallel between the progressive groupings of his synthetic method and genealogical connections, while Carl Fuhlrott (1829:11) suggested that similarity in characters was like both chemical relationships and relationships through descent. As suggested in chapters 3 and 7, too much importance should not be attributed to such remarks.
59. Constance 1964:271.
60. A.-P. de Candolle (1829a:35) observed, "those to whom the revolution in [taxonomic] logic was originally due had paid little attention to making known the fundamental principles and to foreseeing the consequences." It was not entirely the unnamed revolutionary's fault. Jussieu was about as clear as Candolle.
61. Duval-Jouve 1871:475–477.
62. Radlkofer 1883:28, 61, 62. See also Harvey-Gibson 1919:76–79, 112–115; Hocquette 1954:136–139; Metcalfe and Chalk 1979:3. It was fortunate that 1883 was the centenary of the death of the younger Linnaeus. The achievements of the Linnaeuses, father and son, were simultaneously acknowledged and consigned to history; the torch of progress was handed to Radlkofer and the anatomical method. Yet anatomy did not provide much information to Radlkofer in his work on the Sapindaceae, although it did confirm the incorrect placement of some sterile material wrongly placed in the family (Radlkofer 1890:296).
63. Cell- and tissue-level studies were not integrated into systematics (Sprague 1940:450–452; see Stevens 1984b:182).
64. Adanson 1763, 2:390–401; Lawrence 1951:593. See Stafleu 1963:209–211. Lawrence (1953:118) suggested that there had been overmuch reliance in the preceding century on what he called "exomorphic characters of organography.".
65. A.-L. de Candolle 1837b:427.
66. For example, the *Annales des Sciences Naturelles* was started in 1824 to provide a single outlet for findings in physiology, anatomy, comparative studies, botany, zoology, geology, mineralogy (Audouin, Brongniart, and Dumas 1824). Ten years later it split into botanical and zoological parts, shedding most of geology and mineralogy as it did so.
67. A.-P.-C. de Candolle 1879.
68. E.g., Haeckel 1866 and 1874:276. For more on Haeckel, see Gould (1977) and Barabé (1986).
69. A.-L. de Candolle to G. Bentham, 20 Fevr 1879, Bentham *Corr.* 3:1029. K.

70. For example, Sachs 1868.
71. A.-L. de Candolle to J. D. Hooker, 18 Avril 1883, *Dir. Corr.* 135:418. K, emphasis in original. This particular exchange of letters with Hooker started with Hooker's request that Alphonse should produce a sequence of plant families which everybody could follow, rather than having everybody coming up with their own—see also chapter 7.
72. For Julien Vesque and his "epharmonic" characters, see Vesque 1882 and 1883:105–119. He was interested in relating the adaptive and functional importance of characters, especially those taken from cell- and tissue-level anatomical studies, to their importance for systematics.
73. A.-L. de Candolle to J. D. Hooker, 15 Mai 1883, *Dir. Corr.* 135:419. K.
74. Lankester 1873. There is an interesting discussion on ontogeny and phylogeny by the botanist Eduard Adolf Strasburger (1874:5–21); ontogeny did not always yield evidence of relationships.
75. Stevens 1986. Recent discussions in systematics have emphasized that what is important is not only what is observed, but how the observations are evaluated and what classifications represent.
76. Queiroz 1988; cf. Greene 1992:269. Vernon Pratt (1982:210–223) suggests that even Cesalpino's classification was not really about essences, but his assertion remains to be substantiated.
77. The suggestion that there was a "delay" in the systematization of nature after Darwin (Queiroz 1988:241) needs to be modified.
78. In line with this argument, R. J. O'Hara (1993:233) suggests that some zoologists in the first part of the nineteenth century were makers of systems rather than classifications.
79. Turpin 1820:12.
80. Stafleu 1971:46.
81. Daudin 1926a:114–115.
82. Larson 1971:143.
83. Winsor 1976a; Stevens and Cullen 1990, refs. A. J. Cain (1992) has recently suggested that hermeneutic ideas may have affected the patternings Linnaeus saw in nature.
84. Winsor 1985:33, 34, 55.
85. Winsor 1976b.
86. Just how satisfactory the natural system in zoology really was needs close examination (see Wilson 1901:17). By the end of the nineteenth century the search for genealogical trees had become a matter of some ridicule (e.g., Bowler 1989:292–295). M. P. Winsor recently suggested that there were substantial problems within zoology by the end of the nineteenth century; her recent book, *Reading the Shape of Nature* (Winsor 1991), will repay careful reading with this issue in mind.

The situation in zoology is certainly not simple. Several authors were using or mentioning tree diagrams and analogies by the 1840s, e.g., Strickland (1841:190, fn.; cf. 1844) and Wallace (1855:187, 191). See DiGregorio (1982:245–248) and Desmond (1989:369–372) for further references. Phillip Sloan (1992:34–37, 43, 63–64) suggests that trees drawn in 1828 by Joseph Green, and copied by Richard Owen, can be linked to Green's particular concerns that included an integration of the ideas of Kant and Lamarck. Henri Milne-Edwards (1844, fig. facing p. 98) produced a diagram that shows taxa as boxes within boxes; the relationships can be represented as a branching tree (cf. Barsanti 1992a:94). The diagram also shows "collateral affinity" and "recurrent development"; "parenté zoologique" and trees are also mentioned (Milne-Edwards 1844:65, 75–77; cf. p. 82. A.-P. de Candolle 1826, pl. 27 [see figure 14, chapter 7], presents a similar diagram). Milne-Edwards was interested in development (ibid., 66), and his discussion of relationships is compatible with Karl Ernst von Baer's ideas on this subject, but not those of Ernst Haeckel or Étienne Serres (Serres was Milne-Edwards's particular adversary (Milne-Edwards 1851:89–115; see Russell 1916:203–208 for Serres and Ospovat 1976:11–12 for Milne-Edwards's arguments). Milne-Edwards (1844:81) defined a natural group of organisms as all those organisms derived from the same type; in general, his approach is quite compatible with that of Darwin (1859:411–413). Prominent zoologists who believed that there were discrete groups in nature include T. H. Huxley (1868a:357; see DiGregorio 1984:32–33) and Wallace (1856:195).

87. Lamarck 1809:112. See also Cuvier 1812a:73; A.-P. de Candolle 1813b:204–205 and 1829a:53; Lindley 1829a:3; Schwann 1839:v. Schiller (1971a:87–88, 94) and Stemerding (1991:115–123) discuss this point. The problem was apparent to Louis-Jean-Marie Daubenton by 1765 (Daubenton 1765a:227).
88. Buffon 1750:10; Flourens 1838:xxvii; Hooker and Thomson 1855:15–16. Brent Berlin (1992:149) suggests that even at the level of folk classification mid-level groupings of animals are more evident than are those of plants because related animals are more inclined than plants to occupy "a particular adaptive niche."
89. Bessey 1894:244.
90. Such occasions seem to be almost endemic. The lack of discussion over John Ray's attempt to integrate Lockean thinking with classification may be an earlier episode where lack of discussion is noteworthy (Sloan 1972:14–15; cf. Pratt 1982:210–223).
91. Mayr 1982a:204. Mayr appears to include two ideas about synthesis here:

first as upward classification (gradually forming groups of larger extent) and second as the use of many characters.

92. Kuhn (1977 and refs.) provides a good entry into the literature on paradigms; Bernard Cohen (1985) discusses revolutions in science.
93. Gutting 1980:1–2, 13; Greene 1971:3; Kuhn 1970:176–191 and 1977:305–308; Prelli 1989:85–87.
94. The arguments between Henri Baillon and Jean Müller over the virtue of a Jussiaean versus an Adansonian approach to weighting (described in chapter 5) never amounted to much.
95. Kuhn 1977:298, 299, as "taxonomy."
96. Gillispie 1980:152.
97. Foucault 1973:146–148.
98. Foucault 1970:71 and 1973:137; anatomy was not natural history. Foucault thought that plant classification led that of animals because zoologists had for a time dropped their interest in anatomy, and botany presented richer opportunities for those whose gaze was restricted to the surface of things. But his position is partly based on the assumption that dissection of a flower and dissection of a worm are fundamentally different operations. It can even be argued that with Adanson, Jussieu, and others, we see an attempt to reintroduce characters such as those of the qualities of the plant into classification (e.g., Adanson 1764, 1:clv; see Tort 1989:225–233), although the nature of the relationships that could be shown by these qualities had become much restricted from those in which they were involved in earlier times.
99. Foucault 1973:226–232. For Foucault and seeing, consult a useful account in Sheridan 1980:59–60. Insofar as naturalists had started looking at a greater variety of characters much earlier, as well as privileging studies of gross anatomy and restricting the limits of allowable relationships, we should look more closely at the work of people such as Andrea Cesalpino and John Ray for the early stages in these developments.
100. See Nicolson (1987:187–188) for a brief discussion and references and Laudan (1987:86).
101. Bynum 1975:20. See also Schiller 1980:85–89; Caron 1988:223–225.
102. Foucault 1973:230–231.
103. As I suggested in chapter 9 in the context of change in natural history, it is helpful to think of concepts being polythetic.
104. Camille Limoges (1980:236–239) discusses this shift in the context of the Muséum national d'Histoire naturelle in Paris.
105. Daudin 1926a:64, 110–115, 189–195; Baehni 1957:84.
106. Larson 1971:49; Sachs 1875:106. See Linnaeus 1751:62, 98–99.

107. Daudin 1926b, 2:4–5.
108. Thus for Haller, Bonnet, and Buffon all discontinuous systems were artificial (Anderson 1982:34–42; Jahn 1985:282). As Jacques Roger (1989:381–382) observed, Buffon was interested in all the characters, not the classification.
109. Cf. Linnaeus 1736a:10–16. This work on bananas and their relatives is a minor exception.
110. Adolphe Brongniart (1837:9) disapproved of the use of features proper to a key in a natural classification.
111. Systematists mistakenly continue to insist on this latter point (consult Stevens 1990:389–390, 404, refs.). A. H. Gentry (1993:1) provides a recent example.
112. An arrangement in the sense of Lamarck (1809: chap. 5) and Wallace (1856:195); this is equivalent to the natural ordering of Hempel (1965:153).
113. For example, A.-P. de Candolle 1828c:11 and A.-L. de Candolle 1841:149. See also Gray 1836:311.
114. Beckner 1959:24–25; Needham 1976:362.
115. J. É. Planchon 1851:403.
116. I do not know if the "degree of improvement" he saw his method as providing referred to groups, which now had more clearly delimited boundaries, or to nature, the overall shape of which was clearer.
117. Giseke 1792:xv–xvi.
118. A.-L. de Jussieu 1789:lxii.
119. See Rieppel 1987a:127 and 1987b:414, 417; he does not develop the idea.
120. Robinet 1761–1766 and 1768. See Lovejoy 1936:269–283; Roger 1971:642–651; Van der Hammen 1981:7–8; Anderson 1982:48–49. A prototype or archetype might be a unique organization of an atomistic nature (Rieppel 1988:36, 37), but the use of the term organization in this or similar senses for groups above the level of species is incompatible with continuity as it is defined here (cf. Atran 1990:205, fn.22). For the relationship between continuity and essences, see Slaughter (1982:35) and chapter 3.
121. Foucault 1973:146, emphasis in the original.
122. Cf. Stafleu 1964:[xxii]; Bernier 1984:10.
123. A. Smith 1980:37.
124. Pallas 1766; Fries 1821:xvi, xviii–xix; Marchant 1858:214. See Schiller (1980, chap. 11).
125. Robinson 1906:82–83; see also Cook 1898:187.
126. Planchon's work shows that the description of types did not necessari-

ly make the boundaries between groups apparent; Planchon did not believe in the existence of channels.

127. A.-L. de Candolle 1841:149. Alphonse paid much less attention to symmetry, and to comparative morphology in general, than did his father.
128. Candolle 1844:255–258. If variation within a family were reduced to a type, and types of different families were compared, sharp differences might become evident. Interestingly, both the Loganiaceae and the Apocynaceae are paraphyletic, i.e. phylogenetically privative (Bremer and Struwe 1992; Judd, Sanders, and Donoghue 1994).

Epilogue

1. Holton 1975 and 1988, chap. 4.
2. Rieppel 1988, chap. 6.
3. The conflict between Cuvier and Étienne Geoffroy Saint-Hilaire can be interpreted along these lines (Bernier 1988:10). There are similar tensions in morphology as well, as is evident in two articles written a century and a half apart (Turpin 1837:7–10; Hagemann 1982). Synthesis and analysis may ultimately be complementary, as some nineteenth-century commentators realized (e.g., Turpin 1820:10; Sattler 1986), but there has often been tension between them.
4. The synthesis of nature achieved by Linnaeus, the arch classifier and encyclopaedist, is also unexpected. He established systematic and especially morphological continuities that bound nature in a seamless, indivisible whole, even integrating becoming and process within it (Stevens and Cullen 1990).
5. Coleman 1976:168; see also Bather 1927:lxxxv.
6. For references see O'Hara 1988a; Queiroz 1988; Craw 1992.
7. Miller 1939:29.
8. The tradition continues in ethnobiology, where the "nature" in folk classifications is compared with the "nature" in expert evolutionary classifications (Berlin 1992).
9. A.-L. de Candolle 1837b:282.
10. See Kevin De Queiroz (1988:241) and R. J. O'Hara (1988a, 1992) for the consequences of this changed approach to the study of nature. In chapter 10, note 47, recent work is mentioned that emends the limits of some families hitherto considered to be eminently "natural."
11. To compare the development of our understanding of the relationships of flowering plants with that of animals may be inappropriate. We have to compare groups that are really comparable: flowering plants may have

evolved in the Jurassic, perhaps earlier, but they are far younger than the vertebrates, let alone many invertebrate groups. Could the difficulties we find botanists encountering simply be the result of the age of their group? I think not, but this aspect of history, that of the organisms themselves, should not be forgotten.

12. By this I mean that groupings will all be monophyletic, containing all and only the known descendants of a common ancestor. However, endosymbiosis and differential movement of the elements of the cell (nucleus, cytoplasm, and organelles) in what can loosely be called hybridization events make this definition more problematic than it seemed to be fifteen years ago. Genealogies can be reticulating.
13. E.g., Barnes 1990:68–69; see Dean 1979 for a study at lower levels of the classification.
14. Dean (1979:212) opposes classifications that are "made" with those that are "found." There are several comments in the literature from the second half of the nineteenth century onward as to whether taxa (the discussion was about species) were made or discovered (e.g., Powers 1909; Stevens 1991:165, refs.). Although this distinction needs to be rephrased, because no classification is really "found," all being "made," if in different ways, the issue that Dean raises of how competing classifications are evaluated is important.
15. J. D. Hooker (1861) and W. T. Thistleton-Dyer (1905) provide contemporary accounts of these floras. The writing of Bentham and Hooker's *Genera plantarum* (1862–1883) was closely linked to that of the Colonial Floras.
16. Contributors to floras of British dominions certainly felt under nomenclatural constraints (e.g., Kurz 1877, 1:vi); nomenclatural issues cannot be readily separated from more substantive matters (La Vergata 1987; Stevens 1991).
17. I would expect there to have been contemporary discussion of this issue.
18. Crosland 1978:178–179, see also pp. 139–192; Guyton de Morveau, Lavoisier, Berthollet, and Fourcroy 1787.
19. Crosland 1978:141; see also Dagognet 1970:20, 40.
20. Cf. Fleming 1829:301–303.
21. Jean Jacques Rousseau (in Saint Pierre 1784, 3:282–283) and John Bellenden Ker (1823) expressed such hopes.
22. Philosophers in the late eighteenth century saw the species and genera Linnaeus was naming as being no different from a chemical salt or a crystal type, to name substances in two disciplines particularly affected by botany (for mineralogy, see Whewell 1823:xiv–xv; Laudan 1987:70–81).
23. Except in a literal sense, facts as "faits," things made.

24. Gould (1987b:196–197) specifically contrasts the underlying nature of crystals and organisms as illustrative of his comparison between time's arrow (organisms) and time's cycle (crystals).
25. Some species may have more than one origin, resulting from more than one hybridization event (e.g., Wyatt, Odrzykoski, and Stoneburner 1992:536–537).
26. Bentham 1875a:34. If species appear at a level when reticulating events largely cease, varieties and species may be different in nature.
27. Cf. Greene 1971:22. Lincoln Constance (1955:412–414) discusses recurrent problems in systematics; these are in part caused by its misunderstood history.
28. It is partly because of such influences that I think studies of nineteenth- and twentieth-century systems and systematic practice are areas where a Latourian analysis may prove illuminating (cf. Dirk Stemerding [1991, 1993] and his attempt to apply Latourian principles to understand the prevalence of Cuvier's approach to classification over that of Buffon).

Appendix 1

Note: These notes to translations of works by Jussieu are mostly those of Peter F. Stevens. Beginning with note 44, however, some are those of Jussieu himself. Jussieu's own notes are distinguished by an asterisk after the number.

1. E.g., Linnaeus 1751:51.
2. Rey 1981.
3. Jussieu may be thinking of something like the outline of his arrangement that he included at the end of the Introduction to the *Genera plantarum* (see below).
4. The latin "classicus" has been similarly translated.
5. Tournefort (1700, vol. 1), placed *Potentilla* and *Ranunculus* in his 6th class, the herbs and subshrubs with rosaceous flowers, *Aquilegia* in the 11th, the anomalae with polypetalous flowers; Linnaeus (1753) placed *Rumex* and *Colchicum* in his Hexandria trigyna, *Polygonum* in the Octandria trigynia.
6. E.g., Linnaeus 1751:27–36.
7. For this series, see A.-L. de Jussieu 1789:lxiii–lxx.
8. Adanson 1763, vol. 2. See, however, Auguste Jean Baptiste Chevalier (1934:113 and references) for Adanson's own opinion of the resemblance of his system to that of Bernard de Jussieu.
9. Linnaeus 1764:[611]; Adanson 1763, 2:451–460.
10. Adanson 1763, 2:451–460.
11. John Ray (e.g., 1703) in particular distinguished between plants with one

cotyledon and those with two; however, he made this distinction subordinate to his main division of plants, which was by their habit.

12. Linnaeus's Piperitae included *Arum* and *Phytolacca*, Adanson's Aristolochiae, *Asarum*, *Nelumbo*, *Tamus*, and *Vallisneria* (Linnaeus 1751:27; Adanson 1763, 2:71–76).
13. "Port"—"facies" is another word for this. Some authors (e.g., Fries 1825:25–26) distinguished between similarity in habit, which was much more important, and similarity in external appearance. See note 155, following.
14. Adanson (1763, vol. 2) included quite detailed descriptions of seeds in his family descriptions.
15. Linnaeus 1764:[611]. *Fraxinella* is now called *Dictamnus*, and *Harmala*, *Peganum*.
16. Adanson 1763, 2:450.
17. For instance, see Linnaeus 1754:80. But Jussieu himself later used phrases such as "calix superus" and "germen inferus" in his descriptions (e.g., Jussieu 1789:162–163). Whatever Jussieu thought the real situation to be, the appearance that the calyx really was placed on top of the ovary was evidently too striking to be ignored.
18. Vaillant 1718.
19. Adanson 1763, 2:460.
20. Ibid.
21. Tournefort 1694, 1:40–65 and 1700, 1:56–75.
22. Linnaeus 1754:137, 194, 238, 241, 244.
23. Linnaeus (e.g., 1751:73–74) adopted a very broad definition of nectary based more on function than on morphology. This definition was extensively criticized in the ensuing years, and the circumscription of the term was narrowed (e.g., Necker 1790:13–14); see note 90, following.
24. Jussieu (1789:lxiii–lxx) lists the genera and families in Bernard de Jussieu's arrangement.
25. Jussieu (1789:43–48) includes them all in his family of the Junci (Juncaceae).
26. Linnaeus 1735.
27. Linnaeus (1754) placed *Hypecoum* in the Tetrandria digynia, *Chelidonium* in the Polyandia monogynia, *Andromeda* in the Decandria monogynia, *Erica* in the Octandria monogynia, and *Ulmus* and *Apium* in the Pentandria digynia.
28. Linnaeus 1754:137, 197.
29. B. de Jussieu 1741, 1742. See John Farley (1982, chaps. 2–3) for how analogies guided naturalists' thoughts when they compared plants with obvious flowers with those that appeared to lack them. The reasoning

was similar to that which led Bernard de Jussieu to find anthers in pteridophytes.

30. Adanson 1763, 2:455.
31. Geoffroy 1714: pl. 7, figs. 1–20.
32. Adanson 1763, 2:450.
33. Linnaeus 1764:[611].
34. Tournefort 1700, 1:75.
35. The relationships of *Paeonia* are still undecided; unfortunately for Jussieu's argument, a close relationship with the Ranunculaceae is perhaps unlikely.
36. Tournefort 1694, 1700. His primary division was based on habit.
37. See A. G. Morton (1981:213–220) for an account of the discovery of the fertilizing function of pollen.
38. "Corolla" means literally "little crown" or "garland"; if the corolla had no essential function, it could no longer be the crown of the flower.
39. Linnaeus 1737.
40. Antoine-Laurent de Jussieu introduced Linnaean nomenclature to the collections in the Jardin du Roi in 1774, although Buffon, intendant since 1739, had used Tournefort's names up till then (Stafleu 1971:281).
41. See Buffon 1749:21–22.
42. Of course, we saw in chapter 3 that Jussieu did not recognize the Compositae as a family in the *Genera plantarum*.
43. Strictly speaking not orders (= families), but classes.

44*. It would appear that M. Linnaeus had this character in mind when he wrote in his *Classes plantarum*, p. 487, "He who studies how to make the key (to the natural method) knows that no universal part is to be valued more than that of position, especially of the seed, and in the seed the growing point ["punctum vegetans"], which either pierces the length of the seed, or entirely surrounds it, or lies to one side; this either outside the cotyledons or inside; or at the base, near the base, at the side, or at the apex of the seed; at the base of the seed is a small scar where the seed was attached to the pericarp or the actual receptacle." [Linnaeus 1738:487].

45. Linnaeus 1754:xxv–xxvi, referring to his classes Monadelphia, Diadelphia, and Polyadephia.

46*. There is another kind of insertion, sometimes difficult to determine; this is when the stamens are borne on a disc or special fleshy body located between the support and the calyx, which appears to be a production of one or the other, and which one often cannot refer to either: in this case, one decides from analogy following the insertion in congeneric plants.

47. Jussieu 1777:230–232.
48. Jussieu 1777:230.
49. For Bernard de Jussieu's arrangment, see Antoine-Laurent de Jussieu 1789:lxiii–lxx.
50. Bernard de Jussieu probably did not recognize groupings of families, despite Antoine-Laurent's suggestion here that he did (Gidon 1934:299–303). Johann Gottlieb Gleditsch (1751, 1764) seems to have been the first to divide plants using this character; he distinguished four positions—stamens borne on the style, below the ovary, on the ovary, and on the petals.
51*. "Staminal filaments are distinct from a polypetalous corolla, they are inserted on a truly monopetalous corolla; exceptions [occur in plants that have] anthers with two appendages" Linnaeus, *Philos. Bot.* no. 108 [Linnaeus 1751:72].
52. Jussieu is here apparently contrasting Tournefort's method, based largely on the corolla, with Linnaeus's sexual system, based largely on the numbers of stamens and ovaries; he is not comparing systems and methods in general.
53. To the calyx, to the ovary, or to the support.
54. The text reads "dans l'origine."
55. Jussieu 1777:227, 230.
56. See Jussieu 1789:92.
57. Jussieu (like Linnaeus and Ray) described all plants whose fruits had single seeds and in which the pericarp was closely joined to the testa as having naked seeds. However, Jussieu (e.g., 1777:235) realized that many single-seeded plants were distinctive not only in having a single seed, but also because the stamens, etc., were borne on top of the seed.
58. Jussieu is comparing his arrangement with that of Linnaeus in the 6th edition of the *Genera plantarum* (Linnaeus 1764:[606]–[621]). Note Jussieu's reference to "artificial methods."
59. Jussieu now seems to imply that Linnaeus recognized essential characters, contradicting the previous sentence.
60*. M. Linnaeus, in his *Philosophia botanica*, said, no. 134, "all living things arise from eggs, consequently even plants . . ." no. 138, "no egg before fertilization . . ." no. 139, "all species of plants have flowers and fruits, even when sight does not detect them." [Linnaeus 1751:88–89]
61. Jussieu is taking a preformationist position. He also emphasizes that certain organs—stamens, pistil, seed—are essential for reproduction and implies that they will be found in all plants. The search for such organs in mosses, ferns, and even fungi occupied botanists' attention for the next forty years (e.g. Farley 1982, chaps. 2–3; Ainsworth 1976).

62*. One could also add the position of the embryo in the seed, and the position of the latter in the capsule. But, as has been observed above, these characters need examination afresh to determine if they can be admitted to the number of essential [ones].

63. Ipecacuanha comes from *Cephaëlis ipecacuanha*, a member of the Rubiaceae; the "violet" to which Jussieu referred is probably *Hybanthus calceolaria*, of the Violaceae, a species then included in *Viola* itself (see also Linnaeus 1774).

64. Both were placed in the Campanulaceae by Jussieu (1789:165).

65. Resolutives dissolve or disperse morbid matter.

66. Laxative.

67. Inducing sweat.

68. "[H]aving the quality of 'cutting' or loosening viscid humours" (Oxford English Dictionary).

69. Jussieu (1789:136–137) did place them in a separate family.

70. Ibid., 67–69.

71. Placed by Jussieu (ibid., 109) in the Vitaceae.

72. All these family pairs are adjacent in the *Genera plantarum*.

73. Jussieu's negative comments refer to those who proposed systems, rather than looking at nature systematically.

74. Cf. Fourcroy 1794, 1:3–8. The "fruitful art of Synthesis" which Jussieu mentioned can be carried out only after the careful analysis of the problem and its resolution into its simplest elements—see also the previous footnote and chapter 3.

75. As in the title, "historia" means a description, including events of the recent past, but no more.

76. More than simple similarity and difference is involved here, but rather positive attraction ("affinitas") and negative repulsion ("alienatio").

77*. This brief sketch of the structure of plants is taken in part from our early thesis presented at the Medical School of Paris in the year 1770, and entitled, "*An oeconomiam animalem inter et vegetalem analogia.*" [Jussieu 1770]

78. "Concreta" also means "condensed'; in classical descriptions of plant growth, the wood was considered to be a deposit of the inner cortex around the medulla.

79. Jussieu's ideas about plant nutrition, with the roots taking in food from the ground, owe much to Theophrastus.

80. For a diagram of tissue arrangements, see Stevens & Cullen 1990, fig.2, but note that the labels of liber and wood were inadvertently transposed. Jussieu's discussion differs from that of Linnaeus in detail, although remaining very Linnaean in spirit. Note that on p. v Jussieu refers to "cuticulae," although "corticis" seems more appropriate.

81. This group included insects, molluscs, and corals as well as wormlike animals, hence the mention of tracheae, found in insects.
82. That is, whether the flowers are hermaphroditic or not.
83. Jussieu does not distinguish between the endosperm, tissue of maternal and paternal (gametophytic) origin, and perisperm in the strict sense, which is of sporophytic origin.
84. These are the cotyledons, but Jussieu may be drawing a distinction between the cotyledons in the strict sense, found only in the seed, and lobes, the same structures during germination, which eventually change to the seed leaves of the seedling.
85. Here Jussieu probably means all tissue of cortical origin (in the classical sense), i.e., not simply the bark, but also the liber and/or alburnum.
86. Jussieu is making a distinction between prickles, of largely epidermal origin, and thorns, stem structures.
87. Jussieu uses the word "evolutione"; unfolding is another translation.
88. This probably refers to the projections of various shapes found on the leaves.
89. Literally "a multileaf." In general Jussieu thought that all structures that were undivided at the base were unitary; an involucre often consists of several partly fused leaves.

90*. To these Linnaeus adds the *Nectary*, by which name he designates parts sometimes included in the organs mentioned above, sometimes additional organs very different in structure, like glands, setae, squamulae, appendices, tubercula, foveae, sulci, the spurred or horned extensions ["propagines"], etc. Any excessively vague term must be eliminated from the science of botany, descriptions and characters properly being intimately connected, it is preferable that the parts hitherto included in the term "nectary" be renamed as individual organs, and each designated by a separate name [see Linnaeus 1751:73–74. See also note 23].

91*. The calyx is not absent in the fruit as Tournefort has said, nor is it truly superior as Linnaeus has remarked, but covers the whole and grows together with it at the base (see Act. Paris. 1773, p. 223) [Jussieu 1777:223]. Nevertheless, since the Linnaean term is generally accepted in Botany, we will admit it here, while changing the definition.

92*. When a single covering of the flower is present, it is called the calyx by some and the corolla by others in an arbitrary fashion. Tournefort in the Isagoge p. 72 [1700:72] does not solve this difficulty, and is uncertain himself, referring to the corolla in *Tulipa* and *Hyacinthus*, but designating as calyx the equivalent part in *Narcissus* and *Iris*. It is certain, nevertheless, that that part ought to be called the calyx which serves as the special covering of the seed [sic], and that which is shed earlier, petals, whether short- or long-lived. Linnaeus similarly wavers, taking the calyx of the *Helleborus* for the corolla, and he takes the petals of the same [*Helleborus*] to be nectaries, [although] in every way related ["affinia"] to the petals of the Ranunculaceae, now he calls the same organ corolla in *Rheum* and calyx in *Rumex*, failing to consider their affinity. Others have been no more successful in defining the corolla. Nevertheless, if its origin, its close relationships ["cognatio"] with the stamens, its function, its quick decline after fertilization, and its very similar appearance in many flowers are all carefully considered, then the corolla may be defined as that covering of the flower which is surrounded by the calyx and only very rarely naked, continuous with the liber of the peduncle, but not with the epidermis, not persisting, but in most cases shed along with the stamens, enveloping or crowning the fruit, but never fusing with it, and its parts often alternating with the stamens and equal in number with them. Taken together, these signs very clearly distinguish the corolla, which thereby in ambiguous flowers is decided by analogy with congeners. Therefore, we can observe that the perianth of *Narcissus*, which is naked, adnate to the germen, and which does not alternate its lobes with the stamens, is in reality a calyx, as Tournefort has taught us, and that consequently it is of the same character in *Hyacinthus* and other Liliaceae closely related to *Narcissus*.

93. "spiritum"—the fertilizing spirit.
94. Doubles, semidoubles, perhaps simply varieties with ray florets.
95. Jussieu suggests that apparently different parts of the flower are inconvertible; this idea had been more clearly articulated by Linnaeus (Stevens & Cullen 1990).
96. See note 57.
97. Pomes (e.g., an apple) are indehiscent; Jussieu is using "capsulam" in the sense of a hollow, more or less spherical structure containing seeds, whether that structure is dehiscent or not.
98. The exact natures of the various coverings of the "seed" that Jussieu mentions are unclear.
99*. In addition, there are a very few plants that merit the name *Polycotyledones*; in these the seeds have two multipartite or palmate lobes, which together imitate a whorl, as in *Pinus*, which see below, class. 15 ord. 5, p. 411, 415 [a reference to the main body of the *Genera plantarum*].
100. This seems to be a reference to polyembryony, where more than one embryo comes from a single embryo sac.
101. The idea that herbs do not have buds may be found in Ray (1696:3).

102. Linnaeus (1751:274–275) gives an example of a floral clock.
103. Jussieu here does not rank reproduction ("propagationem") above growth ("vegetationem").
104. The meaning is unclear. Jussieu may be suggesting that characters are discontinuous, or, more likely, that *useful* characters are easy to see. This latter meaning links this sentence to the next, where Jussieu talks about identifying plants.
105. These are all horticultural varieties of one sort or another. The "leaves" Jussieu refers to are petals.
106. It was commonly believed that variants reverted to the parental "type."
107. Here and elsewhere (pp. xxxvii, xlii), Jussieu uses the word "fasciculi" to refer to groupings of taxa. "Fasciculus" literally means "a small bundle," "bunch of flowers"; note that Tournefort (1694, 1:13) used a similar analogy.
108. Literally, "apparatus of interpretation."
109. Tournefort 1700, 1:[1]–54.
110. Tournefort 1694, 1:28–33, and 1700, 1:59–60. Secondary genera were based on characters other than those of the fructification. For example, Tournefort recognized three genera: *Quercus*, *Ilex*, and *Suber*, based on characters of the bark and the persistence of the leaves; Linnaeus later placed all three in the single genus, *Quercus*.
111. Literally, "in numerous assembly."
112*. Parts harmful to the flower are drawn out by the stamens as by excretory vessels, and set aside in their apices (the anthers), just as in sewers, *Tournefort, Isagog. p.* 68, 70 [1700, 1:68, 70]. Elsewhere, however, Tournefort seems to get an inkling of the sex of plants when he asks of dioecious plants: "Are small bodies, full of liquid, carried far away from flowers to become delicate fruit, just as if they were awakened from torpor to stimulate their own increase? As for instance of the male and female Palms . . . it is affirmed . . . that the branchlet of the male flower is implanted in the female spathe when the spathe gapes, the unfolding blossom discharges a dust, without whose effect the dates become bitter and disagreeable." *ibid.*, p. 69.
113. Morton 1981:213–220, for references.
114. See note 107.
115*. Nevertheless, he thinks this nomenclature ought to be revised, for he remarks in Isag. p. 63, 64 [Tournefort 1701, 1:63, 64]:"Names are like certain definitions, in which the first word expresses the genus, and the others the differentia. . . . Names ought to be short . . . some of these are so lengthy that they can scarcely be voiced in one breath . . . . It is one thing to name a plant, another to describe it."

116. That is, the polynomial species name that reflected at the same time the essence of the plant (Stevens 1991).
117. "inscripsit." One meaning of this is "furnished with a title," not inappropriate given the parallels that have been drawn between binomials and Linnaeus's abbreviations of book titles (Heller 1959:3–6).
118*. For example, he sometimes clearly rejects all terms not derived from a Latin or Greek source, and on other occasions accepts them very easily. Thus, he excludes any vernacular names that have a barbarian air, not redolent of ancient times, accepted and phrased in Latin, and suppresses *Adhatoda, Calaba, Ceïba, Guazuma, Isora, Ketmia, Mançanilla* [*Hippomane*], *Guanabana, Guaïava, Papaya, Sapota, Tapia* [*Crateva*], etc., for which he substitutes his own names, in another place he diverges from his expressed principles in admitting *Mammea, Basella, Yucca, Hura, Tulipa, Curcuma, Genipa*, etc.
119*. He who makes ["constituit"] a genus must give it a name; this must be a name worthy of the genus since, according to Linnaeus *Phil. Bot. n.* 218, 243 [1751:160, 196], and a chorus of approving botanists, it cannot be changed simply because another may be more suitable. This legislator often breaks his own law, however, as have many others. And there is no apparent reason why new or older names cannot be substituted for many bestowed or consecrated by Tournefort, such as *Abutilon, Acacia, Balsamina, Buglossum, Brunella, Bugula, Caprifolium, Casia, Cassida, Cataria, Elychrysum, Lappa, Mandragora, Pimpinella, Rapuntium, Sphondylium, Stramonium, Syringa, Tamariscus, Tamnus, Terebinthus, Thymelea*, etc. A thoughtful restoration of certain more generally familiar names would be opportune, but must be carried out very sparingly, so that the stain of the crime does not mark the censor himself.
120*. Such as *Corona Solis, Virga aurea, Dens Leonis, Lingua cervina, Lauro-cerasus, Corallodendron.*
121*. Such as *Hypophyllocarpodendron, Stachyarpagophora, Jabotapita.* Certain similar excessively unpolished names we now borrow from Aublet and others; these, however, ought to be made smoother in the future, when the certain constitution ["fabrica"] of those genera is evident.
122*. Inasmuch as the generic name does not have to signify, it can be imposed in an arbitrary manner, 1st. poetic, like *Adonis, Narcissus, Hyacinthus, Amaryllis, Circaea*; 2nd. divine or kingly, like *Serapias, Atropa, Mercurialis, Lysimachia, Artemisia, Helenium*; 3rd. from those who first found or first mentioned plants, like *Nicotiana, Cinchona, Sarracenia*; 4th. from the patrons of Botany, like *Eugenia, Borbonia, Bignonia, Cliffortia*; 5th. from travelers and explorers, like *Banisteria, Lippia,*

Gundelia, Commersonia, Dombeya; 6th. from the best botanists, like *Gesneria, Columnea, Caesalpinia, Bauhinia, Morisonia, Tournefortia, Plumeria, Dillenia, Linnaea, Jussiea, Halleria, Adansonia, Jacquinia*, etc. At first, Tournefort admitted such names only very rarely, but with Linnaeus the number of honorific appellations increased so much that these have now become virtually worthless, they are now regarded as a concession to Botanophiles.

123*. Whence names such as *Gloriosa, Mirabilis, Impatiens* ought to be deleted; such words are really adjectives, and therefore substantives *Phil. Bot. n.* 235 [Linnaeus 1751:167].

124*. Such as *Ephemerum, Onagra, Elephas, Auricula, Sagitta, Bursa.*

125*. Some comparative names prefix a syllable, e.g., *Linagrostis, Pseudodictamnus, Chamaecerasus*, etc.; in others, it is at the end; e.g., *Asteroïdes, Plantaginella, Myrtillus, Salicaria, Alsinastrum, Juncago*, etc.

126*. Such as Fumaria *vesicaria*, Fumaria *bulbosa*, Pyrola *umbellata*, Pyrola *uniflora*, Hordeum *distichum*, Hordeum *hexastichum.* [All specific epithets, whether referring to a "solid character" or not, are now called trivial.]

127*. Such as Veronica *arvensis*, Eryngium *maritimum*, Circaea *Lutetiana*, Leucoïum *vernum*, Iva *annua*, Lamium *album*, Mentha *piperita*, Viola *odorata*, Anchusa *officinalis*, Satureïa *hortensis*, Myagrum *sativum*, Pastinaca *oleracea.*

128. Literally, "reason for existing."

129. I.e., the descriptions are incomprehensible.

130*. He who in his listing of Authors intentionally omits any important or useful ones whatever, does harm to them, to science, and to himself. Linnaeus, like Tournefort a Prince of Botanists, was guilty in this respect when, through a very frequent neglect of Synonyms and changed names, he effectively removed from the eyes of the public those [rivals] whom he did not obscure: such is the love of undivided praise!

131*. Linnaeus calls *Orthodox* those Authors who base their method of arrangement on the fructification, and refers to them as *Fructistae*, or *Calicistae*, or *Corollistae* or *Sexualistae*; he calls *Heterodox* those who distribute Plants on different principles, such as the *Alphabetarii*, who arrange plants in alphabetical order, the *Rhizotomi*, by the structure of roots, the *Phyllophili*, by the different types of leaves, the *Physiognomoni*, by habit, the *Chronici*, by the time of flowering, the *Topophili*, by the native place of each plant, the *Empirici*, by medical use, and finally, the *Seplasiarii*, the arrangement ["dispositione"] of the Pharmacists *Phil. bot. n.* 25–31 [Linnaeus 1751:12–13).

132*. Fifteen were enumerated by Linnaeus (*Class. plant.* [1738]), and arranged chiefly by the organs that determined the methods. Those who drew the characters from the fruit were Cesalpino (1583), Morison (1680), Ray (1682), Christoph. Knaut (1687), [P.] Hermann (1690), Boerhaave (1710); from the calyx Magnol (1720), and Linnaeus (1737); from the corolla Rivinus (1690), Tournefort (1694), Christian Knaut (1716), Ruppius (1718), Pontedera (1720), Ludwig (1737); from the stamens and the sex of plants Linnaeus (1735). In the aforementioned work, each system is separately described, aptly defined, and understood with little trouble. Adanson (*Famil. plant.* [1764]) discusses the same systems and adds others more recently worked out by Royen (1740), Haller (1742), Morandi (1744), Séguier (1745), Wachendorf (1747), Heister (1748), Gleditsch (1749 [= 1751]), Allioni (1762[0]), etc., to these he adds his own numerous ways of distribution, doubtless with the plan of confirming both by example and precept, how very easy it is to multiply [them] by arbitrary arrangements. [Note that Jussieu uses "system" and "method" apparently interchangeably here.]

133. Tournefort 1700, 1:72–74 for a summary. Tournefort's initial division was into trees and shrubs versus herbs, then he divided plants by the kind of corolla the flowers had. Jussieu is puffing Tournefort's method rather more than it deserved, interesting though it is.
134. See note 57.
135. This is a very free translation; Jussieu's meaning is, however, not in doubt.
136. Strictly speaking the Apetalae lack only a corolla, the name of the group under discussion, the *Flore carentes*, implies that there are plants lacking the whole flower, yet still producing fruits.
137. Species that used to be included in *Rubeola* are now placed in *Sherardia* or *Crucianella*.
138. Burckhard 1750.
139. E.g., Linnaeus 1753; see also Linnaeus 1746.
140. Literally "with an armed eye" ("oculo armato").
141. That is, if the luckless botanist actually used a lens and needle.
142. Now included in *Chenopodium* itself. Jussieu's objections in this section are almost all quite reasonable.
143. Now included in *Sibthorpia*.
144*. "The first and most essential step in botany is a correct knowledge of plants. . . . It certainly seems that to know plants is nothing more than to have readily at hand those names which have been suitably given to them." *Tourn. Isag. p.* 1 [1700, 1:1]. . . . "Botany is that part of natural science through whose help, Plants are recognized and retained in

the memory most happily and with the least effort." *Boerh. Hist.* 16 [Boerhaave 1720, 1:16]. . . . "Botany is the natural science that yields a knowledge of Vegetative Beings." *Linn. Phil. Bot. n.* 4 [Linnaeus 1751:(1)].

145*. "Other things being equal, those classes which are more natural, are better. . . . The works of the greatest Botanists today is and ought to be done in this area. The natural method is and will continue to be the ultimate goal of all Botanists." *Linn. Phil. Bot. n.* 206 [1751:137]. "The first and last quest in Botany is for the natural method: little valued by less learned Botanists, it is always highly esteemed among the wiser, even though it has not yet been discovered. . . . For a long time I have worked to find it, and although I have succeeded in making many contributions to the search, I have not been able to complete the task, but shall continue as long as I live." *Linn. Class. pag.* 485 [1738:485].

146*. "Nature makes no leaps. All plants demonstrate their affinities on both sides ['utrinque'] like the Territories on a geographical Map." *Linn. Phil. n.* 77 [1751:27].

147. Linnaeus 1738:485–514.

148*. In order that the whole work of so great a man may become known and compared with that of others, the orders mentioned, displayed both in the gardens of the Trianon and as drawn up in a list in his own hand, are here transcribed at the end of the prologue [see below:lxiii–lxx].

149. Adanson 1763–1764.

150. What Jussieu had in mind is unclear here. He surely was not intending to imply that his method was fundamentally flawed, but he was perhaps being what he saw as realistic about attaining knowledge of plants. Perhaps also he was thinking of Adanson's approach, which did seem to depend on knowing everything about all plants.

151. See note 107.

152. That is, if several plants have this last kind of character in common, they are likely to make a good genus,

153*. Some Authors deny the constancy of this character and argue that monocotyledonous and dicotyledonous plants are sometimes congeneric. Thus, the unilobed *Melocactus* is correctly grouped by Linnaeus with the bilobed *Cactus*, while Adanson places the bilobed *Juncus* together with the unilobed Liliaceae; thus the unilobed *Orobanche* is placed adjacent to the bilobed *Pedicularis*. But these assertions need to be confirmed by repeated observation. It seems to me that *Juncus* is completely unilobed, when it germinates, a simple, primary leaf appears, bearing at its apex the persisting rind of the seed (see below,

Monocotyl. obs. p. 12 [p. 21], and embracing other young leaves in the sheathing base. Similarly, the doubtful germination of Aroids and *Cuscuta* ought to be recognized, as well as that of *Ranunculus glacialis* L., which emit a single seedling leaf, these ought perhaps more correctly to be considered monophyllous, sprouting only in the second year, rather than unilobed, as they appear at first when germinating.

154. Linnaeus 1764:[606]–[621]; Adanson 1763, vol. 2.
155. Note that appearances will not deceive with such a close linkage between inside and outside; see note 13.
156. This is perhaps to be interpreted as a statement of development in the context of preformation.
157*. Meditating casually on these characters, Linnaeus observed:"Those who are anxious to construct the key to the natural method know that no universal part is more valuable than that of position, especially of the seed and within the seed, the growing point, which may vertically perforate the seed, or envelop it completely, or lie at its side; it may be outside or inside the cotyledons; it may be at the base, near the base, or to the side, or to the apex of the seed; the base of the seed is that little scar where the seed was attached to its pericarp or receptacle proper. On these distinctions is built the method of the great Cesalpino." *Linn. Class. plant p.* 487 [1738:487].
158. In terms of importance, not in developmental sequence.
159*. Insertion is ambiguous when the stamens are attached at the same point as the superior calyx to an inferior germen, or when the receptacle of a superior germen separates them from an inferior calyx, in these cases it is sometimes doubtful whether it is epigynous or perigynous, and on other occasions as perigynous or hypogynous. The situation is similar when the stamens are attached at the same point above a slightly projecting disk which, lying between two organs, is, so to speak, of mixed origin. But in both cases, the nature of the insertion is made evident by comparison with neighboring genera in which it is more clearly defined.
160. This is a reference to doubled (luxuriant) flowers.
161. "moreover" would be more appropriate here.
162. This is a difficult passage. Jussieu seems to be saying that where there is variation, it doesn't really matter.
163*. "The filaments of stamens, are separate from a polypetalous corolla, but are inserted on a monopetalous one; [plants with] bicorned anthers being excepted [i.e., the Ericaceae]. Vaillant [1718] observed this in the Monopetalae. From the dissection of 2000 species, Pontedera [1720] revealed that monopetalous flowers bear stamens inserted in

the corolla, while in polypetalous ones they are attached to the receptacle of the flower. *Linn. Phil Bot. n.* 108. [Linnaeus 1751:72].

164★. Any epipetalous insertion of stamens, which is usual in Monopetalae and unusual in Polypetalae, is perhaps nothing but the union of the bases of the stamens and the corolla, these bases, not distinct, but very tightly joined, come together at a common point of insertion, each ministering to the care of the other. Thus the analogy of this form of insertion with others is confirmed, so that the triple insertion of the staminiferous corolla is mutually very distinct.

165★. Polypetalous flowers bear stamens distinct from petals. On very rare occasions, however, exceptions are admitted: *Statice pentapetala* has filaments inserted on the claws of the petals. . . . In *Lychnis* alternate stamens are often attached to the limbs of the claws. *Linn. Phil Bot. n.* 108 [Linnaeus 1751:73]. . . . Certain monopetalous flowers are characterized by stamens separate from the corolla, as in the Bicornes, *Erica, Andromeda* etc. *ibid.* A similar situation exists in Plumbagines and Campanulaceae, but, as will be observed below p. 92, 154, 163, the agreement which exists between polypetalous plants and simply immediate insertion is not undermined by these few exceptions.

166. Note that the "series" is a class of three families (Jussieu 1789:166–192).

167. Jussieu miscounted; this should be the eleventh.

168. That is, the positions of the stamens relative to the pistil.

169★. A table of it follows below p. lxxi. This method was first worked out in 1774 at the School of the Royal Garden of Paris, and its principles were presented at the same time in an academic dissertation (Act. Acad. Paris. 1774 p. 175 [Jussieu 1778:175]), in which a series of arguments rather similar to those presented here was put forward. Although the method has recently been changed a little, its principles, however, have not been changed: since at first [plants were] arranged in fourteen classes, a fifteenth has now been added, of Apetalae with epigynous stamens, and the bilobed corculum of the Aristolochiae confirms this class by germination. In addition, certain orders have moved from one class to another, because the insertion of stamens in them has become better known; certain groups have been further divided on account of a sufficiently distinct character; certain entirely new ones have appeared.

170. See note 110.

171. Jussieu appears to be distinguishing between internally founded characters and external differentiae (see p. xviii).

172. E.g., Linnaeus 1754.

173★. Because of the absence of a corolla and the presence of a superior or inferior germen, Jacquin [e.g., 1788:126] placed a young plant of the Rubiaceae in *Peplis* (see below, p. 333 [*Peplis tetrandra* is a species of *Lucya*]), following in this Linnaeus, who elsewhere had added to *Nyctanthes* a tree congeneric with *Guettarda* (p. 207). His son, in the *Supplementum*, wrongly joined the new *Tabernaemontana* (p. 145) to *Chiococca*, and *Serissa* (p. 209) to *Lycium* [Linnaeus filius 1782:145 (*Chiococca racemosa*) & 150 (*Lycium foetidum*)]. In the *Flora Japonica*, Thunberg [1784:63] named a tree possessing a superior ovary, *Cornus*.

174. That is, they form fascicles, phalanges, or bundles.

175*. "Plants agreeing in genus also agree in virtue; those that are contained in a natural order resemble one another very closely in their virtues; etc." *Linn. Phil Bot. n.* 337. [Linnaeus 1751:178].

176*. "The leaves of grasses provide agreeable food for cattle and horses; their smaller seeds nourish birds, while men find the larger ones good to eat. . . . In dry places, Umbelliferae are aromatic, heat-producing, and repellent; in moist places they are poisonous, etc. . . . Verticillatae (Labiatae) are fragrant, nervine, relaxant, and repellent etc. . . . Siliquosae (Cruciferae), watery, bitter, and binding, have diuretic and drying qualities, and their virtue is decreased on drying. . . . The leaves of the Papilionaceae (Leguminosae) are eaten by cattle and horses, and the seeds, farinaceous and flatulent, by various other animals. . . . The Syngenesia, of the Compositae, much used in medicine, are commonly bitter." *Ibid. n.* 338, 347, 348, 350, 351 [Linnaeus 1751:279, 281].

177*. In his "Fragmenta,' he joins *Commelina* and *Ixia*, *Dioscorea* and *Menispermum*, *Gentiana* and *Hypericum*, *Melianthum* and *Pinguicula*, *Lycium* and *Catesbaea*, *Convolvulus* and *Campanula*, *Gardenia* and *Vinca*, *Guettarda* and *Hippomane*, *Sideroxylum* and *Viburnum*, *Spigelia* and *Oldenlandia*, *Hebenstretia* and *Scabiosa*, etc [Linnaeus 1764:[606]–[621]].

178. The corculum.

179*. In the Linnaean "Fragmenta" are 58 [groups: see Linnaeus 1764:[606]–[621]], as in the *Familles* of Adanson [1763, vol. 2]; the Trianon orders number 65 [see below: lxiii–lxx], and the very recent Parisian ones [i.e., in this book] 100. Lamarck Dict. Encycl. 2:32 [1786, 2:32], arranged these according to the lobes of the corculum, the nature of the corolla, monopetalous, polypetalous, or absent, the location of the stamens, and, finally, on the organization, whether complete or incomplete, simple or more complex; his systematical distribution is truly praiseworthy. [Note that Linnaeus recognized 65 groupings of genera in 1738 and 67 in 1751.]

180. Jussieu here apparently returns to the distinction between (external) differentiae and (internal) characters, see p. xviii.

181*. Just as in any book, an index conveniently and briefly reveals the arrangement of the contents, so a systematic order, or methodical index, is useful in botany, since it produces an easy distribution of plants in which what is wanted is very quickly found. Plants of an uncertain position are here arranged in such a manner (p. 416 below [the key to genera Jussieu could not place in a family]), which is arbitrary until their true affinities become known. I intended to add a similar [index] to the entire work so that those genera that are noted and placed in orders only with difficulty by Beginners, may be quickly named with certainty by the use of an index of a few obvious ["impensis"] signs: but the imminent publication of this work left no time to compile such an index, which however is indispensable and I am working on it continuously so that it will, therefore, be published separately, or be appended to a second edition, if in due course one appears.

182*. The difficulty of establishing affinities and of perfecting the arrangement increases not in the ratio of the number of beings to be classified, but according to the number of organs to be investigated and evaluated in each one: therefore, that similar science, which relates to Animals and especially to Quadrupeds, is more difficult because, embracing many organs, it demands far more extended computations.

183. I am not sure from where Jussieu got the number 60. It is perhaps the number of Bernard de Jussieu's families that have cotyledons, i.e., an embryo (corculum) with obvious structure.

184. Again the distinction between characters and differentiae, see p. xviii.

185. A.-L. de Jussieu 1837. Jussieu's copy of the *Genera plantarum*, with numerous additions and corrections to the body of the text, is in the Laboratorie de Phanérogamie at the Muséum national d'Histoire naturelle in Paris.

186. See chapter 3 for the relationship between Jussieu's work and that of Joseph Gaertner.

187. A similar definition is used by some of Jussieu's followers (e.g., Jaume Saint-Hilaire 1805:xxxvii–xl).

188. Stevens & Cullen 1990:206–209.

189. A.-P. de Candolle 1820.

190. The critical word he used here is "fasciculi"—see note 107.

191. See Magnol 1689.

192. Both *Melastoma* and *Saxifraga* were here described as being perigynous, but Jussieu observed later in this section that their ovaries could be superior or inferior.

193. Jussieu 1789:xviii.

194. See also Jussieu 1789:163.

Appendix 2

1. A.-P. de Candolle 1810a.
2. Miège 1979:10; see also Flourens 1842:x–xi.
3. Candolle 1862.
4. Roger Hahn (1971, chap. 10) discusses the early years of the Institut and its changed position in French science compared to that of the Académie.
5. A.-L. de Jussieu 1803b.
6. Outram 1984:55–56. Jussieu witnessed Cuvier's marriage, as well as supporting him in the Institut. Outram's study provides a fascinating glimpse of the shifting relationships among scientists in Paris during this period; Alfred Moquin-Tandon's letters to Auguste-François Saint-Hilaire (Léotard 1893) mainly cover the period 1837–1851 and are full of comments about botanists and botanical politics. Candolle (1862:98–100) briefly discusses the relations between himself and Cuvier.
7. Candolle 1862:164–169; see Crosland 1967.
8. To Augustin de Candolle: A 212, G-FAC.
9. Of the two botanists who went on that expedition one, André Michaux, left early, while the other, Jean Baptiste Leschenault de la Tour, was often sick; Baudin himself died in Mauritius during the return (Horner 1987).
10. To Augustin de Candolle, 3 Germinal (24 Mars 1800): A 212, G-FAC. Augustin-Pyramus did not capitalize the first letters of his sentences; I have silently capitalized them, but all other changes or comments have been placed in square brackets.
11. Note, however, that Alphonse de Candolle suggested that his father had at this stage not read either botanist very carefully (A.-P. de Candolle 1862:572).
12. A.-L. de Jussieu (1789:xx, xxxvii, xlii) referred to groupings of species and genera as "fasciculi" in the *Genera plantarum*, although he also emphasized the map analogy; see appendix 1.
13. Candolle 1862:121. The murderer was never brought to justice.
14. Candolle 1813b.
15. Candolle 1862:49.
16. Candolle 1813a.
17. To Augustin de Candolle, 15 Février 1813: A 212, G-FAC. Augustin-Pyramus's friends were notably more enthusiastic (Stevens 1984a:58, fn.45).
18. To Augustin de Candolle, 21 Mai 1813: A 212, G-FAC.
19. Candolle 1862:187.
20. Candolle (1862:99) thought that Cuvier was very good at assembling facts, but that he lacked a really inventive mind; he never linked his facts

by a theory that was able to "divine or discover" other facts. As to what Cuvier thought of the *Théorie élémentaire*, I do not know; he later (Cuvier 1834:60, 61) suggested that Candolle's use of analogy, while careful, might induce less critical workers to make mistakes.

21. To Augustin de Candolle, 19 Aout 1802: A 212, G-FAC.
22. Candolle 1862:506.
23. Jussieu 1803b:276–277.
24. The botanists Charles Louis L'Heritier de Brutelle and André Thouin were initially enthusiastic members of the Société Linnéenne in Paris. However, the society was disbanded in 1789, the year after its foundation, ostensibly on the grounds of the protection of naturalists' freedom of expression. Jussieu, along with Adanson and Lamarck, were involved in suppressing the society (Hahn 1971:112–115; Beretta 1992). Pascal Duris (1993:69–77, 149–151) suggests (as did Hahn) that it was seen as a sectarian or partisan group.
25. To Augustin de Candolle, 12 Sept 1806: A 212, G-FAC.
26. Candolle 1862:184–187. Palisot de Beauvois had a very checkered career, and in the early 1800s published on a variety of topics; by the election he had begun his *Flore de Oware* (Palisot 1803–1820; Candolle is not quite fair about Palisot's publications). He was also something of a bryologist and spent considerable effort in defending a mistaken interpretation of the moss capsule (Lamy 1990:260, 266). Cuvier gave a lively account of his work and life in his éloge at the Société Royale (Cuvier 1824).
27. To Augustin de Candolle, 5 Nov 1806; see also letters of 19 Nov 1806 and one dated by his father the 19th of November: A 212, G-FAC.
28. Candolle 1862:95.
29. Candolle 1807a:66–67.
30. Candolle 1862:162, 185. Jussieu was also working on the Rubiaceae at the same time (Jussieu 1807b); I do not know of any friction between Candolle and Jussieu over their presumably overlapping work, but it is a possibility.
31. Candolle 1807a. This is where he criticized Palisot de Beauvois.
32. Candolle 1862:185.
33. Candolle 1807b:220.
34. Ibid., 217. Jussieu (1807b) also divided the Rubiaceae into four groups.
35. Cuvier 1812a.
36. But probably not earlier than 1803 (Coleman 1964:90).
37. Later in his life Candolle (1833b) decided that plants could be divided into four divisions. He played down any parallel with Cuvier's fourfold division of animals, and did not support the "philosophes" who had sug-

gested that groups came in fours. He thought that his paper, although important, had been overlooked; it was the touchstone for both the uncompleted *Prodromus systematis naturalis* on which he and others were working and the likewise uncompleted third edition of the *Theorie élémentaire* (Candolle 1862:398).

38. To Augustin de Candolle, 11 Février 1808: A 212, G-FAC. A seal covers some words.
39. See Jussieu, Desfontaines and Lamarck 1913.
40. For Candolle's opinion of the young Mirbel, see Candolle 1862:91–92. The two seem to have got on well, although Candolle saw a number of his own ideas in the young Mirbel's physiological studies which soon appeared (e.g., Mirbel 1802). Mirbel told Moquin-Tandon that Candolle's *Théorie élémentaire* was "absurd" (Léotard 1893:87), although it should be noted that Mirbel's *Élémens de physiologie végétale et de botanique* appeared three years after the *Théorie élémentaire*, and there was some overlap between the two.
41. To Augustin de Candolle, 1er Novembre 1808: A 212, G-FAC. Mirbel had been "directeur des beaux-arts" to Louis Bonaparte of Holland (for Mirbel's early life, see Payen 1858:11–25).
42. Candolle 1862:95, 184–187, 197–199, 547–548.
43. To Augustin de Candolle, 17 Novembre 1808: A 212, G-FAC.
44. Candolle 1862:369, 522–525.

Bibliography

NOTE: For each author, full surname and all initials are given, even though these may not always appear on the title page of the book. For example, the name of Richard Wettstein von Westersheim may appear simply as Richard Wettstein. In the text, the full name is given the first time a person is mentioned; subsequently, I use only those names by which the person was usually known. The names of some botanists present almost insuperable obstacles to the bibliographer. For example, Frans Stafleu and Richard Cowan (1979, 2:522–523) note under the entry of the botanist John Bellenden Ker, "Born Jon (sic) Gawler; name changed by Royal permission to Ker Bellenden, 5 Nov 1804, but used as Bellenden Ker. Also often referred to as Ker Gawler." He, at least, rarely appears on these pages under any name.

Abbreviations follow those suggested in *B-P-H/S Botanico-Periodicum-Huntianum/Supplementum*, G. D. R. Bridson (compiler and editor), Pittsburgh, Hunt Institute for Botanical Documentation, Carnegie Mellon University, 1991. I have rarely cited reprints or preprints, and parts of journals are indicated only when they are separately paginated.

Adams, J. F. A. 1887. Is botany a suitable study for young men? *Science* 9: 116–117.

Adanson, M. 1763–64. *Familles des plantes.* 2 vols. Paris: Vincent. [Vol. 1, 1764; vol. 2, 1763.]

——. 1777. Premier mémoire sur l'*Acacia* des anciens, et sur quelques autres arbres du Sénégal qui portent la gomme rougeâtre, apellée communément *gomme arabique. Mém. Math. Phys. Acad. Roy. Sci.* (Paris) (1773):1–17.

Agardh, C. A. 1817–1819. *Aphorismi botanici.* Lund: Berlinger. [Pp. 1–16 defended by Johan Forsander, 43–56 by Johannes Brunzelus, and 57–70 by Nicolaus Kihlgren.]

Agardh, J. G. 1858. *Theoria systematis plantarum accedit familarum phanerogamarum in series naturales disposito* . . . . Lund: C. W. K. Gleerup.

Agassiz, J. L. R. 1850. Progressive, embryonic and prophetic types. *Proc. Amer. Assoc. Advancem. Sci.* 2: 432–438.

——. 1859. *An Essay on Classification.* London: Longman, Brown, Green, Longmans and Roberts.

——. 1863. *Methods of Study in Natural History.* Boston: Ticknor and Fields.

Ainsworth, G. C. 1976. *Introduction to the History of Mycology.* Cambridge: Cambridge University Press.

d'Alembert, J. le R. 1757. Géometre, Géométrie. In D. Diderot and J. le R. d'Alembert, eds., *Encyclopédie ou dictionnaire raisonné des sciences, des arts et des métiers.* vol. 7, pp. 627–638. Paris: Briasson, David, Le Breton and Durand.

——. 1765. Nature. In D. Diderot and J. le R. d'Alembert, eds., *Encyclopédie ou dictionnaire raisonné des sciences, des arts et des métiers.* vol. 11, pp. 40–42. Neuchâtel: Samuel Faulche.

——. 1965. *Discours préliminaire de l'Encyclopèdie.* Paris: Éditions Gonthier. [First published in 1751, see *Encyclopédie*, vol. 1].

Alic, M. 1986. *Hypatia's Heritage: A History of Women in Science from Antiquity Through the Nineteenth Century.* Boston: Beacon Press.

Allen, D. E. 1976a. *The Naturalist in Britain: A History.* Harmandsworth: Penguin Books.

——. 1976b. Natural history and social history. *J. Soc. Bibliogr. Nat. Hist.* 7: 509–516.

——. 1980. The women members of the Botanical Society of London, 1836–1856. *Brit. J. Hist. Sci.* 13: 240–254.

Allen, G. E. 1981. Morphology and twentieth century biology: A response. *J. Hist. Biol.* 14: 159–176.

Allioni, C. 1760. *Synopsis methodica stirpium horti regii Taurensis.* Turin.

Allman, W. 1835. Dr Allman's arrangement of plants. *Proc. Fifth Meeting Brit. Assoc. Advancem. Sci.* (1835): [134].

Alston, R. E. and B. L. Turner. 1963. *Biochemical Systematics.* Engelwood Cliffs, New Jersey: Prentice Hall.

Ampère, A. M. 1838–1843. *Essai sur la philosophie des sciences* . . . . 2 vols. Paris: Bachelier.

Anderson, L. 1976. Charles Bonnet's taxonomy and the chain of being. *J. Hist. Ideas* 37: 45–58.
——. 1982. *Charles Bonnet and the Order of the Known.* Dordrecht: Riedel.
Andrews, H. N. 1980. *The Fossil Hunters.* Ithaca: Cornell University Press.
Anker, J. 1938. *Bird Books and Bird Art.* Copenhagen: Levin and Munksgaard.
Anonymous. 1719. Sur les sistèmes de botanique. *Hist. Acad. Roy. Sci.* (Paris) (1718): 45–47.
——. 1765. Plante. In D. Diderot and J. le R. D'Alembert, eds., *Encyclopédie ou dictionnaire raisonné des sciences, des arts et des métiers.* vol. 12, pp. 712–721. Neuchâtel: Samuel Faulche.
——. 1780. *Table analytique et raisonnée des sciences, arts et métiers . . . du dictionnaire des sciences, des arts et des métiers. . .*, vol. 1, chart at beginning. Paris: Panckoucke.
——. 1802–1820. Rees, A. *The Cyclopaedia; or, Universal Dictionary of Arts Science and Literature.* 39 vols. London: Longman, Hurst, Rees, Orme and Brown. [Various unsigned articles.]
——. 1817. *Encyclopaedia Britannica or, a Dictionary of Arts, Sciences and Miscellaneous Literature.* Ed. 5. vol. 14. Edinburgh: Archibald Constable. [Natural history, pp. 628–640; natural philosophy, p. 641.]
——. 1823. *Encyclopaedia Britannica or, a Dictionary of Science, Philosophy, and Miscellaneous Literature.* Ed. 6. vol. 4. Edinburgh: Archibald Constable. [Botany, pp. 63–332.]
——. 1831. Supplement to English Botany Nos VI–XIV. *Mag. Nat. Hist.* 4: 64–65. [Review of Smith and Sowerby 1829–1831.]
——. 1838. History of conchology. *Mag. Zool. Bot.* 2: 238–266.
——. 1842. *Encyclopaedia Britannica or, a Dictionary of Arts, Sciences and Miscellaneous Literature.* Ed. 7. vol. 15. Edinburgh: Archibald Constable. [Natural history, pp. 738–740.]
——. 1854. *Inauguration du buste de A. P. de Candolle dans le Jardin des Plantes de Montpellier le 4 fevrier 1845.* Montpellier: Ricard Frères. [Extracts from the Gazette Médicale de Montpellier, vol. 14.]
——. 1855. Begoniaceen-Gattungen und Arten von J. F. Klotzsch. *Bonplandia* 3: 286–288. [Review of Klotzsch 1854.]
——. 1864. Botanical lesson books. Lessons in elementary botany; the systematic part based on materials left by the late Professor Henslow, with numerous illustrations. By Daniel Oliver, F.R.S., F.L.S. Macmillan and Co., 1864. *Nat. Hist. Rev.* 4: 355–369. [Review.]
——. 1895. Le règne végétal. Aperçu phylogénétique. *Ned. Kruidk. Arch., Sér. 2*, 6: 723–728, pl. 11.
Appel, T. 1980. Henri de Blainville and the animal series: A nineteenth century chain of being. *J. Hist. Biol.* 13: 291–319.

——. 1987. *The Cuvier–Geoffroy Debate.* Oxford: Oxford University Press.

Arnauld, A. 1964. *The Art of Thinking: Port-Royal Logic.* Translated by J. Dickoff and P. James. New York: Dobbs-Merrill.

Arnott, G. A. W. 1831. Botany. In *The Encyclopaedia Britannica or, a Dictionary of Arts, Sciences, and General Literature.* Ed. 7. vol. 5, pp. 30–141. Edinburgh: Archibald Constable. ["D. P."]

Artedi, P. 1792. *Bibliotheca ichthyologica* . . . . vol. 3. [Edited by J. J. Walbaum.] Greifswald: Ant. Ferd. Röse.

Ashworth, W. B., Jr. 1990. Natural history and the emblematic world view. In D. C. Lindberg and R. S. Westman, eds., *Reappraisals of the Scientific Revolution,* pp. 303–332. Cambridge: Cambridge University Press.

Atran, S. 1985. The nature of folk-botanical life forms. *Amer. Anthropol.* 87: 298–315.

——. 1986. *Fondements de l'histoire naturelle* . . . . Paris: Editions Complexe.

——. 1990. *Cognitive Foundations of Natural History.* Cambridge: Cambridge University Press.

Audouin, J. V., A. T. Brongniart, and J. B. Dumas. 1824. Introduction. *Ann. Sci. Nat.* (Paris) 1: v-xvi.

Augier, A. 1801. *Essai d'une nouvelle classification des végétaux.* Lyon: Bruyset Aîné.

Babington, C. C. 1833. On the distinctions between the Linnaean genera *Potentilla* and *Tormentilla. Mag. Nat. Hist. & J. Zool.* 6: 248–249.

Baehni, C. 1957. Les grandes systèmes botaniques depuis Linné. *Gesnerus* 14: 83–93.

Baillon, H. E. 1855. *De la famille des Aurantiacées.* Paris: Rignoux.

——. 1858. *Étude générale du groupe des Euphorbiacées.* Paris: Victor Masson.

——. 1861a. Mémoire sur la symétrie et l'organogénie florale des Marantées. *Adansonia* 1: 306–327, pl. 11.

——. 1861b. Recherches sur l'organisation, le développement et l'anatomie des Caprifoliacées. *Adansonia* 1: 352–380, pl. 12.

——. 1862. Remarques sur l'organisation des Berbéridées. *Adansonia* 2: 268–291.

——. 1863. Mémoire sur la famille des Renonculacées. *Adansonia* 4: 1–57.

——. 1864. Recherches sur l'*Aucuba* et sur les rapports avec les genres analogues. *Adansonia* 5: 179–207.

——. 1891. Taxonomie. In *Dictionnaire de botanique.* vol. 4. Paris: Hachette.

Baker, K. M. 1962. An unpublished essay of Condorcet on technical methods of classification. *Ann. Sci.* (London) 18: 99–123.

——. 1975. *Condorcet: From Natural Philosophy to Social Mathematics.* Chicago: University of Chicago Press.

Balan, B. 1975. Premières recherches sur l'origine et la formation du concept de l'économie animale. *Rev. Hist. Sci. Appl.* 28: 289–326.

——. 1979. *L'ordre et le temps: l'anatomie comparée et l'histoire des vivants au XIXe siècle.* Paris: J. Vrin.

Barabé, D. 1978. Signification du concept de soudure dans les textes de morphologie végétale de A. P. de Candolle. *Candollea* 33: 7–9.

——. 1986. Haeckel et la systématique végétale. *Taxon* 35: 519–525.

Barabé, D. and J. Vieth. 1979. Le concept de fusion en morphologie végétale chez Payer et chez van Tieghem. *Acta Biotheor.* 28: 204–216.

Barber, L. 1980. *The Heyday of Natural History: 1820–1870.* Garden City, New York: Doubleday.

Barber, W. H. 1955. *Leibniz in France, from Arnauld to Voltaire . . . .* Oxford: Clarendon Press.

Barnes, B. 1990. Sociological theories of scientific knowledge. In R. C. Olby, G. N. Cantor, J. R. R. Christie, and M. J. S. Hodge, eds., *Companion to the History of Modern Science,* pp. 60–73. London and New York: Routledge.

Baron, W. 1931. Die idealistische Morphologie Al. Brauns und A. P. de Candolles und ihr Verhältnis zur Deszendenzlehre. *Bot. Centralbl.* Beihefte 48(1): 314–344.

——. 1966. Gedanken über den ursprünglichen Sinn der Ausdrücke Botanik, Zoologie und Biologie. *Sudhoffs Arch.* Beihefte 7: 1–10.

Baron, W. and B. Sticker. 1963. Ansätze zur historischen Denkweise in der Naturforschung an der Wende vom 18. zum 19. Jahrhunderts. *Sudhoffs Arch.* 47: 19–35.

Barrow, M. V. 1992. *Birds and Boundaries: Professionalism, Practice and Preservation in North American Ornithology, 1870–1930.* Ph. D. Thesis, Harvard University.

Barsanti, G. 1984. Linné et Buffon: deux visions différentes de l'histoire naturelle. *Revue de Synthèse* 113–114: 83–107.

——. 1988. Le immagini della natura: scale, mappe, alberi 1700–1800. *Nuncius* 3: 55–125.

——. 1992a. *La scala, la mappa, l'albero: immagini e classificazioni della natura fra sei e ottocento.* Florence: Sansoni.

——. 1992b. Buffon et l'image de la nature: de l'échelle des êtres à la carte géographique et à l'arbre généalogique. In J. Gayon, ed., *Buffon 88.* Paris: VRIN, Libraire Philosophique.

Barthélemy-Madaule, M. 1982. *Lamarck, the Mythical Precursor.* Cambridge: MIT Press.

Bather, F. A. 1927. Biological classifications: Past and future. *Quart. J. Geol. Soc.* 83: lxiii–civ.

Batsch, A. J. G. C. 1786. *Dispositio generum plantarum Ienesium . . . .* Jena: Heller.

——. 1787. *Versuch einer Anleitung zur Kenntnis und Geschichte der Pflanzen . . . .* 2 vols. Halle: Johann Jacob Gebauer.

——. 1802. *Tabula affinitatum regni vegetabilis.* Weimar: Landes-Industrie-Comptoir.

Beal, W. J. 1881. *The New Botany. A Lecture on the Best Method of Teaching.* Transactions 29th Annual Meeting of the Michigan State Teachers' Association.

Beauvois, Palisot de. [See Palisot de Beauvois.]

Beckner, M. 1959. *The Biological Way of Thought.* New York: Columbia University Press.

Bellynck, A. A. A. A. 1869. *La botanique moderne: conférence sur la botanique générale.* Paris: F. Savy.

Benichou, C. 1983. Enquête et reflexions sur l'introduction des termes dégénére(r), dégénération, dégénérescence dans les dictionnaires et encyclopédies scientifiques françaises à partir du 17ème siècle. *Docum. Hist. Vocab. Sci.* 5: 1–83.

Bennett, A. W. 1888. On the affinities and classification of algae. *J. Linn. Soc., Bot.* 24: 49–61.

Bennett, J. J. and R. Brown. 1838–1852. *Plantae javanicae rariores.* London: W. H. Allen.

Bentham, G. 1823. *Essai sur la nomenclature et la classification des principales branches d'art-en-science, ouvrage extrait de "Chrestomathia" de Jérémie Bentham.* Paris: Boussanges Frères.

——. 1827. *Outline of a New System of Logic, with a Critical Examination of Dr Whately's "Elements of Logic."* London: Hunt and Clarke.

——. 1832–1836a. *Labiatarum genera et species.* London: James Ridgway and Sons.

——. 1836b. [Letters to Sir W. J. Hooker from Berlin, Dresden, and Vienna.] *Companion Bot. Mag.* 2: 74–78. [Anonymous.]

——. 1837. On the Eriogoneae, a tribe of the order Polygonaceae. *Trans. Linn. Soc. London* 17: 401–420, pl. 17–20.

——. 1839. Observations on some genera of plants connected with the flora of Guinea. *Trans. Linn. Soc. London* 18: 225–238, pl. 18–20.

——. 1845. De Candolle's Prodromus, vol. IX. *London J. Bot.* 4: 223–248. [Anonymous.]

——. 1847. Notes of a continental tour in the years 1846–7; extracted from letters addressed to the Editor by a botanical friend. *London J. Bot.* 6: 477–482. [Anonymous.]

——. 1856. Notes on Loganiaceae. *J. Proc. Linn. Soc., Bot.* 1: 52–114.

——. 1857. Memorandum on the principles of generic nomenclature in botany as referred to in the preceeding paper. *J. Proc. Linn. Soc., Bot.* 2: 30–33.
——. 1858. *Handbook of the British Flora.* London: Lovell Reeve.
——. 1860. Synopsis of the Dalbergieae, a tribe of Leguminosae. *J. Proc. Linn. Soc., Bot.* 4 (suppl.): 1–134.
——. 1861a. *Flora hongkongensis: a description of the flowering plants and ferns of the island of Hongkong.* London: Lovell Reeve.
——. 1861b. On the species and genera of plants, considered with relation to their practical application to systematic botany. *Nat. Hist. Rev. N. S.* 1: 133–151.
——. 1862. *Address of George Bentham, Esq., President, Read at the Anniversary Meeting of the Linnean Society on Saturday, May 24, 1862.* London: Taylor and Francis.
——. 1863–1878. *Flora australiensis . . . .* 7 vols. London: Lovell Reeve.
——. 1865. *Address of George Bentham, Esq., F.R.S., President, Read at the Anniversary Meeting of the Linnean Society on Wednesday, May 24th, 1865.* London: Taylor and Francis.
——. 1866. *Address of George Bentham, Esq., F.R.S., President, Read at the Anniversary Meeting of the Linnean Society on Thursday, May 24, 1866.* London: Taylor and Francis.
——. 1871. *Address of George Bentham, Esq., F.R.S., President, Read at the Anniversary Meeting of the Linnean Society on Wednesday, May 24, 1871.* London: Taylor and Francis.
——. 1873a. *Address of George Bentham, Esq., F.R.S., President, Read at the Anniversary Meeting of the Linnean Society on Saturday, May 24, 1873.* London: Taylor and Francis.
——. 1873b. Notes on the classification, history, and geographical distribution of Compositae. *J. Linn. Soc., Bot.* 13: 335–577, pl. 8–11.
——. 1875a. On the recent progress and present state of systematic botany. *Rep. Brit. Assoc. Advancem. Sci.* (1874): 27–54.
——. 1875b. Revision of the suborder Mimosoideae. *Trans. Linn. Soc. London* 30: 335–664, pl. 66–70.
Bentham, G. and J. D. Hooker. 1862–1883. *Genera plantarum.* 3 vols. London: Pamplin, Black [vol. 1 only]; Lovell Reeve and Williams and Norgate.
Bentham, J. 1817. *Chrestomathia; Part 2 . . . being an Essay on Nomenclature and Classification.* London: Payne and Foss, R. Hunter.
Beretta, M. 1992. The Société Linnéenne de Paris 1787–1827. *Svenska Linné-Sallsk. Årsskr.* 1990–1991: 151–175.
Berlin, B. 1992. *Ethnobiological Classification: Principles of Categorization of*

Plants and Animals in Traditional Societies. Princeton: Princeton University Press.

Bernier, R. 1975. *Aux sources de la biologie. Tome premier. Les vingt premièrs siècles. La classification.* Montréal: Les presses de l'Université du Québec.

——. 1984. Système et méthode en taxonomie: Adanson, A.-L. de Jussieu et A.-P. de Candolle. *Naturaliste Canad.* 111: 3–12.

——. 1985. La notion de loi en biologie. 1. Lois, principes et règles. *Rivista Biol.* 78: 45–73.

——. 1988. *Aux sources de la biologie. Tome III. L'anatomie.* Frelighsburg, Québec: Orbis.

Bessey, C. E. 1880. *Botany for High Schools and Colleges.* New York: Henry Holt and Co.

——. 1894. Evolution and classification. *Proc. Amer. Assoc. Advancem. Sci.* 42: 237–251.

——. 1897. Phylogeny and taxonomy of the Angiosperms. *Bot. Gaz.* 24: 145–178.

——. 1908. The taxonomic aspect of the species question. *Amer. Naturalist* 42: 218–224.

——. 1915. The phylogenetic taxonomy of flowering plants. *Ann. Missouri Bot. Gard.* 2: 109–164.

Bessey, E. A. 1935. The teaching of botany sixty-five years ago. *Iowa State Coll. J. Sci.* 9: 227–233.

Bicheno, J. E. 1827. On systems and methods in natural history. *Trans. Linn. Soc. London* 15: 479–496. [Reprinted, *Phil. Mag., Ser. 2,* 3: 213–219, 265–271. 1828.]

Bischoff, G. W. 1830–1844. *Handbuch der botanische Terminologie und Systemkunde . . . .* 3 vols. Nurenberg: Leonhard Schrag.

Blackwelder, R. E. 1967. *Taxonomy.* New York: John Wiley.

Blum, A. S. 1993. *Featuring Nature: American Nineteenth Century Zoological Illustration.* Princeton: Princeton University Press.

Blunt, W. 1971. *The Compleat Naturalist: A Life of Linnaeus.* New York: Viking.

Boerhaave, H. 1710. *Index plantarum, quae in horto academico Lugduno-Batavo reperientur.* Leiden.

——. 1720. *Index alter plantarum, quae in horto academico Lugduno-Batavo aluntur.* 2 vols. Leiden: Petrus van der Aa.

Bonnet, C. 1745. *Traité d'insectologie, ou observations sur les pucerons.* 2 vols. Paris: Durand.

——. 1764. *Contemplation de la nature.* 2 vols. Amsterdam: Rey.

Bory de Saint-Vincent, J.-B.-G.-M. de. 1825. Histoire naturelle. In

Dictionnaire Classique d'Histoire Naturelle 8: 244–252. Paris: Rey and Gravier.

Bourdier, F. 1971. Georges Cuvier. In C. C. Gillispie, ed., *Dictionary of Scientific Biography*. vol. 3, pp. 521–528. New York: Charles Scribner's Sons.

Bower, F. O. 1938. *Sixty Years of Botany in Britain (1875–1935): Impressions of an Eyewitness*. London: Macmillan.

Bowler, P. J. 1983. *The Eclipse of Darwinism* . . . . Baltimore: Johns Hopkins University Press.

——. 1988. *The Non-Darwinian Revolution: Reinterpreting a Historical Myth*. Baltimore: Johns Hopkins University Press.

——. 1989. Development and adaptation: evolutionary concepts in British morphology, 1870–1914. *Brit. J. Hist. Sci.* 22: 283–297.

Braun, A. C. H. 1831. Vergleichende Untersuchung über die Ordnung der Schuppen an den Tannenzapfen als Einleitung zur Untersuchung der Blattstellung überhaupt. *Nova Acta Acad. Caes. Leop.-Carol. German. Nat. Cur.* 15: 195–402, pl. 17–50.

——. 1868. Über die Characeen Afrika's. *Monatsber. Königl. Preuss. Akad. Wiss. Berlin*. Dec. 1867: 782–800, 873–944.

Bremekamp, C. E. B. 1953. A re-examination of Cesalpino's classification. *Acta Bot. Neerl.* 1: 580–593.

Bremer, B. and L. Struwe. 1992. Phylogeny of the Rubiaceae and the Loganiaceae: Congruence or conflict between morphological and molecular data? *Amer. J. Bot.* 79: 1171–1184.

Brewster, D. 1833. Life and correspondence of Sir James Edward Smith. *Edinburgh Rev.* 57: 39–69. [Anonymous.]

Broberg, G. 1990. The broken circle. In T. Frängsmyr, J. L. Heilbron, and R. E. Rider, eds., *The Quantifying Spirit*, pp. 45–71. Berkeley: University of California Press.

Brockway, L. H. 1979. *Science and Colonial Expansion: The Role of the British Royal Botanic Gardens*. New York: Academic Press.

Bromhead, Sir E. F. 1836. Remarks on the arrangement of natural botanical families. *Edinburgh. New Philos. J.* 20: 245–254.

Brongniart, A. T. 1822. Sur la classification et la distribution des végétaux fossiles en général, et sur ceux des terrains de sédiment supérieur en particulier. *Mém. Mus. Hist. Nat.* 8: 203–240, pl. 12–15, 297–349, pl. 16–17.

——. 1826. *Mémoire sur la famille des Rhamnacées* . . . . Paris: Didot. [Preprint of *Ann. Sci. Nat. (Paris)* 10: 320–386, pl. 12–17.]

——. 1828a. *Prodrome d'une histoire des végétaux fossiles*. Strasbourg: F. G. Levrault.

——. 1828b(–1837). *Histoire des végétaux fossiles.* 2 vols. Paris: G. Dufour et Ed. D'Ocagne [vol. 1]; Crochard [vol. 2].

——. 1828c. Considerations générales sur la nature de la végétation qui couvrait la surface de la terre aux diverses époques de formation de son écorce. *Ann. Sci. Nat. (Paris)* 15: 225–258.

——. 1829. Recherches sur l'organisation des tiges des Cycadacées. *Ann. Sci. Nat. (Paris)* 16: 389–402.

——. 1837. Notice historique sur Antoine-Laurent de Jussieu. *Ann. Sci. Nat., Bot. Sér. 2,* 7: 5–24.

——. 1846a. Mémoire sur les relations du genre *Noggerathia* avec les plantes vivantes. *Compt. Rend. Hebd. Séances Acad. Sci.* (Paris) 21: 1392–1401.

——. 1846b. *Notice sur A.-P. de Candolle . . . .* Paris: Bouchard Huzard. [From Mem. Soc. Roy. Centr. Agric.]

——. 1850. *Énumeration des genres des plantes . . . .* Ed. 2. Paris: J. B. Baillière.

——. 1868. *Rapport sur les progrès de la botanique phytographique.* Paris: Imprimerie Impériale.

Brongniart, A. T., A.-M.-C. Duméril, R. L. Decaisne, and H. Milne-Edwards. 1853. *Funérailles de M. Adrien de Jussieu.* Paris: Firmin Didot.

Brongniart, A. T., H. Milne-Edwards, and A. Valenciennes. 1854. *Funérailles de M. de Mirbel.* Paris: Firmin Didot.

Brown, R. 1810a. On the Proteaceae of Jussieu. *Trans. Linn. Soc. London* 10: 15–226.

——. 1810b. *Prodromus florae novae hollandiae et insulae van-diemen.* vol. 1. London: Johnson.

——. 1811. On the Asclepiadeae, a natural order of plants separated from the Apocineae of Jussieu. *Mem. Wern. Nat. Hist. Soc.* 1: 12–78.

——. 1814. General remarks, geographical and systematical, on the botany of terra australis. In M. Flinders, *The Voyage to Terra Australis . . . .* vol. 2, pp. 533–613. London: G. and W. Nicol.

——. 1818a. Observations, systematical and geographical, on Professor Christian Smith's collection of plants from the vicinity of the river Congo. In J. K. Tuckey, *Narrative of an Expedition to Explore the River Zaire . . . ,* pp. 420–488. London: John Murray.

——. 1818b. Observations on the natural family of plants called Compositae. *Trans. Linn. Soc. London* 12: 75–142.

——. 1822. An account of a new genus of plants called *Rafflesia. Trans. Linn. Soc. London* 13: 201–234, pl. 15–22.

——. 1826. Botanical Appendix. In Denham, D. and H. Clapperton,

Narrative of Travels and Discoveries in Northern and Central Africa . . ., pp. 208–246. London: John Murray.

——. 1851. Some account of an undescribed fossil fruit. *Trans. Linn. Soc. London* 20: 469–475, pl. 23–24.

Browne, E. J. 1981. Natural History. In W. F. Bynum, E. J. Browne, and R. Porter, eds., *Dictionary of the History of Science,* p. 286. Princeton: Princeton University Press.

——. 1983. *The Secular Ark: Studies in the History of Biogeography*. New Haven: Yale University Press.

Brutelle, L'Héritier de. [See L'Héritier de Brutelle.]

Buchenau, F. G. P. 1890. *Monographia Juncacearum. Bot. Jahrb. Syst.* 12: 1–495, pl. 1–3.

Buffon, G.-L. de L., Comte de. 1749. *Histoire naturelle générale et particulière.* Vol. 1. Paris: Imprimerie Royale.

——. 1750. *Ibid.* Vol. 2.

——. 1753. *Ibid.* Vol. 4.

——. 1755. *Ibid.* Vol. 5.

Buisson, J. P. 1779. *Classes et noms des plantes.* Paris: Hérissant.

Bunge, A. A. von. 1862. Anabasearum revisio. *Mém. Acad. Imp. Sci. St. Pétersbourg, Sér.* 7, 4 (11): 1–102, pl. 1–3.

——. 1872. Die Gattung *Acantholimon* Boiss. *Mém. Acad. Imp. Sci. St. Pétersbourg, Sér.* 7, 18 (2): 1–72, pl. 1–2.

Burckhard, J. H. 1750. *Epistola ad . . . Godofredum Gulielmum Leibnizum . . . cum Laurentii Heisteri praefatione.* Helmstadt: Weygand.

Bureau, L. É. 1856. *De la famille des Loganiacées et des plantes qu'elle fournit à la médicine.* Paris: Rignoux.

——. 1864. *Monographie des Bignoniacées . . . .* Paris: J. B. Ballière et Fils.

Burkhardt, F., D. M. Porter, J. Browne, and M. Richmond. 1993. *The Correspondence of Charles Darwin. Vol 8, 1860.* Cambridge: Cambridge University Press.

Burkhardt, F. and S. Smith. 1985. *The Correspondence of Charles Darwin. Vol. 1. 1831–1836.* Cambridge: Cambridge University Press.

——. 1987. *The Correspondence of Charles Darwin. Vol. 3, 1844–1846.* Cambridge: Cambridge University Press.

——. 1991. *The Correspondence of Charles Darwin. Vol. 7, 1858–1859.* Cambridge: Cambridge University Press.

Burkhardt, R. W., Jr. 1977. *The Spirit of System: Lamarck and Evolutionary Biology.* Cambridge, Massachusetts: Harvard University Press.

Burnett, G. T. 1830. Letter on the philosophy of system. *Quart. J. Lit. Sci. Arts* 29: 368–373.

Burtt, B. L. 1966. Adanson and modern taxonomy. *Notes Roy. Bot. Gard. Edinburgh* 26: 427–431.

Bynum, W. F. 1975. The great chain of being after forty years: An appraisal. *Hist. Sci.* 13: 1–23.

Cain, A. J. 1958. Logic and memory in Linnaeus's system of taxonomy. *Proc. Linn. Soc. London* 169: 144–163.

——. 1959a. Deductive and inductive methods in post-Linnaean taxonomy. *Proc. Linn. Soc. London* 170: 185–217.

——. 1959b. The post-Linnaean development of taxonomy. *Proc. Linn. Soc. London* 170: 234–244.

——. 1981. The development of systematic ideas of variation illustrated by malacology. In A. Wheeler and J. H. Price, eds., *History in the Service of Systematics*, pp. 151–156. London: Society for the Bibliography of Natural History.

——. 1991. *Constantine Samuel Rafinesque-Schmalz on Classification.* Tryonia, no. 20. Department of Malacology, Academy of Natural Sciences, Philadelphia.

——. 1992. Was Linnaeus a Rosicrucian? *Linnean* 8(3): 23–44.

——. 1993. Linnaeus's Ordines naturales. *Arch. Nat. Hist.* 20: 405–415.

——. 1994. Numerus, figura, proportio, situs; Linnaeus's definitory attributes. *Arch. Nat. Hist.* 21: 17–36.

Callot, É. 1965a. Système et méthode dans l'histoire de la botanique. *Rev. Hist. Sci. Applic.* 18: 45–53.

——. 1965b. *La philosophie de la vie au XVIIIe siècle* . . . . Paris: Marcel Rivière.

Camerarius, R. J. 1699. *De convenienta plantarum in fructificatione et viribus.* Tübingen: Joh. Conrad Reis.

Camp, W. H. 1940. Our changing generic concepts. *Bull. Torrey Bot. Club* 67: 381–389.

Candolle, A[lphonse]-L.-P.-P. de. 1830. *Monographie des Campanulacées.* Paris: Desray.

——. 1834. On the history of fossil vegetables. *Edinburgh New Philos. J.* 18: 81–102. [From *Bibl. Universelle Sci. Belles-Lettres Arts, Sci. Arts.*]

——. 1835. *Histoire naturelle des végétaux.* 2 vols. Paris: Roret.

——. 1837a. A review of the natural order Myrsineae. *Trans. Linn. Soc. London* 17: 95–138, pl. 4–8.

——. 1837b. *Introduction à l'étude de la botanique.* Brussels: Meline, Cans and Co.

——. 1841. Troisième mémoire sur la famille des Myrsinacées. *Ann. Sci. Nat., Bot. Sér. 2*, 16: 129–176, pl. 8, 9.

——. 1844. Mémoire sur la familles des Apocynacées. *Ann. Sci. Nat., Bot. Sér. 3*, 1: 235–263.
——. 1856. *Géographie botanique raisonnée* . . . . Paris: Masson.
——. 1862. Etude sur l'espèce à l'occasion d'une révision de la famille des Cupulifères. *Biblioth. Universelle Genève, Arch. Sci. Phys. Nat.* 15: 211–237, 326–365. [Reprinted in *Ann. Sci. Nat., Bot. Sér. 4*, 18: 59–110. 1862]
——. 1867a. *Lois de la nomenclature botanique*. Geneva and Basle: H. Géorg.
——. 1867b. Le Genera plantarum de MM. Bentham et J. D. Hooker. *Biblioth. Universelle Genève, Arch. Sci. Phys. Nat.* 30: 289–305.
——. 1873a. *Prodromus systematis naturalis regni vegetabilis*. vol. 17. Strasbourg and London: Treuttel and Würtz.
——. 1873b. *Histoire des sciences et des savants depuis deux siècles*. Ed. 2. Geneva: H. Georg.
——. 1880. *La phytographie*. Paris: G. Masson.
Candolle, A[nne].-P.-C[asimir] de. 1879. Anatomie comparée des feuilles des quelques familles de Dicotylédons. *Mém. Soc. Phys. Genève* 26: 427–449, pl. 2.
Candolle, A[ugustin]-P. de. 1804. *Essai sur les propriétés médicales des plantes, comparées avec leurs formes extérieures et leur classification naturelle*. Paris: Didot Jeune.
——. 1806. Mémoire sur les pores de l'écorce. *Mém. Inst. Sci. Divers Savans, Sci. Math.* 1: 351–369.
——. 1807a. Sur les champignons parasites. *Mém. Mus. Hist. Nat.* 9: 56–74.
——. 1807b. Mémoire sur le *Cuviera*, genre nouveau de la familles des Rubiacées. *Mém. Mus. Hist. Nat.* 9: 216–222, pl. 16.
——. 1810a. Observations sur les plantes composées, ou syngenèses. Premier mémoire. Sur les Composées et les Cinarocéphales en général. *Mém. Mus. Hist. Nat.* 16: 135–158, pl. 5.
——. 1810b. Ibid. Seconde mémoire. Monographies de quelques genres de Cinarocéphales. *Mém. Mus. Hist. Nat.* 16: 181–208, pl. 6–15.
——. 1812. Ibid. Troisième mémoire. Sur les Composées à corolles labiées, ou Labiatiflores. *Mém. Mus. Hist. Nat.* 19: 59–72, pl. 3–7.
——. 1813a. *Catalogus plantarum horti botanici monspeliensis*. Montpellier: J. Martel.
——. 1813b. *Théorie élémentaire de la botanique*. Paris: Déterville.
——. 1814. *Theoretische Anfangsgrunde der Botanik* . . . . 4 vols. Trans. J. J. Römer. Zurich: Drell, Füssli and Co.
——. 1819. *Théorie élémentaire de la botanique*. Ed. 2. Paris: Déterville.
——. 1820. *Essai élémentaire de géographie botanique*. Strasbourg: F. Levrault.

[From "Géographie botanique," In *Dictionnaire des Sciences Naturelles* 18: 359–422. Strasbourg: F. G. Levrault].

——. 1821a. Mémoire sur la famille des Crucifères. *Mém. Mus. Hist. Nat.* 7: 169–252, pl. 6–7.

——. 1821b. Mémoire sur les affinités naturelles de la famille des Nymphaeacées. *Mém. Soc. Phys. Genève.* 1: 209–244, pl. 1–2.

——. 1825–7. *Mémoires sur la famille des Légumineuses.* Paris: A. Belin.

——. 1827. *Organographie végétale* . . . . 2 vols. Paris: Déterville.

——. 1828a. *Collection de mémoires pour servir à l'histoire du règne végétal . . . I. Melastomatacées.* Paris: A. Belin.

——. 1828b. *Ibid. II. Crassulacées.* Paris: A. Belin.

——. 1828c. *Considérations sur la phytologie ou botanique générale, son histoire et les moyens de la perfectionner.* Paris: J. Tastu. [From "Phytologie ou botanique," In *Dictionnaire Classique d'Histoire Naturelle* 13: 478–491. Paris: Rey and Gravier.]

——. 1829a. De l'état actuelle de la botanique générale. *Rev. Franç.* (Paris, 1828–30, 1837–1839) 8: 33–56.

——. 1829b. Revue de la famille des Cactées, avec des observations sur leur végétation et leur culture, ainsi que sur celles des autres plantes grasses. *Mém. Mus. Hist. Nat.* 17: 1–119, pl. 1–21.

——. 1832. *Physiologie végétale* . . . . 3 vols. Paris: Béchet Jeune.

——. 1833a. Notice sur les progrès de la botanique pendant l'année 1832. *Biblioth. Universelle Sci. Belles-Lettres Arts, Sci. Arts* 52: 142–192.

——. 1833b. Note sur la division du règne végétal en quatre grandes classes ou embranchements. *Biblioth. Universelle Sci. Belles-Lettres Arts, Sci. Arts* 54: 259–268.

——. 1844. *Théorie élémentaire de la botanique.* Ed. 3. Paris: Roret.

——. 1862. *Mémoires et souvenirs de Augustin-Pyramus de Candolle.* Genève: Cherbuliez.

Candolle, A[ugustin]-P. de, and A.-L.-P.-P. de Candolle. 1841. Monstruosités végétales. *Neue Denkschr. Allg. Schweiz. Ges. Gesammten Naturwiss.* 5: 1–23, pl. 1–7.

Candolle, A[ugustin]-P. de, and K. P. J. Sprengel. 1820. *A. P. de Candolle's und K. Sprengel's Grundzüge der wissenschaftlichen Pflanzenkunde zu Vorlesungen.* Leipzig: Kurt Knobloch. [Adaptation of Candolle, 1819: 1821. *Elements of the Philosophy of Plants* . . . . Edinburgh: William Blackwood. Translation.]

Candolle, R. de. 1966. *L'Europe de 1830* . . . . Genève: A. Jullien.

Cannon, S. F. 1978. *Science in Culture: The Early Victorian Period.* New York: Dawson and Science History.

Caron, J. A. 1988. 'Biology' in the life sciences: A historiographical contribution. *Hist. Sci.* 26: 223–268.
Carter, H. B. 1988. *Sir Joseph Banks, 1743–1820.* London: British Museum (Natural History).
Caspary, J. X. R. 1858. Die Hydrilleen (Anacharideen Endl.). *Jahrb. Wiss. Bot.* 1: 377–513, pl. 25–29.
Cassel, F. P. 1817. *Lehrbuch der natürliche Pflanzenordnung.* Frankfort: Andrea.
——. 1820. *Morphonomica botanica, sive observationes circa proportionem et evolutionem partium plantarum.* Cologne: DuMont-Schauberg.
Cassini, A.-H. G. de. 1813. Observations sur le style, et le stigmate des Synanthérées. *J. Phys. Chim. Hist. Nat. Arts* 76: 97–128, 249–275, pl. 1.
——. 1816. De l'influence que l'avortement des étamines paroît avoir sur les périanthes. *J. Phys. Chim. Hist. Nat. Arts* 82: 335–342.
——. 1816–1817. Planches 87–93. *Dict. Sci. Nat.* Planches 2: (Botanique: Végétaux Dicotylédons). Paris: F. G. Levrault.
——. 1817. Quatrième mémoire sur la famille des Synanthérées, contenant l'analyse de l'ovaire et de ses accessoires. *J. Phys. Chim. Hist. Nat. Arts* 85: 5–21.
——. 1822. Inulées. In *Dictionnaire des Sciences Naturelles* 23: 559–582. Paris: F. G. Levrault.
——. 1826. Analyse critique et raisonné des Élémens de Botanique de M. Mirbel. *Opuscules Phytologiques* 2: 287–326. Paris: Levrault.
Cassirer, E. 1951. *The Philosophy of the Enlightenment.* Translated by F. C. A. Koelln and J. P. Pettigrove. Princeton: Princeton University Press.
Castle, T. 1833. *A Synopsis of Systematic Botany as Connected with the Plants Admitted into the Pharmacopoeias . . . .* Southwark: E. Cox.
Cesalpino, A. 1583. *De plantis libri XVI.* Florence: Marescottum.
Charlton, D. G. 1984. *New Images of the Natural in France.* Cambridge: Cambridge University Press.
Chatin, G. A. 1840. *Anatomie comparée végétale appliquée à la classification . . . .* Paris. n.v.
——. 1857. Réponse aux observations présentées par M. R. Caspary sur la division de l'ancienne famille des Hydrocharidées en Ottéliacées et en Hydrocharidées. *Bull. Soc. Bot. France* 4: 156–162.
——. 1866. *Notice sur les travaux scientifique de M. Ad. Chatin.* Paris: Cerf.
——. (1856–)1892. *Anatomie comparée des végétaux—plantes parasites.* Paris: Baillière.
Chavannes, É. L. 1833. *Monographie des Antirrhinées.* Paris: Treuttel and Wurz.
Chevalier, A. J. B. 1934. *Michel Adanson, voyageur, naturaliste et philosophe.* Paris: Larose.

Chodat, R. H. 1891. Monographie Polygalacearum. *Mém. Soc. Phys. Genève*, Suppl. 7: 1–143, pl. 1–12.

Choisy, J. D. 1833. Description des Hydroléacées. *Mém. Soc. Phys. Genève* 6: 95–122.

Churchill, F. B. 1991. The rise of classical descriptive embryology. In S. F. Gilbert, ed., *Developmental Biology, a Comprehensive Synthesis. Volume 7. A Conceptual History of Modern Embryology*, pp. 1–29. New York and London: Plenum.

Clos, D. 1869. Coup d'oeil sur les principes qui servent de base aux classifications botaniques modernes. *Mém. Acad. Sci. Toulouse, Sér.* 7, 1: 125–142.

Cohen, I. B. 1985. *Revolution in Science*. Cambridge, Massachusetts: Belknap Press of Harvard University Press.

Cohn, F. J. 1895. Nathanael Pringsheim. *Ber. Deutsch. Bot. Ges.* 13: (10)–(30).

Colebrooke, H. T. 1829. On dichotomous and quinary arrangements in natural history. *Zool. J.* 4: 43–46.

Coleman, W. 1964. *Georges Cuvier, Zoologist* . . . . Cambridge, Massachusetts: Harvard University Press.

——. 1971. *Biology in the Nineteenth Century*. New York: John Wiley and Sons.

——. 1976. Morphology between type concept and descent theory. *J. Hist. Medicine Allied Sci.* 31: 149–175.

——. 1986. Evolution into ecology? The strategy of Warming's ecological plant geography. *J. Hist. Biol.* 19: 181–196.

Comte, A. 1851. *Système de politique positive*. 4 vols. Paris: L. Mathias.

Condillac, É. B. de. 1749. *Traité des sistèmes* . . . . La Haye: Neaulme.

Condorcet, M.-J.-A.-N. de C., Marquis de. 1777a. Sur les familles naturelles des plantes, et en particulier sur celle des Renoncules. *Hist. Acad. Roy. Sci.* (Paris) 1773: 34–36.

——. 1777b. Sur le gommier rouge du Sénégal. *Hist. Acad. Roy. Sci.* (Paris) 1773: 36–38.

——. 1778. Sur le nouvel ordre de plantes établi dans l'École de Botanique du Jardin du Roi. *Hist. Acad. Roy. Sci.* (Paris) 1774: 27–30.

——. 1780. Éloge de M. de Jussieu. *Hist. Acad. Roy. Sci.* (Paris) 1777: 94–117.

——. 1781. Éloge de M. Linné. *Hist. Acad. Roy. Sci.* (Paris) 1778: 66–84.

——. 1791. Éloge de M. le Comte de Buffon. *Hist. Acad. Roy. Sci.* (Paris) 1788: 50–84.

Conry, Y. 1987. *Darwin en perspective*. Paris: J. Vrin.

Constance, L. 1955. The systematics of the angiosperms. In E. L. Kessel, ed., *A Century of Progress in the Natural Sciences, 1853–1953*, pp. 405–483. San Francisco: California Academy of Sciences.

——. 1964. Systematic botany—an unending synthesis. *Taxon* 13: 257–273.

Cook, O. F. 1898. Stability in generic nomenclature. *Science N. S.* 8: 186–190.

Corner, E. J. H. 1946. Suggestions for botanical progress. *New Phytol.* 45: 185–192.

Cordá, A. K. J. 1842. *Anleitung zum Studium der Mykologia.* Prague: F. Ehrlich.

——. 1845. *Beiträge zur Flora der Vorwelt.* Prague: J. G. Calve.

Corrêa da Serra, J. F. 1805. Observations sur la famille des orangers et sur les limites qui la circonscrivent. *Ann. Mus. Hist. Nat.* 6: 376–387.

Corsi, P. 1983. Models and analogies for the reform of natural history: features of the French debate, 1790–1800. In G. Montalenti and P. Rossi, eds., *Lazzaro Spallanzani e la biologia del Settecento,* pp. 381–396. Florence: Leo S. Olschki.

——. 1988. *The Age of Lamarck: Evolutionary Theories in France 1790–1830.* Berkeley: University of California.

Cournot, A. A. 1872. *Considérations sur la marche des idées et des événements dans les temps modernes.* 2 vols. Paris: Hachette.

Cowles, H. C. 1908. The ecological aspect of the conception of species. *Amer. Naturalist* 42: 265–271.

Craw, R. 1992. Margins of cladistics: Identity, difference and place in the emergence of phylogenetic systematics, 1864–1975. In P. Griffiths, ed., *Trees of Life: Essays in the Philosophy of Biology*, pp. 65–107. Dordrecht: Kluwer Academic.

Crisci, J. V. and T. F. Stuessy. 1980. Determining primitive character states for phylogenetic reconstruction. *Syst. Bot.* 5: 112–135.

Cronk, Q. C. B. 1990. The name of the pea: A quantitative history of legume classification. *New Phytol.* 116: 163–175.

Crosland, M. P. 1967. *The Society of Arcueil: A View of French Science at the Time of Napoleon I.* Cambridge, Massachusetts: Harvard University Press.

——. 1978. *Historical Studies in the Language of Chemistry.* New York: Dover Press.

Cuerrier, A., D. Barabé, and L. Brouillet. 1992. Bessey and Engler: A numerical analysis of their classification of flowering plants. *Taxon* 41: 667–684.

Cullen, J. 1968. Botanical problems of numerical taxonomy. In V. H. Heywood, ed., *Modern Methods in Plant Taxonomy,* pp. 175–183. New York: Academic Press.

Cusset, G. 1982. The conceptual bases of plant morphology. In R. Sattler, ed., *Axioms and Principles of Plant Construction,* pp. 8–86. The Hague: Martinus Nijhoff.

Cuvier, G. 1792. Mémoire sur les cloportes terrestres. *J. Hist. Nat.* 2: 18–31, pl. 26.

——. 1795a. Mémoire sur la structure interne et externe, et sur les affinités des animaux auxquels on a donné le nom de vers. *Décade Philos.* 5: 385–396.

——. 1795b. Second mémoire sur l'organisation et les rapports des animaux à sang blanc, dans lequel on traite de la structure des Mollusques et de leur division en ordre. *Mag. Encycl.* 1795, 2: 433–449.

——. 1795c. Discours prononcé par le citoyen Cuvier, à l'ouverture du cours d'anatomie comparée qu'il fait au Museum national d'histoire naturelle, pour le citoyen Mertrud. *Mag. Encycl.* 1795, 5: 145–155.

——. 1796. Notice sur le squelette d'une très-grande espèce de quadrupède inconnue jusqu'à present, et deposé au cabinet d'histoire naturelle de Madrid. *Mag. Encycl.* 1796, 1: 303–310.

——. 1798. *Tableau élémentaire de l'histoire naturelle des animaux.* Paris: Baudouin.

——. 1799. Sur l'organisation d'un animal nommé méduse. *J. Phys. Chim. Hist. Nat.* 47: 436–440.

——. 1804. Mémoire concernant l'animal de l'Hyale, un nouveau genre de mollusques nus, intermédiaire entre l'Hyale et le Clio, et l'établissement d'un nouvel ordre dans la classe des mollusques. *Ann. Mus. Natl. Hist. Nat.* 4: 223–234, pl. 59.

——. 1805. *Leçons d'anatomie comparée.* 5 vols. Paris: Baudouin.

——. 1810. *Rapport historique sur les progrès des sciences naturelles depuis 1789, et sur leur état actuel . . . .* Paris: Imprimerie Impériale.

——. 1812a. Sur un nouveau rapprochement à établir entre les classes qui composent le Règne animal. *Mém. Mus. Hist. Nat.* 19: 73–84.

——. 1812b. *Recherches sur les ossemens fossiles de quadrupèdes.* Paris: Déterville.

——. 1816. *Le règne animal distribué d'après son organisation . . . .* 4 vols. Paris: Déterville.

——. 1819. *Recueil des éloges historiques.* Vol. 1. Strasbourg: F. G. Levrault.

——. 1824. Éloge historique de M. de Beauvois. *Mém. Acad. Roy. Sci. Inst. France* 4: cccxviii–cccxlvi.

——. 1825. Nature. In *Dictionnaire des Science Naturelles* 34: 261–268. Paris: F. G. Levrault.

——. 1829. On the state of natural history, and the progress which it has made since the return of the maritime peace. *Edinburgh New Philos. J.* 7: 1–14. [Translated.]

——. 1832. Avertissement. *Nouv. Ann. Mus. Hist. Nat.* 1: i–iv.

——. 1834. *Histoire des progrès des sciences naturelles depuis 1789 jusqu'à ce jour.* Vol. 3. Paris: Roret.

——. 1843. *Histoire des sciences naturelles depuis leur origine jusqu'à nos jours.* Vol. 4. Paris: Fortin, Masson.

——. 1845. *Ibid.* Vol. 5. Paris: Fortin, Masson.

Cuvier, G. and É. Geoffroy Sainte-Hilaire. 1795. Mémoire sur les rapports naturels du Tarsier (*Didelphis macrotarsus* Gm.). *Mag. Encycl.* 1795, 3: 147–154.

Dagognet, F. 1970. *Le catalogue de la vie . . . .* Paris: Presses Universitaires de France.

Dahlgren, R. M. T., H. T. Clifford, and P. F. Yeo. 1985. *The Families of Monocotyledons.* Berlin: Springer.

Darwin, C. R. 1859. *On the Origin of Species . . . .* London: John Murray.

Daubenton, L.-J.-M. 1751. Botanique. In D. Diderot and J. le R. d'Alembert, eds., *Encyclopédie ou dictionnaire raisonné des sciences, des arts et des métiers,* vol. 2, pp. 340–345. Paris: Briasson, Durand, Le Breton and Durand. ["I"]

——. 1765a. Histoire naturelle. In D. Diderot and J. le R. d'Alembert, eds., *Encyclopédie ou dictionnaire raisonné des sciences, des arts et des métiers,* vol. 8., pp. 225–230. Neuchâtel: Samuel Faulche.

——. 1765b. Méthode. In D. Diderot and J. le R. D'Alembert, eds., *Encyclopédie ou dictionnaire raisonné des sciences, des arts et des métiers,* vol. 10, pp. 458–460. Neuchâtel: Samuel Faulche.

——. 1797. Observations sur l'organisation et l'accroisement du bois. *Mém. Acad. Sci.* (Paris) 1790: 665–675, pl. 13–14.

Daubeny, C. G. B. 1843. Sketch of the writings and philosophical character of Augustin-Pyramus de Candolle, Professor of Natural Science at the Academy of Geneva, etc. *Edinburgh New Philos. J.* 34: 197–246.

Daudin, H. 1926a. *De Linné à Lamarck: méthodes de la classification et idée de série en botanique et en zoologie (1740–1790).* Paris: Félix Alcan.

——. 1926b. *Cuvier et Lamarck: les classes zoologiques et l'idée de série animale (1790–1830).* 2 vols. Paris: Félix Alcan.

Davis, P. H. and V. H. Heywood. 1963. *Principles of Angiosperm Taxonomy.* Edinburgh: Oliver and Boyd.

Davy de Virville, A. 1954. Botanique des temps anciens. In A. Davy de Virville, ed., *Histoire de la botanique en France,* pp. 17–118. Paris: Société d'Édition d'Enseignement Superieur.

Dean, J. 1979. Controversy over classification; a case study from the history of botany. In B. Barnes and S. Shapin, eds., *Natural Order: Historical Studies of Scientific Culture,* pp. 211–230. Berkeley: Sage.

Decaisne, J. 1842. Essais sur une classification des algues et des polypiers calcifères de Lamouroux. *Ann. Sci. Nat., Bot. Sér. 2*, 17: 297–380, pl. 14–17.

——. 1854. Notice historique sur M. Adrien de Jussieu. *Bull. Soc. Bot. France* 1: 386–399.

De La Rive, A. 1844. Notice sur la vie et les ouvrages de A.-P. de Candolle. *Biblioth. Universelle Sci. Belles-Lettres Arts, Sci. Arts. N. S.* 54: 75–144, 303–377.

Delile, A. R. 1827. Examen de la végétation de l'*Isoetes setacea*, et exposition de ses caractères. *Mém. Mus. Hist. Nat.* 14: 101–119, pl. 6, 7.

Delise, D. F. 1822. *Histoire des lichens. Genre Sticta.* [Caen.]

——. 1825. *Ibid. Atlas.* Caen: T. Chalopin.

Delpino, G. G. F. 1869a. Rivista monografica della famiglia delle Marcgraviaceae precipuamente sotto l'aspetto della biologia ossia delle relazioni di vita esteriore. *Nuovo Giorn. Bot. Ital.* 1: 256–290.

——. 1869b. Breve cenno sulle relazioni biologiche e genealogiche delle Marantacee. *Nuovo Giorn. Bot. Ital.* 1: 293–306.

——. 1888. *Applicazione di nuovi criterii per la classificazione delle pianti. Prima memoria.* Florence: Gamberini and Parmeggiani.

——. 1889. *Ibid. Seconda memoria.* Florence: Gamberini and Parmeggiani.

——. 1890. *Ibid. Quarta memoria.* Florence: Gamberini and Parmeggiani.

Descartes, R. 1962. *Discourse on Method.* Translated by J. Veitch. La Salle: Open Court.

Desfontaines, R. L. 1795. Cours de botanique élémentaire et de physiologie végétale. Discours d'ouverture. *Décade. Philos.* 5: 454–461, 513–527.

——. 1798. Mémoire sur l'organisation des monocotyledons, ou plantes à une feuille seminale. *Mém. Inst. Natl. Sci., Sci. Math.* 1: 478–502, pl. 5.

——. 1807. Précis d'un memoire de M. Mirbel, correspondant de l'Institut, sur l'anatomie des fleurs. *Ann. Mus. Hist. Nat.* 9: 448–468.

Desmond, A. 1982. *Archetypes and Ancestors.* Chicago: University of Chicago Press.

——. 1985. The making of institutional zoology in London, 1822–1836. *Hist. Sci.* 23: 153–185 (Part 1), 223–250 (Part 2).

——. 1989. *The Politics of Evolution: Morphology, Medicine and Reform in Radical London.* Chicago: University of Chicago Press.

Desmond, A. and J. Moore. 1991. *Darwin.* New York: Warner.

Desvaux, N. A. 1813. Théorie élémentaire de botanique. *J. Bot. (Desvaux)* 3: 284–288. ["N. A. D.": Review of Candolle, 1813b.]

Dhombres, N. and J. Dhombres. 1989. *Naissance d'un nouveau pouvoir: sciences et savants en France 1793–1824.* Paris: Puyot.

Dickoff, J., and P. James. 1964. Introduction. In A. Arnauld, *The Art of*

Thinking: Port-Royal Logic, pp. xxvii–li. Translated by J. Dickoff and P. James. New York: Bobbs-Merrill.

DiGregorio, M. A. 1982. In search of the natural system: Problems of zoological classification in Victorian Britain. *Hist. Philos. Life Sci.* 4: 225–254.

——. 1984. *Huxley's Place in Natural Science.* New Haven: Yale University Press.

Dobzhansky, T. 1937. *Genetics and the Origin of Species.* New York: Columbia University Press.

Don, D. 1831. An attempt at a new arrangement of the Ericaceae. *Edinburgh New Philos. J.* 17: 150–160.

——. 1832. A monograph of the genus *Saxifraga. Trans. Linn. Soc. London* 13: 341–452.

Donati, V. 1750. *Della storia naturale marina dell' Adriatico.* Venice: Francesco Storti.

Donoghue, M. J. and J. A. Doyle. 1989. Phylogenetic relationships of angiosperms and the relationships of the Hamamelidae. In P. R. Crane and S. Blackmore, eds., *Evolution, Systematics and Fossil History of the Hamamelidae. Vol. 1. Introduction to the "Lower" Hamamelidae*, pp. 17–45. New York: Clarendon Press.

Dormer, K. J. 1953. A study of phyllotaxis in the genus *Cassia.* New Phytol. 52: 313–315.

Drayton, R. H. 1993. *Imperial Science and a Scientific Empire: Kew Gardens and the Uses of Nature, 1772–1903.* Ph. D. Thesis, Yale University.

Drouin, J.-M. 1990. Un savoir utile et attrayant: René Desfontaines et sa conception de la botanique. *114e Congr. Nat. Soc. Sav. Paris* (1989): 229–240.

——. 1994. Classification des sciences et classification des plantes chez Augustin-Pyramus de Candolle. *Revue de Synthèse.* In press.

Drude, C. G. O. 1879. Über die natürliche Verwandtschaft von *Adoxa* und *Chrysosplenium. Bot. Zeitung* (Berlin) 37: 665–672, pl. 7.

Dubois-Reymond, E. H. 1877. Öffentliche Sitzung der Akadamie zur Feier des Leibnizischen Jahrestages. *Monatsber. König. Preuss. Akad. Wiss. Berlin* 1876: 385–407.

Duby, J. E. 1844. Mémoire sur la famille des Primulacées. *Mém. Soc. Phys. Genève* 10: 395–438, pl. 1–4.

Duchartre, P.-É.-S. 1867. *Éléments de botanique.* Paris: J. B. Baillière et Fils.

Duchesne, A.-N. 1766. *Histoire naturelle des fraisiers . . . .* Paris: Didot and C. J. Pancoucke.

——. 1795. Sur les rapports entre les êtres naturels. *Mag. Encycl.* 1795, 6: 289–294, chart.

Duhamel du Monceau, H. L. 1755. *Traité des arbres et arbustes qui se cultivent en France* . . . . 2 vols. Paris: Guérin and Delatour.

——. 1758. *La physique des arbres, où il est traité de l'anatomie des plantes et de l'économie végétale* . . . . 2 vols. Paris: Guérin and Delatour.

Duméril, A. M. C. 1806. *Zoologie analytique ou méthode naturelle.* Paris: Allais.

Dumortier, B. C. J. 1822. *Observations botaniques* . . . Tournay: Casterman-Dieu.

——. 1864. Discours sur la marche de la classification générale des plantes, depuis Jussieu jusqu'à nos jours. *Bull. Soc. Roy. Bot. Belgique* 3: 155–213.

Dunal, M. F. 1817. *Monographie de la famille des Anonacées.* Paris: Treuttel and Wurz.

——. 1829. *Considérations sur la nature et les rapports de quelques-uns des organes de la fleur.* Montpellier: Gabon.

——. 1842. *Éloge historique du A.-P. de Candolle.* Montpellier: J. Martel, ainé.

Dupree, H. 1959. *Asa Gray 1810–1888.* Cambridge, Massachusetts: Belknap Press of Harvard University Press.

Durande, J. F. 1781. *Notions élémentaires de botanique.* Dijon: L. N. Frantin.

Duris, P. 1993. *Linné et la France, 1780–1850.* Genève: Libraire Droz.

Duval-Jouve, J. 1864. *Histoire naturelle des Equisetum de France.* Paris: J. B. Baillière.

——. 1865. Variations parallèles des types congénères. *Bull. Soc. Bot. France* 12: 196–211.

——. 1871. Des comparaisons histotaxiques et de leur importance dans l'étude critique d'espèces végétales. *Mém. Acad. Sci. Montpellier* 7: 471–526.

Edgar, R. K. 1991. Farlow's students. *Newslett. Friends Farlow* 19: 1–3. [R. K. E.]

Eichler, A. W. 1875–1878. *Blüthendiagramme construirt und erlaütert.* 2 vols. Leipzig: Wilhelm Engelmann.

Elliott, B. 1992. Victorian gardens and botanical nomenclature. *Bot. J. Linn. Soc.* 109: 473–483.

——. 1993. The origins of floral diagrams. *Garden (London 1975+)* 118(1): 24–27.

Endress, P. K. 1993. Federico Delpino and early views on angiosperm evolution and macroevolution. *Diss. Bot.* 196: 77–83.

Engler, H. G. A. 1874a. Studien über die Verwandtschaftsverhältnisse des Rutaceae, Simarubaceae und Burseraceae nebst Beiträgen zur Anatomie und Systematik dieser Familien. *Abh. Naturf. Ges. Halle* 13: 109–158, pl. 12–13.

——. 1874b. Ueber Begrenzung und systematische Stellung der natürliche

Familie der Ochnaceae. *Nova Acta Phys.-Med. Acad. Caes. Leop.-Carol. Nat. Cur.* 37(2): 1–28, pl. 12–13.

——. 1881. Über die morphologische Verhältnisse und die geographische Verbreitung der Gattung *Rhus* wie der mit ihr verwandten, lebenden und ausgestorbenen Anacardiaceen. *Bot. Jahrb. Syst.* 1: 364–426, pl. 4.

——. 1893. Die botanische Centralstelle für die deutschen Colonien am Kön. botanisch Garten der Universität Berlin und die Entwickelung botanischer Versuchsstationen in der Colonien. *Bot. Jahrb. Syst.* 15, Nachtr. 35: 10–14.

——. 1897. Principien der systematischen Anordnung, insbesondere der Angiospermen. In A. Engler and K. Prantl, eds., *Die natürlichen Pflanzenfamilien* . . . . Nachträge zum II–IV Teil, pp. 5–14. Leipzig: Wilhelm Engelmann.

English, M. P. 1987. *Mordecai Cubitt Cooke: Victorian Naturalist, Mycologist, Teacher and Eccentric.* Bristol: Biopress.

Eriksson, G. 1969. *Romantikens värdsbild speglad i 1800-talets svenska vetenskap.* Stockholm: Wallström and Widstrand.

Esposito, J. L. 1977. *Schelling's Idealism and Philosophy of Nature.* Lewisburg: Bucknell University Press.

Farber, P. L. 1975. Buffon and Daubenton: Divergent traditions within the *Historie Naturelle. Isis (Sarton)* 66: 63–74.

——. 1976. The type concept in zoology during the first half of the nineteenth century. *J. Hist. Biol.* 9: 93–119.

——. 1982a. *The Emergence of Ornithology as a Scientific Discipline: 1760–1850.* Dordrecht: D. Riedel.

——. 1982b. The transformation of natural history in the nineteenth century. *J. Hist. Biol.* 15: 145–152.

Farley, J. 1982. *Gametes and Spores.* Baltimore: Johns Hopkins University Press.

Farlow, W. G. 1887. The task of American botanists. *Popular Sci. Monthly* 31: 305–314.

——. 1913. The change from the old to the new botany in the United States. *Science N. S.* 37: 79–86.

Fenzl, E. 1841. Darstellung und Erläuterung vier minder bekannter, ihrer Stellung im natürlichen System nach bisher zweifelhaft gebliebener, Pflanzen-Gattungen. *Denkschr. Königl.-Baier. Bot. Ges. Regensburg* 3: 153–270, pl. 1–5.

Figlio, K. M. 1976. The metaphor of organization: An historiographical perspective on the bio-medical sciences of the early nineteenth century. *Hist. Sci.* 14: 17–53.

Fleming, J. 1829. On systems and methods in natural history. By J. E.

Bicheno, Esq. 1829 (Linn. Trans., xv, part 2.). *Quart. Rev.* 41: 302–327. [Anonymous: review of Bicheno, 1829.]

Flourens, M. J. P. 1838. Éloge historique de Antoine-Laurent de Jussieu. *Mém. Acad. Roy. Sci. Inst. France.* 17: i–lx.

——. 1842. Éloge historique de Pyramus de Candolle. *Mém. Acad. Roy. Sci. Insti. France* 19: i–xlviii.

Foucault, M. 1970. La situation de Cuvier dans l'histoire de la biologie. *Rev. Hist. Sci. Applic.* 23: 63–92.

——. 1973. *The Order of Things.* New York: Vintage Books.

Fourcroy, A. F. de 1794. *Élémens d'histoire naturelle et de chimie.* Ed. 5, 5 vols. Paris: Cuchet.

Frankel, H. 1981. The paleobiological debate over the problem of disjunctively distributed life forms. *Studies Hist. Philos. Sci.* 12: 211–259.

Fries, E. M. 1821–1832. *Systema mycologicum.* 3 vols. Lund: Berlinger [vols. 1, 2], Griefswald: Moritz [vol. 3].

——. 1825. *Systema orbis vegetabilis . . . Pars 1. Plantae homonemeae.* Lund: Typographica Academica.

——. 1849. *Summa vegetabilium scandinaviae.* Stockholm and Leipzig: Bonnier.

Frodin, D. G. 1984. *Guide to the Standard Floras of the World.* Cambridge: Cambridge University Press.

Fuhlrott, C. 1829. *Jussieu's und De Candolle's natürliche Pflanzen-Systeme nach ihren Grundsätzen entwickelt . . . .* Bonn: Eduard Weber.

Gaertner, C. F. von. 1805–1807. *Supplementum carpologicae . . . .* Leipzig: Richter.

Gaertner, J. 1788–1792. *De fructibus et seminibus plantarum . . . .* Stuttgart: the author; Tübingen: Schramm.

Gage, A. T. and W. T. Stearn. 1988. *A Bicentenary History of the Linnean Society of London.* London: Academic Press.

Galton, Sir D. 1895. [Presidential] address. *Rep. Brit. Assoc. Advancem. Sci.* 1895: 1–35.

Gandoger, M. 1883–1891. *Flora europae . . . .* 27 vols. London: Bernard Quaritch.

Gaudry, A. 1866. *Considérations générales sur les animaux fossiles des Pikermi.* Paris: F. Savy.

Gentry, A. H. 1993. *A Field Guide to the Families and Genera of Woody Plants of Northwest South America . . . .* Washington, D.C.: Conservation International.

Geoffroy, C. J. 1714. Sur la structure et l'usage des principales parties des fleurs. *Mém. Math. Phys. Acad. Roy. Sci.* (Paris) (1711): 210–234, pl. 7.

Geoffroy, É. L. 1762. *Histoire abregée des insectes qui se trouvent aux environs de Paris . . . .* 2 vols. Paris: Durand.

Geoffroy Saint-Hilaire, É. 1794. Extrait d'un mémoire sur un nouveau genre des quadrupèdes, de l'ordre des rongeurs (Glires L.). *Décade Philos.* 4: 193–206, pl.

——. 1796. Mémoire sur les rapports naturels de Makis *Lemur,* L., et description d'une espèce nouvelle des mammifères. *Mag. Encycl.* 1796, 1: 20–50.

——. 1798. Sur un prétendu orang-outang des Indies, publié dans les actes de la société de Batavia. *J. Phys. Chim. Hist. Nat. Arts* 46: 342–346.

Geoffroy Saint-Hilaire, É. and G. Cuvier. 1795a. Mémoire sur une nouvelle division des mammifères, et sur les principes qui doivent server de base dans cette sorte de travail. *Mag. Encycl.* 1795, 2: 164–190.

Geoffroy Saint-Hilaire, É. and G. Cuvier. 1795b. Histoire naturelle des orangs-outangs. *Mag. Encycl.* 1795, 3: 451–463, pl. [See also: 1798. Mémoire sur les orangs-outangs. *J. Phys. Chim. Hist. Nat. Arts* 46: 185–191, pl.]

Geoffroy Saint-Hilaire, I. 1827. Monstre. In *Dictionnaire Classique d'Histoire Naturelle* 11: 108–150. Paris: Rey and Gravier.

——. 1832a. Considérations sur les caractères employés en ornithologie pour la distinction des genres, des familles, et des ordres, et détermination de plusieurs genres nouveaux. *Nouv. Ann. Mus. Hist. Nat.* 1: 356–397.

——. 1832b-36. *Histoire générale et particulière des anomalies d'organisation chez l'homme et les animaux* . . . . 3 vols. Paris: J. B. Baillière.

——. 1841. *Essai de zoologie générale* . . . . Paris: Roret.

——. 1842. Sur la classification des animaux en séries parallèles; par M. Brullé. *Compt. Rend. Acad. Sci.* (Paris) 14: 226–228.

——. 1845. Classification parallélique des Mammifères. *Compt. Rend. Acad. Sci.* (Paris) 20: 757–761.

Germain de Saint-Pierre, J. N. E. 1855. *Histoire iconographique des anomalies de l'organisation dans le règne végétal.* . . . Paris: F. Klincksieck.

——. 1870. *Nouveau dictionnaire de botanique* . . . . Paris: J.-B. Baillière et Fils.

Ghiselin, M. 1967. *The Triumph of the Darwinian Method.* Berkeley: University of California Press.

Gidon, F. 1934. Linné, Jussieu ou Adanson? (a propos d'un texte de 1747). *Mém. Acad. Nat. Sci. Caen, N. S.* 7: 287–309.

Gillispie, C. C. 1960. *The Edge of Objectivity.* Princeton: Princeton University Press.

——. 1971. Condillac, Étienne Bonnot Abbé de. In C. C. Gillispie, ed., *Dictionary of Scientific Biography*, vol. 3., pp. 380–383. New York: Charles Scribner's Sons.

——. 1980. *Science and Polity in France at the End of the Old Regime.* Princeton: Princeton University Press.

Gilmour, J. S. L. 1989. Two early papers on classification. *Plant Syst. Evol.* 167: 97–107.

Ginzburg, C. 1983. Clues. Morelli, Freud and Sherlock Holmes. In U. Eco and T. A. Sebeok, eds., *The Sign of Three: Dupin, Holmes, Pierce*, pp. 81–118. Bloomington: Indiana University Press.

Giseke, P. D. 1792. *Praelectiones in ordines naturales plantarum* . . . . Hamburg: Hoffmann.

Gleditsch, J. G. 1751. Système des plantes fondé sur la situation et la liaison des étamines. *Hist. Acad. Roy. Sci.* (Berlin) (1749): 109–136, pl. 4.

——. 1764. *Systema plantarum a staminum situ* . . . . Berlin: Haude and Spener.

Goebel, K. I. E. von. 1926. *Wilhelm Hofmeister: The Work and Life of a Nineteenth Century Botanist.* London: The Ray Society.

Göppert, [J.] H. R. 1864–1865. Die fossile Flora der Permischen Formation. *Palaeontographia* 12: 1–316, pl. 1–64.

——. 1881. Revision meiner Arbeiten über die Stämme der fossilen Coniferen, insbesondere der Araucariten, und über die Descendenzlehre. *Bot. Centralbl.* 5: 378–385, 6: 27–30, 98–101, 170–174, 207–212.

Goethe, J. W. von. 1790. *Versuch die Metamorphose der Pflanzen zu erklären.* Gotha: Ettinger.

Gooday, G. 1991. "Nature" in the laboratory: Demonstration and discipline with the microscope in Victorian life science. *Brit. J. Hist. Sci.* 24: 307–341.

Gould, S. J. 1977. *Ontogeny and Phylogeny.* Cambridge, Massachusetts: Harvard University Press.

——. 1984. The Rule of Five. *Nat. Hist.* 93(10): 14–23.

——. 1987a. Hatracks and theory. *Nat. Hist.* 96(3): 12–23.

——. 1987b. *Time's Arrow, Time's Cycle: Myths and Metaphor in the Discovery of Geological Time.* Cambridge, Massachusetts: Harvard University Press.

Gram, M. S. and R. M. Martin. 1980. The perils of plenitude: Hitinkka contra Lovejoy. *J. Hist. Ideas* 41: 497–511.

Grandmaison, Millin de. [See Millin de Grandmaison.]

Grant, V. 1957. The plant species in theory and practice. *Publ. Amer. Assoc. Advancem. Sci.* 50: 39–80.

Gray, A. 1836. *Elements of Botany.* New York: G. and C. Carvill.

——. 1837. Art. VIII.—A natural system of botany . . . by John Lindley. *Amer. J. Sci. Arts* 32: 292–303. [Review of Lindley, 1836]

——. 1842. *The botanical textbook.* New York: Wiley and Putnam.

——. 1846. Scientific intelligence II. Botany. *Amer. J. Sci. Arts, Ser. 2*, 21: 134–137. [A. G.]

——. 1850. On the composition of the plant by phytons, and some applications of phyllotaxis. *Proc. Amer. Assoc. Advancem. Sci.* 2: 438–444.
——. 1858a. Scientific intelligence III. Botany and Zoology. 1. Monographie de la famille des Urticées, par H. A. Weddell. *Amer. J. Sci. Arts, Ser. 2*, 25: 109–111. ["A. G."; review of Weddell 1856.]
——. 1858b. Ibid. 8. Dr. J. D. Hooker on the structure and affinities of the Balanophoreae. *Amer. J. Sci. Arts, Ser. 2*, 25: 116–120. ["A. G."; review of Hooker 1856.]
——. 1861. *Amer. J. Sci. Arts, Ser. 2*, 32: 124–127. [Review of Bentham 1861a.]
——. 1879. *Gray's Botanical Text Book*, ed. 6, vol. 1. *Structural Botany, or Organography on the Basis of Morphology, to which is added the Principles of Taxonomy and Phytography*. New York: American Book Company.
——. 1880. Botany for high schools and colleges. *Amer. J. Sci. Arts, Ser. 3*, 20: 337. ["A. G."; review of Bessey 1880.]
Green, E. R. 1914. *A History of Botany in the United Kingdom from the Earliest Times to the End of the 19th Century*. London and Toronto: Dent.
Greene, E. L. 1983. *Landmarks of Botanical History*. 2 vols. Stanford: Stanford University Press.
Greene, J. C. 1971. The Kuhnian paradigm and the Darwinian revolution in natural history. In D. Roller, ed., *Perspectives in the History of Science and Technology*, pp. 3–25. Norman: University of Oklahoma Press.
——. 1992. From Aristotle to Darwin: Reflections on Ernst Mayr's interpretation in *The Growth of Biological Thought*. *J. Hist. Biol.* 25: 257–284.
Greenman, J. M. 1940. Genera from the standpoint of morphology. *Bull. Torrey Bot. Club* 67: 371–375.
Gregg, J. R. 1950. Taxonomy, language and reality. *Amer. Naturalist* 84: 421–433.
Grew, N. 1673. *An Idea of a Phytological History Propounded*. London: Richard Chiswell.
Griffiths, G. C. D. 1974. On the foundations of biological systematics. *Acta Biotheor.* 23: 85–131.
Grisebach, A. H. R. 1838. *Genera et species Gentianearum*. Stuttgart: J. G. Cotta.
——. 1848. Bericht über den Leistungen in der Pflanzengeographie und systematische Botanik während des Jahres 1847. *Arch. Naturgesch.* 14(2): 257–350.
——. 1854. *Grundriss der systematischen Botanik für akademische Vorlesungen*. Göttingen: Dieterich.
Gruber, J. W. 1991. Does the platypus lay eggs? The history of an event in

science. *Arch. Nat. Hist.* 18: 61–123.
Guédès, M. 1967. La méthode taxonomique de Adanson. *Rev. Hist. Sci. Applic.* 20: 361–386.
——. 1971. Le botaniste Lamarck. In J. Schiller, ed., *Colloque international "Lamarck"*, pp. 47–85. Paris: A. Blanchard.
——. 1972. La théorie de la métamorphose en morphologie végétale: A.-P. de Candolle et P.-J.-F. Turpin. *Rev. Hist. Sci. Appl.* 25: 253–270.
——. 1973. Duchesne, Buisson, Durande, early followers of the natural method of the Jussieus. *Taxon* 22: 211–219.
——. 1982. Nothing new with cladistics. *Taxon* 31: 95–96.
Gusdorf, G. 1972. *Les sciences humaines et la pensée occidentale. V. Dieu, la nature, l'homme au siècle des lumières.* Paris: Payot.
Gutting, G. 1980. Introduction. In G. Gutting, ed., *Paradigms and Revolution*, pp. 1–21. Notre Dame: University of Notre Dame.
Guyton de Morveau, L.-B., A.-L. Lavoisier, C.-L. Berthollet, and A. F. de Fourcroy. 1787. *Méthode de la nomenclature chimique.* Paris: Cuchet.
Haeckel, E. H. P. A. 1866. *Generelle Morphologie der Organismen.* 2 vols. Berlin: Georg Reimer.
——. 1874. *Anthropogenie oder Entwicklungsgeschichte der Menschen . . . .* Leipzig: W. Engelmann.
Hagemann, W. 1982. Vergleichende Morphologie und Anatomie—Organismus und Zelle, ist ein Synthese möglich? *Ber. Deutsch. Bot. Ges.* 95: 45–56.
Hagen, J. B. 1986. Ecologists and taxonomists: Divergent traditions in twentieth-century plant geography. *J. Hist. Biol.* 19: 197–214.
Hahn, R. 1971. *The Anatomy of a Scientific Institution: The Paris Academy of Sciences.* Berkeley: University of California.
Haines, J. H. and I. F. Salkin, eds. 1986. History of North American mycology. *Mycotaxon* 26: 1–79.
Haldane, E. G. and G. R. T. Ross. 1931. *The Philosophical Works of Descartes.* 2 vols. Cambridge: Cambridge University Press.
Hallé, F., R. A. A. Oldeman, and P. B. Tomlinson. 1978. *Tropical Trees and Forests.* Berlin: Springer.
Haller, A. von. 1742. *Enumeratio methodica stirpium Helvetiae indigenarum . . . .* Goettingen: Abram Vandenhoek.
——. 1983. *The Correspondence between Albrecht von Haller and Charles Bonnet.* Edited by O. Sonntag. Bern: Hans Huber.
Hallier, H. 1893. Versuch einer natürlichen Gliederung der Convolvulaceen auf morphologischer und anatomischer Grundlage. *Bot. Jarhb. Syst.* 16: 453–591.
——. 1903. Über eine Zwischenform zwischen Apfel und Pflaume. *Verhand.*

Naturwiss. Vereins Hamburg 1903, 3, Folge X: 8–19.
Hamy, E.-T. 1893. *Les derniers jours du Jardin du Roi et la fondation du Muséum d'Histoire Naturelle*. Paris: Imprimerie Nationale.
——. 1909. *Les debuts de Lamarck, suivis des recherches sur Adanson, Jussieu, Pallas, etc*. Paris: E. Guilmoto.
Harvey-Gibson, R. J. 1919. *Outlines of the History of Botany*. London: A. and C. Black.
Haworth, A. H. 1819. *Supplementum plantarum succulentarum* . . . . London: J. Harding.
——. 1821. *Saxifragearum enumeratio*. London: Wood.
——. 1823. A few observations on the natural distribution of animated nature. *Philos. Mag*. 62: 200–202.
——. 1825a. A new binary arrangement of the macrurous crustacea. *Philos. Mag*. 65: 183–184.
——. 1825b. Observations on the dichotomous distribution of animals: together with a binary arrangement of the natural order Saxifrageae. *Philos. Mag*. 65: 428–430.
Hayata, B. 1921. The natural classification of plants according to the dynamic system. *Icon. Plant. Formosan*. 10: 97–234.
Hays, T. 1976. An empirical method for the detection of covert categories in ethnobiology. *Amer. Ethnol*. 3: 489–507.
Heckel, É. M. 1894. *Étude monographique de la famille des Globulariées*. Marseille: J. Cayer.
Hedge, I. C. 1967. The specimens of Paul Dietrich Giseke in the Edinburgh herbarium. *Notes Roy. Bot. Gard. Edinburgh* 28: 73–86.
Hedwig, J. 1785–1797. *Descriptio et adumbratio microscopico-analytica muscorum frondosum* . . . . Leipzig: I. G. Müller.
Heilbron, J. L. 1990. Introductory Essay. In T. Frängsmyr, J. L. Heilbron, and R. E. Rider, eds., *The Quantifying Spirit*, pp. 1–23. Berkeley: University of California Press.
Heister, L. 1748. *Systema plantarum generale ex fructificatione* . . . . Helmstadt: Ioannis Drimbornii.
Heller, J. L. 1959. Index auctorum et librorum a Linnaeo (*Species plantarum*, 1753) citatorum. In C. Linnaeus, *Species plantarum*, vol. 2, pp. 3–60. London: Ray Society. [Reprint.]
Hempel, C. G. 1965. *Aspects of Scientific Explanation and Other Essays in the Philosophy of Science*. New York: The Free Press, Macmillan.
Henderson, L. J. 1917. *The Order of Nature: An Essay*. Cambridge, Massachusetts: Harvard University Press.
Henslow, J. S. 1835. *The Principles of Descriptive and Physiological Botany*. London: Longman, Rees, Orme, Brown and Green, and John Taylor.

——. 1837. On the requisites necessary for the advance of botany. *Mag. Zool. Bot.* 1: 113–125.

Hermann, J. 1777. *Tabula affinitatum animalium.* Strasbourg: J. G. Treuttel.

Hermann, P. 1690. *Florae lugduno-batavae flores* . . . . Leiden: Haaring.

Hesse, M. 1974. *The Structure of Scientific Inference.* Berkeley: University of California Press.

Heywood, V. H. 1988. The structure of systematics. In D. C. Hawksworth, ed., *Prospects in Systematics*, pp. 44–56. Oxford: Clarendon Press.

——. 1989. Nature and natural classification. *Plant Syst. Evol.* 167: 87–92.

Hiern, W. P. 1873. A monograph of Ebenaceae. *Trans. Cambridge Philos. Soc.* 22: 27–300, pl. 1–11.

Hilgard, T. C. 1857. Exposition of a natural series by immediate catholic affinities in the Vegetable Kingdom. *Trans. Acad. Sci. St. Louis* 1: 125–156, pl. 6–7.

Hill, J. 1759. *The Vegetable System or, a Series of Experiments, and Observations Tending to Explain the Internal Structure, and the Life of Plants* . . . . vol. 1. Printed for the author, London.

Hoare, M. E. 1976. *The Tactless Philosopher: Johann Reinhold Forster (1729–1798).* Melbourne: Hawthorn Press.

Hocquette, M. 1954. Morphologie, anatomie, cytologie. In A. Davy de Virville, ed., *Histoire de la botanique en France*, pp. 120–152. Paris: Société d'Édition d'Enseignement Supérieur.

Hofmeister, W. F. B. 1851. *Vergleichende Untersuchungen der Keimung, Entfaltung und Fruchtbildung höherer Kryptogamen* . . . . Regensberg: Hinrichs.

Holman, E. W. 1985. Evolutionary and psychological effects in pre-evolutionary classifications. *J. Classific.* 2: 29–39.

——. 1992. Statistical properties of large published classifications. *J. Classific.* 9: 187–210.

Holton, G. 1975. On the role of themata in scientific thought. *Science* 188: 328–338.

——. 1988. *Thematic Origins of Scientific Thought.* Ed. 2. Cambridge, Massachusetts: Harvard University Press.

Hooker, J. D. 1853. *The Botany of the Antarctic Voyage . . . II. Flora novae zelandiae. Part 1. Flowering plants.* Vol. 1. London: Lovell Reeve.

——. 1856a. On the structure and affinities of the Balanophoreae. *Trans. Linn. Soc. London* 22: 1–68, pl. 1–16.

——. 1856b. *Géographie botanique raisonnée* . . . par M. Alp. de Candolle . . . . *Hooker's J. Bot. Kew Gard. Misc.* 8: 54–64, 82–88, 112–121, 151–157, 181–191, 248–256.

——. 1859. *The Botany of the Antarctic Voyage . . . Part III. Flora tasmaniae. Vol. 1. Dicotyledones*. London: Lovell Reeve.

——. 1861. Colonial floras. *Nat. Hist. Rev., Ser. 2*, 1: 255–266. [Anonymous.]

——. 1862. Outlines of the distribution of arctic plants. *Trans. Linn. Soc. London* 23: 251–348.

——. 1863. On *Welwitschia*, a new genus of Gnetaceae. *Trans. Linn. Soc. London* 24: 1–48, pl. 1–14.

——. 1873. Appendix on the classification of plants by the natural method . . . . In E. Le Maout and J. Decaisne, *A general System of botany . . .* , pp. 985–1023. Translated by Mrs Hooker. London: Longmans, Green, Reader and Dyer.

——. 1888. Eulogium on Robert Brown. *Proc. Linn. Soc. London* (1887–1888): 54–67.

Hooker, J. D. and T. Thomson. 1855. *Flora indica*. vol. 1. London: W. Pamplin.

Hooker, W. J. 1835. *The British Flora* . . . . Ed. 3, vol. 1. London: Longman, Rees, Orme, Brown, Green and Longman.

Hooker, W. J. and G. A. W. Arnott. 1855. *Ibid*. Ed. 7. London: Longman, Brown, Green and Longman.

Hopkirk, T. 1817. *Florae anomalae. A General View of the Anomalies in the Vegetable Kingdom*. Glasgow: John Smith.

Hoppe, B. 1971. Le concept de biologie chez G. R. Treviranus. In J. Schiller, ed., *Colloque international "Lamarck"*, pp. 199–237. Paris: A. Blanchard.

Horner, F. 1987. *The French Reconnaisance: Baudin in Australia*. Melbourne: Melbourne University Press.

Howard, R. A. 1988. *Charles Wright in Cuba, 1856–1867*. Alexandra, Vermont: Chadwyck-Healey.

Hull, D. L. 1965. The effect of essentialism on taxonomy: Two thousand years of stasis. *Brit. J. Philos. Sci.* 15: 314–326, and 16: 1–18.

——. 1970. Contemporary systematic philosophies. *Annual Rev. Ecol. Syst.* 19–51.

——. 1988. *Science as a Process: An Evolutionary Account of the Social and Conceptual Development of Science*. Chicago and London: University of Chicago Press.

Hutchinson, J. 1923. *The Families of Flowering Plants 1. Dicotyledons*. London: Macmillan.

Huxley, L. 1901. *Life and Letters of Thomas Henry Huxley*. 2 vols. New York: D. Appleton.

——. 1918. *Life and Letters of Sir Joseph Dalton Hooker O.M., G.C.S.I.* New York: D. Appleton.

Huxley, T. H. 1868a. On the animals which are most nearly intermediate between birds and reptiles. *Geol. Mag.* 5: 357–365.

——. 1868b. Letters, Announcements, etc. *Ibis N. S.* 4: 357–362. [Letter addressed to Alfred Newton.]

——. 1888. The gentians: Notes and queries. *J. Linn. Soc., Bot.* 24: 101–124, pl. 2.

Hyman, A. 1982. *Charles Babbage, Pioneer of the Computer.* Princeton: Princeton University Press.

Jackson, B. D. 1906. *George Bentham.* London: J. M. Dent.

Jacob, F. 1976. *The Logic of Life.* New York: Random House.

Jacobs, M. 1980. Revolutions in plant description. In J. C. Arends, G. Boelema, C. T. de Groot, and A. J. M. Leeuwenberg, eds., *Liber gratulatorius in honorem H. C. D. de Wit*, pp. 155–181. Wageningen: Veenman.

Jacquin, N. J., Baron von. 1785. *Anleitung zur Pflanzenkenntnis nach Linné's Methode.* Vienna: Christian Friedrich Wappler.

——. 1788. *Selectarum stirpium americanum historia* . . . . Mannheim: Bibliopolo Novo Aul. & Acad.

Jacyna, L. S. 1984. Principles of general physiology: The comparative dimension to British neuroscience in the 1830s and 1840s. *Stud. Hist. Biol.* 7: 47–92.

Jäger, G. F. von. 1814. *Ueber die Missbildungen der Gewächse* . . . . Stuttgart: J. F. Steinkopf.

Jahn, I. 1985. Die Ordnungswissenschaften und die Begründung biologischer Disziplinen im 18. und zu Beginn des 19. Jahrhunderts. In I. Jahn, R. Löther, and K. Senglaub, eds., *Geschichte der Biologie*, pp. 263–323. Jena: Gustav Fischer.

Jaume Saint-Hilaire, J. H. 1805. *Exposition des familles naturelles et de la germination des plantes.* 2 vols. Paris and Strasbourg: Treuttel and Würtz.

Jenyns, L. 1837. Some remarks on the study of zoology, and on the present state of the science. *Mag. Zool. Bot.* 1: 1–31.

Jevons, W. S. 1874. *The principles of science.* New York: Macmillan.

Joly, P. 1986. Les classifications botaniques. In P. Tassy, ed., *L'ordre et la diversité du vivant: quel statut scientifique pour les classifications biologiques?*, pp. 52–66. Paris: Fondation Diderot, Fayard.

Jolyclerc, N.-M.-T. 1797. *Élémens de botanique.* Lyon: Pierre Bernuset.

Jordanova, L. J. 1984. *Lamarck.* Oxford: Oxford University Press.

Judd, W. S., R. W. Sanders, and M. J. Donoghue. 1994. Angiosperm family pairs—preliminary phylogenetic analyses. *Harvard Pap. Bot.* 5: 1–51.

Junker, T. 1989. Darwinismus und Botanik. Kritik und theoretische

Alternativen in Deutschland des 19. Jahrhunderts. *Quellen Stud. Gesch. Pharm.* 54: 1–367.

Jussieu, A[drien]-H.-L. de. 1823. Considérations sur la famille des Euphorbiacées. *Mém. Mus. Hist. Nat.* 10: 317–355.

——. 1825. Sur le group des Rutacées. *Mém. Mus. Hist. Nat.* 12: 384–542, pl. 14–29.

——. 1834. A MM. les rédacteurs des Annales des Sciences Naturelles, sur un point de l'histoire de la botanique. *Ann. Sci. Nat., Bot. Sér.* 2, 2: 302–308.

——. 1843a. Monographie de la famille des Malpighiacées. *Arch. Mus. Hist. Nat.* 3: 5–151, 255–616, 23 pl.

——. 1843b. *Cours élémentaire d'histoire naturelle . . . Botanique.* Paris: Fortin Masson and Langlois & Leclerc.

——. 1848. *Taxonomie: coup d'oeil sur l'histoire et les principes des classifications botaniques.* Paris. [Reprint of "Taxonomie," in d'Orbigny, *Dictionnaire Universal d'Histoire Naturelle* 12: 368–431.]

Jussieu, A[ntoine] de. 1719. Examen des causes des impressions de plantes marquées sur certaines pierres des environs de Saint-Chaumont dans le Lionnois. *Mém. Math. Phys. Acad. Roy. Sci.* (Paris) 1718: 287–297, pl. 13–14.

Jussieu, A[ntoine]-L. de. 1770. *Quaestio medica, . . . Carolo-Jacobo-Ludovico Coquereau, Doctore Medico, Praeside. An oeconomiam animalem inter et vegetalem analogia?* Paris: Quillau.

——. 1777. Examen de la famille des Renoncules. *Mém. Math. Phys. Acad. Roy. Sci.* (Paris) (1773): 214–240.

——. 1778. Exposition d'un nouvel ordre des plantes adopté dans les démonstrations du Jardin Royal. *Mém. Math. Phys. Acad. Roy. Sci.* (Paris) (1774): 175–197.

——. 1784. *Rapport de l'un des commissaires chargés par le roi, de l'examen du magnétisme animal.* Paris: Veuve Hérissant and Théophile Barrois.

——. 1787. Extrait d'un mémoire de M. Cusson sur les plantes ombellifères. *Hist. Soc. Roy. Méd.* 1785: 275–285.

——. 1788. Sur les rapports existans entre les caractères des plantes, et leurs vertus. *Hist. Soc. Roy. Méd.* 1786: 188–197.

——. 1789. *Genera plantarum.* Paris: Hérissant and Barrois. [Reprint: 1965, Weinheim: Cramer.]

——. 1802. Mémoire sur la plante nommé par les botanistes *Erica daboecia*, et sur la nécessité de la rapporter à un autre genre et à une autre famille. *Ann. Mus. Natl. Hist. Nat.* 1: 52–56.

——. 1803a. Observations sur la famille des plantes Amarantacées. *Ann. Mus. Natl. Hist. Nat.* 2: 131–135.

——. 1803b. Observations sur la famille des plantes Nyctaginées. *Ann. Mus. Natl. Hist. Nat.* 2: 269–279.

——. 1804a. Mémoire sur quelques espèces du genre *Hypericum*. *Ann. Mus. Natl. Hist. Nat.* 3: 159–162.

——. 1804b. Observations sur la famille des plantes Onagraires. *Ann. Mus. Natl. Hist. Nat.* 3: 315–325, pl. 30.

——. 1804c. Mémoire sur l'*Opercularia*, genre de plantes voisin de la famille des Dipsacées. *Ann. Mus. Natl. Hist. Nat.* 4: 418–428, pl. 70–71.

——. 1804d. Mémoire sur le *Loasa*, genre de plantes qui devra constituer avec le *Mentzelia*, une nouvelle famille. *Ann. Mus. Hist. Nat.* 5: 18–27, pl. 1–5.

——. 1805a. Premier mémoire sur quelques nouvelles espèces du genre *Passiflora*, et sur la nécessité d'établir une famille des Passiflorées. *Ann. Mus. Hist. Nat.* 6: 102–116.

——. 1805b. Mémoire sur la réunion de plusieurs genres de plantes en un seul dans la famille des Laurinées. *Ann. Mus. Hist. Nat.* 6: 197–213.

——. 1805c. Quatrième mémoire sur les caractères généraux de familles, tirés des graines, et confirmés ou rectifiés par les observations de Gaertner. Corolles monopétales épigynes, à anthères réunies. Premier partie. *Ann. Mus. Hist. Nat.* 6: 307–324.

——. 1806a. *Ibid.* Seconde partie. *Ann. Mus. Hist. Nat.* 7: 373–391.

——. 1806b. *Ibid.* Troisième partie. *Ann. Mus. Hist. Nat.* 8: 170–186.

——. 1807a. Mémoire sur le *Dicliptera* et le *Blechum*, genres nouvaux de plantes, composés de plusieurs espèces auparavant réunies au *Justicia*. *Ann. Mus. Hist. Nat.* 9: 251–271, pl. 21.

——. 1807b. Septième mémoire sur les caractères généraux des familles, tirés des graines, et confirmés ou rectifiés par les observations de Gaertner. *Ann. Mus. Hist. Nat.* 10: 307–332.

——. 1808. Suite des observations sur quelques genres de la Flore de Cochinchine de Loureiro, avec quelques réflexions sur l'*Elaeocarpus* et les genres qui doivent s'en rapprocher dans l'ordre naturel. *Ann. Mus. Hist. Nat.* 11: 231–240.

——. 1809a. Mémoire sur les Monimiées, nouvel ordre de plantes. *Ann. Mus. Hist. Nat.* 14: 116–135.

——. 1809b. Mémoire sur une nouvelle espèce de *Marcgravia*, et sur les affinités botaniques de ce genre. *Ann. Mus. Hist. Nat.* 14: 397–411, pl. 25.

——. 1815. Douzième mémoire sur les caractères généraux des familles tirés des graines. Aurantiacées - Théacées. *Mém. Mus. Hist. Nat.* 2: 436–443.

——. 1819a. Treizième mémoire . . . Meliacées - Tiliacées. *Mém. Mus. Hist. Nat.* 5: 226–246.

——. 1819b. Note sur quelques genres anciens de plantes non classés antérieurement, et maintenant rapportés à leurs familles. *Mém. Mus. Hist. Nat.* 5: 247–248.

——. 1820. Sur la famille des plantes Rubiacées. *Mém. Mus. Hist. Nat.* 6: 365–409.

——. 1823. Marcgraviacées. In *Dictionnaire des Sciences Naturelles* 29: 109–110. Strasbourg: F. G. Levrault. ["J."]

——. 1824. *Principes de la méthode naturelle des végétaux.* Paris: F. G. Levrault. [From *Dictionnaire des Science Naturelles* 30: 426–468. Strasbourg: F. Levrault.]

——. 1837. Introductio in historiam plantarum. *Ann. Sci. Nat., Bot. Sér. 2,* 8: 97–160, 193–239. [Also separately printed, Paris: Paul Renouard.]

Jussieu, A[ntoine]-L. de., R. L. Desfontaines, and J.-B.-P.-A. de M. de Lamarck. 1913. Rapport sur le mémoire de M. DeCandolle, relatif à la distribution des Composées. *Procès-Verbaux Séances Acad. Sci.* (Paris) 4: 27–29.

Jussieu, B. de. 1741. Histoire d'une plante, connu par les botanistes sous le nom de *Pilularia. Mém. Math. Phys. Acad. Roy. Sci.* (Paris) (1739): 240–256, pl. 11.

——. 1742. Histoire du *Lemma. Mém. Math. Phys. Acad. Roy. Sci.* (Paris) (1740): 263–275, pl. 15.

Keeney, E. B. 1992. *The Botanizers: Amateur Scientists in Nineteenth Century America.* Chapel Hill: University of North Carolina Press.

Ker, J. B. 1817. *Aster novae-angliae. Edwards Bot. Reg.* 3: pl. 183. [Anonymous.]

——. 1823. *Erythrina caffra. Edwards Bot. Reg.* 9: pl. 736. [Anonymous.]

Kieser, D. G. 1814. *Mémoire sur l'organisation des plantes . . . .* Haarlem: J. J. Beets.

Kirby, W. and W. Spence. 1815–1826. *An Introduction to Entomology: Or Elements of the Natural History of Insects.* 4 vols. London: Longman, Hurst, Rees, Orme and Brown.

Klotzsch, J. F. 1855. Begoniaceen Gattungen und Arten. *Abh. Königl. Akad. Wiss.* (Berlin) (1854): 121–255, pl. 1–12.

——. 1860. Linné's natürliche Pflanzenklasse Tricoccae. *Abh. Köngigl. Akad. Wiss.* (Berlin) (1859): 1–108.

Knaut, C[hristian]. 1716. *Methodus plantarum genuina . . .* Leipzig: Sell.

Knaut, C[hristoph]. 1687. *Enumeratio plantarum . . . .* Leipzig: Lanckisch.

Knight, D. 1985. William Swainson: Types, circles and affinities. In J. D. North and J. J. Roche, eds., *The Light of Nature: Essays in the History and Philosophy of Science Presented to A. C. Crombie,* pp. 83–94. Dordrecht: Martinus Nijhoff.

Knight, I. F. 1968. *The Geometric Spirit: The Abbé de Condillac and the French Enlightenment.* New Haven: Yale University Press.

Knowlson, J. 1975. *Universal Language Schemes in England and France 1600–1800.* Toronto: University of Toronto Press.

Kohlstedt, S. G. 1988. Curiosities and cabinets: Natural history museums and education on the Antebellum campus. *Isis (Sarton)* 79: 405–426.

Kuhn, T. 1970. *The Structure of Scientific Revolutions.* Ed. 2. Chicago: University of Chicago Press.

——. 1977. *The Essential Tension: Selected Studies in Scientific Tradition and Change.* Chicago: University of Chicago Press.

Kunth, C. S. 1815. Considérations générales sur les Graminées. *Mém. Mus. Hist. Nat.* 2: 62–75.

——. 1831. *Handbuch der Botanik.* Berlin: Duncker and Humblot. [As K. S. Kunth.]

——. 1833. *Zwei botanische Abhandlungen.* Berlin: Königlichen Akademie der Wissenschaften.

Kuntz, M. L., and P. G. Kuntz. 1987. *Jacob's Ladder and the Tree of Life.* New York: Peter Lang.

Kurz, W. S. 1877. *Forest Flora of British Burma.* Calcutta: Government Printer.

Kusnezow, N. I. 1896–1904. Subgenus Eugentiana Kusnez. generis *Gentiana* Tournef. *Acta Horti Petrop.* 15: 1–507, pl. 1–4.

Kützing, F. T. 1843. *Phycologia generalis oder Anatomie, Physiologie und Systemkunde der Tange . . . .* Leipzig: F. A. Brockhaus.

——. 1844. *Die Sophisten und Dialektiker.* Norhausen: Ferd. Förstermann.

Lacepède, B.-G.-É. de. 1795. Introduction au cours d'ichthyologie, donné dans les galeries du Muséum d'histoire Naturelle. *Mag. Encycl.* 1795, 1: 448–457.

Laissus, J. 1965. Antoine-Laurent de Jussieu, "l'aimable professeur." In *Comptes-Rendus 89e Congr. Soc. Sav., Lyons, Sect. Sci. Hist. Sci.* (1964): 27–39.

——. 1967. La succession de Le Monnier au Jardin du Roi: Antoine-Laurent de Jussieu et René Louiche Desfontaines. In *Comptes-Rendus du 91e Cong. Soc. Sav., Lyons, Sect. Sci. Hist. Sci.* (1966): 137–152.

Lam, H. J. 1936. Phylogenetic symbols, past and present. *Acta Biotheor.* 2: 153–194.

Lamarck, J.-B.-P.-A. de M. de. 1778. *Flore françoise.* 3 vols. Paris: Imprimerie Royale.

——. 1783. Analyses. In J.-B.-P.-A. de M. de Lamarck and J.-L.-M. Poiret, *Encyclopédie méthodique. Botanique.* vol. 1(1), pp. 142–143. Paris: Panckoucke.

——. 1785. Botanique. *Ibid.* vol. 1(2), pp. 439–449. Paris: Panckoucke.

——. 1786. Classes. *Ibid.* vol. 2(1), pp. 29–36. Paris: Panckoucke.
——. 1788a. Genres. *Ibid.* vol. 2(2), pp. 630–634. Paris: Panckoucke.
——. 1788b. Sur les classes les plus convenables à établir parmi les végétaux et sur l'analogie de leur nombre avec celles determinés dans le règne animal, ayant égard de part et d'autre à la perfection graduée des organes. *Mém. Acad. Roy. Sci.* (Paris) (1785): 437–453.
——. 1791. *Tableau encyclopédique et méthodique des trois règnes de la nature.* vol. 1. Paris: Panckoucke.
——. 1792a. Sur les systèmes et les méthodes de Botanique, et sur l'analyse. *J. Hist. Nat.* 1: 300–307.
——. 1792b. Sur l'étude des rapports naturels. *J. Hist. Nat.* 1: 361–371.
——. 1792c. Observations sur les coquilles, et sur quelques-uns des genres qu'on a établis dans l'ordre des vers testacées. *J. Hist. Nat.* 2: 269–280.
——. 1792d. Lichen. In J.-B.-P.-A. de M. de Lamarck and J.-L.-M. Poiret, *Encyclopédie méthodique. Botanique.* vol. 3(2), pp. 470–508. Paris: Panckoucke.
——. 1802. *Recherches sur l'organisation des corps vivans.* Paris: Maillard.
——. 1809. *Philosophie zoologique, ou exposition des considérations relatives à l'histoire naturelle des animaux . . . .* Paris: Dentu.
Lamarck, J.-B.-P.-A. de M. de and A.-P. de Candolle. 1805. *Flore Française.* Ed. 3, vols. 1–5. Paris: Desray.
Lamarck, J.-B.-P.-A. de M. de and A.-P. de Candolle. 1815. *Ibid.* [Includes vol. 6.]
Lamarck, J.-B.-P.-A. de M. de and C.-F. B. de Mirbel. 1803. *Histoire naturelle des végétaux, classés par familles.* 15 vols. Paris: Crapelat.
Lamy, D. 1990. A. M. F. J. Palisot de Beauvois et la classification des mousses. *114e Congr. Nat. Soc. Sav., Paris* (1989): 259–273.
——. 1992. Introduction à la cryptogamie de J. B. Lamarck. In *La botanique de Lamarck.* vol. 4, pp. 4–17. Albertville: Alzieu.
Lankester, E. R. 1873. On the primitive cell-layers of the embryo as the basis of genealogical classification of animals, and on the origin of vascular and lymph systems. *Ann. Mag. Nat. Hist., Ser. 4,* 11: 321–338.
Larson, J. L. 1971. *Reason and Experience.* Berkeley: University of California Press.
——. 1980. Linné's French critics. In G. Broberg, ed., *Linnaeus: Progress and Prospects in Linnaean Research,* pp. 67–79. Stockholm: Almqvist and Wiksell.
——. 1986. Not without a plan: geography and natural history in the late eighteenth century. *J. Hist. Biol.* 19: 447–488.
Laudan, R. 1982. Tensions in the concept of geology: natural history or natural philosophy. *Earth Sci. Hist.* 1: 7–13.

——. 1987. *From Mineralogy to Geology* . . . . Chicago: University of Chicago Press.

Laurent, G. 1975. Lamarck: de la philosophie du continu à la science du discontinu. *Rev. Hist. Sci. Applic.* 28: 327–360.

——. 1987. *Paléontologie et évolution en France de 1800 a 1860: une histoire des idées de Cuvier et Lamarck à Darwin.* Paris: Comité des Travaux Historiques et Scientifiques.

La Vergata, A. 1987. Au nom de l'espèce: classification et nomenclature au XIXe siècle. In S. Atran, et al., *Histoire du concept de l'espèce dans les sciences de la vie*, pp. 193–225. Paris: Fondation Singer-Polignac.

Lawrence, G. H. M. 1951. *Taxonomy of Vascular Plants.* New York: Macmillan.

——. 1953. Plant genera, their nature and definition: The need for an expanded outlook. *Chron. Bot.* 14: 117–125.

Layton, D. 1973. *Science for the People* . . . . London: Science History Publ.

Lecomte, P. H. and J. Leandri. 1930. L'oeuvre botanique de Lamarck. *Arch. Mus. Natl. Hist. Nat., Sér. 6*, 6: 31–44.

Leenhouts, P. W. 1966. Keys in biology: A survey and a proposal of a new kind. *Proc. Kon. Ned. Akad. Wetensch., C*, 64: 571–596.

Lefébure, L. F. H. 1817. *Atlas botanique ou clef du jardin de l'univers d'après les principes de Tournefort et Linné réunis.* Paris: Treuttel and Wurz.

——. 1835. *Flore de Paris.* Paris: Casimir.

Légée, G. and M. Guédès. 1981. Lamarck, botaniste et évolutionniste. *Hist. & Nat.* 17–18: 19–31.

Le Grand, H. E. 1988. *Drifting Continents and Shifting Theories: The Modern Revolution in Geology and Scientific Change.* Cambridge: Cambridge University Press.

Le Guyader, H. 1986. Objectivité et taxinomie: des systèmes et méthodes à la classification naturelle. In P. Tassy, ed., *L'ordre et la diversité du vivant: quel statut scientifique pour les classifications biologiques?*, pp. 70–82. Paris: Fondation Diderot, Fayard.

——. 1988. *Théories et histoire en biologie.* Paris: J. Vrin.

Léman, D. S. 1810. Observations sur les plantes composées ou syngénèses; par M. de Candolle. *Nouv. Bull. Sci. Soc. Philom. Paris* 2: 223–227. ["S. L."]

Le Maout, J. E. M. 1844. *Leçons élémentaires de botanique* . . . . Paris: Fortin, Masson.

——. 1852. *Botanique organographie et taxonomie* . . . . Paris: L. Curmer.

Le Maout, J. E. M. and J. Decaisne. 1868. *Traité générale du botanique.* Paris: Firmin Didot.

Lennox, J. G. 1980. Aristotle on genera, species, and "the more and the less." *J. Hist. Biol.* 13: 321–346.

Lenoir, T. 1978. Generational factors in the origin of Romantische naturphilosophie. *J. Hist. Biol.* 11: 57–100.

——. 1981. The Göttingen school and the development of transcendental Naturphilosophie in the Romantic Era. *Stud. Hist. Biol.* 5: 110–205.

——. 1982. *The Strategy of Life*. Dordrecht: D. Riedel.

Léotard, S. 1893. *Lettres inédites de Moquin-Tandon à Auguste de Saint-Hilaire*. Clermont-L'Herault: Saturnin Léotard.

Lepenies, W. 1976. *Das Ende der Naturgeschichte*. Munich and Vienna: Hanser.

Lesch, J. E. 1990. Systematics and the geometrical spirit. In T. Frängsmyr, J. L. Heilbron, and R. E. Rider, eds., *The Quantifying Spirit*, pp. 73–111. Berkeley: University of California Press.

Lestiboudois, F. J. 1781. *Botanographie belgique*. Lille: J. B. Henry.

——. 1804. *Botanographie universelle ou tableau général des végétaux*. Lille: Vanackere. [Ed. 2, vol. 3(1) of *Botanographie belgique*, 1800.]

L'Héritier de Brutelle, C.-L. 1795. Mémoire sur un nouveau genre de plant appelé *Cadia*. *Mag. Encycl.* 1795, 5: 20–31, pl.

Limoges, C. 1980. The development of the Muséum d'Histoire Naturelle of Paris, c. 1800–1914. In R. Fox and G. Weisz, eds., *The Organization of Science and Technology in France 1808–1914*, pp. 211–240. Cambridge: Cambridge University Press.

Lincoln, A. H. 1832. *Familiar Lectures on Botany* . . . . Hartford: F. J. Huntington.

Lindley, J. 1826. Some account of the spherical and numerical system of nature of M. Elias Fries. *Philos. Mag.* 68: 81–91.

——. 1829a. *Synopsis of the British Flora*. London: Longman, Rees, Orme, Brown and Green.

——. 1829b. *An Introductory Lecture Delivered on the Opening of the University of London*. London: Richard Taylor.

——. 1830. *An Introduction to the Natural System of Botany* . . . . London: Longman, Rees, Orme, Brown and Green.

——. 1831. *Introduction to the Natural System of Botany* . . . . American Edition. New York: Carvill.

——. 1832. *An Introduction to Botany* . . . . London: Longman, Rees, Orme, Brown, Green and Longman.

——. 1834. On the principal questions at present debated in the philosophy of botany. *Proc. Brit. Assoc. Advancem. Sci.* (1833): 27–57. [Separately printed, London: Richard Taylor, 1833.]

——. 1835a. *A Key to Structural, Physiological, and Systematic Botany for the Use of Classes*. London: Longman, Rees, Orme, Brown, Green and Longman.

——. 1835b. *A Synopsis of the British Flora*. Ed. 2. London: Longman, Rees, Orme, Brown, Green, and Longman.

——. 1836. *A Natural System of Botany; or a Systematic View of the Organisation, Natural Affinities, and Geographical Distribution of the Whole Vegetable Kingdom* . . . . London: Longman, Rees, Orme, Brown, Green and Longman.

——. 1839. *School botany* . . . . London: Longman, Orme, Brown, Green and Longmans.

——. 1844. *Alona coelestis*. *Edwards Bot. Reg.* 30: pl. 46.

——. 1846. *Vegetable Kingdom*. London: Bradbury and Evans.

——. 1853. *Ibid.* Ed. 3. London: Bradbury and Evans.

——. 1854. *School Botany* . . . . Ed. 2. London: Bradbury and Evans.

Lindley, J. and W. Hutton. 1831–1837. *The Fossil Flora of Great Britain: Or Figures and Descriptions of the Vegetable Remains Found in the Fossil State in this Country*. London: James Ridgway. [Dates of publication of the various parts uncertain; cited here as 1831.]

Lindroth, C. H. 1973. Systematics specialises between Fabricius and Darwin: 1800–1859. In R. F. Smith, T. E. Mitter, and C. N. Smith, eds., *History of Entomology*, pp. 119–154. Palo Alto: Annual Reviews.

Link, J. H. F. 1798. *Philosophiae botanicae novae seu institutionum phytographicarum prodromus* . . . . Göttingen: Dieterich.

——. 1824. *Elementa philosophiae botanicae* . . . . Berlin: Haude and Spener.

Linnaeus, C. 1735. *Systema naturae* . . . . Leiden: Theodor Haak.

——. 1736a. *Musa cliffortiana* . . . . Leiden.

——. 1736b. *Fundamenta botanica*. Amsterdam: Schouten.

——. 1737. *Genera plantarum* . . . . Leiden: Conrad Wishoff.

——. 1738. *Classes plantarum* . . . . Leiden: Conrad Wishoff.

——. 1744a. *Peloria*. Uppsala. [Defended by D. Rudberg.]

——. 1744b. *Systema naturae* . . . . Ed. 4. Paris: Michael Anton David.

——. 1746. *Sponsalia plantarum*. Stockholm: Salvius. [Defended by J. G. Wahlbom.]

——. 1747. *Vires plantarum*. Uppsala. [Defended by F. Hasselquist.]

——. 1751. *Philosophia botanica* . . . . Stockholm: Kiesewetter.

——. 1753. *Species plantarum* . . . . Stockholm: Salvius.

——. 1754. *Genera plantarum* . . . . Ed. 5. Stockholm: Salvius.

——. 1758. *Systema naturae* . . . . Ed. 10. Stockholm: Salvius.

——. 1764. *Genera plantarum* . . . Ed. 6. Stockholm: Salvius.

——. 1774. *De Viola ipecacuanha*. Uppsala: Edmann. [Defended by D. Wickman.]

Linnaeus, C. von, fils. 1782. *Supplementum plantarum* . . . . Brunswick: Impensis orphanotrophei.

Lonsdale, K. 1970. Women in science: Reminiscences and reflections. *Impact Sci. Soc.* 20: 45–59.

Lovejoy, A. O. 1936. *The Great Chain of Being.* Cambridge, Massachusetts: Harvard University Press.

Lowe, P. D. 1976. Amateurs and professionals: The institutional emergence of British plant ecology. *J. Soc. Bibliogr. Nat. Hist.* 7: 517–535.

Ludwig, C. G. 1737. *Definitiones generum plantarum in usum auditorum.* Leipzig: Gleditsch.

Lurie, E. 1960. *Louis Agassiz: A Life in Science.* Chicago: University of Notre Dame Press.

Lyon, J., and Sloan, P. R. 1981. *From Natural History to the History of Nature.* Notre Dame: University of Notre Dame Press.

Mabberley, D. J. 1985. *Jupiter botanicus: Robert Brown of the British Museum.* Braunschweig: Cramer.

Macleay, W. S. 1819–1821. *Horae entomologicae: Or Essays on Annulose Animals.* Bagster: London.

——. 1825. Remarks on the identity of certain general laws which have been lately observed to regulate the natural distribution of insects and fungi. *Trans. Linn. Soc. London.* 14: 46–68.

——. 1829. A letter to J. E. Bicheno, Esq., F.R.S., in examination of his paper "On systems and methods" in the Linnean Transactions. *Philos. Mag., Ser. 2*, 6: 199–212. [See also *Zool. J.* (1829) 4: 402–415.]

——. 1830. On the dying struggle of the dichotomous system. *Philos. Mag., Ser. 2*, 7: 431–445; 8: 53–57, 134bis–140bis, 200–207.

MacMillan, C. 1893. On the emergence of "sham" biology in America. *Science* 21: 184–186.

McRae, R. F. 1961. *The Problem of the Unity of the Sciences: Bacon to Kant.* Toronto: University of Toronto Press.

Mägdefrau, K. 1992. *Geschichte der Botanik* . . . . Ed. 2. Stuttgart: Gustav Fischer.

Madriñán, S. 1994. Richard Spruce's pioneering work on tree architecture. In M. R. D. Seward, ed., *Richard Spruce, Botanist and Explorer.* London: Royal Botanic Gardens, Kew, and the Linnean Society. In press.

Magnol, P. 1689. *Prodromus historiae generalis plantarum, in quo familiae plantarum per tabulas disponuntur.* Montpellier: Pech.

——. 1720. *Novus character plantarum* . . . . Montpellier: Pech.

Malesherbes, C. G. de L. de 1798. *Observations sur l'histoire naturelle générale et particulière du Buffon et Daubenton.* 2 vols. Paris: Charles Pugens. [Slatkin Reprints, Genève, 1975.]

Marchand, N. L. 1867. *Des classifications et des méthodes en botanique.* Angers: Lachèse, Bellevue and Dolbeau.

Marchant, L. 1858. *Lettres de Georges Cuvier à C. M. Pfaff . . . .* Paris: Victor Masson.

Marcus, S. 1974. *Engels, Manchester, and the Working Class.* New York: Random House.

Maroske, S. and H. M. Cohn. 1992. "Such ingenious birds": Ferdinand Mueller and William Swainson in Victoria. *Muelleria* 7: 529–553.

Martin, H. N. 1877. The study and teaching of biology. *Popular Sci. Monthly* 10: 298–309.

Martius, C. F. P. von. 1822a. De plantis nonnullis antediluvianis ope specierum inter tropicos nunc viventium illustrandis. *Denkschr. Königl.-Baier. Bot. Ges. Regensburg* 2: 121–147, pl. 2–3.

——. 1822b. Novum plantarum genus. *Denkschr. Königl.-Baier. Bot. Ges. Regensburg* 2: 148–160, pl. 4–10.

——. 1825. On certain antediluvian plants susceptible of being illustrated by means of species now living in the tropics. *Edinburgh New Philos. J.* 12: 47–56, 270–281.

——. 1826. Beitrag zur Kenntnis der natürlichen Familie der Amarantaceen. *Nov. Act. Phys.-Med. Acad. Caes. Leop.-Carol. Nat. Cur.* 13: 211–322, pl. 14A, 14B.

——. 1835. *Conspectus regni vegetabilis secundum characteres morphologicas praesertim carpicos et in classes ordines et familias digesti.* Nürnberg: Joh. Leonhard Schrag.

——. 1843. Notice of the life and labors of de Candolle. *Amer. J. Sci. Arts* 44: 217–239. ["K. F. P.": translation.]

Masters, M. T. 1877. Remarks on the superposed arrangement of parts of the flower. *J. Linn. Soc., Bot.* 15: 456–478.

Maury, L. F. A. 1864. *L'ancienne académie des sciences.* Paris: Didier.

Mayr, E. 1982a. *The Growth of Biological Thought: Diversity, Evolution and Inheritance.* Cambridge, Massachusetts: Belknap Press of Harvard University Press.

——. 1982b. Comments on David Hull's paper on exemplars and type specimens. *Philos. Sci. Assoc.* 2: 504–511.

——. 1987. The species as category, taxon and population. In S. Atran, et al., *Histoire du concept d'espèce dans les sciences de la vie*, pp. 303–320. Paris: Fondation Singer-Polignac.

——. 1988. *Towards a New Philosophy of Biology.* Cambridge, Massachusetts: Belknap Press of Harvard University Press.

Mayr, E and D. H. Colless. 1991. *Principles of Systematic Zoology.* Ed. 2. New York: McGraw Hill.

McIntosh, R. P. 1975. H. A. Gleason, "individualistic ecologist" 1882–1975: His contribution to ecological theory. *Bull. Torrey Bot. Club* 102: 253–273.

McMullin, E. 1990. The development of philosophy of science 1600–1900. In R. C. Olby, G. N. Cantor, J. R. R. Christie, and M. J. S. Hodge, eds., *Companion to the History of Modern Science*, pp. 816–852. London and New York: Routledge.

McNeill, J. 1979. Structural value: A concept used in the construction of taxonomic classifications. *Taxon* 28: 481–504.

McOuat, G. 1992. *Species, Names and Things, from Darwin to the Experimentalists.* Ph. D. Thesis, University of Toronto.

Meeuse, A. D. J. 1975. Floral evolution in the Hamamelidae. 1. General assessment of the probable phylogeny and taxonomic position of the group. *Acta Bot. Neerl.* 24: 155–164.

Meissner, C. F. 1826. *Monographiae generis polygoni prodromus.* Geneva: A. Lador.

Merriam, E. H. 1893. Biology in our colleges: A plea for a broader and more liberal biology. *Science* 21: 352–355.

Merrill, E. D. 1954. *The Botany of Cook's Voyages and its Unexpected Significance in Respect to Anthropology, Biogeography, and History.* Waltham: Chronica Botanica.

Merrill, L. L. 1989. *The Romance of Victorian Natural History.* Oxford: Oxford University Press.

Merxmüller, H. 1972. Systematic botany—an unachieved synthesis. *Biol. J. Linn. Soc.* 4: 311–321.

Metcalfe, C. R. and L. Chalk. 1979. *Anatomy of the Dicotyledons.* Ed. 2, vol. 1. Oxford: Clarendon Press.

Miège, J. 1979. Augustin-Pyramus de Candolle: sa vie, son oeuvre, son action à travers la Société de Physique et d'Histoire Naturelle de Genève. *Mém. Soc. Phys. Genève* 43: 1–45.

Miers, J. 1845. Contribution to the botany of South America. *London J. Bot.* 4: 319–371.

——. 1851–1852. Observations on the affinities of the Olacaceae. *Ann. Mag. Nat. Hist., Ser. 2*, 8: 161–184, 9: 128–132.

——. 1858a. On the Canellaceae. *Ann. Mag. Nat. Hist., Ser. 3*, 1: 342–353.

——. 1858b. On the Winteraceae. *Ann. Mag. Nat. Hist., Ser. 3*, 2: 33–50, 109–115.

Miller, G. A. 1956. The magical number seven, plus or minus two: Some limits on our capacity for processing information. *Psychol. Rev.* 63: 81–97.

Miller, H. 1939. *History and Science.* Berkeley: University of California Press.

Millin de Grandmaison, A.-L. 1792. Discours sur l'origine et les progrès de l'histoire naturelle, en France. *Actes Soc. Hist. Nat. Paris* 1: I–XVI.

Milne, C. 1770. *A Botanical Dictionary*. London: William Griffin.
——. 1771. *Institutes of Botany*. London: W. Griffin.
Milne-Edwards, A. 1864. *De la famille des Solanacées*. Paris: E. Martinet.
Milne-Edwards, H. H. 1844. Considérations sur quelques principes relatifs à la classification naturelle des animaux et plus particulièrement sur la distribution méthodique des mammifères. *Ann. Sci. Nat., Zool. Sér. 3*, 1: 65–99.
——. 1851. *Introduction à la zoologie générale*. Paris: Victor Masson.
Minelli, A., G. Fusco, and S. Sartori. 1991. Self-similarity in biological classifications. *Biosystems* 26: 89–97.
Miquel, F. A. W. 1843. *Systema Piperacearum*. Rotterdam: H. A. Kramers.
Mirbel, C.-F. B. de. 1801. Suite de l'anatomie des végétaux. Fin de l'anatomie des monocotyledons. *J. Phys. Chim. Hist. Nat. Arts* 53: 200–213.
——. 1802. *Traité de anatomie et de physiologie végétales . . . .* 2 vols. Paris: F. Dufart.
——. 1806. Mémoire sur les glandes contenus dans les végétaux suivi du note sur l'organisation des plantes. *Ann. Mus. Hist. Nat.* 7: 274–300, pl. 17.
——. 1809a. Observations anatomiques et physiologiques sur la croissance et le développement des végétaux. *Mém. Sci. Math. Phys. Inst. France* (1808): 303–330, 8 fig.
——. 1809b. Oeuvre sur un système d'anatomie comparée des végétaux fondé sur l'organisation de la fleur. *Mém. Sci. Math. Phys. Inst. France* (1808): 331–362, 2 fig.
——. 1809c. *Exposition de la théorie de l'organisation végétale . . . .* Paris: Dufart.
——. 1810a. Considérations sur la manière d'étudier l'histoire naturelle des végétaux, servant d'introduction à un travail anatomique, physique et botanique sur la famille des Labiées. *Ann. Mus. Hist. Nat.* 15: 110–141.
——. 1810b. Mémoire sur l'anatomie et physiologie des plantes de la famille des Labiées. *Ann. Mus. Hist. Nat.* 15: 213–260.
——. 1814. Précis de plusieurs leçons de physiologie végétale sur les plantes cryptogames ou agames. *J. Bot. (Desvaux)* 5: 31–65.
——. 1815. *Élémens de physiologie végétale et de botanique.* 3 vols. Paris: Magimel.
——. 1833. Discours prononcé par M. de Mirbel aux funérailles de M. Desfontaines, le lundi 18 Novembre. *Arch. Bot.* (Paris) 2: 568–572.
——. 1843. Recherches anatomiques et physiologiques sur quelques végétaux monocotylés. *Compt. Rend. Hebd. Séances Acad. Sci.* (Paris) 16: 1213–1235.

——. 1844. Suite des recherches anatomiques et physiologiques sur quelques végétaux monocotylés. *Compt. Rend. Hebd. Séances Acad. Sci.* (Paris) 19: 689–699. [Reprinted in 1845, *Ann. Sci. Nat., Bot. Sér. 3*, 3: 321–337, pl. 13–15.]

Mirbel C.-F. B. de. and N.-M.-T. Jolyclerc. 1802–1806. *Histoire naturelle, générale et particulière, des plantes* . . . . 18 vols. Paris: F. Dufart. [C. S. Sonnini (ed.), *Suites à Buffon*.]

Mohl, H. von. 1831. De palmarum structura. In C. F. P. von Martius, ed., *Historia naturalis palmarum*, vol. 1., pp. I–LII, pl. A–Q. Leipzig: T. O. Weigel.

——. 1851. *Grundzüge der Anatomie und Physiologie der vegetabilische Zelle.* Braunschweig: Friedrich Vieweg.

Moldenhawer, J. J. P. 1812. *Beyträge zur Anatomie der Pflanzen.* Kiel: C. L. Wäser.

Monk. J. 1818. Considerations regarding Cambridge, more particularly relating to its botanical professorship, by Sir James Edward Smith. *Quart. Rev.* 19: 434–446. [Anonymous; review of Smith 1818.]

Montgomery, W. M. 1970. The origins of the spiral theory of phyllotaxy. *J. Hist. Biol.* 3: 299–323.

Moquin-Tandon, C.-H.-B.-A. 1826. *Essai sur les dédoublemens ou multiplications d'organes dans les végétaux.* Montpellier: Jean Martel le Jeune.

——. 1832. Considérations sur les irrégularités de la corolle dans les Dicotylédones. *Ann. Sci. Nat.* (Paris) 27: 225–290.

——. 1841. *Élémens de tératologie végétale.* Paris: P. J. Loss.

Morandi, G. 1744. *Historia botanica practica* . . . . Milan: Malatesta.

Morison, R. 1680. *Plantarum historiae universalis* . . . . Oxford: e theatro Sheldoniano.

Morton, A. G. 1981. *Outlines of Botanical History.* London: Academic Press.

Morveau, Guyton de. [See Guyton de Morveau.]

Mueller, F[erdinand] J. H. von. 1882. *Systematic Census of Austral Plants with Chronologic, Literary and Geographic Annotations.* Melbourne: The Victorian Government.

Müller, F[ritz] J. F. T. 1864. *Für Darwin.* Leipzig: Wilhelm Engelmann. [Translated by W. S. Dallas as *Facts and Arguments for Darwin*, London: John Murray, 1869.]

Müller, H. L. H. 1881. *Alpenblümen, ihre Befruchtung durch Insekten und ihrer Anpassungen an dieselben.* Leipzig: Wilhelm Engelmann.

Müller [Argoviensis], J. 1857. *Monographie de la famille des Résédacées.* Zürich: Zurcher and Furrer.

——. 1866. Nachschrift zu meiner systematischen Arbeit über die Euphorbiaceen. *Bot. Zeitung* (Berlin) 24: 333–345.

——. 1875. Replik auf Dr. Baillon's "Nouvelles observations sur les Euphorbiacées." *Bot. Zeitung* (Berlin) 33: 223–239, 254–256.

Naef, A. 1919. *Idealistische Morphologie und Phylogenetik*. Jena: Gustav Fischer.

Nägeli, C. W. 1847. *Die neuern Algensysteme* . . . . Zürich: Friedrich Schulthess.

——. 1865. *Entstehung und Begriff der naturhistorischen Art* . . . . Munich: Köningl. Akademie.

Naudin, C. V. 1852. Considerations philosophiques sur l'espèce et la variété. *Rev. Hort., Sér. 4*, 1: 102–109.

Necker, N. J. de. 1790. *Elementa botanica* . . . . Neuwied: Societas typographica.

Needham, R. 1976. Polythetic classification: convergence and consequences. *Man (N. S.)* 10: 349–369.

Nees von Esenbeck, C. G. D. 1817. *Das System der Pilz und Schwämme*. Ed. 2. Würzburg: Stahel.

——. 1820–1821. *Handbuch der Botanik*. 2 vols. Berlin: Johann Leonhard Schrag.

——. 1831. Laurinae indiae orientalis. In N. Wallich, *Plantae asiaticae rariores* . . . . vol. 2., pp. 58–76. London: Treuttel and Würtz.

——. 1833–1838. *Naturgeschichte der Europäischen Lebermoose* . . . . 4 vols. Berlin: August Rücker [vols. 1–2]; Breslau: Grass, Barth and Comp. [vols. 2–4].

——. 1836. *Systema laurinarum*. Berlin: Veit.

Nees von Esenbeck, T. F. L. 1820. *Radix plantarum mycetoidearum*. Bonn: A. Marius.

Nelson, G. 1978. From Candolle to Croizat: Comments on the history of biogeography. *J. Hist. Biol.* 11: 269–305.

——. 1979. Cladistic analysis and synthesis: Principles and definitions, with a historical note on Adanson's *Familles des plantes* (1763–1764). *Syst. Zool.* 28: 1–21.

Nelson, G. and Platnick, N. 1981. *Systematics and Biogeography: Cladistics and Vicariance*. New York: Columbia University Press.

Nelson, G. and Platnick, N. 1984. Systematics and evolution. In M.-W. Ho and P. T. Saunders, *Beyond Neo-Darwinism*, pp. 143–158. London and New York: Academic Press.

Nicolson, M. 1987. Alexander von Humboldt, Humboldtian science and the origin of the study of vegetation. *Hist. Sci.* 25: 167–194.

Northrop, F. S. C. 1947. *The Logic of the Sciences and the Humanities*. New York: Macmillan.

Nuttall, T. 1818. *The Genera of North American Plants, and a Catalogue of Their Specimens, to the Year 1817*. Philadelphia: D. Heartt.

Oeder, G. C. 1764–1766. *Elementa botanicae . . . .* Copenhagen: Philibert.

O'Brian, P. 1987. *Joseph Banks: A Life.* London: Collins Harvill.

O'Hara, R. J. 1988a. Homage to Clio, or, toward an historical philosophy for evolutionary biology. *Syst. Zool.* 37: 142–155.

——. 1988b. Diagrammatic classifications of birds, 1819–1901: Views of the natural system in 19th-century British ornithology. In H. Ouellet, ed., *Acta XIX congressus internationalis ornithologici,* pp. 2746–2759. Ottawa: National Museum of Natural Sciences.

——. 1991. Representations of the natural system in the nineteenth century. *Biol. & Philos.* 6: 255–274.

——. 1992. Telling the tree: Narrative representation and the study of evolutionary history. *Biol. & Philos.* 7: 135–160.

——. 1993. Systematic generalization, historical fate, and the species problem. *Syst. Biol.* 42: 231–246.

Oliver, F. W. 1906. Botany in England: A reply. *New Phytol.* 5: 173–176.

Oppenheimer, J. M. 1987. Haeckel's variations on Darwin. In H. M. Hoenigswald and L. F. Wiener, eds., *Biological Metaphor and Cladisitic Classification,* pp. 123–135. Philadelphia: University of Pennsylvania Press.

Osborne, D. V. 1963. Some aspects of the theory of dichotomous keys. *New Phytol.* 62: 144–161.

Ospovat, D. 1976. The influence of Karl Ernst von Baer's embryology, 1828–1859, a reappraisal in the light of Richard Owen's and William B. Carpenter's "palaeontological application of 'Von Baer's Law.'" *J. Hist. Biol.* 9: 1–28.

——. 1981. *The Development of Darwin's theory.* Cambridge: Cambridge University Press.

Outram, D. 1984. *Georges Cuvier: Vocation, Science and Authority in Post-Revolutionary France.* Manchester: Manchester University Press.

——. 1986. Uncertain legislator: Georges Cuvier's laws of nature in their intellectual context. *J. Hist. Biol.* 19: 323–368.

Owen, R. 1992. *The Hunterian Lectures in Comparative Anatomy, May–June 1837.* [Edited by P. R. Sloan.] Chicago: University of Chicago Press.

Palisot de Beauvois, A. M. F. J. 1803–1820. *Flore de Oware et de Bénin, en Afrique.* 2 vols. Paris: Fain.

——. 1812. *Essai d'une nouvelle agrostographie; ou nouvelles genres des Graminées . . . .* Paris: Fain.

Pallas, P. S. 1766. *Elenchus zoophytorum . . . .* Hagae-Comitum: Petrus van Cleef.

Panchen, A. L. 1992. *Classification, Evolution and the Nature of Biology.* Cambridge: Cambridge University Press.

Paris, C. A. and D. S. Barrington. 1990. William Jackson Hooker and the generic classification of ferns. *Ann. Missouri Bot. Gard.* 27: 228–238.

Parks, S. 1821. On the employment of common salt for the purposes of horticulture. *Quart. J. Lit. Sci. Arts* 10: 52–72.

Parlatore, F. 1843. *Lezioni di botanica comparata.* Florence: Societá Tipografica.

Patterson, C. 1977. The contribution of paleontology to teleostean phylogeny. In M. K. Hecht, P. C. Goody, and B. M. Hecht, eds., *Major Patterns of Vertebrate Evolution*, pp. 579–643. New York: Plenum.

——. 1982. Morphological characters and homology. In K. A. Joysey and A. E. Friday, eds., *Problems of Phylogenetic Reconstruction*, pp. 21–74. London and New York: Academic Press.

Paul, C. B. 1980. *Science and Immortality: The Eloges of the Paris Academy of Sciences (1699–1791).* Berkeley: University of California Press.

Pauly, P. J. 1984. The appearance of academic biology in late nineteenth-century America. *J. Hist. Biol.* 17: 369–397.

Pax, F. A. 1884. Die Anatomie der Euphorbiaceen in ihrer Beziehung zum System derselben. *Bot. Jarhb. Syst.* 5: 384–421, pl. 6–7.

——. 1888. Monographishe Übersicht über die Arten der Gattung *Primula*. *Bot. Jahrb. Syst.* 10: 75–241.

Payen, A. 1858. *Éloge historique de M. de Mirbel.* Paris: Bouchard-Huzard. [Reprint from Mém. Soc. Imp. Centr. d'Agriculture, 1858.]

Payer, J.-B. 1844. *Des classifications et des méthodes en histoire naturelle.* Paris: Lacour and Maistrasse.

——. 1854–1857a. *Traité d'organogénie comparée de la fleur.* Paris: Victor Masson.

——. 1857b. *Éléments de botanique.* Paris: Langlois and Leclercq & Victor Masson.

Pennell, F. W. 1930. Genotypes of the Scrophulariaceae in the first edition of Linné's *Species plantarum. Proc. Acad. Nat. Sci. Philadelphia* 82: 9–26.

Perrot, É. C. 1899. *Anatomie comparée des Gentianacées.* Paris: Masson and Cie. [Also, *Ann. Sci. Nat., Bot. Sér. 8*, 7: 105–292, pl. 1–9.]

Petit-Thouars, L.-M. A. A. du. 1811. Dissertation sur l'enchaînement des êtres. In his *Mélanges de botanique et de voyages* . . . . Paris: Arthrus Bertrand. [Reprint of article first published in 1788.]

——. 1817. Botanique. In *Dictionnaire des Sciences Naturelles* 15: 168–241. Strasbourg: F. G. Levrault.

Pfister, D. H. 1986. Mycology, universities and 19th century American science. *Mycotaxon* 26: 65–74.

Pictet, F. G. 1839. On the writings of Goethe relative to natural history. *Ann. Nat. Hist.* 2: 313–322.

Piveteau, J. 1981. Lamarck et Cuvier. L'échelle des êtres. In Centre de Recherche sur l'Histoire des Idées de l'Université de Picardie, eds., *Lamarck et son temps. Lamarck et notre temps*, pp. 189–197. Paris: J. Vrin.

Planchon, [F.] G. 1860. *Les principes de la méthode naturelle.* Montpellier: Boehm.

Planchon, J.-É. 1846–1847a. Sur le genre *Godoya* et ses analogues, avec des observations sur les Ochnacées, et une revue des genres et espèces de ce group. *London J. Bot.* 5: 584–600, pl. 21–24, and 6: 1–31.

——. 1847b–1848. Sur la famille des Linées. *London J. Bot.* 6: 588–603, and 7: 165–186, 473–501, 507–528.

——. 1851. Meliantheae, a new natural order, proposed and defined. *Trans. Linn. Soc. London* 20: 403–418, pl. 20.

Plée, F. 1842–1864. *Types de chaque famille et des principaux genres des plantes . . . .* 2 vols. Paris: the author.

Poiret, J.-L.-M. 1797. Méthode. In J-B.-P.-A. de M. de Lamarck and J.-L.-M. Poiret, *Encyclopédie méthodique. Botanique.* vol. 4, pp. 128–133. Paris: Panckoucke.

——. 1804. Rapports. *Ibid.* vol. 6, pp. 76–77. Paris: Panckoucke.

——. 1808. Trèfle, *Trifolium. Ibid.* vol. 8, pp. 1–31. Paris: Panckoucke.

——. 1816. Organisation. *Ibid.* Suppl. vol. 4, pp. 183–184. Paris: Panckoucke.

Pontedera, G. 1720. *Anthologia, sive de floris natura libri tres . . . .* Padua: J. Manfré.

Pool, R. J. 1935. The evolution and differentiation of laboratory teaching in the botanical sciences. *Iowa State Coll. J. Sci.* 9: 235–251.

Potez, H. 1909. *Pages choisis des grands écrivains. Fontenelle.* Paris: Colin.

Powers, J. 1909. Are species realities, or concepts only? *Amer. Naturalist* 43: 598–610.

Pratt, V. 1982. Aristotle and the essence of natural history. *Hist. Philos. Life Sci.* 4: 203–223.

——. 1985. System-building in the eighteenth century. In J. D. North and J. J. Roche, eds., *The Light of Nature: Essays in the History and Philosophy of Science Presented to A. C. Crombie*, pp. 421–431. Dordrecht and Boston: Martinus Nijhoff.

Prelli, L. J. 1989. *A Rhetoric of Science.* Columbia, South Carolina: University of South Carolina Press.

Pujoulx, J. B. 1803. *Promenades au Jardin des Plantes.* Paris: Libraire Économique.

Queiroz, K. de. 1988. Systematics and the Darwinian revolution. *Philos. Sci.* 55: 238–259.

Radlkofer, L. [A. T.] 1883. *Ueber die Methoden in botanischer Systematik, ins-*

besondere die anatomische Methode. Munich: Königliche Bayerische Akademie.

——. 1890. Ueber die Gliederung der Familie den Sapindaceen. *Sitzungsber. Math.-Phys. Cl. Königl. Bayer. Akad. Wiss. München* 1890: 105–379.

Rafinesque[-Schmaltz], C. S. 1814. *Principes fondamentaux de Somiologie* . . . . Palermo: F. Abate.

——. 1815. *Analyse de la nature* . . . . Palermo.

——. 1836. *Flora telluriana* . . . . Philadelphia.

Ramsbottom, J. 1938. Linnaeus and the species concept. *Proc. Linn. Soc. London* 150: 192–219.

Ray, J. 1682. *Methodus plantarum nova* . . . . London: Henry Faithorne and John Kersey.

——. 1696. *De variis plantarum methodis dissertatio brevis* . . . . London: S. Smith and B. Walford.

——. 1703. *Methodus plantarum emendata et aucta* . . . . London: Smith and Walford.

Redtenbacher, L. 1843. *Illustrationes et descriptiones coleoptorum novorum Syriaci.* Stuttgart: Schweizerbart.

Rehbock, P. F. 1983. *The Philosophical Naturalist.* Madison: University of Wisconsin Press.

Reichenbach, H. G. L. 1828. *Conspectus regni vegetabilis per gradus naturales evoluti* . . . . Leipzig: Cnobloch.

——. 1837. *Handbuch des natürlichen Pflanzensystems* . . . . Dresden and Leipzig: Arnold.

Reif, W.-E. 1983. Evolutionary theory in German paleontology. In M. Grene, ed., *Dimensions of Darwinism: Themes and Counter-Themes in Twentieth-Century Evolutionary Theory*, pp. 173–203. Cambridge: Cambridge University Press.

Renault, B. 1879. Structure comparée de quelques types de la flore Carbonifère. *Nouv. Arch. Mus. Hist. Nat., Sér. 2*, 2: 217–348, pl. 10–17.

Rey, R. 1981. Aspects du vocabulaire de la classification dans l'Encyclopédie. *Docum. Hist. Vocabul. Sci.* 2: 45–63.

Richard, A. 1822. Anatomie végétale. In *Dictionnaire Classique d'Histoire Naturelle* 1: 340–341. Paris: Rey and Gravier.

——. 1826. Méthode. In *Dictionnaire Classique d'Histoire Naturelle* 10: 493–511. Paris: Rey and Gravier.

——. 1828. *Nouveaux élémens de botanique et physiologie végétale.* Ed. 4. Paris: Bichat Jeune.

——. 1834. Mémoire sur la famille des Rubiacées, contenant la description générale de cette famille et les caractères des genres qui la composent. *Mém. Soc. Hist. Nat. Paris* 5: 81–305, pl. 11–25.

Rider, R. E. 1990. Measure of idea, rule of language: Mathematics and language in the eighteenth century. In T. Frängsmyr, J. L. Heilbron, and R. E. Rider, eds., *The Quantifying Spirit*, pp. 114–140. Berkeley: University of California Press.

Ridley, M. 1986. *Evolution and Classification: The Reformation of Cladism*. London and New York: Longmans.

Rieppel, O. 1987a. "Organization" in the Lettres Philosophiques of Louis Bouguet compared to the writings of Charles Bonnet. *Gesnerus* 44: 125–132.

——. 1987b. Pattern and process: The early classification of snakes. *Biol. J. Linn. Soc.* 31: 405–420.

——. 1988. *Fundamentals of Comparative Biology*. Basel: Birkhauser.

Ritvo, H. 1990. New presbyter or old priest? Reconsidering zoological taxonomy in Britain, 1750–1840. *Hist. Human. Sci.* 3: 259–276.

Rivinus, A. Q. 1690. *Ordo plantarum, quae sunt florae irregulari monopetalo*. Leipzig: the author.

Robinet, J.-B. R. 1761–1766. *De la nature*. 4 vols. Amsterdam: E. van Harrevelt.

——. 1768. *Vue philosophique de la gradation naturelle des formes de l'être . . . .* Amsterdam: E. van Harrevelt.

Robinson. B. L. 1906. The generic concept in the classification of flowering plants. *Science N. S.* 23: 81–92.

Rodgers, A. D. 1944. *American Botany, 1873–1892: Decades of Transition*. Princeton: Princeton University Press.

Rodrigues C., R. L. 1950. A graphic representation of Bessey's taxonomic system. *Madroño* 10: 214–218.

Roger, J. 1970. Buffon, Georges-Louis Leclerc, Comte de. In C. C. Gillispie, ed., *Dictionary of Scientific Biography*, vol. 2, pp. 576–582. New York: Charles Scribner's Sons.

——. 1971. *Les sciences de la vie dans la pensée française du xviiie siècle*. Ed. 2. Paris: Colin.

——. 1989. *Buffon, un philosophe au Jardin du Roi*. Paris: Fayard.

——. 1990. L'histoire naturelle au XVIIIe siècle: de l'échelle des êtres à l'évolution. *Bull. Soc. Zool. France* 115: 245–254.

Roscoe, W. 1813. On artificial and natural arrangements of plants: and particularly on the systems of Linnaeus and Jussieu. *Trans. Linn. Soc. London* 11: 50–78. [See also: *Philos. Mag., Ser. 2*, 7: 15–23, 97–104, 180–185.]

Roth, V. L. 1991. Homologies and hierarchies: Problems solved and unresolved. *Evol. Biol.* 4: 167–192.

Roule, L. 1930. L'esprit Lamarckien dans les classifications actuelles. *Arch. Mus. Natl. Hist. Nat., Sér. 6*, 6: 25–30.

Rousseau, J. J. 1785. *Letters on the Elements of Botany Addressed to a Lady*. Trans. T. Martyn. London: B. and J. White.

Royen, A. van. 1740. *Florae Leydensis prodromus* . . . . Leiden: Samuel Luchtmans.

Rudolph, E. D. 1973. How it developed that botany was the science thought most suitable for Victorian young ladies. *Children's Lit.* 2: 92–97.

——. 1984. Almira Hart Lincoln Phelps (1793–1884) and the spread of botany in nineteenth century America. *Amer. J. Bot.* 71: 1161–1167.

Rudwick, M. J. S. 1976. *The Meaning of Fossils: Episodes in the History of Palaeontology*. London: Macdonald.

Rüling, J. P. 1766. *Commentatio botanica de ordinibus naturalibus plantarum*. Göttingen: Rosenbusch. [Reprinted in P. Usteri, ed., 1793. *Delectus opusculorum botanicorum*. vol. 2, pp. 431–462 and fig.]

Ruppius, H. B. 1718. *Flora Jenensis* . . . . Frankfurt and Leipzig: Balliar.

Ruse, M. 1988. Particles to man: Evolutionary biology and thoughts of progress. In M. H. Nitecki, ed., *Evolutionary Progress*, pp. 97–126. Chicago: University of Chicago Press.

Russell, E. S. 1916. *Form and Function*. London: John Murray.

Russo, R. P. F. 1981. La notion d'organisation chez Lamarck. In Centre de Recherche sur l'Histoire des Idées de l'Université de Picardie, eds., *Lamarck et son temps. Lamarck et notre temps*, pp. 119–141. Paris: J. Vrin.

Rutishauser, R. and R. Sattler. 1985. Complementarity and heuristic value of contrasting models in structural botany. 1. General considerations. *Bot. Jahrb. Syst.* 107: 415–455.

Sachs, [F. G.] J. 1868. *Lehrbuch der Botanik*. Leipzig: Wilhelm Engelmann.

——. 1875. *Geschichte der Botanik*. Munich: Oldenbourg. [Transl. H. E. F. Garnsey and I. B. Balfour, 1890, *History of Botany*, Oxford: Clarendon Press; reprinted 1967, New York: Russell and Russell.]

Saint-Hilaire, A. F. C. P. de. 1819. Mémoire sur les Cucurbitacées et les Passifloracées. Première partie. *Mém. Mus. Hist. Nat.* 5: 304–350, pl. 24, 25.

——. 1823. Premier mémoire sur le gynobase. Du gynobase considéré dans les polypétales. *Mém. Mus. Hist. Nat.* 10: 129–164.

——. 1825a. *Histoire des plantes les plus remarquables du Brésil et du Paraguay*. Paris: A. Belin.

——. 1825b. *Flora brasiliae meridionalis*. vol. 1. Paris: A. Belin.

——. 1827. Sur la série linéaire des plantes polypétales, et en particulier de celles qui font partie de la flore Brasilienne. *Mém. Mus. Hist. Nat.* 14: 120–130.

——. 1837. *Deuxième mémoire sur les Résédacées* . . . . Montpellier: Jean Martel Ainé.

——. 1838a. Rapport sur la traduction de la partie botanique des oeuvres de Goethe, publiée par M. Martins. *Compt. Rend. Hebd. Séances Acad. Sci.* 7: 434–440.

——. 1838b. Mémoire sur les Myrsinées, les Sapotées et les embryons parallèles au plan de l'ombilic. *Mém. Acad. Roy. Sci. Inst. France* 16: 117–167.

——. 1840. *Leçons de botanique contenant principalement la morphologie végétale* . . . . Paris: P. J. Loss.

Saint-Hilaire, A. F. C. P. de and C.-H.-B.-A. Moquin-Tandon. 1830a. Deuxième mémoire sur la famille des Polygalées. *Mém. Mus. Hist. Nat.* 19: 305–339.

Saint-Hilaire, A. F. C. P. de and C.-H.-B.-A. Moquin-Tandon. 1830b. Mémoire sur la symétrie des Capparidées et des familles qui ont le plus de rapports avec elles. *Ann. Sci. Nat., Bot.* 20: 318–326.

Saint-Hilaire, Geoffroy. [See Geoffroy Saint-Hilaire.]

Saint-Hilaire, Jaume. [See Jaume Saint-Hilaire.]

Saint-Pierre, Germain de. [See Germain de Saint-Pierre.]

Saint Pierre, J. H. B. 1784–1788. *Études de la Nature.* 4 vols. Paris: l'Imprimerie de Monsieur.

Saint-Vincent, Bory de. [See Bory de Saint-Vincent.]

Sattler, R. 1986. *Biophilosophy: Analytic and holistic perspective.* Berlin: Springer.

Schama, S. 1989. *Citizens: A Chronicle of the French Revolution.* New York: Knopf.

Schiller, J. 1971a. L'échelle des êtres et la série chez Lamarck. In J. Schiller, ed., *Colloque international 'Lamarck'*, pp. 87–103. Paris: A. Blanchard.

——. 1971b. A propos de la diffusion du terme 'biologie.' In J. Schiller, ed., *Ibid*, pp. 239–242.

——. 1978. *La notion d'organisation dans l'histoire de biologie.* Paris: Maloine.

——. 1980. *Physiology and classification: historical relations.* Paris: Maloine.

Schleiden, M. J. 1842–1843. *Grundzüge der wissenschaflichen Botanik* . . . . 2 vols. Leipzig: Wilhelm Engelmann.

——. 1844. *Schellings und Hegels Verhältniss zur Naturwissenschaft.* Leipzig: Wilhelm Engelmann.

Schleiden, M. J. and T. Vogel. 1838. Beiträge zur Entwickelungsgeschichte des Blüthenteiles bei den Leguminosen. *Acta Acad. Caes.-Leop. Carol. Nat. Cur.* 19: 61–84, pl. 9–11.

Schultz[-Schultzenstein], C. H. 1832. *Natürliches System des Pflanzenreichs nach seiner inneren Organisation* . . . . Berlin: August Hirschwald.

——. 1847. *Neues System der Morphologie der Pflanzen nach den organischen Bildungsgesetzen* . . . . Berlin: August Hirschwald.

Schwann, T. 1839. *Mikroscopische Untersuchungen über Uebereinstimmung in der Struktur und dem Wachsthum der Thiere und Pflanzen*. Berlin: Sanders.

Scott, D. H. 1896. Present position of morphological botany. *Rep. Brit. Assoc. Advancem. Sci.* (1896): 992–1010.

Séguier, J. F. 1745. *Plantae Veronenses* . . . . 2 vols. Verona: Typis Seminarii.

Senebier, J. 1775. *L'art d'observer*. 2 vols. Genève: C. Philibert and B. Chirol.

Sepkoski, J. J. Jr. 1993. Ten years in the library: New data confirm paleontological patterns. *Paleobiology* 19: 43–51.

Seringe, N. C. 1815. *Essai d'une monographie des saules de la Suisse*. Berne: Maurhofer and Dellenbach.

Seringe, N. C. and J. C. A. Guillard. 1835. *Essai de formules botaniques représentant les caractères des plantes par des signes analytiques* . . . . Paris: J. Albert Mercklein.

Shapere, D. 1989. Evolution and continuity in scientific change. *Philos. Sci.* 56: 419–437.

Shaw, E. A. 1987. *Charles Wright at the Boundary, 1849–1852*. Westport: Meckler.

Shaw, G. 1800. *General Zoology or Systematic Natural History*. London: G. Kearsley.

Sherff, E. E. 1940. The delimitations of genera from the conservative point of view. *Bull. Torrey Bot. Club* 67: 375–380.

Sheridan, A. 1980. *Michel Foucault, the Will to Truth*. London and New York: Tavistock Publications.

Shteir, A. B. 1987. Botany in the breakfast room: Women and early nineteenth-century British plant study. In P. G. Abir-Am and D. Outram, eds., *Uneasy Careers and Intimate Lives: Women in Science, 1789–1979*, pp. 31–42. New Brunswick and London: Rutgers University Press.

Simpson, G. G. 1961. *Principles of Animal Taxonomy*. New York: Columbia University Press.

Skinner, A. S. 1979. *A System of Social Science. Papers Relating to Adam Smith*. Oxford: Clarendon Press.

Slaughter, M. M. 1982. *Universal Languages and Scientific Taxonomy in the Seventeenth Century*. Cambridge: Cambridge University Press.

Sloan, P. R. 1972. John Locke, John Ray, and the problem of the natural system. *J. Hist. Biol.* 5: 1–53.

——. 1979. Buffon, German biology, and the historical interpretation of biological species. *Brit. J. Hist. Sci.* 12: 109–153.

——. 1987. From logical universals to historical individuals: Buffon's idea of biological species. In S. Atran, et al., *Histoire du concept d'espèce dans les sciences de la vie*, pp. 101–140. Paris: Fondation Singer-Polignac.

——. 1990. Natural history, 1670–1802. In R. C. Olby, G. N. Cantor, J.

R. R. Christie, and M. J. S. Hodge, eds., *Companion to the History of Modern Science*, pp. 295–313. London and New York: Routledge.
——. 1992. Introductory essay on the edge of evolution. In R. Owen, *The Hunterian Lectures in Comparative Anatomy, May–June 1837*, pp. 3–72 [edited by P. R. Sloan]. Chicago: University of Chicago Press.
Sloane, H. 1707. *A Voyage to the Islands of Madera, Barbados, Nieves, S. Christophers and Jamaica* . . . . 2 vols. London: B. M.
Smith, A. 1980. *Essays on Philosophical Subjects*. [Edited by W. P. D. Wightman and J. C. Bryce.] Oxford: Clarendon Press.
Smith, Sir J. E. 1793. *A Specimen of the Botany of New Holland*. London: J. Davis.
——. 1798. *Tracts Relating to Natural History*. London: J. Davis. [See also: Introductory discussion on the rise and progress of natural history. *Trans. Linn. Soc. London* 1: 1–55. 1790.]
——. 1811. An account of a new genus of New Holland plants named *Brunonia*. *Trans. Linn. Soc. London* 10: 365–370, pl. 28–29.
——. 1813. Natural orders. In A. Rees, ed., *The Cyclopaedia; or, Universal Dictionary of Arts, Sciences and Literature*. vol. 24. London: Longman, Hurst, Rees, Orme and Brown.
——. 1817. Botany. *Supplement to the Fourth, Fifth, and Sixth Editions of the Encyclopaedia Britannica*. vol. 2, pp. 376–422. Edinburgh: Archibald Constable. ["J. J."]
——. 1818. *Considerations Respecting Cambridge, More Particularly Relating to its Botanical Professorship* . . . . London: Longman, Hurst, Rees, Orme and Brown.
——. 1821. *A Selection of the Correspondence of Linnaeus, and other Naturalists*. 2 vols. London: Longman, Hurst, Rees, Orme and Brown.
——. 1822. *A Grammar of Botany* . . . . New York: James V. Seaman.
Smith, Sir J. E. and J. Sowerby. 1829–1831. *Supplement to the English Botany of the late Sir. J. E. Smith and Mr Sowerby*. [Contributions by W. J. Hooker and others, illustrations by J. de C. Sowerby.] London: J. D. C. and C. E. Sowerby.
Smith, L. H. J. 1852. *Parks and Pleasure Grounds*. London: Reeve and Co.
Smith, Lady P. 1832. *Memoir and Correspondence of the Late Sir James Edward Smith* . . . . 2 vols. London: Longman, Rees, Orme, Brown, Green, and Longman.
Smocovitis, V. B. 1992a. Unifying biology: The evolutionary synthesis and evolutionary biology. *J. Hist. Biol.* 25: 1–65.
——. 1992b. Disciplining botany: A taxonomic problem. *Taxon* 41: 459–470.
Sneath, P. H. A. 1962. The construction of taxonomic groups. In G. S.

Ainsworth and P. H. A. Sneath, eds., *Microbial Classification*, pp. 289–332. Cambridge: Cambridge University Press.

——. 1964. Mathematics and classification from Adanson to the present. In G. H. M. Lawrence, ed., *Adanson: The Bicentennial of Michel Adanson's "Familles des plantes," Part Two*, pp. 471–498. Pittsburgh: Hunt Botanical Library, Carnegie Institute of Technology.

Sokal, R. R. 1962. Typology and empiricism in taxonomy. *J. Theor. Biol.* 3: 230–267.

Sokal, R. R. and P. H. A. Sneath. 1963. *Principles of Numerical Taxonomy.* San Francisco and London: W. H. Freeman.

Solereder, H. 1898–1899. *Systematische Anatomie der Dicotyledonen* . . . . Stuttgart: Ferdinand Enke.

Solms-Laubach, H. 1887. *Einleitung in die Paläophytologie.* Leipzig: Arthur Felix. [Translated by H. E. F. Garnsey as *Fossil Botany*, Oxford: Clarendon Press, 1891.]

Sprague, T. A. 1925. The classification of dicotyledons. 1. General principles. *J. Bot.* 63: 9–13.

——. 1940. Taxonomic botany, with special reference to the angiosperms. In J. Huxley, ed., *The New Systematics*, pp. 435–454. Oxford: Oxford University Press.

Sprengel, K. [P. J.] 1817. *Geschichte der Botanik.* 2 vols. Altenburg and Leipzig: E. A. Brockhaus.

Spring, A. F. 1842. Monographie de la famille des Lycopodiacées. *Nouv. Mém. Acad. Roy. Sci. Bruxelles* 15: 1–358.

Spruce, R. 1861. On the mode of branching of some Amazon trees. *J. Linn. Soc., Bot.* 5: 3–14.

Stafford, B. M. 1984. *Voyage into Substance* . . . . Cambridge, Massachusetts: Massachusetts Institute of Technology Press.

Stafford, R. A. 1989. *Scientist of Empire, Sir Roderick Murchison: Scientific Exploration and Victorian Imperialism.* Cambridge: Cambridge University Press.

——. 1990. Annexing the landscapes of the past: British imperial geology in the nineteenth century. In J. Mackenzie, ed., *Imperialism and the Natural World*, pp. 67–89. Manchester: Manchester University Press.

Stafleu, F. A. 1963. Adanson and his "Familles des plantes." In G. H. M. Lawrence, ed., *Adanson: The Bicentennial of Michel Adanson's "Familles des plantes," Part One*, pp. 123–264. Pittsburgh: Hunt Botanical Library, Carnegie Institute of Technology.

——. 1964. Introduction to Jussieu's *Genera plantarum.* In Antoine Laurent de Jussieu, *Genera plantarum*, pp. [v]–[xlvii]. Weinheim: Cramer. [Reprint.]

——. 1966a. Adanson's familles des plantes. In M. Adanson, *Familles des plantes*, pp. [v]–[xv]. Lehre: J. Cramer. [Reprint.]

——. 1966b. Brongniart's Histoire des Végétaux Fossiles. *Taxon* 15: 320–324.

——. 1966c. F. A. W. Miquel, Netherlands botanist. *Wentia* 10: 1–95.

——. 1969a. A historical review of systematic biology. In *Systematic Biology*, pp. 16–44. Washington: National Academy of Sciences.

——. 1969b. Joseph Gaertner and his Carpologia. *Taxon* 18: 216–223.

——. 1971. *Linnaeus and the Linnaeans*. Utrecht: Oosthoek.

Stafleu, F. A. and R. S. Cowan. 1976–1988. *Taxonomic Literature: A Selective Guide to Botanical Publications with Dates, Commentaries and Types*. Ed. 2, 7 vols. Utrecht: Bohn, Scheltema and Holkema.

Stafleu, F. A. and E. A. Mennega. 1992. *Ibid. Supplement 1*. Königstein: Koelz.

Stearn, W. T. 1957. An introduction to the *Species plantarum* and cognate botanical works of Carl Linnaeus. In C. Linnaeus, *Species plantarum*. vol. 1., pp. vii, 1–176. London: Ray Society. [Facsimile reprint.]

——. 1958. Botanical exploration to the time of Linnaeus. *Proc. Linn. Soc. London* 169: 173–196.

——. 1959. Four supplementary Linnaean publications: Methodus (1736), Demonstrationes plantarum (1753), Genera plantarum (1754), Ordines naturales (1764). In C. Linnaeus, *Species plantarum*. vol. 2., pp. 73–102. London: Ray Society. [Facsimile reprint.]

——. 1961. Botanical gardens and botanical literature in the eighteenth century. In A. H. Stevenson (compiled), *Catalogue of Botanical Books in the Library of Rachel McMasters Miller Hunt*. vol. 2., part 1, *Introduction to Printed Books 1701–1800*, pp. xli–cxl. Pittsburgh: Hunt Foundation.

——. 1962. An introduction to Robert Brown's "Prodromus florae novae hollandiae." In W. T. Stearn, *Three Prefaces on Linnaeus and Robert Brown*, pp. [v]–[lii]. Weinheim: Cramer.

——. 1990. John Lindley (1799–1865), a sketch of the life and work of a pioneer British orchidologist and gardener-botanist. In S. Sprunger, ed., *Orchids from the Botanical Register, 1815–1847*. vol. 1., pp. 15–44. Basel: Birkhäuser.

Stemerding, D. 1991. *Plants, Animals, and Formulae*. Twente: Universiteit Twente.

——. 1993. How to make oneself nature's spokesman? A Latourian account of classification in eighteenth- and early nineteenth-century natural history. *Philos. Sci*. 8: 199–223.

Sternberg, K. M. von. 1820–1838. *Versuch einer geognostisch-botanischen Darstellung der Flora der Vorwelt* . . . . 2 vols. Regensburg: Christoph Ernst Brenck.

Stevens, P. F. 1980. Evolutionary polarity of character states. *Annual Rev. Ecol. Syst.* 11: 333–358.

——. 1983. Augustin Augier's "Arbre botanique" (1801), a remarkable early botanical representation of the natural system. *Taxon* 32: 203–211.

——. 1984a. Haüy and A.-P. de Candolle: Crystallography, botanical systematics and comparative morphology, 1780–1840. *J. Hist. Biol.* 17: 49–82.

——. 1984b. Metaphors and typology in the development of botanical systematics 1690–1960, or the art of putting new wine in old bottles. *Taxon* 33: 169–211.

——. 1986. Evolutionary classification in botany, 1960–1985. *J. Arnold Arbor.* 67: 313–339.

——. 1990. Nomenclatural stability, taxonomic instinct, and flora writing—a recipe for disaster? In P. Baas, K. Kalkman, and R. Geesink, eds., *The Plant Diversity of Malesia*, pp. 387–410. Dordrecht: Kluwer.

——. 1991. George Bentham and the Kew Rule. In D. L. Hawksworth, ed., *Improving the Stability of Names: Needs and Options*, pp. 157–168. Königstein: Koeltz.

——. 1992. Species: Historical perspectives. In E. F. Keller and E. A. Lloyd, eds., *Keywords in Evolutionary Biology*, pp. 302–311. Cambridge, Massachusetts: Harvard University Press.

Stevens, P. F. and S. P. Cullen. 1990. Linnaeus, the cortex-medulla theory, and the key to his understanding of plant form and natural relationships. *J. Arnold Arbor.* 71: 179–220.

Strasburger, E. A. 1872. *Die Coniferen und die Gnetaceen. Eine morphologische Studie.* 2 vols. Leipzig: Ambr. Abel.

——. 1874. *Ueber die Bedeutung phylogenetischer Methoden fur die Erforschung lebender Wesen.* Jena: Mauke.

——. 1895. The development of botany in Germany during the nineteenth century. *Bot. Gaz.* 20: 193–204, 249–257.

Stresemann, E. 1975. *Ornithology from Aristotle to the Present.* Cambridge, Massachusetts: Harvard University Press.

Strickland, H. E. 1840. Observations on the affinities and analogies of organised beings. *Mag. Nat. Hist. N. S.* 4: 219–226.

——. 1841. On the true method of describing the natural system in zoology and botany. *Ann. Mag. Nat. Hist.* 6: 184–194.

——. 1844. Description of a chart of the natural affinities of the Insessorial Order of birds. *Rep. Brit. Assoc. Advancem. Sci.* (1843): 69.

——. 1845. Report on the recent progress and present state of ornithology. *Rep. Brit. Assoc. Advancem. Sci.* (1844): 170–221.

Stuessy, T. F. 1990. *Plant Taxonomy: The Systematic Evaluation of Comparative Data.* New York: Columbia University Press.

Suringar, W. F. R. 1895. *Het plantenrijk (regnum vegetabile). Phylogenetische schets.* Leeuwarden: Hugo Suringar.

Swainson, W. 1834. *A Preliminary Discourse on the Study of Natural History.* London: Longman, Rees, Orme, Brown, Green and Longman.

Tassy, P. 1991. *L'arbre à remonter le temps: les rencontres de la systématique et de l'évolution.* Paris: C. Bourgois.

Tchouproff, O. 1895. Quelques notes sur l'anatomie systématique des Acanthacées. *Bull. Herb. Boissier* 3: 550–560.

Tesi, D. 1982. Augustin-Pyramus de Candolle: essai d'élaboration d'une taxonomie théorique du XIXe siècle. *Gesnerus* 39: 295–303.

Thistleton-Dyer, W. T. 1880. *The Botanical Enterprise of Empire.* London: George E. Eyre and William Spottiswoode.

——. 1889. Section D, Biology. *Proc. Brit. Assoc. Advancem. Sci.* (1888): 686–701. [Presidential Address.]

——. 1895. Section K, Botany. *Proc. Brit. Assoc. Advancem. Sci.* (1895): 836–850. [Presidential Address.]

——. 1905. Botanical survey of the empire. *Bull. Misc. Info. Kew* (1905): 9–43.

Thompson, J. A. 1899. The study of natural history. *Nat. Sci.* 14: 437–449.

Thunberg, C. P. 1784. *Flora japonica* . . . . Leipzig: J. G. Müller.

Tort, P. 1989. *La raison classificatoire. Quinze études.* Aubier: Res.

Tournefort, J. P. de. 1694. *Élémens de botanique.* 3 vols. Paris: Imprimerie Royale.

——. 1700. *Institutiones rei herbariae.* 4 vols. Paris: Typographia Regia.

Trécul, A. [A. L.] 1845. Recherches sur la structure et développement du *Nuphar lutea. Ann. Sci. Nat., Sér. Bot. 3*, 4: 286–345, pl. 10–13.

Treviranus, G. R. 1832. On the fundamental types of organisation. *Edinburgh New Philos. J.* 13: 75–86.

Triana, J. 1872. Les Melastomatacées. *Trans. Linn. Soc. London* 27: 1–188, pl. 1–28.

Trigger, B. G. 1989. *A History of Archaeological Thought.* Cambridge: Cambridge University Press.

Tufte, E. R. 1983. *The Visual Display of Quantitative Information.* Cheshire, Connecticut: Graphics Press.

Turpin, P. J. F. 1819. Mémoire sur l'inflorescence des Graminées et des Cyperacées, comparée avec celles des autres végétaux sexifères; suivi de quelques observations sur les disques. *Mém. Mus. Hist. Nat.* 5: 426–492, pl. 30–31.

——. 1820. *Essai d'une iconographie élémentaire et philosophique des végétaux* . . . . Paris: C. L. F. Panckoucke. [vol. 3 of J. L. M. Poiret, *Leçons de flore.*]

——. 1837. Esquisse d'organographie végétale, fondée sur le principe d'unité de composition organique et d'évolution rayonnante ou centrifuge . . . . In C. F. Martins, ed., *Oeuvres d'histoire naturelle de Goethe . . .* , col. 5–70, pl. 3–5. Paris: Ch. Fr. Martins.

Turrill, W. B. 1942. Taxonomy and phylogeny. *Bot. Rev. (Lancaster)* 8: 247–270, 473–532, 655–707.

——. 1963. *Joseph Dalton Hooker: Botanist, Explorer and Administrator.* London: Thomas Nelson and Sons.

Tutin, T. G. 1952. Phylogeny of flowering plants: Fact or fiction? *Nature* 169: 126–127.

Tutt, J. W. 1898. Some considerations of natural genera, and incidental references to the nature of species. *Proc. S. London Entomol. Nat. Hist. Soc.* (1898): 20–30.

Uschmann, G. 1967. Zur Geschichte der Stammbaum-Darstellungen. In M. Gersch, ed., *Gesammelte Vorträge über moderne Probleme der Abstammlungslehre.* vol. 2, pp. 9–30. Jena: Friedrich-Schiller-Universität.

Vahl, M. 1804. *Enumeratio plantarum.* 2 vols. Copenhagen: Mölleri.

Vaillant, S. 1718. *Discours sur la structure des fleurs . . . les differences et l'usage de leurs parties.* Leiden: Van der Aa.

——. 1719. Établissement des nouveaux caractères de trois familles ou classes des plantes à fleurs composées; sçavoir des Cynarocéphales, des Corymbifères, et des Cichoracées. *Mém. Math. Phys. Acad. Roy. Sci.* (Paris) (1718): 143–191, pl. 5–6.

Valmont de Bomare, J. C. 1768. Histoire naturelle. In *Dictionnaire raisonné universel d'histoire naturelle* 2: 474–483. Paris: Lacombe.

Ventenat, É. P. 1794–1795. *Principes de botanique . . . .* Paris: Sallior.

——. 1795. Extrait d'un mémoire sur la précision avec laquelle on doit observer et déterminer les différens organes du végétal pour l'accélération des progrès de la botanique, et particulièrement sur les meilleurs moyens de distinguer la calyce de le corolle. *Mag. Encycl.* 1795, 3: 303–313.

——. 1797–1798. [Icones et descriptiones plantarum . . . .] In A. J. Cavanilles, *Icones et descriptiones plantarum . . . .* vol. 4, pp. [72]-[77]. Madrid: Regia Typographia. [Reprint of a review of earlier volumes of Cavanilles's work, originally appearing in the *Magasin Encyclopédique* of 1797.]

——. 1799. *Tableau du règne végétal selon la méthode de Jussieu . . . .* 4 vol. Paris: J. Drissonier.

Vernon, K. 1993. Desperately seeking status: Evolutionary systematics and the taxonomists' search for respectability, 1940–60. *Brit. J. Hist. Sci.* 26: 207–227.

Vesque, J. 1882. L'espèce végétal consideré au point du vue de l'anatomie comparée. *Ann. Sci. Nat., Bot. Sér. 6,* 13: 5–46.

——. 1883. Contribution à l'histologie systématique de la feuille des

Caryophyllinées précedées de remarques complémentaires sur l'importance des caractères anatomiques en botanique descriptive. *Ann. Sci. Nat., Bot. Sér. 6*, 15: 105–148, pl. 7–8.

——. 1889a. *Epharmoses sive materiae ad instruendam anatomiam systematis naturalis. 1. Folia capparearum.* Vincennes: Delapierre.

——. 1889b. *Ibid. 2. Genitalia foliaque Garcinearum et Calophyllearum.* Vincennes: Delapierre.

Virville, Davy de. [See Davy de Virville.]

Voigt, F. S. 1806. *Darstellung der natürlichen Pflanzensystems von Jussieu, nach seinem neuesten Verbesserungen.* Leipzig: C. H. Reclam.

Voss, E. 1952. The history of keys and phylogenetic trees in systematic biology. *J. Sci. Lab. Denison Univ.* 43: 1–25.

Vriese, W. H. de. 1854. Goodenovieae. *Natuurk. Verh. Holl. Maatsch. Wetensch. Haarlem, Ser. 2*, 10: viii, 1–198, pl. 1–38.

Wachendorff, E. J. van. 1747. *Horti ultrajectini index.* Utrecht: Van Vucht.

Walker, M. 1988. *Sir James Edward Smith . . . .* London: Linnean Society of London.

Wallace, A. R. 1855. On the law which has regulated the introduction of new species. *Ann. Mag. Nat. Hist., Ser. 2*, 16: 184–196.

——. 1856. Attempts at a natural arrangement of birds. *Ann. Mag. Nat. Hist., Ser. 2*, 18: 193–216.

Walters, S. M. 1961. The shaping of angiosperm taxonomy. *New Phytol.* 60: 74–84.

——. 1986. The name of the rose: A review of ideas on the European bias in angiosperm classification. *New Phytol.* 104: 527–546.

——. 1988. The purposes of systematic botany. *Symb. Bot. Upsal.* 28(3): 13–20.

Warburg, O. 1897. Monographie der Myristicaceen. *Nova Acta Acad. Caes. leop.-Carol. German. Nat. Cur.* 68: 1–680, pl. 1–25.

Watson, H. C. 1833. Letter to the Editor, on the relation between cerebral development and the tendency to pursuits;—and on the heads of botanists. *Phrenological J. & Misc.* 8: 97–108.

Webster, G. 1987. The saga of the spurges: A review of classification and relationships in the Euphorbiales. *Bot. J. Linn. Soc.* 94: 3–46.

Weddell, H. A. 1856. Monographie de la famille des Urticacées. *Arch. Mus. Hist. Nat.* 9: 1–592, pl. 1–20.

Westfall, R. H., H. F. Glen, and M. G. Panagos. 1986. A new identification aid combining features of a polyclave and an analytical key. *Bot. J. Linn. Soc.* 92: 65–73.

Wettstein [von Westersheim], R. 1895. Globulariaceen-Studien. *Bull. Herb. Boissier* 3: 271–290.

——. 1896. Die Europäischen Arten der *Gentiana* aus der Section *Endotriche* Froel. und ihr entwicklungsgeschichtlicher Zusammenhang. *Denkschr. Math.-Nat. Cl. Akad. Wiss. Wien* 64: 309–382, maps 1–3, fig. 1–4.

——. 1898. *Grundzüge de geographisch-morphologischen Methode der Pflanzensystematik* . . . . Jena: Gustav Fischer.

——. 1901. Die Entwicklung der Morphologie, Entwicklungsgeschichte und Systematik der Phanerogamen in Österreich von 1850–1900. In A. Handlirsch and R. von Wettstein, eds., *Botanik und Zoologie in Österreich in den Jahren 1850 bis 1900*, pp. 198–218. Vienna: Alfred Hölder.

Whewell, W. 1823. *An Essay on Mineralogical Classification and Nomenclature* . . . . Cambridge: J. Smith.

——. 1840. *The Philosophy of the Inductive Sciences.* 2 vols. London: John W. Parker.

——. 1847. *Ibid.* Ed. 2. 2 vols. London: John W. Parker.

——. 1848. *The History of the Inductive Sciences.* Ed. 2. 3 vols. London: John W. Parker.

White, G. 1836. *The Natural History and Antiquities of Selbourne* . . . . Ed. 6. Edinburgh: Fraser and Co.

Wigand, J. W. A. 1846. *Kritik und Geschichte der Lehre von der Metamorphose der Pflanzen.* Leipzig: Wilhelm Engelmann.

Wight, R. 1840. *Illustrations of Indian Botany.* 2 vols. Madras: J. B. Pharaoh.

——. 1845. *Spicilegium neilgherrense.* Madras: Franck and Co.

Willdenow, C. L. 1792. *Grundriss der Kraüterkunde zu vorlesungen.* Berlin: Haude and Spener.

Williams, C. B. 1964. *Patterns in the Balance of Nature and Related Problems in Quantitative Ecology.* London: Academic Press.

Williams, R. L. 1988. Gerard and Jaume: Two neglected figures in the history of Jussiaean classification. *Taxon* 37: 2–34, 233–271.

Willis, J. C. 1973. *A Dictionary of Flowering Plants and Ferns.* Ed 8, revised by H. K. Airy Shaw. Cambridge: Cambridge University Press.

Wilson, E. B. 1901. Aims and methods of study in natural history. *Science N. S.* 13: 14–23.

Winsor, M. P. 1976a. The development of Linnaean insect classification. *Taxon* 25: 57–67.

——. 1976b. *Starfish, Jellyfish, and the Order of Life.* New Haven and London: Yale University Press.

——. 1979. Louis Agassiz and the species question. *Stud. Hist. Biol.* 13: 89–117.

——. 1985. The impact of Darwin on the Linnaean enterprise, with special reference to the work of T. H. Huxley. In J. Weinstock, ed.,

Contemporary Perspectives on Linnaeus, pp. 55–84. Lanham, Maryland: University Press of America.

——. 1991. *Reading the Shape of Nature*. Chicago: University of Chicago Press.

Withers, C. W. J. 1991. The Rev. Dr John Walker and the practice of natural history in late eighteenth century Scotland. *Arch. Nat. Hist.* 18: 201–220.

——. 1992. Natural knowledge as cultural property: Disputes over the 'ownership' of natural history in late eighteenth-century Edinburgh. *Arch. Nat. Hist.* 19: 289–303.

Wolfsohl, C. 1991. Gilbert White's natural history and history. *Clio* 20: 271–281.

Woods, J. 1838. Observations on the genera of European grasses. *Trans. Linn. Soc. London* 18: 1–57.

Wyatt, R., I. J. Odrzykoski, and A. Stoneburner. 1992. Isozyme evidence of reticulate evolution in mosses: *Plagiomnium medium* is an allopolyploid of *P. ellipticum* x *P. insigne*. *Syst. Bot.* 17: 532–550.

Yates, F. A. 1966. *The Art of Memory*. London: Routledge and Kegan Paul.

Yeo, R. R. 1986. The principle of plenitude and natural theology in nineteenth century Britain. *Brit. J. Hist. Sci.* 19: 263–282.

——. 1991. Reading encyclopedias: Science and the organisation of knowledge in British dictionaries of arts and sciences, 1730–1850. *Isis (Sarton)* 82: 24–49.

Zahlbruckner, A. 1901. Die Entwicklung der Morphologie, Entwicklungsgeschichte und Systematik der Kryptogamen in Österreich von 1850–1900. In A. Handlirsch and R. von Wettstein, eds., *Botanik und Zoologie in Österreich in den Jahren 1850 bis 1900*, pp. 155–194. Vienna: Alfred Hölder.

Zaunick, R. 1960. Kützing in wissenschaftshistorischer Sicht. In R. H. W. Müller and R. Zaunick, eds., *Friedrich Traugott Kützing 1807–1893: Aufzeichnungen und Erinnerungen*, pp. 12–22. Leipzig: Johann Ambrosius Barth.

Zimmermann, W. 1930. *Die Phylogenie der Pflanzen*. Jena: Gustav Fischer.

Subject and Author Index

Authors mentioned by Jussieu in the works translated in appendix 1 are not included below, and details of the subjects he covered there are also not mentioned; references to these translations can be found throughout the text.

Organism Index

The variable endings of plant names at the rank of family and above present problems. Names are grouped at the appropriate taxon level and, disregarding the termination, under the name that is currently used for the taxon (or the closest equivalent mentioned in the text above). Thus Ericae, Ericaceae, and Ericacées—all names for a taxon at the family level—are indexed together. Cross references to modern names for Jussieu's groups are given, e.g., the Gramineae are cross-referenced under the Poaceae. Names of the major groups in the Linnaean sexual system and those in the lists illustrating nomenclatural points in the notes to Jussieu's ``Introduction'' to the Genera plantarum are not indexed.

Designer: Linda Secondari
Text: Bembo 11/13
Compositor: Columbia University Press
Printer: Book Crafters
Binder: Book Crafters